Occurrence, Detection, and Fate of Microplastics in Freshwater Ecosystems

Kiran Bala • Regina Nogueira
Gopala Krishna Darbha • Dirk Weichgrebe
Editors

Occurrence, Detection, and Fate of Microplastics in Freshwater Ecosystems

Springer

Editors
Kiran Bala
Department of Biosciences and Biomedical Engineering
Indian Institute of Technology Indore
Simrol, Madhya Pradesh, India

Gopala Krishna Darbha
Department of Earth Sciences
Indian Institute of Science Education and Research Kolkata
Mohanpur, West Bengal, India

Regina Nogueira
Institute of Sanitary Engineering and Waste Management
Leibniz University Hannover
Welfengarten, Hannover, Germany

Dirk Weichgrebe
Institute of Sanitary Engineering and Waste Management
Leibniz University Hannover
Welfengarten, Hannover, Germany

ISBN 978-981-95-0224-0 ISBN 978-981-95-0225-7 (eBook)
https://doi.org/10.1007/978-981-95-0225-7

This Springer imprint is published by the registered company Springer Nature Singapore Pte Ltd.
The registered company address is: 152 Beach Road, #21-01/04 Gateway East, Singapore 189721, Singapore

Preface

Microplastic pollution is an emerging environmental challenge, which is one of the most complex and urgent challenges threatening ecosystems, biodiversity, and human health. Microplastics—plastic particles smaller than 5 mm—are generated through the fragmentation of larger plastic debris or are intentionally added in everyday consumer products such as textiles, personal care items, and cleaning agents. These microplastic particles, persistent pollutants, once introduced into freshwater systems infiltrate water bodies (river, lakes), sediments, soil, biota, and food chain leading to widespread contamination. Aquatic organisms, including plankton and fish, can ingest these particles, resulting in bioaccumulation and causes potential biomagnification through the food web, ultimately posing indirect risks and long-term impact to human health.

The pervasive nature of microplastics calls for a comprehensive and interdisciplinary approach to understanding their origin, pathways, persistence, and mitigation. The proposed book, *Occurrence, Detection, and Fate of Microplastics in Freshwater Ecosystems*, aims to provide a comprehensive literature on sources, distribution, detection, ecological and health consequences, and mitigation of microplastics in freshwater environments.

The book covers several aspects of microplastic contamination such as firstly, it introduces microplastics in freshwater systems and examines its interactions with microbes and terrestrial sources further contributing to aquatic contamination. The book also explores various pathways through which microplastics can enter and move within freshwater and aquatic ecosystems, including their subsequent movement into soil and its biota.

The book also provides insight into different conventional and advanced techniques used for identifying and characterizing microplastics in water, sediments, and biological tissue. This insight is very critical for understanding the distribution of microplastics and designing effective interventions to mitigate it. The book addresses impact of microplastics ingestion and long-term exposure on human health. Further, the book provides outlook on how microplastics interfere with aquatic biogeochemical processes such as carbon and other nutrient cycling. The book also attempts to evaluate the ecological risks posed by microplastic pollution, with a detailed case study from the Saurashtra coastline.

As plastic has become now an inevitable part of human civilization, so apart from detection and characterization, we need to deduce strategies for degradation of microplastics for mitigation of this menace. This book explores different degradation mechanisms and remediation strategies for microplastics and manufactured nanoparticles (MNPs). The book offers practical insights into their removal from environment and further ecological restoration.

By integrating insights from environmentalists, chemist, biologists, engineers, and public health workers, this book aims to foster collaborative discussions and drive innovative solutions. It aims to serve as a valuable resource addressing this pressing environmental issue. Furthermore, it will provide comprehensive information leading to actionable change and contribute to devising innovations and policy for protection of freshwater ecosystems for future generations.

Simrol, Madhya Pradesh, India — Kiran Bala
Welfengarten, Hannover, Germany — Regina Nogueira
Mohanpur, West Bengal, India — Gopala Krishna Darbha
Welfengarten, Hannover, Germany — Dirk Weichgrebe

Acknowledgments

All the editors (Dr. Kiran Bala, Dr. Regina Nogueira, Dr. Gopala Krishna Darbha, Dr. Dirk Weichgrebe) acknowledge the Scheme for Promotion of Academic and Research Collaboration - SPARC (Ministry of Education, India) to support the idea and meetings held to accomplish this book, Occurrence, Detection, and Fate of Microplastics in Freshwater Ecosystems.

Contents

Editors and Contributors

About the Editors

Kiran Bala has been serving as Professor in the Department of Biosciences & Biomedical Engineering at the Indian Institute of Technology Indore since October 2023. Prior to this, she held the position of Associate Professor and Assistant Professor from February 2017. With a robust academic journey, she began as INSPIRE Faculty at the Indian Institute of Technology Indore in November 2012, following a stint as a DBT scientist at DAVV, Indore, from July 2012 to October 2012. Her research focuses on microplastics, which includes monitoring and assessment, and developing enzyme-based biodegradation for emerging contaminants. She has more than 65 papers in international journals, along with 11 book chapters. She has also edited two books with reputed publishers.

Regina Nogueira is a distinguished academician with extensive experience and expertise in water and wastewater biology. Since January 2012, she held the esteemed position of Professor for Water and Wastewater Biology at the Leibniz Universität Hannover, Germany. Prior to this, she served as an Assistant Professor at the University of Minho in Braga, Portugal, from February 2003 to January 2012. In the field of microplastics, she has developed a method for extraction and analysis of microplastics from complex biological matrices. She has experience in characterizing the composition and function of biofilms growing on the surface of microplastics/biopolymers using molecular tools and microscopy. Her commitment to sustainability is evident through research grants and third-party funding, including projects like LegioGrowth and Intelligente Systeme

Wasser, which focus on critical issues such as the impact of climate change on *Legionella pneumophila* and the spread of antibiotic-resistant germs in aquatic systems. To date, 80 research papers have been published in international peer-reviewed journals.

Gopala Krishna Darbha is an accomplished academic with a diverse educational background and a wealth of experience in the field of environmental sciences. He completed his PhD in Chemistry at Jackson State University, USA, in 2008. He is currently serving as Associate Professor in the Department of Earth Sciences at IISERK. He has garnered significant recognition for his contributions, such as the prestigious Ramanujan Fellowship award from SERB and being highlighted by Nature (India) for research on nano and microplastics in 2021. In the field of micro- and nanoplastics, he has made significant contributions to (1) understanding their presence in various environmental compartments (air, water, soil, and sea food) and finding the human exposure pathways, (2) mechanisms of eco-accumulation and mobility, ecological toxicity, and (3) sustainable methods for their remediation from wastewaters. With a prolific publication record comprising 80 scientific publications and a patent.

Dirk Weichgrebe currently serves as Associated Professor, Senior Scientist, and Head of the Research Field Waste Management, Circular Economy, and Bioenergy at the Institute for Sanitary Engineering and Waste Management (ISAH) at Leibniz University Hannover since 2015. He is a distinguished academician with a rich educational background and significant achievements in the field of engineering and environmental science. He obtained his Habilitation at Leibniz University Hannover in 2015, focusing on Compendium Biogas, and earned the degree of Privat Dozent (Dr.-Ing. habil.), akin to an Associate Professor. His expertise and dedication to the field were further recognized with the EU Madame-Curie Fellowship from 2011 to 2015. He brings research experience in the fields of analysis and characterization of MPs, material and energy flow analysis, life cycle assessment as well as wastewater and waste management and processing.

Contributors

Irédon Adjama School of Doctoral Studies & Research (SDSR), National Forensic Sciences University, Gandhinagar, Gujarat, India

Ebrahim Ahmadisharaf Department of Civil and Environmental Engineering, FAMU-FSU College of Engineering, Tallahassee, FL, USA

Resilient Infrastructure and Disaster Response Centre, FAMU-FSU College of Engineering, Tallahassee, FL, USA

Paromita Banerjee Department of Cardiology, All India Institute of Medical Sciences, Rishikesh, Uttarakhand, India

Gayatri Barabde Department of Environmental Science, The Institute of Science, Dr. Homi Bhabha State University, Mumbai, India

Department of Analytical Chemistry, The Institute of Science, Dr. Homi Bhabha State University, Mumbai, India

Abhishek Biswas Environmental Nanoscience Laboratory, Department of Earth Sciences, Indian Institute of Science Education and Research Kolkata, Mohanpur, West Bengal, India

Akshay Botle Department of Environmental Science, The Institute of Science, Dr. Homi Bhabha State University, Mumbai, Maharashtra, India

Gopala Krishna Darbha Environmental Nanoscience Laboratory, Department of Earth Sciences, Indian Institute of Science Education and Research Kolkata, Mohanpur, West Bengal, India

Centre for Climate and Environmental Studies, Indian Institute of Science Education and Research Kolkata, Mohanpur, West Bengal, India

Siddhant Dash Department of Civil Engineering, School of Engineering and Sciences, SRM University, AP, Amaravati, Andhra Pradesh, India

Escuela de Ingeniería y Ciencias, Tecnologico de Monterrey, Campus Monterey, Monterrey, Nuevo Leon, Mexico

Surya Narayan Dash Institute of Biotechnology, Viikinkaari, University of Helsinki, Helsinki, Finland

Hemen Dave School of Pharmacy, National Forensic Sciences University, Gandhinagar, Gujarat, India

Shubha Dixit Environmental Toxicology Group, FEST Division, CSIR-Indian Institute of Toxicology Research (CSIR-IITR), Lucknow, Uttar Pradesh, India

Academy of Scientific and Innovative Research (AcSIR), Ghaziabad, Uttar Pradesh, India

Bhanu Duggal Department of Cardiology, All India Institute of Medical Sciences, Rishikesh, Uttarakhand, India

Khalid Muzamil Gani Department of Civil Engineering, National Institute of Technology, Srinagar, Jammu, and Kashmir, India

Sudha Goel Environmental Engineering and Management, Department of Civil Engineering, Indian Institute of Technology Kharagpur, Kharagpur, West Bengal, India

School of Environmental Science and Engineering, Institute of Technology Kharagpur, Kharagpur, West Bengal, India

Abdul Mobin Ibna Hafiz Department of Civil and Environmental Engineering, FAMU-FSU College of Engineering, Tallahassee, FL, USA

Resilient Infrastructure and Disaster Response Centre, FAMU-FSU College of Engineering, Tallahassee, FL, USA

Mihir Herlekar Symbiosis Centre for Management Studies, Pune, Symbiosis International (Deemed University), Pune, India

Mayur Shirish Jain Department of Civil Engineering, Indian Institute of Technology Indore, Indore, Madhya Pradesh, India

Farhad Jazaei Department of Civil Engineering, The University of Memphis, Memphis, TN, USA

Ritu Kothari Department of Civil Engineering, Indian Institute of Technology Indore, Indore, Madhya Pradesh, India

Ghanshyam Kumar Department of Cardiology, All India Institute of Medical Sciences, Rishikesh, Uttarakhand, India

Manish Kumar Escuela de Ingeniería y Ciencias, Tecnologico de Monterrey, Campus Monterey, Monterrey, Nuevo Leon, Mexico

Sustainability Cluster, School of Advanced Engineering, UPES, Dehradun, Uttarakhand, India

Manoj Kumar Environmental Toxicology Group, FEST Division, CSIR-Indian Institute of Toxicology Research (CSIR-IITR), Lucknow, Uttar Pradesh, India

Academy of Scientific and Innovative Research (AcSIR), Ghaziabad, Uttar Pradesh, India

Ayan Lodh School of Environmental Science and Engineering, Institute of Technology Kharagpur, Kharagpur, West Bengal, India

Arpita Maurya Environmental Toxicology Group, FEST Division, CSIR-Indian Institute of Toxicology Research (CSIR-IITR), Lucknow, Uttar Pradesh, India

Academy of Scientific and Innovative Research (AcSIR), Ghaziabad, Uttar Pradesh, India

Payal Mazumder Centre for the Environment, Indian Institute of Technology Guwahati, Guwahati, Assam, India

Nishigandha Muduli Environmental Science Laboratory, Department of Zoology, Ravenshaw University, Cuttack, Odisha, India

Pratyusha Nayak Environmental Science Laboratory, Department of Zoology, Ravenshaw University, Cuttack, Odisha, India

Regina Nogueira Institute of Sanitary Engineering and Waste Management, Leibniz University Hannover, Welfengarten, Hannover, Germany

Pavan Hirenbhai Pandya School of Pharmacy, National Forensic Sciences University, Gandhinagar, Gujarat, India

Dinesh Parida Department of Biosciences and Biomedical Engineering, Indian Institute of Technology Indore, Simrol, Madhya Pradesh, India

Institute of Sanitary Engineering and Waste Management, Leibniz University Hannover, Welfengarten, Hannover, Germany

Lipika Patnaik Environmental Science Laboratory, Department of Zoology, Ravenshaw University, Cuttack, Odisha, India

Smruti Prajna Pradhan Environmental Science Laboratory, Department of Zoology, Ravenshaw University, Cuttack, Odisha, India

Maryam Salehi Department of Civil and Environmental Engineering, University of Missouri, Columbia, MO, USA

Missouri Water Centre, Columbia, MO, USA

Sayli Salgaonkar Department of Environmental Science, The Institute of Science, Dr. Homi Bhabha State University, Mumbai, Maharashtra, India

Mozim Shafi Environmental Engineering and Management, Department of Civil Engineering, Indian Institute of Technology Kharagpur, Kharagpur, West Bengal, India

Sheha Shaji Research Scholar, Department of Civil Engineering, National Institute of Technology, Tadepalligudem, Andhra Pradesh, India

Adarsh Singh Department of Civil Engineering, Indian Institute of Technology Indore, Indore, Madhya Pradesh, India

Ashutosh Singh Department of Biochemistry, University of Lucknow, Lucknow, Uttar Pradesh, India

Rajesh Singh Environmental Hydrology Division, National Institute of Hydrology, Roorkee, Uttarakhand, India

Deborshee Sinha Department of Civil Engineering, Indian Institute of Technology Indore, Indore, Madhya Pradesh, India

Baranidharan Sundaram Department of Civil Engineering, National Institute of Technology, Tadepalligudem, Andhra Pradesh, India

Divyanka Thakur Department of Environmental Science, The Institute of Science, Dr. Homi Bhabha State University, Mumbai, Maharashtra, India

Sheetal Verma Department of Microbiology, King George's Medical University, Lucknow, India

Microbes' and Microplastics' Interactions in Freshwater Ecosystems: Fate and Implications for Environmental Health

1

Shubha Dixit, Arpita Maurya, Ashutosh Singh, Sheetal Verma, Rajesh Singh, and Manoj Kumar

Abstract

Microorganisms play a pivotal role in the biogeochemical cycles of aquatic environment, mediating interactions between biotic and abiotic components and forming complex associations with plastic debris. Their widespread distribution is a global concern due to their movement through food chains, which poses risks to ecosystems and human health, and their capacity to facilitate the spread of infectious microbes in contaminated environments. Interactions between microorganisms and microplastics (MPs) can yield both beneficial and harmful effects. Microorganisms, including bacteria, fungi, algae, and viruses are indispensable for nutrient cycling, organic matter decomposition, microplastic degradation, and maintaining overall ecosystem health. Certain bacterial groups, such as Proteobacteria, Cyanobacteria, and Bacteroidetes, are commonly associated with MPs like polyethylene (PE) and polypropylene (PP), while plastics like polyvinyl chloride (PVC) and polystyrene (PS) are less frequently colonized. These bacteria play a crucial role in breaking down large polymer molecules into

S. Dixit · A. Maurya · M. Kumar (✉)
Environmental Toxicology Group, FEST Division, CSIR-Indian Institute of Toxicology Research (CSIR-IITR), Lucknow, Uttar Pradesh, India

Academy of Scientific and Innovative Research (AcSIR), Ghaziabad, Uttar Pradesh, India
e-mail: manojkumar@iitr.res.in; manoj.kumar13@csir.res.in

A. Singh
Department of Biochemistry, University of Lucknow, Lucknow, India

S. Verma
Department of Microbiology, King George's Medical University, Lucknow, India

R. Singh
Environmental Hydrology Division, National Institute of Hydrology, Roorkee, Uttarakhand, India

K. Bala et al. (eds.), *Occurrence, Detection, and Fate of Microplastics in Freshwater Ecosystems*, https://doi.org/10.1007/978-981-95-0225-7_1

smaller components and monomers, ultimately leading to mineralization into CO_2, H_2O, and biomass via enzymatic processes. The intricate relationships between microorganisms and microplastics, including microbial colonization, biofilm formation on MPs (commonly referred to as the plastisphere), biodegradation mechanisms, and their broader implications for pollution dynamics and ecological health, underscore the complexity of the issue. While microbial degradation of microplastics offers potential solutions, it faces significant challenges. Despite these challenges, understanding the mechanisms of microbe-MP interactions and their environmental implications remains critical for developing strategies to combat microplastic contamination in freshwater ecosystems. This chapter explores these interactions in detail, focusing on the dual role of microorganisms in mitigating and exacerbating MP-related pollution. By advancing research on microbial processes and environmental variables that influence plastic degradation, it is possible to work toward innovative solutions for managing microplastic contamination and safeguarding environmental health.

Keywords

Microplastics · Freshwater ecosystem · Microbe-plastic interaction · Plastisphere · Environmental health

1.1 Introduction

The biogeochemical cycle is an integral part of an ecosystem that governs the interaction and movement of vital elements like carbon, nitrogen, phosphorus, and sulfur across Earth's biosphere, lithosphere, atmosphere, and hydrosphere (Odum and Barrett 2005). This biogeochemical cycle plays a crucial role in mediating the exchange of elements among physical, chemical, and biological components of the environment, including terrestrial inputs and atmospheric depositions (Schneider et al. 2017). In aquatic environment, the biogeochemical cycle operates as a complex dynamic system that transports essential elements, maintaining the ecological balance of lakes, rivers, wetlands, and seas (Reddy et al. 2022; Chakraborty and S.K. 2021). Microorganisms are key drivers of biogeochemical cycles at both local and global scales, actively transporting and transforming elements and minerals, such as carbon, nitrogen, phosphorus, and sulfur, within soils, sediments, and the aquatic environment. Through continuous processes like oxidation and reduction, these microorganisms facilitate essential transformations that are both perpetual and periodic. These processes play a pivotal role in shaping ecological characteristics by maintaining and sustaining global biogeochemical cycles. Furthermore, it support ecosystem processes that are essential for maintaining biodiversity and climatic stability, including energy flow, biological interactions, and ecological succession (Smith et al. 2015).

The introduction of harmful substances (pollution) or contaminants into the environment disrupts its natural balance and adversely affects ecosystem functioning

(Özkara and Akyıl 2018). Pollution arises from various sources, including industrial emissions, agricultural runoff, urban waste, and deforestation, significantly impacting the ecological processes that maintain biodiversity, nutrient cycling, and climatic stability, threatening the health and sustainability of ecosystems globally (Ali and Rahman 2024). Pollution affects ecosystems primarily through its interference with essential ecological processes. Contributions of industrial and vehicular emissions (greenhouse gases (GHGs) like carbon dioxide (CO_2) and methane (CH_4)) to global warming and climate change are already well-established, and besides this, contaminants like heavy metals, plastics, and nutrients from agricultural runoff and improper waste disposal negatively impact soil health and fertility (Geyer et al. 2017). These contaminants disrupt soil microbial communities, which are critical for nutrient cycling, organic matter decomposition, and overall soil productivity. Loss of key microbial functions can trigger cascading effects across ecosystems, compromising plant health, reducing crop productivity, and impacting herbivore populations (Shi et al. 2023). Moreover, contaminated soil can serve as a source of pollutents for water bodies through leaching and runoff, perpetuating the cycle of ecosystem degradation (Akhtar et al. 2021). Additionally, pollutants such as pesticides and heavy metals harm aquatic organisms directly by causing toxicity and indirectly by altering habitat structures and nutrient availability (Amoatey et al. 2023). The sources and impact of common pollutants, such as pesticides, heavy metals, nutrients, and contaminants like microbial pathogens, have been characterized and well-established, but in the case of emerging pollutant such as microplastics, there is very limited data and research available for its sources and impact (Upadhyay et al. 2024).

While microorganisms are the backbone of aquatic ecosystems and their functioning, the impact of plastic pollutants on the aquatic food chain is presently gaining more attention, alongside studies on microplastics. Most of the waste that ends up in water bodies originates from land-based sources and is transported to the aquatic environment either through wind, direct human disposal, dumping, transportation or disaster damage, and overflow from sewage treatment plants during periods of heavy monsoon season (Oberbeckmann et al. 2014). Aquatic life across all the trophic levels is at risk from plastics and plastic fragments, as aquatic organisms often perceive, by mistake, plastic and its particles as food, leading to ingestion or entanglement, which can severely impact their health. In addition, persistent organic contaminants from the aquatic environment can adhere to plastic surfaces or leach out of plastic components, creating plastic-associated toxins that are highly toxic to living organisms. These toxins may bioaccumulate in organisms that ingest plastics, potentially causing widespread and unpredictable effects on the food chain. There is also a significant risk of microplastic transmission to humans, as they are the ultimate consumers of aquatic foods that are heavily affected by microplastics (Saha et al. 2021). This chapter explores these interactions in detail, focusing on the dual role of microorganisms in mitigating and exacerbating MP-related pollution. By advancing research on microbial processes and environmental variables that influence plastic degradation, it is possible to work toward innovative solutions for managing microplastic contamination and safeguarding environmental health.

1.2 Microplastics: Definition, Types, and Sources

1.2.1 Definition and Classification of Microplastics (MPs) and Their Characteristics

Microplastics (MPs) are extensive worldwide environmental pollutants as these are ubiquitous, persistent, and have the potential to affect the environment and human health (Yuan et al. 2022). These have been classified based on their occurrence, chemical composition, and environmental persistence (Wang et al. 2021) and broadly categorized into primary and secondary microplastics (da Costa et al. 2017). Primary microplastics are directly released into the environment as small particles without breakdown, in the form of microbeads, granules, microfibers, and powder, generally through personal care products, industrial activities, and other usages (Duis et al. 2016), whereas secondary microplastics are formed through the weathering and breakdown of primary MPs and large plastic products, such as bottles, plastic packaging, and disposed fishing nets and tools, through natural processes like biodegradation, ultraviolet (UV) degradation, and mechanical scratching (Bostan et al. 2023). These MPs generally appear in the form of fragments, fibers, and films (Sooriyakumar et al. 2022). Their chemical composition varies, and they are classified as polyethylene (PE), polypropylene (PP), polystyrene (PS), and polyvinyl chloride (PVC) (Silva et al. 2018). These materials differ in properties like density, crystallinity, and surface features, influencing their interactions with microorganisms (Restrepo-Flórez et al. 2014). Lightweight polymers like PE and PP can float on the aquatic surface and are mostly available on the top surface, increasing their exposure to breeding fish and aquatic birds (Lusher et al. 2017). These polymers are carried long distances through water tides and flow. Dense polymers like PVC sink to the bottom of water bodies, accumulate in sediments, and degrade slowly due to limited exposure to UV light and microbial activity. Microplastic debris can threaten benthic organisms and be transferred through the food web via bottom-up trophic pathways (Setälä et al. 2014). The crystal property of polymers also impacts their degradation. Polymers with high crystalline properties, like PE, degrade slowly due to their arranged molecular structures, which are less susceptible to enzymatic breakdown and microbial attack. Polymers with a high amorphous nature are more susceptible to environmental degradation. Additionally, surface properties like hydrophobicity and surface area play an important role in their environmental interactions. Hydrophobic polymers like PS and PP can adsorb hydrophobic contaminants, including persistent organic pollutants (POPs), serving as vectors for spreading contamination and hazardous materials (Agboola, et al. 2021). Microplastics with larger surface areas increase interactions with aquatic organisms, potentially affecting multiple trophic levels (Wright et al. 2013). The environmental behavior and adsorption properties of microplastics are influenced by factors such as surface functional groups, particle size and shape, surface charge, buoyancy, and hydrophobicity (Anderson et al. 2016). Plastics can be categorized into thermoplastics or thermosetting plastics (Bîrca et al. 2019). Thermoplastics can melt when heated and solidify when cooled. Ultimately, the persistence, transport, and ecological interactions of microplastics highlight their role as vectors for environmental and biological risks (Syberg et al. 2015).

1.3 Sources of MPs in Aquatic Environments

Microplastics (MPs) are rising as a global concern in both fresh- and marine water (Mallik et al. 2021). While earlier studies focused on marine environments, it has been evidenced that microplastic contamination has extended to freshwater systems like rivers, streams, lakes, ponds, creeks, and estuarine rivers (Peller et al. 2020). The sources of MPs in these environments are diverse and heterogeneous and stem from industrial discharge, urban runoff, agronomic practices, and other anthropogenic activities (Kumar et al. 2020). Primary microplastics are manufactured and used in industries such as synthetic clothing, personal care products, and air-blasting facilities, contributing to this pollution (see Fig. 1.1). Previous studies suggested that over 94,500 microbeads are discharged per wash from cosmetics and personal care products (Anagnosti et al. 2021). Secondary microplastics are formed from the fragmentation of larger plastic debris due to mechanical degradation, UV radiation, and other environmental factors (Huber et al. 2022). Both primary and secondary MPs enter aquatic systems through land-based sources, ocean-based activities, and

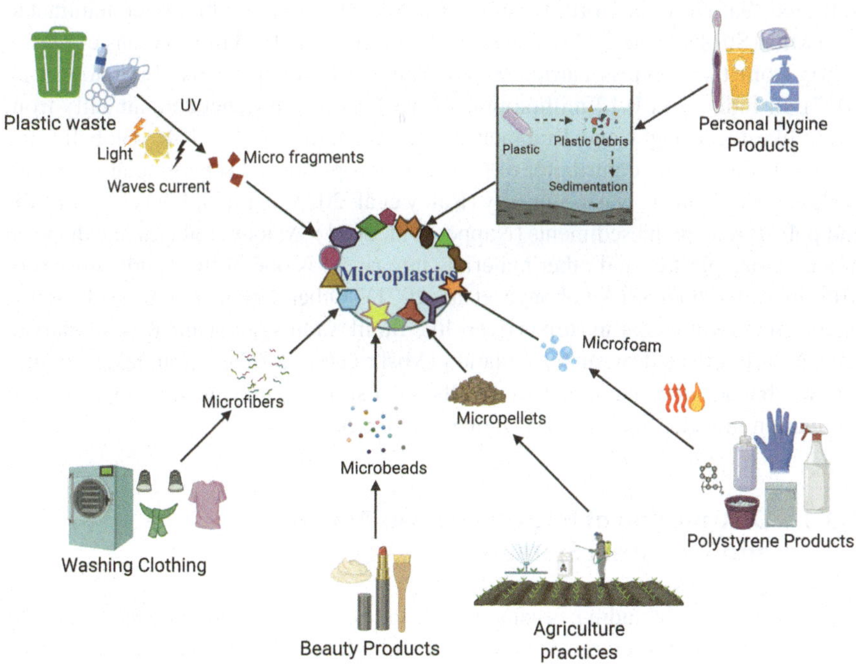

Fig. 1.1 Sources and characteristics of different types of microplastics in the environment

atmospheric transport (Dris et al. 2015). Land-based sources include urban runoff, agricultural practices using plastic mulch and sewage sludge, and effluents of wastewater treatment plants (WWTPs), which directly release microfibers and microbeads. Ocean-based sources of MPs include fishing gear, paint flakes, and maritime activities. Wind transport further disperses fibers and fragments, depositing them into aquatic environments. Microplastics are extensively used in various daily products, including cosmetics, personal care items, synthetic textiles, and industrial applications. Since 1990s, microbeads have been detected as a major source of MPs in marine ecosystems. MPs are consumed by marine organisms like plankton, leading to adverse ecological impacts (Cole et al. 2013). Industrial sources include waste from plastic manufacturing and processing plants and discharge materials such as polyethylene (PE) and polypropylene (PP) being released into the environment (Akhbarizadeh et al. 2021, Hamidian et al. 2021). Synthetic textile fibers shed during laundering, tire wear particles generated by road friction, and plastic waste resulting from littering and improper disposal are the sources of MPs in urban areas. Additionally, agronomic practices introduce MPs through sewage sludge, plastic covering, and recycled wastewater for irrigation. Poorly conserved landfills also play an important role in plastic degradation mediated by UV radiation and microbial activities, leading to MPs' discharge into the surrounding soil and water (Lin et al. 2023).

Wastewater treatment plants (WWTPs) capture a substantial quantity of MPs, but they still release them into freshwater systems through treated effluents. Studies indicated that rivers in North America, Europe, and Asia are highly contaminated with MPs (Strokal et al. 2023). Annually, approximately 0.33 million tons of plastic debris from the ocean is carried by the Yangtze River in China (Lebreton et al. 2017). In India, over 0.12 million tons of plastic waste is generated annually from rivers like the Ganges and Brahmaputra (Priyadarsini et al. 2024). Globally, the Ganges is the second contributor of plastics/microplastics, discharging around 0.12 million tons of plastic waste annually (Rafey et al. 2023), with high levels of PE, PP, and polystyrene in the sediments (Napper et al. 2021). At tourist places, the disposal of single-use plastics and other materials into rivers is one of the major sources of MPs in water bodies (Amobonye et al. 2021). Other Indian rivers, such as the Netravathi in Karnataka and urban rivers like Tamil Nadu's Adyar and Kosasthalaiyar, are similarly affected by such a situation (Mishra et al. 2023). Indian lakes, including Vembanad, Veeranam, and Red Hills, also show significant microplastic contamination in water bodies (Neelavannan & Sen 2023) .

1.3.1 Distribution of Microplastics in Aquatic and Terrestrial Systems

Aquatic systems, including oceans, rivers, lakes, and even groundwater, serve as significant microplastic reservoirs due to their high connectivity and ability to transport contaminants across large spatial scales (Du et al. 2020). Oceans, in particular, are heavily impacted by microplastics. A study estimates that approximately 14 million tons of microplastics currently reside on the ocean floor, far exceeding the quantities detected on the surface (Bohdan et al. 2022).

The global concentration of microplastics in surface oceans ranges 10^{-4} to 10^{4} particles per cubic meter (median 205 particles per cubic meter) (Zhao et al. 2025), with hotspots in heavily industrialized and densely populated coastal regions, such as the North Pacific Gyre, known as the "Great Pacific Garbage Patch." Rivers are critical conduits for microplastic transport into oceans, contributing up to 2.4 million metric tons of plastic annually (Kurniawan et al. 2023). For instance, the Yangtze, Ganges, and Nile rivers alone account for a significant proportion of this influx due to inadequate waste management. Freshwater lakes are also significant reservoirs of microplastics. A study on Lake Victoria found microplastic concentrations exceeding 400 particles per cubic meter, primarily originating from urban runoff and untreated wastewater (Egessa et al. 2020). Aquatic organisms, from zooplankton to fish, often ingest microplastics in rivers and lakes, causing physical and biochemical damages. In a recent assessment, microplastic ingestion was detected in >70% of commercially important fish species, raising concerns about bioaccumulation and human consumption (Makhdoumi et al. 2023). Groundwater systems, once considered less susceptible, are increasingly reported to contain microplastics. A study revealed microplastics contamination in groundwater in China (44 MPs/m^3), Australia (38 MPs/m^3), Bangladesh (24 MPs/m^3), United States (15 MPs/m^3), India (4 MPs/m^3) and United Kingdom (<1 MPs/m^3) (Jahan Rakib et al. 2025). These contaminants likely infiltrate groundwater through soil leaching, irrigation with contaminated water, and landfills, underscoring the systemic distribution of microplastics in hydrological cycles. Terrestrial environments, often overlooked compared to aquatic ecosystems, are now recognized as significant sinks for microplastics. Recent research estimates that terrestrial systems contain 4 to 23 times more microplastics than marine environments (Horton et al. 2017). Agricultural practices, industrial activities, and plastic waste mismanagement drive this vast accumulation. Microplastics in soils originate from multiple sources, including the degradation of plastic mulching films, wastewater irrigation, atmospheric deposition, and the application of sewage sludge as fertilizer. Sewage sludge, a byproduct of wastewater treatment plants, is a major contributor to soil microplastic pollution. A recent study (Shang and Chi 2023) reported that sewage sludge can contain up to 300,000 microplastic particles per kilogram. In agricultural fields, these particles are in the form of microbeads and micro-fragments integrated with soils; due to inadequate biodegradation, they remain there for decades (Briassoulis 2023). This practice introduces microplastics into food crops, as demonstrated in a study (Silva et al. 2021) that detected microplastics in the roots, stems, and leaves of edible plants, such as lettuce and wheat. Plastic covering, which is a common agricultural practice to improve crop yields, contaminates the soil badly. These covering plastic films degrade into micro- and nanoplastics (MNPs) with time. Khalid et al. (2023) estimated that over 70 kg of plastic residues per hectare accumulate annually in intensively cultivated fields. These residues of MNPs change the soil structure, reduce water retention, and interfere with nutrient cycling by disrupting microbial communities. Microplastics can easily be transported by wind and rainfall, leading to their deposition in terrestrial and aquatic systems (Rezaei et al. 2019). Evangeliou et al. (2020) detected microplastics in the Arctic snow and the Himalayas, indicating their long-range atmospheric movement. Microplastic concentrations show particularly high deposition rates in urban areas in the air, ranging from 0.2 to 1.5 particles per cubic meter (Dehghani et al. 2017). These particles settle down and remain on terrestrial and aquatic surfaces, contributing to soil, water, and vegetation contamination (Smith and Jones 2000).

1.4 Monitoring of MP Pollution and Its Diversity

1.4.1 Monitoring Microplastics in Aquatic Systems

Detection is the primary requirement for the monitoring of the smaller MPs and nanoplastics (<1 μm). Advanced filtration systems and imaging technologies, including Raman spectroscopy and Scanning Electron Microscopy (SEM), have helped enhanced and refine detection methods to capture MPs of varying sizes and compositions (Hale et al. 2020). In addition to surface monitoring, deep-sea sediments harbor approximately 14 million tons of MPs globally, primarily transported via marine snow aggregation and sinking processes (Xu et al. 2024). Rivers serves as major pathways for MP transport, channeling an estimated 1.15–2.41 million tons of plastics annually into marine systems. Studies indicated that ten major rivers, including the Yangtze, Ganges, and Mekong, contribute to approximately 80% of global riverine plastic discharge (Lebreton et al. 2017). Real-time monitoring via remote sensing and autonomous water-sampling drones has identified seasonal variations, with higher MP concentrations during monsoon periods (Möller et al. 2020). Furthermore, hydrological models integrated with in situ data have demonstrated that urban runoff contributes the largest share of riverine MPs, dominated by fibers and fragments derived from textiles, packaging, and vehicle tires. Freshwater lakes and reservoirs are increasingly being recognized as significant reservoirs for MPs. For instance, Lake Victoria and Lake Michigan exhibit MP concentrations of 1568.6 ± 1473.8 particles per cubic meter, dominated by polyethylene (PE) and polypropylene (PP) fragments (Mason et al. 2016). Emerging concerns highlight the bioaccumulation of MPs in lake ecosystems, with a recent study reported that 70% of sampled fish species contained detectable MPs in their gastrointestinal tracts (Sun and Wang 2023).

Monitoring MPs is inherently complex due to their heterogeneity in size, shape, and chemical composition. Size ranges from visible fragments (>1 mm) to nanoplastics (<1 μm), which require specialized imaging techniques such as Atomic Force Microscopy and separation techniques like Field Flow Fractionation (Rochman et al. 2016). Shapes such as fibers, films, spheres, and fragments are associated with specific sources, such as textiles, packaging, and industrial processes (Rillig and Lehmann 2020). Polymer identification using techniques such as pyrolysis–GC/MS (Gas Chromatography-Mass Spectrometry) reveals that MPs frequently co-occur with toxic additives (phthalates and bisphenol A) or adsorbed pollutants such as polychlorinated biphenyls (PCBs) and polycyclic aromatic hydrocarbons (PAHs). A report indicates that MPs can adsorb ten times their weight in hydrophobic pollutants, increasing their toxicity when ingested by organisms (Li et al. 2019, Santos et al. 2021).

1.5 Innovations in Monitoring Technologies

1.5.1 Machine Learning and Artificial Intelligence (AI)

Recent advancement in AI have enabled automatic MP detection in spectroscopic datasets, reducing processing time and human error. Moreover, 95% of machine learning algorithms trained on polymer spectra can accurately differentiate between varoius types of polymers and particle shapes (Ramanna et al. 2022, Li et al. 2023).

1.5.2 Remote Sensing and Drones

The accumulation of plastic on shorelines and marine bodies is increasingly being detected via remote sensing technology, such as satellite imagery and drone-based surveillance. To forecast MP dispersion patterns, these systems provide real-time geographical data that may be combined with atmospheric and hydrological models (Gorman et al. 2020, Garaba and Dierssen 2018).

1.5.3 Nanofiber Filters

Recently, researchers have developed high-efficiency nanofiber filters for nanoplastic monitoring that are capable to detect particles smaller than 20 nm, making it simple and easy to measure and classify them using advanced imaging methods (Enyoh et al. 2021).

1.5.4 Global Monitoring Networks and Databases

Global monitoring networks and databases play an important role in tracking microplastic pollution worldwide. To generate dynamic maps of MP pollution, international projects like the United Nations Clean Seas Campaign (UNCSC) and Global Plastic Watch incorporate information from the research studies, satellite imaging, and citizen science (Nyadjro et al. 2023). These networks support international policies like the Intergovernmental Negotiating Committee on Plastic Pollution (https://www.unep.org/inc-plastic-pollution) by promoting standardized data collection and monitoring protocols to ensure effective implementation and enabling coordinated responses. Monitoring MP pollution and its diversity requires interdisciplinary approaches and technological innovations (Xue et al. 2024). Strong and enhanced collaborations among researchers, policymakers, and industries are essential for developing actionable mitigation strategies (Miller et al. 2025, Bakir et al. 2024). Enhanced monitoring is important to raise global awareness and safeguard environmental and human health (Lamichhane et al. 2023).

1.6 Microbial Interactions with Microplastics

The association between aquatic plastics, microbes, and associated toxins has been a topic of recent interest. This concern dates back to 1972, when the first report highlighted bacteria and diatoms colonizing plastic (Carpenter and Smith 1972). However, it took over four decades for this issue to regain significant attention. Dang and Lovell (2016) studied the initial phases of biofilm formation on various plastic surfaces, including polyethylene terephthalate (PET), polymethyl methacrylate (PM), and polyvinyl chloride (PVC), during 24 and 72 h of exposure (Hadjiev et al. 2007). Their findings revealed that Rhodobacterales were the most dominant primary surface-colonizing group. Studies indicate that microbes interact with microplastics to varying extents, depending on their surface properties and chemical composition.

1.6.1 Processes of Microbial Colonization and Biofilm Formation on MPs

Microplastics (MPs) are first colonized by microbes when pioneer species interact with the polymer surface to form a conditioning layer (Debroy et al. 2022). The initial attachment of biofilm occurs through reversible interactions like hydrophobic interactions and van der Waals forces. At this stage, extracellular polymeric substances (EPSs) secreted by microorganisms provide a sticky matrix that promotes adherence and leads to a long-lasting, irreversible attachment. Environmental factors, such as temperature, pH, salinity, and nutrient availability, influence microbial attachment. Additionally, the chemical composition and surface energy of different polymers affect microbial colonization, with polymers like polyethylene and polypropylene being particularly prone to colonization due to their hydrophobic nature (Bazaka et al. 2011). Surface properties can also be altered by coatings, additives, or degradation products. After the initial attachment, the phase of microcolony formation begins. During this stage, additional microbes are attracted, and cells proliferate, resulting in the development of microcolonies. The maturation stage follows, where these microcolonies evolve into a three-dimensional biofilm structure with stratified microbial communities and distinct nutrient pathways. Secondary colonizers attach during this process, further diversifying the microbial community. This community's successional dynamics are shaped by environmental conditions, competition, and interactions. Eventually, during the dispersion phase, components of the biofilm may separate, enabling microbes to migrate to other surfaces (Kaplan 2010). Microplastics occur in various shapes, such as fibers, spheres, and fragments,

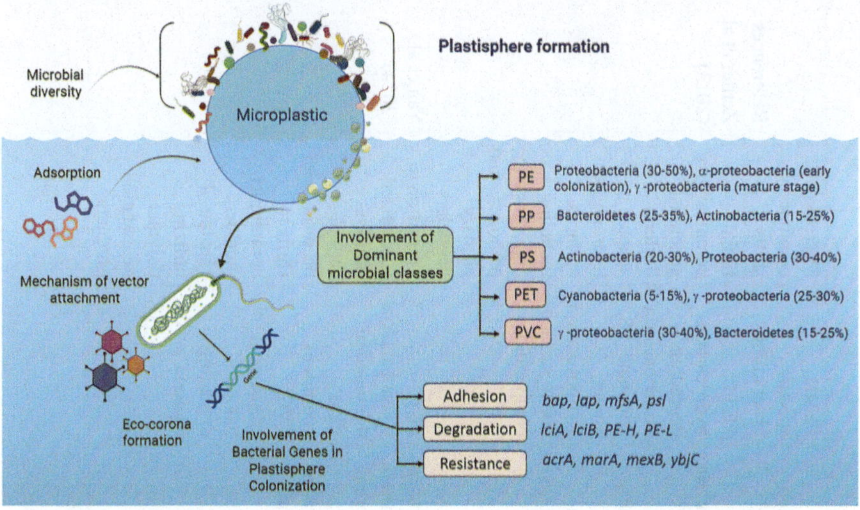

Fig. 1.2 Interaction of microbes with microplastics, eco-corona formation, and involvement of microbial diversity and genes

with smaller particles having a higher surface-area-to-volume ratio that promotes stronger microbial interactions. Irregular surfaces, as opposed to smooth ones, provide microhabitats and enhance cell adhesion. Environmental aging mechanisms, including UV exposure, oxidation, and mechanical abrasion, lead to surface cracks and breaks, further facilitating microbial attachment and biofilm growth (Fig. 1.2 and Table 1.1). These biofilms, through their developmental phases, modify the physical and chemical properties of MPs, enhancing their role as vectors for spreading impurities and pathogens. The formation of biofilms on MPs also has ecological implications. Biofilms influence the movement of hydrophobic organic pollutants (HOCs) between plastic waste and water due to their sorptive properties and their capacity to metabolize them (Edwards and Kjellerup 2013). Many bacteria, fungi, and algae within biofilms can break down HOCs, making them valuable for use in constructed bioreactors or in situ bioremediation of surface waters (Mishra et al. 2023). Furthermore, biofilms may be supported by the discharge of chemicals, which provide nutrients that encourage microbial growth. Age-related changes in the shape and size of MPs and naturally occurring dispersed particles also affect biofilm formation. The small size, resilience, and hydrophobic surface of MPs make them suitable substrates for microbial colonization. After being released into the environment, MPs interact with different microbial communities, leading to biofilm development.

Table 1.1 Factors involved in colonization and biofilm formation on microplastics

Stage	Key processes	Dominant microbial classes	Phylum	Dominant genera	Associated genes	Plastic-specific observations	Microbial community shifts	References
Early stage (0–10 days)	Adhesion to plastic surfaces, eco-corona formation	Alphaproteobacteria (40–50%)	Proteobacteria	*Pseudomonas, Alcanivorax, Marinobacter*	fadL (hydrocarbon uptake), ompA (adhesion), luxR (quorum sensing (QS))	PE/PP: hydrocarbon-degrading Proteobacteria dominate. PS: higher Actinobacteria due to surface compatibility	Initial colonizers are generalist bacteria adapted to hydrophobic surfaces, including opportunistic pathogens and primary degraders	Zettler et al. (2013)
Mature biofilm (10–30 days)	EPS production, microbial interactions, and biofilm stabilization	Gammaproteobacteria (20–40%), Bacteroidetes (25–35%)	Proteobacteria, Bacteroidota	*Flavobacterium, Rhodococcus, Streptomyces, Nocardia*	bcsA (biofilm stabilization), lasI (quorum sensing), yfiR (biofilm regulation)	PET: Cyanobacteria dominate in light conditions. PVC: enrichment of halogen-degrading bacteria	Transition to specialized biofilm-forming bacteria, with shifts to community members capable of utilizing polymer degradation products	Yan et al. (2024)

Long-term colonization (>30 days)	Enhanced microbial diversity, organic matter decomposition, and persistent colonization	Actinobacteria and Cyanobacteria increase	Actinomycetota, Cyanobacteria	*Synechococcus, Mycobacterium, Ideonella, Comamonas*	phaC (PHA synthesis), sodA (oxidative stress tolerance), PETase (polymer degradation)	PS and PET: diverse communities due to eco-corona PVC: persistent stress-tolerant species dominate	Stable, diverse communities with high specialization, including secondary degraders and syntrophic interactions for efficient polymer breakdown	Hou et al. (2025), Hou et al. (2021)
Eco-corona formation	Adsorption of organic matter, heavy metals, and pollutants enhances microbial attachment and colonization	Hydrocarbon-degrading bacteria (PE/PP), photoautotrophs (PET), and specialized degraders (PVC)	Varies	*Pseudomonas, Burkholderia, Halomonas*	tolC (toxin efflux), groEL (stress response), haloalkane dehalogenase (halogen compound degradation)	PE/PP: hydrocarbon-rich eco-corona. PET: complex organics support Cyanobacteria PVC: heavy metal adsorption selects metal-tolerant species	Selects bacteria with metabolic pathways for detoxification, stress tolerance, and utilization of adsorbed organic compounds	Yan et al. (2024)

(continued)

Table 1.1 (continued)

Stage	Key processes	Dominant microbial classes	Phylum	Dominant genera	Associated genes	Plastic-specific observations	Microbial community shifts	References
Plastic-specific biofilm composition	Variations based on surface chemistry and polymer structure influence microbial succession and community composition	PE: Proteobacteria dominate PP: higher Bacteroidetes and Actinobacteria PET: Cyanobacteria under light exposure	Varies	Depends on the plastic type	bphK (biofilm formation), csrA (carbon storage regulation), pahZ (polymer degradation)	Biofilm composition shifts over time, with variations influenced by polymer type and environmental conditions	Community shifts reflect adaptation to polymer type, surface roughness, and environmental factors, such as light, temperature, and nutrient availability	Tang et al. (2024)

1.6.2 Genetic Factors Involved in Colonization and Biofilm Formation on Microplastics

Microplastics (MPs) in freshwater ecosystems create a unique microenvironment called the plastisphere, where diverse microbial communities interact with environmental factors and plastic surfaces (Li et al. 2021). These interactions are significantly influenced by the type of plastic substrate, which affects microbial colonization patterns, biofilm composition, and functional dynamics (Jacquin et al. 2019). The genetic mechanisms and microbial community shifts that occur during colonization offer critical insights into the ecological impact of plastics and their potential applications in biotechnology. Different plastic types, such as polyethylene (PE), polypropylene (PP), polystyrene (PS), polyethylene terephthalate (PET), and polyvinyl chloride (PVC), serve as distinct habitats for microbial colonization due to their varied surface chemistries, hydrophobicity, and polymer structures (Cai et al. 2019). PE, widely used in packaging, is hydrophobic and primarily colonized by Proteobacteria, particularly Alphaproteobacteria during initial adhesion and Gammaproteobacteria in later stages of biofilm maturation (Zhang et al. 2022a, 2022b). The colonization is facilitated by genes such as *fadL*, involved in hydrocarbon uptake, and *ompA*, critical for adhesion (Waters et al. 2024). PP, with similar characteristics but a higher melting point, supports the growth of Bacteroidetes and Actinobacteria, with genes like *bcsA* aiding biofilm stabilization and *cutinase* playing a role in polymer hydrolysis (Augimeri et al. 2015). PS, commonly found in disposable items, promotes the dominance of Actinobacteria and Proteobacteria, leveraging genes such as *lipA* for lipase production and *katG* for oxidative stress protection (Hou and Majumder 2021). PET, used in bottles and textiles, attracts Cyanobacteria in illuminated conditions and Gammaproteobacteria, which utilize genes like *petase PETase* for PET hydrolysis and *epsA* for exopolysaccharide production (Basuki et al. 2020). PVC, being durable and used in pipes, is colonized by Gammaproteobacteria and Bacteroidetes, with specialized genes like *haloalkane dehalogenase* for dehalogenation and *hsp70* for stress response.

Biofilm composition on microplastics evolves, influenced by environmental conditions and the plastic type (Tu et al. 2020). During the early stage (0–10 days), Alphaproteobacteria dominate, particularly on PE and PP, due to rapid adhesion facilitated by their hydrophobic nature (Mammo et al. 2020). By the biofilm maturation stage (10–30 days), Gammaproteobacteria and Bacteroidetes become dominant, with significant contributions from Actinobacteria and Cyanobacteria, depending on the substrate (Sushmitha et al. 2021). PET, under light exposure, fosters Cyanobacteria, while PVC biofilms are enriched with halogen-degrading bacteria. Long-term colonization (>30 days) sees increased diversity, with Actinobacteria and Cyanobacteria thriving as organic matter and photosynthetic byproducts accumulating (Lemonnier et al. 2022). Plastic-specific variations persist, with PS and PET supporting diverse communities due to eco-corona formation, while PVC sustains stress-resilient species. The eco-corona, a layer of adsorbed organic matter and pollutants, further modulates colonization (Schefer et al. 2023). On PE and PP, hydrocarbon-rich layers attract hydrocarbon-degrading bacteria. PET's eco-corona supports cyanobacteria and photoautotrophs due to its photosynthetic ability that lead to enhanced microbial diversity to the surface of MPs composed of complex organic molecules (Tang 2024). PVC, known for adsorbing heavy metals and halogenated compounds, selects bacteria capable of specialized degradation (Yu et al. 2016).

The genetic repertoire of microbes colonizing microplastics is central to their ecological functions and plastic degradation capabilities. In PE and PP, genes like *ompA* drive initial adhesion, while *fadL* and *lipA* supports hydrocarbon uptake and degradation respectively (Pinto et al. 2022). For PS, oxidative stress response and polymer breakdown are facilitated by genes such as *katG* and *cutinase* (Cai et al. 2023). PET degradation is uniquely driven by *PETase*, which hydrolyzes PET, and *bcsA*, which stabilizes biofilm formation (Huang et al. 2024). PVC-degrading microbes utilize *haloalkane dehalogenase* for dehalogenation and *groEL* for stress tolerance. These genetic mechanisms not only enable colonization but also enhance the potential for bioremediation by breaking down polymers into simpler compounds.

The ecological and biotechnological implications of plastisphere colonization are vast (Jacquin et al. 2019). The microbial communities forming on plastics influence nutrient cycling, pollutant bioaccumulation, and the dissemination of pathogens (Ormsby et al. 2023). Plastisphere communities act as vectors for contaminants, raising concerns about ecosystem health and human safety (Gkoutselis et al. 2021). From a biotechnological perspective, leveraging plastic-specific bacterial genes offers opportunities for improving waste management and pollution mitigation. For instance, engineering microbes with enhanced PETase activity can expedite PET degradation, while microbial consortia can be designed for multi-polymer breakdown (Tournier et al. 2020). Long-term shifts in biofilm composition highlight the potential for microbial adaptation and natural biodegradation of plastics in the environment. This understanding can inform the development of targeted strategies to mitigate microplastic pollution and leverage microbial capabilities for sustainability.

Plastic type significantly determines microbial colonization dynamics in the plastisphere, influencing both ecological interactions and potential biotechnological applications (Wright et al. 2013). As different plastics host distinct microbial communities with specialized genes, this knowledge provides a foundation for advancing bioremediation strategies and addressing the environmental challenges posed by microplastics (Wani et al. 2023). Through a comprehensive understanding of these interactions, coupled with the development of innovative approaches, we can mitigate the adverse effects of microplastics and harness microbial potentials for ecological and industrial benefits. These biofilms play a structural role as interfaces between sediments and surrounding water, with significant applications in ecotoxicology for assessing the impact of chemicals in aquatic environments (Sabater et al. 2007).

1.6.3 Factors Influencing Microbial Colonization: MP Type, Size, and Environmental Conditions

Microorganisms play a crucial role in the biogeochemical cycle of aquatic environments by interconnecting biotic and abiotic factors by influencing the intricate association with plastic debris (Rogers et al. 2020). Polyethylene (PE), polypropylene (PP), polystyrene (PS), and polyvinyl chloride (PVC) are among the several polymers that make up microplastics (Dey et al. 2022). Microbial colonization is affected by the unique physical and chemical properties associated with different types of polymers and has become a prominent field of research on these interactions, providing information on microbial community dynamics, pollutant transfer, and ecological effects (Zhang et al. 2018). Microbial adhesion is significantly affected by the hydrophobicity and surface characteristics of plastics. Many polymers are hydrophobic, which promotes the growth of biofilms (Do Sul and Costa 2014). Compared to hydrophilic polymers like polyamide (PA), PE, and PP, which are extremely hydrophobic, they frequently show more widespread microbial colonization. However, on the other hand, hydrophilic polymers could draw particular microbial species that are suited to less hydrophobic surfaces. Microplastic chemical composition, including leachates and additives, has a major effect on microbial colonization (Table 1.1). Depending on their chemical composition, additives like plasticizers, stabilizers, and flame retardants either facilitate or restrict microbial development. Common plasticizers like phthalates, for example, may act as carbon sources for certain microorganisms, promoting colonization (Gu 2007). The condition of breakdown of microplastics affects microbial colonization as well. Rough surfaces as well as increased surface area are common features of weathered microplastics, which promote microbial adsorption. The chemistry of the polymer is changed by processes such as oxidation and photodegradation, which increases the possibility that particular microbial populations may colonize it. Microbial colonization patterns are impacted by the size of microplastics, which can range from millimeter-scale fragments to nanoplastics (plastic particles with at least one dimension in the nanometer scale, ranging from 1 to 100 nm). Because of their larger surface area with volume ratio, smaller particles like nanoplastics provide space to facilitate microbial adhesion. This property allows the formation of thick biofilms on their surfaces, resulting in a plastisphere (Fig. 1.2). Furthermore, the mobility around small microplastics allows them to spread throughout all types of environments and encounter a wide range of microbial communities. When microplastics aggregate with organic matter and other particles, they can create microenvironments that are favorable to microbial growth, which can further affect colonization dynamics. Additionally, the size of microplastics has an enormous impact on how aquatic life consumes them. Ingesting microplastics can change the colonization process by interacting with microbial communities in the gut of an organism, which is suitable for specific environmental niches. Many environmental factors, like temperature, salinity, pH, nutrient availability, and the level of contaminants, affect microbial colonization on microplastics (Tang et al. 2024).

Microbial communities are greatly affected by temperature since it influences their metabolic rates and the formation of biofilms (Villanueva et al. 2011). In general, higher temperatures increase microbial activity and the colonization rate. On the other hand, psychrophilic microorganisms (microbes such as bacteria, archaea, and fungi that grow and reproduce at temperatures ranging between 0 °C and 15 °C or in extremely cold environments) can survive in cold environments, leading to distinct colonization patterns.

The presence of salinity gradients in aquatic environments directly impacts microbial colonization. Freshwater systems sustain a variety of microbial species, while marine environments with high salt concentrations commonly contain halotolerant (organisms that survive and grow in environments with different salt concentrations but do not require high salt levels for growth) or halophilic microorganisms (microbes that require high concentrations of salt (sodium chloride) to grow and function). Microplastics are the surfaces that assist these communities in interacting with each other. The pH of the environment affects the growth and adherence of microorganisms. The structure and variety of microbial biofilms on microplastics can be changed by alkaline or acidic environments, which promotes acidophiles (microbes that grow in low pH, generally below 7) and basophiles (microbes that survive in highly alkaline conditions, or pH above 7), species that can tolerate different pH levels (Lowe et al. 1993)).

1.7 Impact of Microplastics on Microbe-Assisted Environmental Processes and Ecological Implications of Microplastic Interactions

1.7.1 Impact of MPs on Microbial Communities and Their Metabolic Functions

Microplastics (MPs) are ubiquitous pollutants that have major impacts on biological and ecological systems. MPs have a major effect on microbial communities and are the main agents driving biogeochemical cycles, organic matter breakdown, and environmental health in microbial populations. For microbial colonization, microplastics (MPs) offer a special substrate that promotes the development of biofilms, sometimes referred to as "plastispheres" (Zhai et al. 2023). These biofilms include diverse microbial communities from the surrounding environments (Yin et al. 2019, Di Cesare et al. 2024) and have shown that the taxonomic composition of biofilm communities on MPs varies, with particular bacterial taxa, including *Pseudomonas, Vibrio,* and *Alteromonas,* commonly enriched (Fig. 1.2). The type of MP polymer, surface roughness, and environmental factors, including temperature, salinity, and nutrition availability, all affect the development of biofilms. MPs may also alter the variety of microorganisms by promoting particular microbial groups that possess the ability to adhere to or break down plastic surfaces (Debroy et al. 2022). Bohdan (2022) have shown that plastispheres have a greater diversity of microbes than ambient water or sediment, while others have shown that selective factors have led to a reduction in diversity.

MPs can disrupt microbial carbon metabolism by providing an artificial carbon supply, which has a significant impact on the carbon cycle. Energy can be converted from natural carbon sources to MP degradation by microbes that can break down MPs into monomers, such as certain strains of *Rhodococcus* and *Ideonella sakaiensis* (Danso et al. 2019) and the simplified monomers (artificial carbon supply) can be utilized by the other microbes, leading to artificial changes in the microbial diversity, which may alter the function of the ecological niche. MPs have an adverse impact on nitrification and denitrification processes, which, in turn, affect nitrogen cycling, leading to decreased nitrogen cycling efficiency in sediments contaminated with MPs (Wang et al. 2023).

1.7.2 MPs as a Vector for Microbial Pathogens and Pollutants

The vector impact of microplastics (MPs) is especially concerning because it may alter their behavior in the environment by interacting with the chemical or biological agents they adsorb or attach to, thus complicating the risk associated with their presence (Huang et al. 2021). MPs can adsorb chemical pollutants and attach to biological organisms, potentially acting as vectors for these harmful agents (Cao et al. 2021). This phenomenon is known as the "vector effect." MPs are capable of transmitting metals, medications, and persistent organic pollutants (POPs). These MPs are found in many environmental components, including air, freshwater, wastewater, oceans, bottled water, tap water, and food (Fig. 1.3). Plastic surfaces can support abundant growth of the Bacteroidetes phylum, with the order Flavobacteriales, in particular, identified as colonizing polystyrene (PS), low-density polyethylene (LDPE), polypropylene (PP), and polyethylene terephthalate (PET) (Roager and Sonnenschein 2019).

1.7.3 Microplastics and Antimicrobial Resistance and Dissemination

The increase in antimicrobial resistance (AMR) is currently one of the biggest public health concerns (Ferri et al. 2017). The mechanisms of microbial resistance are complex, involving both acquired and innate strategies. A common path for AMR is genetic adaptation, where microbes acquire resistance genes through horizontal gene transfer (HGT) (Michaelis and Grohmannet al. 2023). This process, facilitated by integrons, transposons, and plasmids, allows microbes to inherit components such as transformed target proteins, enzymes that degrade antibiotics (ATs), and efflux pumps that reduce the effectiveness of drugs (Aminov 2011). Another significant mechanism of resistance is biofilm-mediated resistance. Biofilms help bacteria evade antimicrobial treatment by preventing drug penetration, altering microenvironments, and promoting the expression of resistance genes (Uruén et al. 2020). Strategies to combat AMR include antibiotic stewardship programs, the development of novel drugs, and the exploration of alternative therapies like phage therapy

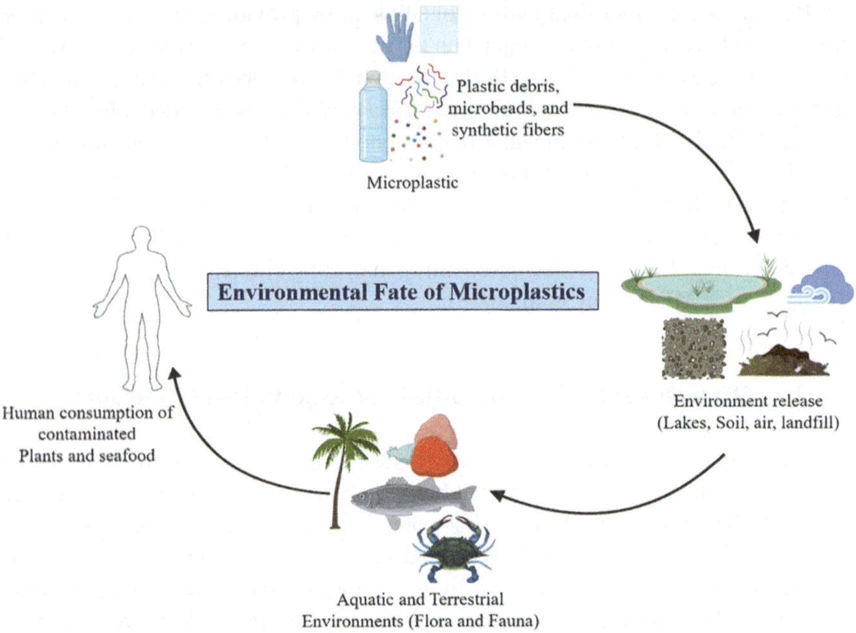

Fig. 1.3 The *environmental cycle of microplastics* illustrates the movement and impact of microplastics in the ecosystem. Plastic debris, microbeads, and synthetic fibers contribute to the formation of microplastics, which are released into various environmental compartments, including lakes, soil, air, and landfills. These microplastics infiltrate aquatic and terrestrial ecosystems, affecting flora and fauna. Contaminated plants and seafood that humans consume reintroduce microplastics into the food chain, highlighting plastic pollution's potential health and ecological risks

and antimicrobial peptides. In addition to antimicrobial resistance, antibiotics (ATs) and MPs have emerged as pollutants that harm the freshwater environment (Li et al. 2018). It has been shown that common antibiotics, such as amoxicillin (AMX), ciprofloxacin (CIP), tetracycline (TC), trimethoprim (TMP), and sulfadiazine (SDZ), are adsorbed by various types of microplastics, including polyamide (PA), polyethylene (PE), polypropylene (PP), polystyrene (PS), and polyvinyl chloride (PVC), in freshwater systems (Baruah et al. 2024). Among these, PA has been recorded to show the highest adsorption capacity for antibiotics. Tetracycline, in particular, is of great concern due to its high biotoxicity, hydrophilicity, solubility, and abundance of functional groups. The presence of both microplastics and antibiotics in aquatic ecosystems can synergistically affect aquatic organisms. Research indicates that the prevalence of antibiotic resistance genes (ARGs) is often higher in MP-associated biofilms compared to the surrounding water or sediments. Microplastics have also been shown to enhance quorum sensing in multidrug-resistant strains of *Escherichia coli* and *Pseudomonas aeruginosa*, further stabilizing biofilms and contributing to the persistence of ARGs (Bohdan 2022).

1.7.4 Implications for Microbial Resilience and Resistance to Environmental Stressors (UV Radiation and pollutants)

Microorganisms inhabit a wide range of ecological niches and are essential to life on Earth due to their remarkable ability to adapt to external environmental stresses. Ultraviolet (UV) radiation and antimicrobial agents are some of the major risks to microbial survival. UV radiation, in particular, is the most common environmental stress that can harm DNA, denature proteins, and disrupt cellular membranes. To survive and grow in such conditions, microbes have developed several defense mechanisms, with DNA repair being one key process. Photoreactivation, nucleotide excision repair (NER), and base excision repair (BER) are employed to fix UV-induced DNA damage, such as cyclobutane pyrimidine dimers (CPDs). Many prokaryotic and eukaryotic microorganisms possess enzymes called photolyases that use light energy to reverse CPDs. In addition to DNA repair, many microorganisms synthesize protective pigments, such as carotenoids and mycosporine-like amino acids (MAAs), to defend against UV radiation. These UV-absorbing compounds act as natural sunscreens, minimizing UV penetration and protecting cellular components from damage. Interestingly, MPs can influence microbial UV resistance by providing a physical shield against direct UV exposure. Microbes colonizing the surface of MPs often reside in biofilms, where the matrix and the plastic surface attenuate UV penetration. Moreover, the hydrophobic surface of MPs enhances the persistence of UV-resistant microorganisms, allowing them to thrive in high-UV environments. In addition to serving as substrates, MPs leach hazardous chemicals such as bisphenol A (BPA), phthalates, and other endocrine disruptors into the environment. They also adsorb and concentrate persistent organic pollutants (POPs) like polycyclic aromatic hydrocarbons (PAHs) and polychlorinated biphenyls (PCBs), which interact with microbial communities. Some microbes possess metabolic pathways to degrade these compounds, further highlighting their adaptive potential (Wang et al. 2021).

1.8 Biodegradation of Microplastics by Microorganisms

Plastics, especially synthetic polymers like polyethylene, polypropylene, and polystyrene, are commonly used in several industry sectors. Their extensive usage, along with their durability in the environment, has raised concerns regarding the accumulation of microplastics in marine, terrestrial, and freshwater habitats. Because of their small size, these microplastics represent substantial environmental and health risks, enabling organisms across the food chain to ingest them easily. While the environmental impact of microplastics is widely established, their removal remains a difficult task. Conventional methods of plastic waste management, such as physical and chemical processes, are sometimes resource- and energy-intensive and insufficient. Biological degradation, particularly enzymatic biodegradation, is an effective strategy for reducing microplastic pollution (Liu et al. 2022) (Fig. 1.4). Enzymatic degradation utilizes naturally existing enzymes to degrade polymer

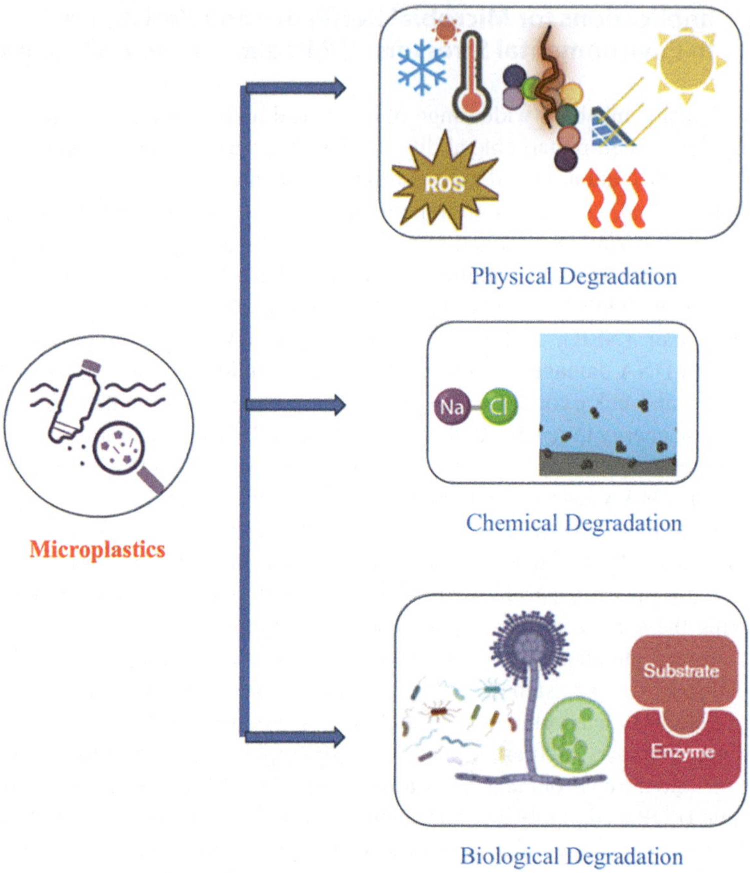

Fig. 1.4 Overview of microplastic degradation pathways, including physical, chemical, and biological processes. Physical degradation involves factors like temperature, radiation, and reactive oxygen species (ROS). Chemical agents facilitate chemical degradation, while biological degradation relies on different types of microbes and their microbial enzymes and substrates

chains, potentially converting plastic waste into sustainable and eco-friendly products and electricity (Kalathil et al. 2022, Mohanan et al. 2020).

1.8.1 Microbial Diversity in MP Degradation: Proteobacteria, Cyanobacteria, and Bacteroidetes

In 1972, the first report on microbial colonization of microplastic waste was recorded. Since then, the structure, composition, morphology, and diversity of microbial communities concerning their taxonomic units have become a subject of increasing research. The bacterial groups that exist on the surfaces of freshwater plastic litter are quite distinct from those that survive on other types of particles or in the surrounding pelagic oceans (De Tender et al. 2015; Oberbeckmann et al. 2018; Ogonowski et al. 2018; Zettler et al. 2013). Plastic biofilms are commonly enriched with bacterial groups belonging to the phyla Proteobacteria, Bacteroidetes, Firmicutes, and Cyanobacteria, indicating that plastic could serve as a favorable ecological niche for particular taxa (Zhai et al. 2023). This increase in bacterial diversity indicates that these species may have undergone metabolic adaptations that facilitate their colonization and survival on plastic substrates, such as improved adhesion, chemotaxis, additive resistance, and degradation capacities. PET can be degraded and absorbed by novel bacterial species *Ideonella sakaiensis* 201-F6, which was isolated in 2016 by Yoshida et al. (2016) from a PET bottle–recycling plant. In comparison to seawater, the most ubiquitous phylum of Proteobacteria on marine plastic waste is also the most prominent phylum worldwide. The abundance of this phylum was found to be more on plastic surfaces than on glass or organic surfaces. Alphaproteobacteria, Gammaproteobacteria, and sometimes Betaproteobacteria and Deltaproteobacteria are the classes of this phylum that are concentrated on plastic-based substances. PP, PS, and PE are often colonized by members of the Hyphomonadaceae alphaproteobacterial family, which also includes organisms classified as methylotrophs, hydrocarbon decomposers, and species isolated from hydrocarbon-enriched environments. PE and PS were substantially correlated with the Hyphomonadaceae family in contrast to the colonization of wooden pellets (Oberbeckmann et al. 2018).

Microplastic biodegradation is primarily dependent on microbial diversity. Bacteria and fungi that produce plastic-degrading enzymes are commonly found in areas that have extensive plastic deposits, such as landfills, leachates, compost heaps, and oceans. These microorganisms degrade plastics through a range of enzymatic methods, generally in conjunction with other microbial processes, including oxidation and fermentation. Symbiotic associations between different microbial species may improve the overall degradation rate because specific bacteria produce enzymes that break down plastics into simpler monomers that are subsequently metabolized by other microbes. Recent advancements in metagenomics and microbial genomes have resulted in the identification of novel plastic-degrading bacteria. For example, many strains of *Ideonella sakaiensis* have been reported to produce PETase, an enzyme capable of degrading PET into its constituent monomers (Taniguchi et al. 2019). Similarly, *Pseudomonas* species have been identified for producing esterases, enzymes capable of degrading polyester (Wilkes and Aristilde 2017). Fungal species such as *Aspergillus* and *Penicillium* have also been investigated for their potential in the degradation of plastics, especially those containing aromatic compounds (Sánchez 2020).

Additionally, microplastics may act as a vehicle for several harmful contaminants, such as pharmaceuticals, personal care items, polyfluoroalkyl substances (PFAs), heavy metals, polycyclic aromatic hydrocarbons (PAHs), polychlorinated biphenyls (PCBs), dichlorodiphenyltrichloroethane (DDT), and polybrominated diphenyl ethers (PBDEs) (Ahmed et al. 2023). Variations in plastic composition affect how pollutants interact with and may spread across freshwater habitats. How plastic materials interact with environmental pollutants, particularly those of a chemical nature, can vary and include van der Waals forces, hydrophobic interactions, hydrogen bonding, electrostatic interaction-repulsion forces, halogen bonding, π-π interaction, or partition effect (Tumwesigye et al. 2023) In addition to chemical pollutants, environmental microorganisms can colonize MP surfaces and produce biofilms. MPs and environmental bacteria interplay to generate extracellular polymeric substances (EPSs). The PETase enzyme initiates the aerobic decomposition of PET, while styrene monooxygenase initiates the most prominent biodegradation in route for PS (Jacquin et al. 2019). Few instances of esterases, lipases, and cutinases that may hydrolyze PET are present (Zettler et al. 2013).

1.8.2 Mechanisms of Microplastic Biodegradation: Enzymatic Breakdown

The primary process by which microorganisms degrade MPs is the synthesis of extracellular enzymes that fragment polymer chains into simpler monomers and oligomers. Hydrolases (such as cutinases and esterases) and oxidases (such as laccases and peroxidases) are enzymes involved in the degradation of plastic polymers (Dhali et al. 2024). Cutinase enzyme works against polyethylene terephthalate (PET), which breaks ester bonds within the polymer (Fig. 1.5), whereas the mechanism of hydrolase enzymes, such as lipases and carboxylesterases, depends on the type, character and variety of polymers. The class of oxidase enzymes, like laccases and peroxidases, initiates the oxidative cleavage of carbon-carbon bonds in polymers such as polyethylene (PE) and polystyrene (PS). Enzymes mainly belong to the class of hydrolases, such as esterases, lipases, and laccases, and have gained focus for their capacity to degrade plastics. Hydrolases are enzymes that catalyze the hydrolysis of chemical bonds with water. Esterases, a class of hydrolases, are especially crucial for the biodegradation of polyesters, such as polyethylene terephthalate (PET), which is widely used in beverage bottles, textiles, and packaging. PETase and mono(2-hydroxyethyl) terephthalic acid hydrolase (MHETase) enzymes have been found in a variety of microorganisms and have shown potential in the degradation of PET. Yoshida et al. (2016) found PETase for its capability to depolymerize PET into its monomers, terephthalic acid (TPA) and ethylene glycol (EG), providing an extremely effective enzymatic process that has prompted interest in biotechnological applications in plastic waste recycling (Fig. 1.5). Lipase, an enzyme that hydrolyzes ester linkages in fats and oils, has been proven to degrade polyolefins like polypropylene and polyethylene (Tournier et al. 2023). However, the process is sometimes slow and needs certain environmental conditions. Lipase activity is regulated by temperature, pH, and the presence of cofactors; hence, optimizing lipase-based degradation systems is a current research area.

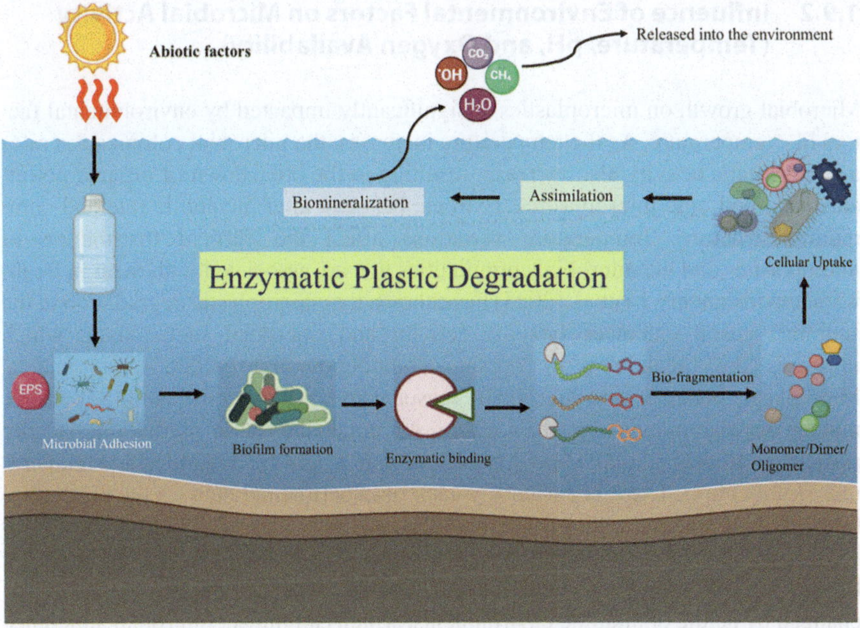

Fig. 1.5 An overview of the enzymatic degradation through microorganisms that break down the plastic polymer into small monomers and fragments, which are then processed by intracellular enzymes

1.9 Challenges in Microbial Degradation of Microplastics

1.9.1 The MP Degradation Efficiency of Microbes

The persistence of MPs and their harmful impact on biodiversity creates MP pollution, a severe environmental problem. The distribution, ingestion, density, behavior, quantity, and impact of MPs have all been the subject of research in the past decade. However, the efficiency of strategies developed for MP removal and repair, however, is still unknown. MPs can resist degradation and remain in the environment, but specific microorganisms can break them down to reduce MP pollution. Factors like environmental conditions, temperature, pH, surface properties, and MP composition influence the effectiveness of degradation. Microbes can break down several compounds, including MPs, and are naturally able to adapt to any environmental condition. Utilizing bacteria to break down MPs will improve biodegradation without affecting the environment, making it an appropriate and safe method to support natural bioremediation and help in the detoxification of natural ecosystems without having a negative effect. Many viable microorganisms have been identified to date, but the links between microbes and MPs are still unidentified, and there is still a lack of information about MP removal (Bacha et al. 2021). The breakdown of MPs is a comprehensive process that includes microbiological and physicochemical degradation components in different circumstances.

1.9.2 Influence of Environmental Factors on Microbial Activity (Temperature, pH, and Oxygen Availability)

Microbial growth on microplastics is significantly impacted by environmental factors like temperature. As the temperature increases, the microbial colonization rates and enzymatic activity also increase, resulting in the breakdown of organic pollutants adsorbed onto microplastics. Extreme heat can alter metabolic rate and community structure. Temperature variations affect the vertical distribution of microplastics and the microbial communities that are associated with them in freahwater environments. Li et al. (2023) have shown that microplastic degradation in the summer season enhances enzyme activity and microbial metabolism, which improves the breakdown of organic compounds and certain contaminants. Microplastic behavior is influenced by seasonal and diurnal stratification variations, making regular monitoring challenging. Accurate dispersion predictions require associated models that take into consideration microplastic dynamics and temperature stratification. There is currently a lack of experimental data on the concentrations of microplastics in thermal layers (Caniato et al. 20212021).

The metabolism and adhesion of microorganisms to microplastics are impacted by the pH of the environment. Surface characteristics of microplastics can be changed by acidic or alkaline environments, which can impact microbial adherence and biofilm development. In acidic conditions, heavy metals like lead and cadmium adsorb to the surface of microplastics. Organic contaminants can be promoted in basic conditions by changing the surface charge and chemical bonds. Microplastics may break down more quickly in extremely basic or acidic pH environments because of higher rates of oxidation and hydrolysis. Smaller particles, or nanoplastics, may be produced as a result, and these might have distinct and perhaps more detrimental impacts on the environment and biodiversity. Plastics break down in aquatic environments due to an interaction between mechanical forces, UV rays, and pH variations. These substances may be toxic to aquatic life, especially in low-pH conditions, when leaching is more likely to occur (Li et al. 2024a, b).

It is possible to forecast the environmental behavior of microplastics and minimize their negative impact on wildlife and human health by keeping an eye on pH levels in environments where they are common. Changes in pH affect the solubility of pollutants that have been adsorbed, which, in turn, affects the routes of microbial breakdown (Wu et al. 2012). The fate of microplastics and the activity of the microbial communities that are linked to them are significantly influenced by the availability of water. Water currents and turbulence in aquatic environments have an effect on the circulation and accumulation of microplastics, which modifies the dynamics of microbial colonization. Microplastics lack organic nutrients but enable surfaces for microbial colonization (Sheridan et al. 2022). On the one hand, nutrient-poor environments may restrict microbial activities and enhance oligotrophic or specialized degrading agents, whereas nutrient-rich environments enhance biofilm development and metabolic rates.

Microplastics provide hotspots for microbial associations with contaminants by acting as transporters of hydrophobic organic substances, heavy metals, and pesticides. On the other hand, phthalates are hormonally active because their constituents resemble and/or have hormonal actions that cause endocrine disruption and may pose health risks similar to those of MPs (Mihai et al. 2021). These pollutants can either activate degradative pathways in resilient microbial communities or generate toxicity, inhibiting microbial activity (Zhu et al. 2017). Microplastics and contaminants occur together, indicating their coupled microbial and ecological effects (Mammo et al. 2020). Ecological factors rarely functions independently. For instance, the metabolic process of bacteria correlated with microplastics is regulated by both temperature and the accessibility of nutrients (Sheridan et al. 2022).

1.10 Limitations and Gaps in Current Research on MP Degradation

The interaction between these microplastic particles and biotic factor in the different ecological niche are still part of the gap in research, even though no studies have yet identified any harmful effects of MPs on primary consumers except few studies like on microalgae in freshwater ecosystem (Nava and Leoni 2021). Although MPs were effectively detected in biofilms using flow cytometry in a recent investigation, but no toxicological or morphological effects were noted on the biofilm(Sgier et al. 2016). The direct negative impacts of plastic waste and related pollutants on the biofilms, however, gain less attention. Most of the study focuses on the sources/ occurrence, fate, and effects of microplastics in the aquatic ecosystem, but there is still a gap in the fate and effects of microplastics in terrestrial ecosystem, which is a potent source of aquatic MPs (; Horton et al. 2017).

To understand the ecological effects of microplastic contamination, it is essential to examine these interrelationships of environmental factors. Establishing practical methods to reduce plastic pollution requires an understanding of the microbial impact on microplastics. Microbes linked to microplastics may contribute to the breakdown of pollutants or act as carriers of pathogens (Mammo et al. 2020). Understanding their activity may assist in plastic waste management practices and bioremediation techniques. MPs are more prevalent in developed countries and areas with poor waste management practices (Mihai et al. 2021). Additionally, wind may carry and leave behind lightweight plastic particles from synthetic grass, construction materials, and household dust through air fallout in both terrestrial and aquatic habitats.

1.11 Future Directions in Research and Mitigation Strategies

1.11.1 Need for Comprehensive Monitoring of MP Pollution

The degradation of plastic in many environmental sectors results in MP pollution, leading to the need for demanding monitoring to formulate mitigation and control measures (Singh et al. 2023). Interestingly, human feces have also been found to contain microplastics. The ability to accurately detect these microscopic plastic particles is essential for estimating the amount present in tap and bottled water. Standardized techniques for evaluating microbial activity on microplastics in various environments have to be developed. The potential of microorganisms linked to microplastics for bioremediation applications has not been yet investigated well, resulting in an extensive gap in their methods and capacities to combat environmental contamination. Research in understudied habitats, such as fresh water, deep-sea and polar regions, is noticeably lacking, and more investigation is required to fill this knowledge gap. A crucial knowledge gap about the global environmental effects of microplastics is highlighted by the lack of research on the long-term ecological consequences of the microbial activity linked to these plastic fragments.

1.11.2 Analytical Techniques for Microplastic Detection

Microplastics have become a significant environmental issue due to their widespread presence in waterbodies, soil, air, and even the food chain. As the environmental risks of microplastics grow, effective techniques for their detection are essential. To identify and quantify MPs in water, many analytical methods were implemented. Their detection is crucial for environmental monitoring, prompting the development of various analytical techniques. Visual inspection, combined with microscopy, is one of the simplest methods used for MP detection. This technique allows for the direct observation of plastic particles under an optical microscope, and fluorescent dyes can enhance the contrast of microplastics. This method is labor-intensive, time-consuming, and less effective for detecting microscopic particles.

Fluorescent dyes can improve visibility, though the method is labor-intensive. FTIR (Fourier Transform Infrared Spectroscopy) spectroscopy identifies polymer types based on absorption spectra, detecting MPs as small size as 20 μm. Raman spectroscopy offers high-resolution molecular identification but is costly and sensitive to fluorescence interference (Picó & Barceló. 2019). Scanning electron microscopy, often paired with energy-dispersive X-ray analysis, reveals particle morphology and elemental composition. Density separation and filtration aid in isolating MPs by size and density. Near-infrared spectroscopy is nondestructive and effective for certain polymers. Pyrolysis-GC/MS thermally degrades MPs to identify polymer types with high sensitivity, though it is destructive. Thermal methods like DSC (Differential Scanning Calorimetry) and TGA (Thermogravimetric Analysis) distinguish plastics via thermal behavior but are less suitable for complex samples. Flow cytometry rapidly assesses particle shape, size, and fluorescence but struggles with particles under 1 μm. Microwave digestion with ICP-OES (Inductively Coupled Plasma - Optical Emission Spectroscopy) or ICP-MS (Inductively Coupled Plasma - Mass Spectrometry) analyzes both plastic and inorganic contaminants, though it alters sample integrity. Nanoparticle Tracking Analysis (NTA) provides size distribution data in liquids but lacks polymer specificity. Each method varies in sensitivity, resolution, sample compatibility, and equipment requirements. No single technique suffices alone; thus, combining complementary methods enhances accuracy. The choice depends on particle characteristics, sample type, and analytical goals. Advancing these technologies is key to mitigating microplastic pollution effectively.

1.11.3 Exploration of Microbial Consortia and Bioengineering for Enhanced Degradation

In the twentieth century, the first biodegradable plastic bottles made from polyethylene terephthalate (PET) were manufactured (Nisticò 2020). During the past 20 years, synthetic biology has engineered microorganisms to perform certain functions for various uses, such as biosensing, biocomputing, and biomanufacturing (Khalil and Collins 2010). This research has both advanced and benefited from the growing range of biological tools. With less power consumption and low-temperature requirements, PET biodegradation has gained attention as an eco-friendly solution. Recycling the byproducts in an effective and timely manner is an easy and efficient way to reduce plastic pollution. Tokiwa Suzuki et al. (1977) first postulated the enzyme degradation of polymers. Numerous enzymes, such as lipases, cutinases, and esterases, that degrade PET have now been identified and described from various microbes. PET degradation involves hydrophilizing the biofilm layer, where the hydrolysis of polymer chains produces carboxylic acid and hydroxyl group residues at the end, forming loops (Kushwaha et al. 2023). PETase production and secretion have been observed in some host organisms, notably microalgae, *Escherichia coli*, *Bacillus* sp., and *Pichia pastoris* (Soong et al. 2022). Quorum sensing (QS) and cellular communication mechanisms are frequently used in engineering microbial consortia (Scott and Hasty 2016). Genetic engineering can improve the ability of microbes to break down plastic contaminants. However, systems biology uses several omics techniques to understand how DNA, RNA, and metabolites are expressed and interact with other molecules throughout the degradation process (Hasan et al. 2024). Due to their ability to acquire enzymes that enable them to use plastic as a nutrient substrate, microbes are an ideal model for reducing plastic contaminants.

1.11.4 Development of Mitigation Strategies and Technologies to Reduce MP Pollution

In today's world, the excessive use, unsustainable utilization, and poor disposal of plastic products cause plastic pollution and pose risks to ecosystems, the economy, and human health. Although there are ongoing efforts to reduce the harmful impacts of plastic pollution, existing remediation techniques cannot keep pace with the increasing amount of plastic entering the environment. Reducing plastic inputs into the environment requires a global, interdisciplinary strategy. An integrated waste management system can improve the production, use, and disposal of plastics, minimizing uncontrolled garbage, which is a major land-based source of plastic pollution. Microbial activity, which is strongly influenced by environmental conditions, plays a significant role in shaping the ecological functions and implications of microplastics. To overcome the challenges caused by microplastic contamination, an integrated strategy should influence modern technologies and multidisciplinary research. We can better predict and mitigate the ecological impacts of microplastics by understanding the interactions between microplastics and microbial populations. Many techniques are recommended to establish an extensive structure for microplastic (MP) monitoring.

Initially, it is important to set global standards. International organizations like UNEP (United Nations Environment Programme) and ISO (International Organization for Standardizations) should lead efforts to develop standardized protocols for MP sampling, analysis, and reporting. These guidelines will ensure consistent and reliable results across studies and areas. Second, developing cross-sector partnerships between governments, academia, business, and nongovernmental organizations (NGOs) is important, as it can help gather resources, exchange expertise, and set up goals to effectively reduce MP pollution. Investing in technology and innovation is an important additional critical step. Cutting-edge detection technologies and modern data management systems may simplify monitoring operations and improve data accuracy, permitting improved decision-making. Geographic expansion should also be prioritized. Targeted financing and capacity-building programs are required to close gaps in underrepresented countries and provide a fair global representation of data. Furthermore, combining multimedia monitoring strategies is important. A comprehensive approach that considers water, soil, air, and biota will help in capturing all aspects of MP contamination and its impact on different ecosystems. Finally, data normalization and open access are important. Establishing centralized databases and open-access platforms will allow for worldwide data exchange, trend analysis, and evidence-based policy formation.

1.12 Summary

Microplastic particles are generally smaller than 5 μm, mainly originating from anthropogenic activities involving industrial sources and daily consumer products. These can cause risk to the environment and human health. This chapter highlights the types of microplastics, their sources, and vectors for the contamination of freshwater ecosystems. Further, it explains the interaction between MPs and microorganisms in freshwater ecosystems. Microplastics serve as substrates for microbial colonization, particularly for bacterial groups like Proteobacteria, Cyanobacteria, and Bacteroidetes, which are commonly associated with plastics such as polyethylene (PE) and polypropylene (PP). In contrast, plastics like polyvinyl chloride (PVC) and polystyrene (PS) show lower levels of microbial colonization. These microorganisms colonize and form a biofilm (plastisphere) that enhances microbial survival and resistance to environmental stressors. This microplastic-microbe association, or eco-corona, is influenced by various environmental factors, such as temperature, pH, water flow, and nutrient availability. Furthermore, microbial degradation of

MPs into small fragments through an enzymatic process may lead to further fragmentation, and due to expansion in surface area, other pollutants are co-transported through MPs after adsorption. Microbial breakdown and environmental factors, like temperature, pH, flow of water, and availability of nutrients, decide the fate of MPs and the environmental health.

Acknowledgments Shubha Dixit and Arpita Maurya are thankful to DST and UGC, respectively, for providing Junior Research Fellowship. Manoj Kumar is thankful to CSIR-IITR for providing funds. The CSIR-IITR manuscript communication number for this chapter is **IITR/SRC/MS/2024/87.**

Declarations The authors report no financial interests or personal relationships that could have affected the results of this study.

Competing Interest All the authors declare no conflicts of interest.

References

Agboola OD, Benson NU (2021) Physisorption and chemisorption mechanisms influencing micro (nano) plastics-organic chemical contaminants interactions: a review. Front Environ Sci 9:678574

Ahmed AS, Billah MM, Ali MM, Bhuiyan MKA, Guo L, Mohinuzzaman M et al (2023) Microplastics in aquatic environments: a comprehensive review of toxicity, removal, and remediation strategies. Sci Total Environ 876:162414

Akhbarizadeh R, Dobaradaran S, Torkmahalleh MA, Saeedi R, Aibaghi R, Ghasemi FF (2021) Suspended fine particulate matter (PM2. 5), microplastics (MPs), and polycyclic aromatic hydrocarbons (PAHs) in air: their possible relationships and health implications. Environmental Research, 192:110339

Akhtar N, Syakir Ishak MI, Bhawani SA, Umar K (2021) Various natural and anthropogenic factors responsible for water quality degradation: a review. Water 13(19):2660

Ali I, Rahman A (2024) Environmental degradation: causes, effects and solutions. Int J Multidiscip Res 6(3):1–10

Aminov RI (2011) Horizontal gene exchange in environmental microbiota. Front Microbiol 2:158

Amoatey P, Baawain MS (2019) Effects of pollution on freshwater aquatic organisms. Water Environ Res 91(10):1272–1287

Amobonye A, Bhagwat P, Raveendran S, Singh S, Pillai S (2021) Environmental impacts of microplastics and nanoplastics: a current overview. Front Microbiol 12:768297

Anagnosti L, Varvaresou A, Pavlou P, Protopapa E, Carayanni V (2021) Worldwide actions against plastic pollution from microbeads and microplastics in cosmetics focusing on European policies. Has the issue been handled effectively? Mar Pollut Bull 162:111883

Anderson JC, Park BJ, Palace VP (2016) Microplastics in aquatic environments: implications for Canadian ecosystems. Environ Pollut 218:269–280

Augimeri RV, Varley AJ, Strap JL (2015) Establishing a role for bacterial cellulose in environmental interactions: lessons learned from diverse biofilm-producing Proteobacteria. Front Microbiol 6:1282

Bacha AUR, Nabi I, Zhang L (2021) Mechanisms and the engineering approaches for the degradation of microplastics. ACS ES & T Eng 1(11):1481–1501

Bakir A, McGoran AR, Silburn B, Russell J, Nel H, Lusher AL et al (2024) Creation of an international laboratory network towards global microplastics monitoring harmonisation. Sci Rep 14(1):12714

Baruah J, Singh LS, Salvia T, Sarma J (2024) Antimicrobial resistance a continued global threat to public health–a perspective and mitigation strategies. J Lab Physicians, 16(1)

Basuki H, Chaurasia I, Husen K, Kasena G, Kim YE, Park HS, Wanandi R (2020) PET-degrading cyanobacteria for marine environments. Biotreks: Peer Reviewed High School Synthetic Biology

Bazaka K, Jacob MV, Crawford RJ, Ivanova EP (2011) Plasma-assisted surface modification of organic biopolymers to prevent bacterial attachment. Acta Biomater 7(5):2015–2028

Bîrcă A, Gherasim O, Grumezescu V, Grumezescu AM (2019) Introduction in thermoplastic and thermosetting polymers. In Materials for biomedical engineering (pp 1-28). Elsevier

Bohdan K (2022) Estimating global marine surface microplastic abundance: systematic literature review. Sci Total Environ 832:155064

Bostan N, Ilyas N, Akhtar N, Mehmood S, Saman RU, Sayyed RZ et al (2023) Toxicity assessment of microplastic (MPs); a threat to the ecosystem. Environmental Research, 234, 116523

Briassoulis D (2023) Agricultural plastics as a potential threat to food security, health, and environment through soil pollution by microplastics: problem definition. Sci Total Environ 892:164533

Cai L, Wu D, Xia J, Shi H, Kim H (2019) Influence of physicochemical surface properties on the adhesion of bacteria onto four types of plastics. Sci Total Environ 671:1101–1107

Cai Z, Li M, Zhu Z, Wang X, Huang Y, Li T et al (2023) Biological degradation of plastics and microplastics: a recent perspective on associated mechanisms and influencing factors. Microorganisms 11(7):1661

Caniato M, Cozzarini L, Schmid C, Gasparella A (2021) Acoustic and thermal characterization of a novel sustainable material incorporating recycled microplastic waste. Sustainable Materials and Technologies, 28

Cao Y, Zhao M, Ma X, Song Y, Zuo S, Li H, Deng W (2021) A critical review on the interactions of microplastics with heavy metals: Mechanism and their combined effect on organisms and humans. Sci Total Environ 788:147620

Carpenter EJ, Smith KL Jr (1972) Plastics on the Sargasso Sea surface. Science 175(4027):1240–1241

Chakraborty SK, (2021) Trophic interactions and biogeochemical cycles in river ecosystem. In: Riverine ecology volume 1: eco-functionality of the physical environment of rivers, pp 167–234. Cham: Springer International Publishing.

Cole M, Lindeque P, Fileman E, Halsband C, Goodhead R, Moger J, Galloway TS (2013) Microplastic ingestion by zooplankton. Environ Sci Technol 47(12):6646–6655

da Costa JP, Duarte AC, Rocha-Santos TA (2017) Microplastics–occurrence, fate and behaviour in the environment. In: Comprehensive analytical chemistry, vol 75. pp 1–24, Elsevier.

Dang H, Lovell CR (2016) Microbial surface colonization and biofilm development in marine environments. Microbiol Mol Biol Rev 80(1):91–138

Danso D, Chow J, Streit WR (2019) Plastics: environmental and biotechnological perspectives on microbial degradation. Appl Environ Microbiol 85(19):e01095–e01019

De Tender CA, Devriese LI, Haegeman A, Maes S, Ruttink T, Dawyndt P (2015) Bacterial community profiling of plastic litter in the Belgian part of the North Sea. Environ Sci Technol 49(16):9629–9638

Debroy A, George N, Mukherjee G (2022) Role of biofilms in the degradation of microplastics in aquatic environments. J Chem Technol Biotechnol 97(12):3271–3282

Dehghani S, Moore F, Akhbarizadeh R (2017) Microplastic pollution in deposited urban dust, Tehran metropolis, Iran. Environ Sci Pollut Res 24(25):20360–20371

Dey S, Rout AK, Behera BK, Ghosh K (2022) Plastisphere community assemblage of aquatic environment: plastic-microbe interaction, role in degradation and characterization technologies. Environ Microbiome 17(1):32

Dhali SL, Parida D, Kumar B, Bala K (2024) Recent trends in microbial and enzymatic plastic degradation: a solution for plastic pollution predicaments. Biotechnol Sustain Mater 1(1):11

Di Cesare A, Sathicq MB, Sbaffi T, Sabatino R, Manca D, Breider F et al (2024) Parity in bacterial communities and resistomes: microplastic and natural organic particles in the Tyrrhenian Sea. Mar Pollut Bull 203:116495

Do Sul JAI, Costa MF (2014) The present and future of microplastic pollution in the marine environment. Environ Pollut 185:352–364

Dris R, Imhof H, Sanchez W, Gasperi J, Galgani F, Tassin B, Laforsch C (2015) Beyond the ocean: contamination of freshwater ecosystems with (micro-) plastic particles. Environ Chem 12(5), 539-550

Du J, Xu S, Zhou Q, Li H, Fu L, Tang J et al (2020) A review of microplastics in the aquatic environmental: distribution, transport, ecotoxicology, and toxicological mechanisms. Environ Sci Pollut Res 27(11):11494–11505

Duis K, Coors A (2016) Microplastics in the aquatic and terrestrial environment: sources (with a specific focus on personal care products), fate and effects. Environ Sci Eur 28(1):2

Edwards SJ, Kjellerup BV (2013) Applications of biofilms in bioremediation and biotransformation of persistent organic pollutants, pharmaceuticals/personal care products, and heavy metals. Appl Microbiol Biotechnol 97(2):9909–9921

Egessa R, Nankabirwa A, Ocaya H, Pabire WG (2020) Microplastic pollution in surface water of Lake Victoria. Sci Total Environ 741:140201

Enyoh CE, Wang Q, Chowdhury T, Wang W, Lu S, Xiao K, Chowdhury MAH (2021) New analytical approaches for effective quantification and identification of nanoplastics in environmental samples. PRO 9(11):2086

Evangeliou N, Grythe H, Klimont Z, Heyes C, Eckhardt S, Lopez-Aparicio S, Stohl A (2020) Atmospheric transport is a major pathway of microplastics to remote regions. Nat Commun 11(1):3381

Ferri M, Ranucci E, Romagnoli P, Giaccone V (2017) Antimicrobial resistance: a global emerging threat to public health systems. Crit Rev Food Sci Nutr 57(13):2857–2876

Garaba SP, Dierssen HM (2018) An airborne remote sensing case study of synthetic hydrocarbon detection using shortwave infrared absorption features identified in marine-harvested macro- and microplastics. Remote Sens Environ 205:224–235

Geyer R, Jambeck JR, Law KL (2017) Production, use, and fate of all plastics ever made. Sci Adv 3(7):e1700782

Gkoutselis G, Rohrbach S, Harjes J, Obst M, Brachmann A, Horn MA, Rambold G (2021) Microplastics accumulate fungal pathogens in terrestrial ecosystems. Sci Rep 11(1):13214

Gorman D, Gutiérrez AR, Turra A, Manzano AB, Balthazar-Silva D, Oliveira NR, Harari J (2020) Predicting the dispersal and accumulation of microplastic pellets within the estuarine and coastal waters of South-Eastern Brazil using integrated rainfall data and Lagrangian particle tracking models. Frontiers in Environmental Science, 8:559405

Gu JD (2007) Microbial colonization of polymeric materials for space applications and mechanisms of biodeterioration: a review. Int Biodeterior Biodegradation 59(3):170–179

Hadjiev D, Dimitrov D, Martinov M, Sire O (2007) Enhancement of the biofilm formation on polymeric supports by surface conditioning. Enzym Microb Technol 40(4):840–848

Hale RC, Seeley ME, La Guardia MJ, Mai L, Zeng EY (2020) A global perspective on microplastics. Journal of Geophysical Research: Oceans, 125(1):e2018JC014719

Hamidian AH, Ozumchelouei EJ, Feizi F, Wu C, Zhang Y, Yang M (2021) A review on the characteristics of microplastics in wastewater treatment plants: a source for toxic chemicals. J Clean Prod 295:126480

Hasan, M., Islam, M. S., & Islam, M. S. (2024). Bioremediation of Microplastics by Microorganisms: Trends, Challenges, and Perspectives. Microbial Bioremediation and Multiomics Technologies for Sustainable Development: Recent Trends, 13, 49.

Horton AA, Walton A, Spurgeon DJ, Lahive E, Svendsen C (2017) Microplastics in freshwater and terrestrial environments: Evaluating the current understanding to identify the knowledge gaps and future research priorities. Sci Total Environ 586:127–141

Hou L, Majumder ELW (2021) Potential for and distribution of enzymatic biodegradation of polystyrene by environmental microorganisms. Materials 14(3):503

Hou S, Wang J, Feng D, Wu D, Chen S, Fan J et al (2025) Colonization time of plastisphere drives the dynamics of organic carbon stability and microbial communities in seagrass bed sediments. J Hazard Mater 492:138078

Huang W, Song B, Liang J, Niu Q, Zeng G, Shen M et al (2021) Microplastics and associated contaminants in the aquatic environment: a review on their ecotoxicological effects, trophic transfer, and potential impacts to human health. J Hazard Mater 405:124187

Huang Y, Wu Y, Hu H, Tong B, Wang J, Zhang S et al (2024) Accelerating the design of pili-enabled living materials using an integrative technological workflow. Nat Chem Biol 20(2):201–210

Huber M, Archodoulaki VM, Pomakhina E, Pukánszky B, Zinöcker E, Gahleitner M (2022) Environmental degradation and formation of secondary microplastics from packaging material: A polypropylene film case study. Polymer Degradation and Stability, 195:109794

Jacquin J, Cheng J, Odobel C, Pandin C, Conan P, Pujo-Pay M et al (2019) Microbial ecotoxicology of marine plastic debris: a review on colonization and biodegradation by the "Plastisphere". Front Microbiol 10:865

Jahan Rakib MR, Mohiuddin M, Al Masud MA, Murshed MF, Senapathi V, Islam MA, Jakariya M (2025) Exploring the impact of microplastics in groundwater on human health: A comprehensive review. Current Opinion in Environmental Science and Health, 44:100605

Kaplan JÁ (2010) Biofilm dispersal: mechanisms, clinical implications, and potential therapeutic uses. J Dent Res 89(3):205–218

Kalathil S, Miller M, Reisner E (2022) Microbial fermentation of polyethylene terephthalate (PET) plastic waste for the production of chemicals or electricity. Angewandte Chemie International Edition, 61(45):e202211057

Khalid N, Aqeel M, Noman A, Rizvi ZF (2023) Impact of plastic mulching as a major source of microplastics in agroecosystems. J Hazard Mater 445:130455

Khalil AS, Collins JJ (2010) Synthetic biology: applications come of age. Nat Rev Genet 11(5):367–379

Kumar M, Xiong X, He M, Tsang DC, Gupta J, Khan E et al (2020) Microplastics as pollutants in agricultural soils. Environ Pollut 265:114980

Kurniawan TA, Haider A, Ahmad HM, Mohyuddin A, Aslam HMU, Nadeem S et al (2023) Source, occurrence, distribution, fate, and implications of microplastic pollutants in freshwater on environment: a critical review and way forward. Chemosphere 325:138367

Kushwaha A, Goswami L, Singhvi M, Kim BS (2023) Biodegradation of poly (ethylene terephthalate): mechanistic insights, advances, and future innovative strategies. Chem Eng J 457:141230

Lamichhane G, Acharya A, Marahatha R, Modi B, Paudel R, Adhikari A et al (2023) Microplastics in environment: global concern, challenges, and controlling measures. Int J Environ Sci Technol 20(4):4673–4694

Lebreton LC, Van Der Zwet J, Damsteeg JW, Slat B, Andrady A, Reisser J (2017) River plastic emissions to the world's oceans. Nat Commun 8(1):15611

Lemonnier C, Chalopin M, Huvet A, Le Roux F, Labreuche Y, Petton B et al (2022) Time-series incubations in a coastal environment illuminates the importance of early colonizers and the complexity of bacterial biofilm dynamics on marine plastics. Environ Pollut 312:119994

Li J, Zhang K, Zhang H (2018) Adsorption of antibiotics on microplastics. Environ Pollut 237:460–467

Li X, Mei Q, Chen L, Zhang H, Dong B, Dai X, Zhou, J (2019) Enhancement in adsorption potential of microplastics in sewage sludge for metal pollutants after the wastewater treatment process. Water Research, 157:228-237

Li C, Wang L, Ji S, Chang M, Wang L, Gan Y, Liu J (2021) The ecology of the plastisphere: microbial composition, function, assembly, and network in the freshwater and seawater ecosystems. Water Res 202:117428

Li J, Shan E, Zhao J, Teng J, Wang Q (2023) The factors influencing the vertical transport of microplastics in marine environment: a review. Sci Total Environ 870:161893

Li Y, Chen L, Zhou N, Chen Y, Ling Z, Xiang P (2024a) Microplastics in the human body: a comprehensive review of exposure, distribution, migration mechanisms, and toxicity. Sci Total Environ, 946: 174215

Li Y, Liu C, Yang H, He W, Li B, Zhu X et al (2024b) Leaching of chemicals from microplastics: a review of chemical types, leaching mechanisms and influencing factors. Sci Total Environ 906:167666

Lin X, Zhang S, Yang S, Zhang R, Shi X, Song L (2023) A landfill serves as a critical source of microplastic pollution and harbors diverse plastic biodegradation microbial species and enzymes: study in large-scale landfills, China. J Hazard Mater 457:131676

Liu L, Xu M, Ye Y, Zhang B (2022) On the degradation of (micro) plastics: degradation methods, influencing factors, environmental impacts. Sci Total Environ 806:151312

Lowe SE, Jain MK, Zeikus JG (1993) Biology, ecology, and biotechnological applications of anaerobic bacteria adapted to environmental stresses in temperature, pH, salinity, or substrates. Microbiol Rev 57(2):451–509

Lusher A, Hollman P, Mendoza-Hill J (2017) Microplastics in fisheries and aquaculture: status of knowledge on their occurrence and implications for aquatic organisms and food safety. FAO

Makhdoumi P, Hossini H, Pirsaheb M (2023) A review of microplastic pollution in commercial fish for human consumption. Reviews on environmental health, 38(1):97-109

Mallik A, Xavier KM, Naidu BC, Nayak BB (2021) Ecotoxicological and physiological risks of microplastics on fish and their possible mitigation measures. Sci Total Environ 779:146433

Mammo FK, Amoah ID, Gani KM, Pillay L, Ratha SK, Bux F, Kumari S (2020) Microplastics in the environment: Interactions with microbes and chemical contaminants. Sci Total Environ 743:140518

Mason SA, Kammin L, Eriksen M, Aleid G, Wilson S, Box C et al (2016) Pelagic plastic pollution within the surface waters of Lake Michigan, USA. J Great Lakes Res 42(4):753–759

Michaelis C, Grohmann E (2023) Horizontal gene transfer of antibiotic resistance genes in biofilms. Antibiotics 12(2):328

Mihai FC, Gündoğdu S, Markley LA, Olivelli A, Khan FR, Gwinnett C et al (2021) Plastic pollution, waste management issues, and circular economy opportunities in rural communities. Sustainability 14(1):20

Miller D, Adams QH, Wellenius GA (2025) Urban Collaborations: Bridging the Gap Between Research and Action on Climate Change and Health. J of City Climate Policy and Economy, 3(2/3):204-214

Mishra AK, Singh J, Mishra PP (2023) Microplastics in freshwater ecosystem: a serious threat for freshwater environment. Int J Environ Sci Technol 20(8):9189–9204

Mohanan N, Montazer Z, Sharma PK, Levin DB (2020) Microbial and enzymatic degradation of synthetic plastics. Front Microbiol 11:580709

Möller JN, Löder MG, Laforsch C (2020) Finding microplastics in soils: a review of analytical methods. Environ Sci Technol 54(4):2078–2090

Napper IE, Bakir A, Rowland SJ, Thompson RC (2015) Characterisation, quantity and sorptive properties of microplastics extracted from cosmetics. Marine pollution bulletin, 99(1-2):178-185

Napper IE, Baroth A, Barrett AC, Bhola S, Chowdhury GW, Davies BF et al (2021) The abundance and characteristics of microplastics in surface water in the transboundary Ganges River. Environ Pollut 274:116348

Nava V, Leoni B (2021) A critical review of interactions between microplastics, microalgae and aquatic ecosystem function. Water Res 188:116476

Neelavannan K, Sen IS (2023) Microplastics in freshwater ecosystems of India: current trends and future perspectives. ACS omega, 8(38):34235-34248

Nisticò R (2020) Polyethylene terephthalate (PET) in the packaging industry. Polym Test 90:106707

Nyadjro ES, Webster JA, Boyer TP, Cebrian J, Collazo L, Kaltenberger G et al (2023) The NOAA NCEI marine microplastics database. Sci Data 10(1):726

Oberbeckmann S, Loeder MG, Gerdts G, Osborn AM (2014) Spatial and seasonal variation in diversity and structure of microbial biofilms on marine plastics in Northern European waters. FEMS Microbiol Ecol 90(2):478–492

Oberbeckmann S, Kreikemeyer B, Labrenz M (2018) Environmental factors support the formation of specific bacterial assemblages on microplastics. Front Microbiol 8:2709

Odum EP, Barrett GW (2005) Fundamentals of ecology, 5th edn. Thomson Brooks/Cole

Ogonowski M, Motiei A, Ininbergs K, Hell E, Gerdes Z, Udekwu KI et al (2018) Evidence for selective bacterial community structuring on microplastics. Environ Microbiol 20(8):2796–2808

Ormsby MJ, Akinbobola A, Quilliam, RS (2023) Plastic pollution and fungal, protozoan, and helminth pathogens–A neglected environmental and public health issue?. Science of the Total Environment, 882:163093

Özkara A, Akyıl D (2018) Environmental pollution and the effects of the pollutants on the ecosystem. Türk bilimsel derlemeler dergisi 11(2):11–17

Peller JR, Nelson CR, Babu BG, Iceman C, Kostelnik E (2020) A review of microplastics in freshwater environments: locations, methods, and pollution loads. In: Contaminants in our water: identification and remediation methods, 65–90

Pinto M, Zhao Z, Klun K, Libowitzky E, Herndl GJ (2022) Microbial consortiums of putative degraders of low-density polyethylene-associated compounds in the ocean. Msystems, 7(2):e01415-21

Priyadarsini P, Barik D, Barik B (2024) Plastic pollution and associated emerging contaminant across Indian subcontinent river catchments: A grave concern of anthropogenic epoch. In River basin ecohydrology in the Indian sub-continent (pp 169-181). Elsevier

Picó Y, Barceló D (2019) Analysis and prevention of microplastic pollution in water: current perspectives and future directions. ACS Omega 4(4):6709–6719

Rafey A, Siddiqui FZ (2023) A review of plastic waste management in India–challenges and opportunities. Int J Environ Anal Chem 103(16):3971–3987

Ramanna S, Morozovskii D, Swanson S, Bruneau J (2022) Machine Learning of polymer types from the spectral signature of Raman spectroscopy microplastics data. arXiv preprint arXiv:2201.05445.

Restrepo-Flórez JM, Bassi A, Thompson MR (2014) Microbial degradation and deterioration of polyethylene–a review. Int Biodeterior Biodegradation 88:83–90

Reddy KR, DeLaune RD, Inglett PW (2022) Biogeochemistry of wetlands: science and applications. CRC Press

Rezaei M, Riksen MJ, Sirjani E, Sameni A, Geissen V (2019) Wind erosion as a driver for transport of light density microplastics. Sci Total Environ 669:273–281

Rillig MC, Lehmann A (2020) Microplastic in terrestrial ecosystems. Science 368(6498):1430–1431

Roager L, Sonnenschein EC (2019) Bacterial candidates for colonization and degradation of marine plastic debris. Environ Sci Technol 53(20):11636–11643

Rochman CM, Browne MA, Underwood AJ, Van Franeker JA, Thompson RC, Amaral-Zettler LA (2016) The ecological impacts of marine debris: unraveling the demonstrated evidence from what is perceived. Ecology 97(2):302–312

Rogers KL, Carreres-Calabuig JA, Gorokhova E, Posth NR (2020) Micro-by-micro interactions: how microorganisms influence the fate of marine microplastics. Limnol Oceanogr Lett 5(1):18–36

Sabater S, Guasch H, Ricart M, Romaní A, Vidal G, Klünder C, Schmitt-Jansen M (2007) Monitoring the effect of chemicals on biological communities. The biofilm as an interface. Anal Bioanal Chem 387(4):1425–1434

Saha M, Naik A, Desai A, Nanajkar M, Rathore C, Kumar M, Gupta P (2021) Microplastics in seafood as an emerging threat to marine environment: a case study in Goa, west coast of India. Chemosphere 270:129359

Sánchez C (2020) Fungal potential for the degradation of petroleum-based polymers: an overview of macro-and microplastics biodegradation. Biotechnol Adv 40:107501

Schefer RB, Armanious A, Mitrano DM (2023) Eco-corona formation on plastics: adsorption of dissolved organic matter to pristine and photochemically weathered polymer surfaces. Environ Sci Technol 57(39):14707–14716

Schneider B, Dellwig O, Kuliński K, Omstedt A, Pollehne F, Rehder G, Savchuk O (2017) Biogeochemical cycles. In: Biological oceanography of the Baltic Sea, pp 87–122. Dordrecht: Springer Netherlands

Scott SR, Hasty J (2016) Quorum sensing communication modules for microbial consortia. ACS Synth Biol 5(9):969–977

Setälä O, Fleming-Lehtinen V, Lehtiniemi M (2014) Ingestion and transfer of microplastics in the planktonic food web. Environ Pollut 185:77–83

Sgier L, Freimann R, Zupanic A, Kroll A (2016) Flow cytometry combined with viSNE for the analysis of microbial biofilms and detection of microplastics. Nat Commun 7(1):11587

Shang Q, Chi J (2023) Impact of biochar coexistence with polar/nonpolar microplastics on phenanthrene sorption in soil. J Hazard Mater 447:130761

Sheridan EA, Fonvielle JA, Cottingham S, Zhang Y, Dittmar T, Aldridge DC, Tanentzap AJ (2022) Plastic pollution fosters more microbial growth in lakes than natural organic matter. Nature Communications, 13(1):4175

Shi J, Wang Z, Peng Y, Zhang Z, Fan Z, Wang J, Wang X (2023) Microbes drive metabolism, community diversity, and interactions in response to microplastic-induced nutrient imbalance. Science of the Total Environment, 877:162885

Silva AB, Bastos AS, Justino CI, da Costa JP, Duarte AC, Rocha-Santos TA (2018) Microplastics in the environment: challenges in analytical chemistry-a review. Anal Chim Acta 1017:1–19

Silva GC, Galleguillos Madrid FM, Hernández D, Pincheira G, Peralta AK, Urrestarazu Gavilán M et al (2021) Microplastics and their effect in horticultural crops: food safety and plant stress. Agronomy 11(8):1528

Singh S, Malyan SK, Maithani C, Kashyap S, Tyagi VK, Singh R et al (2023) Microplastics in landfill leachate: occurrence, health concerns, and removal strategies. J Environ Manag 342:118220

Smith KE, Jones KC (2000) Particles and vegetation: implications for the transfer of particle-bound organic contaminants to vegetation. Sci Total Environ 246(2–3):207–236

Smith P, Cotrufo MF, Rumpel C, Paustian K, Kuikman PJ, Elliott JA et al (2015) Biogeochemical cycles and biodiversity as key drivers of ecosystem services provided by soils. Soil 1(2):665–685

Soong YHV, Sobkowicz MJ, Xie D (2022) Recent advances in biological recycling of polyethylene terephthalate (PET) plastic wastes. Bioengineering 9(3):98

Santos LH, Rodríguez-Mozaz S, Barceló D (2021) Microplastics as vectors of pharmaceuticals in aquatic organisms–an overview of their environmental implications. Case Studies in Chemical and Environmental Engineering, 3:100079

Sooriyakumar P, Bolan N, Kumar M, Singh L, Yu Y, Li Y et al (2022) Biofilm formation and its implications on the properties and fate of microplastics in aquatic environments: a review. J Hazard Mater Adv 6:100077

Strokal M, Vriend P, Bak MP, Kroeze C, van Wijnen J, van Emmerik T (2023) River export of macro-and microplastics to seas by sources worldwide. Nat Commun 14(1):4842

Sun A, Wang WX (2023) Human exposure to microplastics and its associated health risks. Environ Health 1(3):139–149

Sushmitha TJ, Rajeev M, Sriyutha Murthy P, Ganesh S, Toleti SR, Karutha Pandian S (2021) Bacterial community structure of early-stage biofilms is dictated by temporal succession rather than substrate types in the southern coastal seawater of India. PLoS One 16(9):e0257961

Syberg K, Khan FR, Selck H, Palmqvist A, Banta GT, Daley J et al (2015) Microplastics: addressing ecological risk through lessons learned. Environ Toxicol Chem 34(5):945–953

Tang KHD (2024) Terrestrial and aquatic plastisphere: formation, characteristics, and influencing factors. Sustainability 16(5):2163

Taniguchi I, Yoshida S, Hiraga K, Miyamoto K, Kimura Y, Oda K (2019) Biodegradation of PET: current status and application aspects. ACS Catal 9(5):4089–4105

Tokiwa Y, Suzuki T (1977) Purification and some properties of polyethylene adipate-degrading enzyme produced by Penicillium sp. Strain 14–3. Agricultural and Biological Chemistry, 41(2):265-274

Tournier V, Topham CM, Gilles A, David B, Folgoas C, Moya-Leclair E et al (2020) An engineered PET depolymerase to break down and recycle plastic bottles. Nature 580(7802):216–219

Tournier V, Duquesne S, Guillamot F, Cramail H, Taton D, Marty A, André I (2023) Enzymes' power for plastics degradation. Chem Rev 123(9):5612–5701

Tu C, Chen T, Zhou Q, Liu Y, Wei J, Waniek JJ, Luo Y (2020) Biofilm formation and its influences on the properties of microplastics as affected by exposure time and depth in the seawater. Sci Total Environ 734:139237

Tumwesigye E, Nnadozie CF, Akamagwuna FC, Noundou XS, Nyakairu GW, Odume ON (2023) Microplastics as vectors of chemical contaminants and biological agents in freshwater ecosystems: current knowledge status and future perspectives. Environ Pollut 330:121829

Upadhyay S, Sharma PK, Dogra K, Bhattacharya P, Kumar M, Tripathi V, Karmakar R (2024) Microplastics in freshwater: Unveiling sources, fate, and removal strategies. Groundwater for Sustainable Development, 26:101185

Uruén C, Chopo-Escuin G, Tommassen J, Mainar-Jaime RC, Arenas J (2020) Biofilms as promoters of bacterial antibiotic resistance and tolerance. Antibiotics 10(1):3

Villanueva VD, Font J, Schwartz T, Romani AM (2011) Biofilm formation at warming temperature: acceleration of microbial colonization and microbial interactive effects. Biofouling 27(1):59–71

Wang C, Zhao J, Xing B (2021) Environmental source, fate, and toxicity of microplastics. J Hazard Mater 407:124357

Wang X, Zhang X, Yao C, Shan E, Lv X, Teng J, Wang Q (2023) Impact of aged and virgin microplastics on sedimentary nitrogen cycling and microbial ecosystems in estuaries. Sci Total Environ 878:162977

Wani AK, Akhtar N, Naqash N, Rahayu F, Djajadi D, Chopra C et al (2023) Discovering untapped microbial communities through metagenomics for microplastic remediation: recent advances, challenges, and way forward. Environ Sci Pollut Res 30(34):81450–81473

Waters JK, Eijkelkamp BA (2024) Bacterial acquisition of host fatty acids has far-reaching implications on virulence. Microbiology and Molecular Biology Reviews, 88(4):e00126-24

Wieczorek AM, Morrison L, Croot PL, Allcock AL, MacLoughlin E, Savard O, Doyle, TK (2018) Frequency of microplastics in mesopelagic fishes from the Northwest Atlantic. Frontiers in Marine Science, 5, 339138

Wilkes RA, Aristilde L (2017) Degradation and metabolism of synthetic plastics and associated products by Pseudomonas sp.: capabilities and challenges. J Appl Microbiol 123(3):582–593

Wright SL, Thompson RC, Galloway TS (2013) The physical impacts of microplastics on marine organisms: a review. Environ Pollut 178:483–492

Wu Y, Li T, Yang L (2012) Mechanisms of removing pollutants from aqueous solutions by microorganisms and their aggregates: a review. Bioresource technology, 107:10-18 ISBN: 978-81-965020-4-1

Xu J, Zuo R, Wu G, Liu J, Liu J, Huang C, Wang Z (2024) Global distribution, drivers, and potential hazards of microplastics in groundwater: a review. Sci Total Environ 954: 176194

Xue Y, Lu H, Feng S, Kang J, Guan Y, Li H et al (2024) Standardization of monitoring data reassesses spatial distribution of aquatic microplastics concentrations worldwide. Water Res 254:121356

Yan X, Chio C, Li H, Zhu Y, Chen X, Qin W (2024) Colonization characteristics and surface effects of microplastic biofilms: Implications for environmental behavior of typical pollutants. Sci Total Environ 937, 173141

Yin W, Wang Y, Liu L, He J (2019) Biofilms: the microbial "protective clothing" in extreme environments. Int J Mol Sci 20(14):3423

Yoshida S, Hiraga K, Takehana T, Taniguchi I, Yamaji H, Maeda Y et al (2016) A bacterium that degrades and assimilates poly (ethylene terephthalate). Science 351(6278):1196–1199

Yu J, Sun L, Ma C, Qiao Y, Yao H (2016) Thermal degradation of PVC: a review. Waste Manag 48:300–314

Yuan Z, Nag R, Cummins E (2022) Human health concerns regarding microplastics in the aquatic environment-From marine to food systems. Sci Total Environ 823:153730

Zettler ER, Mincer TJ, Amaral-Zettler LA (2013) Life in the "plastisphere": microbial communities on plastic marine debris. Environ Sci Technol 47(13):7137–7146

Zhai X, Zhang XH, Yu M (2023) Microbial colonization and degradation of marine microplastics in the plastisphere: a review. Front Microbiol 14:1127308

Zhang S, Yang X, Gertsen H, Peters P, Salánki T, Geissen V (2018) A simple method for the extraction and identification of light density microplastics from soil. Sci Total Environ 616:1056–1065

Zhang X, Lin T, Jiang F, Zhang X, Wang S, Zhang S (2022a) Impact of pipe material and chlorination on the biofilm structure and microbial communities. Chemosphere, 289:133218

Zhang SJ, Zeng YH, Zhu JM, Cai ZH, Zhou J (2022b) The structure and assembly mechanisms of plastisphere microbial community in the natural marine environment. J Hazard Mater 421:126780

Zhao S, Kvale KF, Zhu L, Zettler ER, Egger M, Mincer TJ, Stubbins A (2025) The distribution of subsurface microplastics in the ocean. Nature, 641(8061):51-61

Zhu X, Chen B, Zhu L, Xing B (2017) Effects and mechanisms of biochar-microbe interactions in soil improvement and pollution remediation: a review. Environ Pollut 227:98–115

ISBN: 978-81-965020-4-1

Release of Microplastics by Terrestrial Sources into Freshwater Environments

2

Maryam Salehi, Abdul Mobin Ibna Hafiz, Ebrahim Ahmadisharaf, and Farhad Jazaei

Abstract

This chapter examines the significant role of various terrestrial sources in releasing microplastics (MPs) into freshwater systems. Key contributors include stormwater runoff, agricultural activities, and wastewater treatment plants (WWTPs), each facilitating the transport and accumulation of microplastics in aquatic environments. We will explore the mechanisms through which these sources contribute to microplastic pollution, highlighting the implications for water quality and ecosystem health. The chapter details the types of microplastics emitted from each source, their size distribution, and their abundance, providing a clearer picture of the scale and impact of this environmental issue. Additionally, we will discuss other important sources, such as roadway tire wear particles (TWPs), landfill leachate, and industrial discharges, to comprehensively understand the various terrestrial origins contributing to microplastic pollution. By analyzing these factors, this chapter aims to elucidate the pathways through which microplastics enter freshwater ecosystems, emphasizing the need

M. Salehi (✉)
Department of Civil and Environmental Engineering, University of Missouri, Columbia, MO, USA

Missouri Water Center, Columbia, MO, USA
e-mail: mshfp@missouri.edu

A. M. I. Hafiz · E. Ahmadisharaf
Department of Civil and Environmental Engineering, FAMU-FSU College of Engineering, Tallahassee, FL, USA

Resilient Infrastructure and Disaster Response Center, FAMU-FSU College of Engineering, Tallahassee, FL, USA

F. Jazaei
Department of Civil Engineering, The University of Memphis, Memphis, TN, USA

K. Bala et al. (eds.), *Occurrence, Detection, and Fate of Microplastics in Freshwater Ecosystems*, https://doi.org/10.1007/978-981-95-0225-7_2

for effective management strategies to mitigate their environmental impact. Understanding these sources and their characteristics is crucial for developing targeted interventions to reduce microplastic contamination and protect freshwater resources for future generations.

Keywords

Terrestrial · Microplastics · Sources · Pollution

2.1 Introduction

Despite numerous studies investigating plastic pollution in marine and coastal environments, this global challenge has been understudied for freshwater systems. Considering the limited availability of freshwater resources and their critical role in providing drinking water and supporting aquatic ecosystems, it is essential to protect them from microplastic (MP) pollution (Salehi 2022). Although the long-term impacts of MPs on human health have remained largely unexplored, recent investigations have reported their presence in human tissues, such as blood and the liver (Bocker and Silva 2025). Additionally, MPs also reported altering the biodiversity and reducing both biomass and productivity in aquatic flora and fauna (Sridharan et al. 2021). Primary MPs (e.g., microbeads, microcapsules, and pellets) are generated intentionally. However, the disintegration of plastic-containing products during their usage and after disposal due to exposure to external environmental forces can create secondary MPs (SMPs). The size distribution, shape, surface chemistry, morphology, and bulk characteristics of SMPs vary depending on the plastics' intrinsic characteristics, exposure durations, and environmental conditions (Salehi et al. 2024). These characteristics collectively influence their subsequent distribution within the natural and built environmental systems. Land-based sources, such as wastewater treatment plants (WWTPs), roadways, agricultural activities, disposal sites, and commercial and recreational activities, release approximately 20 times more MP compared to aquatic sources (Horton et al. 2017). For instance, the annual discharge of MPs by Chinese municipal wastewater treatment plants has been estimated to reach 220–950 tons per year (Ren et al. 2023). Moreover, the annual inputs of 44,000–300,000 and 63,000–430,000 tons of MPs to North American and European farmlands have been estimated through sludge application (Nizzetto et al. 2016). On the other hand, the annual global emission of tire wear particles has been estimated to be around six million tons (Kole et al. 2017). Major pathways for transport of MPs from land to freshwater are recognized as surface runoff, wastewater effluent, wind, and streams (Ibna Hafiz et al. 2025). It is critical to elucidate factors influencing the extent of MP transport via these pathways to more accurately predict their accumulation within aquatic systems such as ponds and wetlands and their potential for reaching drinking water resources and assess their subsequent transport to estuaries, oceans, and marine environments. Furthermore, effective mitigation strategies and control measures can be developed using this knowledge to ensure drinking water safety and environmental protection. In this chapter, we provide an overview of the significant terrestrial sources that release MPs into freshwater environments. Next, we discuss the transport of MPs from the terrestrial systems to the freshwater environment through surface runoff, erosion, wind, atmospheric deposition, and streams. In the last section, we discuss mitigation and management strategies, focusing on urban stormwater runoff management and sustainable agricultural practices.

2.2 Major Terrestrial Sources Contributing to MPs' Release into the Freshwater System

Primary (intentionally manufactured) and secondary (disintegrated from larger plastic materials) MPs are released to the natural and built terrestrial environments through a wide range of human-driven activities (Salehi et al. 2024). Major land-based contributors to MP release are wastewater treatment plants, roadways, littering and illegal dumping, agricultural practices, and industrial activities. In this section, we discuss the extent and characteristics of the MPs released by these major sources.

2.2.1 Domestic and Industrial Wastewater Effluents

The daily consumed items, such as packaging materials, synthetic textiles, and personal care products, can release MPs into domestic wastewater. Approximately 78–98% of MPs can be removed from influent wastewater through the primary wastewater treatment processes; however, a further reduction by 7–20% may occur via secondary treatments (Prata 2018). Thus, a significant portion of MPs (80–90%) that were initially present in the wastewater influent is retained in the generated sludge called biosolid. The total mass of MPs retained in biosolids and the overall MP load released by effluents vary by the type and capacity of each treatment facility. The biosolid can be landfilled or used for soil amendments and agricultural practices (Bretas Alvim et al. 2020). The small percentage of MPs left in wastewater effluent can result in a substantial MP release to freshwater systems due to an elevated volume of daily wastewater effluent discharges (Zubris and Richards 2005). A survey of 17 wastewater treatment facilities by Mason et al. (2016) revealed an average effluent concentration of 0.05 MP/L and an estimated daily discharge of four million MPs. Moreover, other studies reported MP concentrations in wastewater effluent to vary within a wide range from <0.5 to 297 MPs/L (Hidayaturrahman and Lee 2019; Lv et al. 2019). The wide variability in the concentration, size distribution, polymer type, and shape of MPs present in wastewater effluents from different domestic wastewater treatment plants can be driven by variations in the population composition and density, water consumption patterns, and social characteristics of the community they serve. The most common shapes of MPs in wastewater effluent have been reported to be fibers and fragments. The high length-to-width ratio of fibers makes their removal challenging compared to the other shapes of MPs (e.g., fragments and films) (Cristaldi et al. 2020).

In addition to the effluent from municipal wastewater treatment facilities, industrial wastewater effluent, particularly from textile and other manufacturing industries, also contributes significantly to MP release into freshwater systems. Table 2.1 summarizes major studies that examined the MP pollution resulting from industrial effluent and industrial wastewater effluent. Although the critical role of textile industries in MP release into the freshwater systems is underscored in recent studies, the contribution of other sectors, such as pharmaceutical, resin/paint, and plastic industries, should not be ignored. Carr (2017) estimated that the global production of synthetic fibers exceeded 5.4 million tons in 2016. Unfortunately, many of these fibers are released from clothing material into domestic wastewater. Fibers' small size and high length-to-width ratio facilitate their passage through wastewater treatment processes and subsequent release into streams. Additionally, a significant level of microfibers is released from the textile industry wastewater compared to municipal sewage treatment plants. Therefore, it makes textile industries substantially contribute to microfiber pollution in aquatic environments (Gambino et al. 2025). Zhou et al. (2020) reported a maximum concentration of 54,100 microfibers per liter (MFs/L) in textile printing and dyeing wastewater influent. Although the removal efficiency exceeded 85%, the treated effluent still contained 537.5 MFs/L. This process released approximately 430 billion MFs per day. In addition to the textile industry, the wastewater effluents from other sectors such as paint, resin, pharmaceutical, and PVC manufacturing also contribute to MP pollution in freshwater systems, as suggested by Magalhães et al. (2022). This study reported alkyd resin and poly(acrylate/styrene) as examples of MPs released by the paint and coating industries. Despite the persistence of MPs in effluents from industry and municipal wastewater treatment plants, regulations limiting their discharge levels have not been established yet. Another study by Karakurt et al. (2025) reported the MP types and their abundance in the effluents from the paint industry pretreatment, the food industry, the textile industry, and the metal industry pretreatment and suggested a 95% reduction in MPs' abundance through industrial wastewater treatment processes, capturing 127 million particles per day but still emitting 7.6 million particles.

2.2.2 Littering and Illegal Dumping

Littering is a substantial source of plastic pollution in urban areas. The extent of littering is influenced by population density, land use (e.g., commercial, residential, and recreational), and waste management practices. Generally, more significant waste and litter generation is observed in areas with denser populations. Moreover, parks and recreational sites often serve as hotspots for littering (Cohen et al. 2022). Recent estimations for the United States indicated 152 items of litter per resident. This involves over 2000 pieces of litter per mile (1.61 km), 80 items per capita in waterways, and 73 items per capita along roadways. Most of the littered debris in the United States are plastics (38.6%) and cigarette butts (19.6%) (Keep America

Table 2.1 Summary of major studies investigating MP pollution in industrial and treated industrial effluents

Industry	Location	Sample type	Dominant plastic type	Dominant size range	Abundance	References
Paint industry pretreatment	Türkiye	Effluent	HDPE[a]	38–435 μm	112, 94 MP/L	Karakurt et al. (2025)
Food industry			PS[b]		39, 82 MP/L	
Textile			PA[c]		48 MP/L	
Metal industry pretreatment			HDPE, PS		24–33 MP/L	
PVC[d] production	Germany	Plant effluent	PVC	10–100 μm	2×10^5–4×10^5 MPP[e]/m^3	Barkmann-Metaj et al. (2023)
PE[c] production		Plant effluent	PE	100–500 μm	8×10^5–13×10^5 MPP/m^3	
PE production		Plant effluent	PE	50–100 μm	8.2×10^5 MPP/m^3	
Paint	Portugal	WWTP[f] effluent	PET[g]	10–600 μm	89.0 g MPs/ton effluent	Magalhães et al. (2022)
Textile					15.6 g MPs/ton effluent	
Resins					41.7 g MPs/ton effluent	
PVC					70.6 g MPs/ton effluent	
Pharmaceutical		WWTP effluent	PE, HDPE, PET		24.7 g MPs/ton effluent	
Textile	Portugal	WWTP effluent	PET	10–50 μm	15.6 g MPs/ton effluent	Magalhães et al. (2024)
Textile		WWTP effluent		500–600 μm	33.6 g MPs/ton effluent	
Textile	China		NR[h]	100–300 μm	5–1800 MFs/L	Zhou et al. (2020)

(continued)

Table 2.1 (continued)

Industry	Location	Sample type	Dominant plastic type	Dominant size range	Abundance	References
Textile	China	WWTP effluent	PP, PS, PE	0.1–0.5 mm	6–12 MP/L	Wang et al. (2020)
Machine manufacturing			PP (polypropylene)			
Chemical			PE			
Steel			PS			
Chemical		Plant effluent	PE, PS, PET	0.1–0.2 mm	8–23 MP/L	
Electroplating			PE, PS	0.2–0.5 mm		
Machine manufacturing			PE, PET, PS	0.1–0.5 mm		

[a]*HDPE* high-density polyethylene
[b]*PS* polystyrene
[c]*PA* polyamide
[d]*PVC* polyvinyl chloride
[e]*MPP* MP particle
[f]*WWTP* wastewater treatment plant
[g]*PET* polyethylene terephthalate
[h]*NR* not reported

Beautiful 2021). Litter, such as discarded plastic materials, packaging, food containers, bags, bottles, and cups, is commonly found in mismanaged waste disposal sites, along roadways, and in open spaces such as parks and parking lots. PET, HDPE, low-density polyethylene (LDPE), and PS are the major plastic types reported for urban plastic litter (Esquinas et al. 2020). Plastic litter could be simply transported by runoff to urban drainage systems, clog those, and intensify the negative impact of flooding, further impacting environmental safety and public health (Anggraini et al. 2020). Plastic litter exposure to solar radiation and mechanical forces can result in alterations to their surface chemistry and gradual disintegration into MPs (Battacharjee et al. 2023; Hadiuzzaman et al. 2022; Herath and Salehi 2022). These generated MPs can accumulate in road deposits, pavement surfaces, vegetation, and sediments during dry conditions. This process is known as buildup. Then, during rainfall events, runoff transports these MPs into ponds, rivers, lakes, and oceans, bypassing traditional waste treatment systems. This process is known as wash-off. The low weight and small size of MPs facilitate their transport via surface runoff and wind action into the adjacent freshwater resources (Aghilinasrollahabadi et al. 2021; Beheshtimaal et al. 2024).

Illegal dumping of solid waste may occur in poorly monitored areas, such as roadsides, canals, and abandoned properties. Plastic debris left on dumping sites can be easily transported to waterways or the surrounding environment. Moreover, they can gradually be disintegrated into MPs through environmental weathering (e.g., photodegradation) and exposure to external mechanical forces (e.g., abrasion). Individuals and industries would engage in illegal dumping due to the cost of waste collection and disposal. As reported by Lopatta et al. (2016), firms under financial pressure would be more willing to increase their profits by making compromises. A study by Sotamenou et al. (2019) for households in Cameroon showed that accessibility to a centralized waste collection system is a major parameter that limits illegal waste dumping. Initiatives such as return and recycling programs and producers' extended responsibility (PER) systems for waste have been applied to mitigate illegal dumping and littering, as well as reduce collection costs (Prata et al. 2019). In terms of plastic pollution, PER systems emphasize the circular economy and strengthening of corporate social responsibility, as suggested by Zhou and Luo (2024). However, they were not able to keep up with the extent of plastic litter released into the environment globally. Recent studies reported plastic waste as the main component of illegally dumped materials, alongside household waste and construction debris (Šedová 2016; Niyobuhungiro and Schenck 2021). Advanced remote sensing technologies and satellite imagery have successfully been used for land surface monitoring. For instance, Sakti et al. (2023) conducted a study in Indonesia, which utilized Sentinel-2 for the detection of illegally dumped plastic waste around the Rancamanyar River (Sakti et al. 2023). The information collected from the computer-aided identification of the dumping sites can be utilized for the development of effective solid waste compliance and enforcement practices. The gradual disintegration of plastic waste results in the generation of MPs and their distribution around the dump sites. Thus, a greater abundance of MPs has been reported around the dumping sites (Mandal et al. 2025). As reported by Priya et al.

(2023), an average of 12 MP items per liter in samples were collected from the groundwater wells near dumping sites in Kollam, India. In addition to the plastic waste associated with household items, industrial (e.g., packaging and abrasives) and construction (e.g., pipes, fittings, insulation materials, and flooring) waste can also gradually disintegrate into MPs, as they are left within the environment. The location of dumping sites and proximity to waterways influence the extent of MPs from those sites being transported into the adjacent aquatic systems (Bharath et al. 2021).

2.2.3 Roadways

Surface abrasion of motor vehicles' tires with the road surface results in the generation of tire and road wear particles (TWRPs). They play a critical role in MP pollution in the terrestrial environments. The generated TWRPs can accumulate on the pavement surface, road dust, adjacent soil, and vegetation or be dispersed in the air (Siegfried et al. 2017). Several studies underscored the critical health risks associated with TWRP's contribution to $PM_{2.5}$ and PM_{10} pollution. Giechaskiel et al. (2024) reported that PM_{10} emission is 2.5% of the total abrasion for TWPs, and the ratio of $PM_{2.5}$ to PM_{10} was found to be 40%. Traffic volume, vehicle type, road surface characteristics (e.g., roughness), soil moisture, and weather conditions (e.g., precipitation, temperature, wind, and relative humidity) influence the extent of their generation and further distribution. A greater abundance of TWRPs is reported around bridges and highways that experience a greater traffic volume (Leads and Weinstein 2019). Liu et al. (2022b) reported a greater rate of TWRPs for hybrid vehicles than conventional vehicles, and the wear rate was found to be 1.8 times greater for left-front tires than left-rear tires. Moreover, the tire composition also influences the extent of tire wear particles' (TWPs') generation. The study by Yan et al. (2021) revealed that tread rubber with more silica content emits a greater abundance of TWPs but fewer coarse TWPs. However, the complex composition of TWRPs and variability in sampling procedures make it challenging to accurately and reliably determine the flux of TWRPs into the environment (Baensch-Baltruschat et al. 2020).

Tamis et al. (2021) reported a wide concentration of TWPs in urban runoff, ranging from 975 to 58,500 μg/L, with an average of 29,737 μg/L in highway runoff. The black rubbery fragments have been reported to make up more than 83% of all fragment-shaped MPs and around 48% of all shapes of MPs in stormwater samples collected from an urban watershed (Zhu et al. 2021). Braga Moruzzi et al. (2020) reported the tire particles to be around 30% of all fragments quantified in stormwater samples collected in stormwater ponds. This high proportion of TWPs, relative to other MP types, underscores their significance as a primary contributor to urban MP pollution. The distribution of TWPs varies across different urban sites. For instance, in industrial areas, rubber particles, including TWPs, make up around 8% of total MPs in stormwater runoff and road dust (Kabir et al. 2024). Studies investigating runoff from hiking trails have found rubber particles comprising approximately 3% of larger MPs (>100 µm) and 1% of smaller particles (<100 µm), indicating that even recreational areas are subjected to TWP pollution (Forster et al. 2023). Moreover, road marking materials, typically polymer-based paints, contribute to additional MP pollution through degradation caused by ultraviolet radiation and mechanical wear (Turner et al. 2024).

2.2.4 Agricultural Activities

The use of plastics in agricultural practices is called "plasticulture" and has led to an annual global consumption of approximately 12.5 million tons of plastics in the agricultural sector (FAO 2021). The primary purpose of plastics in agriculture is to enhance crop yields and conserve water. Typical applications include plastic mulch, greenhouses, silage films, irrigation pipes, drippers, plant protectors, liners, nursery pots, agrochemical containers, water reservoirs, and controlled-release fertilizers, as shown in Fig. 2.1.

Plastic mulch, one of the most widely used agricultural plastic products, is projected to reach a market value of $5.7 billion by 2026 (Research and Market 2021). It provides multiple benefits, such as increasing soil temperature, resulting in weed suppression, and conserving moisture, leading to increased plastic mulch application for the cultivation of specialty crops (e.g., vegetables and berries) (Kasirajan and Ngouajio 2012). Linear low-density polyethylene (LLDPE) dominates the conventional plastic mulch market due to its superior mechanical and agronomic performance (Yu et al. 2024). Despite these advantages, end-of-life management of agricultural plastics presents a significant challenge. Residual plastics, such as

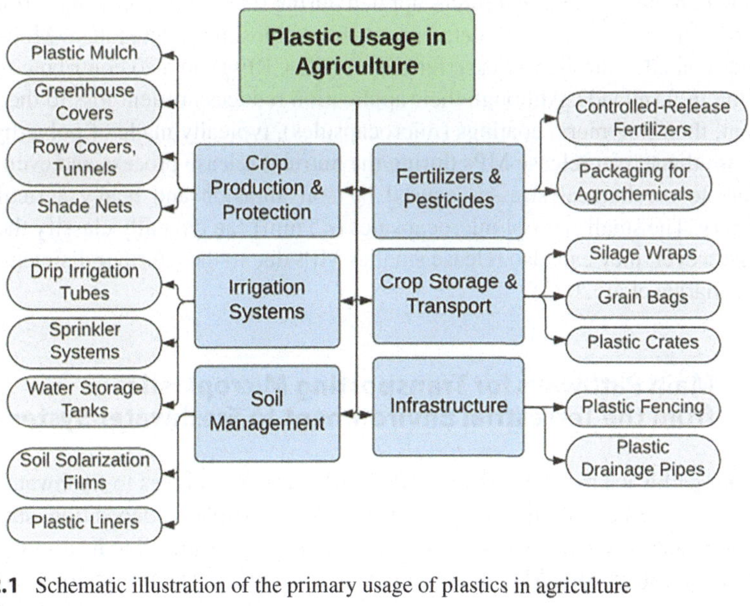

Fig. 2.1 Schematic illustration of the primary usage of plastics in agriculture

micro- and nanoplastics, are left in the soil after retrieving plastic products like mulch, silage films, and irrigation pipes. They can penetrate the soil column, negatively affecting soil biota, structural properties, and geochemistry. Moreover, runoff can transport those plastic residuals to adjacent water bodies and streams, resulting in declining freshwater quality and negatively impacting aquatic life. Those MPs retained within agricultural soil could release their additives and degradation byproducts as they further degrade within the soil environment through a combination of physical, chemical, and biological processes. These released chemicals can penetrate into root zones and potentially be adsorbed by crops, posing risks to food safety (Hurley and Nizzetto 2018; Piehl et al. 2018). Huang et al. (2020) reported a significant linear correlation between using plastic mulching in farmlands across China (2012–2016) and macroplastic abundance. This study examined soil samples collected from 19 provinces in China and reported that plastic residual abundance increased significantly from 80.3 ± 49.3 to 308 ± 138.1 and 1075.6 ± 346.8 pieces/kg soil, following 5, 15, and 24 years of continuous plastic mulching, respectively (Huang et al. 2020). Moreover, plastic mulch residuals' exposure to solar radiation can result in their surface chemical oxidation, making them more brittle and susceptible to further disintegration, leading to the generation of smaller MP fragments (Battacharjee et al. 2023; Bonyadinejad et al. 2022).

As mentioned earlier, treated wastewater sludge, known as biosolid, is another major contributor to MP pollution in the terrestrial systems. In the United States, approximately 25% of generated biosolids are applied to agricultural land (Environmental Protection Agency 2021). The MPs added to the soil matrix via biosolid application can be further washed away by runoff and irrigation water and/or be blown away by wind, further dispersed within the environment, and contaminate surface and groundwater resources (Kumar et al. 2023). Biosolid composition, moisture content, and external forces applied during rain events and wind action can influence the extent of MP detachment from biosolid aggregates. Moreover, polymer-coated controlled-release fertilizers (PC-CRFs) can also contribute to MP pollution in farmlands. Although their application reduces nutrient loss to the environment, their polymeric coatings (microcapsules), typically made of polyethylene or polyurethane, can release MPs during the nutrient release process and even after that due to mechanical stresses caused by soil abrasion and heavy agricultural machinery. The small sizes of microcapsules (<5 mm) can typically classify them as MPs. However, they can also release smaller MPs due to their further disintegration (Bhattacharjee et al. 2025).

2.3 Main Pathways for Transporting Microplastics from the Terrestrial Environment to Freshwater Systems

The main pathways for carrying the MPs from terrestrial sources to freshwater systems involve surface runoff, wind action, erosion, atmospheric deposition, streams, and wastewater discharge. Urban stormwater runoff is considered to be a significant pathway for transporting MPs to adjacent water resources. The impervious surfaces of urban areas, such as roadways and parking lots, facilitate the rapid movement of MPs during rain events. Moreover, wind action can blow the MPs and transport them over long distances.

As discussed earlier, both municipal and industrial wastewater treatment plants discharge millions of MPs into freshwater systems (Murphy and Quinn 2018). The low efficiency of these treatment plants in removing smaller MPs results in a significant number of MPs being released into freshwater systems (Afkarina et al. 2020; Zhao et al. 2022). Moreover, after MPs enter the waterways, they can be further carried by streams and rivers to lakes and marine environments, impacting aquatic life and threatening drinking water safety. In this section, we discuss the roles of runoff, wind, atmospheric deposition, and streams in transporting MPs to freshwater systems.

2.3.1 Urban Stormwater Runoff

Urban stormwater runoff is a primary route for carrying MPs to freshwater systems, with concentrations ranging from 1.3 to 6000 particles per liter, affected by precipitation characteristics (e.g., rainfall intensity and duration) and land use/land cover patterns (Ross et al. 2023). The recent studies that quantified MPs in urban runoff owing to various land uses are summarized in Table 2.2. As suggested by Beheshtimaal et al. (2024), MPs' characteristics, such as type, size distribution, density, and shape, can influence the extent and mechanism of their transport by stormwater runoff. For small settling MPs, the wash load was dominated, regardless of their shape and density. However, for larger settling MPs, flow depth, size, and shape were effective (Beheshtimaal et al. 2024). Synthetic fibers, fragments, and TRWPs have been reported as the major MPs found in urban stormwater (Siegfried et al. 2017). Polypropylene (34%), rubber particles (28.5%), and PE (17.9%) were reported as the major chemical types for MPs found in urban stormwater. The evaluated percentage of rubber particles underscores the significant contribution of TRWPs to urban MP pollution. On the other hand, PET (7.5%), PVC (4%), and PS (1.8%) MPs made up smaller percentages of those found in urban stormwater (Braga Moruzzi et al. 2020; Forster et al. 2023; Kabir et al. 2024; Olubusoye et al. 2024; Sugiura et al. 2021). This distribution pattern of major plastic types aligns with findings from multiple studies that reported the prevalence of tire-derived MPs in stormwater runoff (Leads and Weinstein 2019; Jayalakshmamma et al. 2023).

MPs transported by urban runoff can accumulate within stormwater retention infrastructure (e.g., ponds, wetlands, and swales). A study conducted in Denmark revealed a greater concentration of MPs in retention ponds within industrial and commercial areas compared to stormwater samples collected from highway zones or residential sites. The PP, PVC, polyester, PE, and PS MPs were abundant in stormwater retention ponds, with their relative abundances varying by land use and/or land cover (Liu et al. 2019). The characteristics of impervious surfaces (e.g., slope and roughness) within the urban environment significantly affect the pattern of MP transport. By limiting water infiltration, these surfaces result in enhanced runoff volumes

Table 2.2 Summary of studies quantifying microplastics (MPs) in urban stormwater

Study area, location	Studied size range	MP quantity	MP shapes/types	References
Highways in Germany, Sweden, and the Netherlands	4–350 μm	975–58,500 μg/L	Shapes: NR[a]; rubber polymers	Tamis et al. (2021)
San Francisco Bay, California, United States	0.1–195 mm (median: 1.0 mm)	BDL[b] to 0.056 MP/L (mean 0.0015 MP/L)	Fragments (42%), fibers (42%), and foam (9%); non-synthetic fibers (30%), PE (18%), and PP (8%)	Zhu et al. (2021)
Expressway in Tokyo, Japan	10 μm–5 mm	First flush: 292 MP/L Steady-state: 81 MP/L	Mostly flakes; PE (25–49%), polyethylene-polypropylene diene (13–30%), and polyethylene-polypropylene copolymer (29–35%)	Sugiura et al. (2021)
A farm in the Mississippi Delta, United States	40–1000 μm	Range: 27–609 particles/L Average: 237 particles/L	Fragments: 95.7% Films: 2.7% Fibers: 1.6% Mostly PE (21%), polyarylamide (20%), PA (19%), PVC (11%) (site BC) ABS (32%), PU (21%), PE (18%), PS (10%) (site NSD)	Olubusoye et al. (2024)
Stormwater retention pond in Viborg, Denmark	10–500 μm	270 MPs per liter, corresponding to 4.2 μg/L	Fragments (dominant shape) and fibers; PP (dominant plastic type)	Olesen et al. (2019)
A car park, a road site, and a stormwater wetland in Queensland, Australia	>25 μm	Raw stormwater: 3.8–59 MPs/L Treated stormwater: 1.8–32 MPs/L	Fragments and fibers, dominated by TWPs, poly(butadiene) rubber and styrene-butadiene rubber (SBR)	Ziajahromi et al. (2023)
Hiking trails, New South Wales, Australia	>100 μm	9–55 MPs per plot	Fibers (85%) and irregular-shaped fragments (15%) composed of PU, PP, and PET	Forster et al. (2023)

(continued)

Table 2.2 (continued)

Study area, location	Studied size range	MP quantity	MP shapes/types	References
Town center of Madeán, Lima region, Peru	6 μm–5 mm	Minimum 144.44 to maximum 622.22 MP items/kg of dry sediment (average abundance of 292.91 ± 149.32 MP items/kg of sediment)	Fibers: 60.66% Fragments: 38.39% Film: 0.95% Plastic types: NR	Canchari and Iannacone (2023)
Residential, commercial, and industrial zones in Melbourne, Australia	125–960 μm	17–43 MP/L	Fragments, fibers, films, and microbeads Mostly PE, PP, PET, and polyester	Kabir et al. (2024)
An urban center in Calgary, Canada	37 μm–>1000	0.7–200.4 MP/L	Fibers, fragments, films, and spheres Plastic type: NR	Ross et al. (2023)
Bioretention cells, Ontario, Canada	25 μm–1.3 mm	Average concentration of 324 ± 284 MPs/L	Fragments and films; PP, PE, PS, polyester polyamide, and PVC	Smyth et al. (2024)

[a]Not reported
[b]Below detection limit

and consequently contribute to the rapid transport of MPs to adjacent freshwater systems. During dry periods, MPs accumulate on impervious surfaces, such as roads, highways, parking lots, and sidewalks. They can later be mobilized during precipitation events (Ross et al. 2023). The accumulation and subsequent mobilization of MPs following rain events is more significant in densely populated urban areas due to a combination of extensive impervious surfaces and concentrated human activities (Jayalakshmamma et al. 2023). Moreover, urban drainage networks rapidly transport the mobilized MPs to the receiving water bodies, creating point sources for MP release into urban waterways. In addition to runoff, soil erosion further facilitates the MP transport process through the displacement of soil and sediment particles. As reported by Rehm et al. (2021), soil erosion results in lateral loss of coarse MPs (250–300 μm) from agricultural soils; however, for fine MPs (53–100 μm), vertical transport into the soil column was more significant. The hydrodynamic forces applied to MPs within the environment not only facilitate their transport but also result in their further fragmentation. For instance, a study of MPs in glacial environments has revealed that erosive processes not only transport existing MPs but also result in the formation of secondary MPs by the disintegration of larger plastic particles (Dong et al. 2021a). The spatial distribution of MP particles in both runoff and receiving waters showed a clear linkage to these erosional processes.

The hydrologic factors can influence the extent of MPs' transport and their subsequent spatial and temporal distribution in receiving aquatic systems (Beheshtimaal et al. 2024). Recent studies have reported elevated levels of MPs in urban rivers during wet weather conditions (Ross et al. 2023; Townsend et al. 2019). This phenomenon is particularly evident in a study conducted in Southern California, where elevated MP abundances correlate strongly with precipitation events. This finding emphasizes the role of stormwater systems as primary conveyance pathways (Ross et al. 2023). Another study conducted for the Great Lakes region also reported the significant role of urban runoff in releasing MPs into aquatic systems through stormwater infrastructure (Baldwin et al. 2016). Moreover, studying the land use influence on MPs' abundance by Soltani et al. (2022) revealed a greater MP pollution in both water and sediment samples collected from an urban area in Iran than in the studied industrial, agricultural, and natural sites.

2.3.2 Microplastic Transport via Wind and Atmospheric Deposition

Atmospheric deposition and transport significantly influence the continental distribution of MPs. It was suggested by Chen et al. (2023) that MPs smaller than 25 μm are globally transportable. The suspension and transport of MPs through atmospheric processes are mainly controlled by the characteristics of MPs such as their shape, size, and density. Recent research by Tatsii et al. (2024) suggested that smaller particles and fibers are more susceptible to long-range transport due to their aerodynamic properties. The fibers, due to their elongated shapes, generally move more quickly compared to spherical particles, which may settle earlier (Xiao et al. 2023). Moreover, it was reported that they have high mobility and potential for transporting lightweight MPs, especially synthetic fibers, by wind (Conard et al. 2023; Dris et al. 2016; Wright et al. 2020). Polymer type influences the MP density and their suspension and settling characteristics in air (Chen et al. 2023).

Additionally, meteorological parameters like wind speed can influence MP deposition within the environment. It was reported that there is a seasonal variation of MP deposition rates and that it is correlated with wind patterns and rainfall events (Brahney et al. 2020; Szewc et al. 2021). The MPs' abundance is reported to decrease from densely populated areas to remote sites (Yang et al. 2025). Heavy traffic and synthetic textile usage in densely populated urban centers are also suggested to result in greater deposition rates for MPs (Cakaj et al. 2023). The extent of atmospheric transport of MPs can vary for highly populated urban regions, suburban regions, and remote areas, each requiring specific monitoring and management strategies. Recent studies reported a strong correlation between MP concentrations and land cover/land use characteristics in receiving aquatic systems (Tamminga et al. 2022). Often, urbanized regions consistently show higher MP deposition rates than rural areas (Tanentzap et al. 2021; Zhang et al. 2019). Radford et al. (2023) reported that MP atmospheric deposition rates for rural areas vary from 136.5 to 512.0 MP particles/m^2/day. However, Wright et al. (2020) reported deposition rates of 575–1008 particles/m^2/day for an urban area. The suspended MPs in the atmosphere are deposited into the terrestrial or aquatic environment during precipitation events as they contact rain droplets or snowflakes (Dris et al. 2016; Yücel et al. 2023). Chen et al. (2023) reported a greater abundance of MPs in the first rainfall following a long-term dry period in Beijing, China, where the wet deposition rate was found to vary from 2 to 463 μg/m^2. Aves et al. (2022) investigated the MPs' abundance and characteristics in snow samples collected from 19 sites in Ross Island in Antarctica. They reported an average concentration of 29 MP/L, with fibers being the most abundant. Identification of MPs in snow samples from various terrestrial systems, such as remote polar regions, demonstrates the significant role of atmospheric deposition in the long-distance transport of MPs (Bergmann et al. 2019; Kanhai et al. 2020). Additionally, MPs' interactions with other atmospheric pollutants, such as primary air pollutants, particulate matter, and biological particles, can affect their potential health impacts. Ortega and Cortés-Arriagada (2023) reported the adsorption of primary air pollutants such as CO, CO_2, NO, NO_2, N_2O, NH_3, and SO_2 by polyethylene terephthalate (PET) micro- and nanoplastics.

2.3.3 Microplastic Transport via Streams

Rivers and streams are considered the key pathways for connecting the terrestrial sources of MP pollution to freshwater and marine environments (Ashiq et al. 2023; Kukkola et al. 2023). Several interconnected components are present within the stream or river systems that influence the flow and particle transport dynamics. Among those, the main channels are the key components, generally with a high flow rate and minimal aquatic vegetation to facilitate the sediment transport process. The riverbanks stabilize against erosion, hosting grasses, shrubs, and trees that reduce runoff and trap solid particles. Dense and moisture-loving vegetation, such as willows and reeds, that are present in riparian zones act as a natural buffer and enhance the deposition of sediments. During high flows, floodplains provide natural storage for water, reduce flow velocity, and enhance the deposition of fine sediments. Thus, they support seasonal grasses and nutrient-rich vegetation. Floodplains can also transition into wetlands, which remain saturated for extended periods or year-round. These wetlands trap sediments and provide unique hydrologic and ecological functions, such as nutrient cycling, habitat for diverse species, and additional filtration of MPs and other pollutants. Beneath the surface, the hyporheic zone serves as a transition between surface water and groundwater, facilitating the trapping of fine particles such as MPs and supporting nutrient cycling and microbial activity, which influence the broader hydrodynamics and ecological processes of the river system (Bennett and Simon 2004; Boon and Raven 2012; Chicharo et al. 2015; National Research Council 2002).

There exists a common misconception, particularly among the general public and decision-makers, that all MPs transported through streams inevitably reach the marine environment (Boos et al. 2024). In reality, rivers and streams can serve as long-term sinks for MPs, in which particles become entrapped in sediments and vegetation (e.g., aquatic plants and riparian grasses) or along riverbanks (e.g., floodplain deposits), often remaining within freshwater systems for extended periods. There are several factors that affect this entrapment, including hydrodynamic processes (e.g., changes in flow velocity and turbulence), sediment composition (e.g., fine-grained sediments and organic-rich sediments), and local geomorphological features (e.g., meanders and pools) (Shamskhany et al. 2021; Boos et al. 2024; Kukkola et al. 2023). Recent studies showed that MP uptake by benthic organisms and alteration of their physical characteristics by biofilm growth on their surface can enhance their retention in freshwater ecosystems (Drummond et al. 2020; Lambert and Wagner 2018).

A number of researchers have also highlighted that structures, such as dams and ponds, can influence the capture and retention of MPs (Lebreton et al. 2017; Talbot and Chang 2022). However, there is a significant knowledge gap regarding the MP transport mechanisms, including capture, storage, release, and degradation dynamics within the streams (Du et al. 2020; Lebreton et al. 2017; Niu et al. 2023). Microplastic discharge into riverine systems is closely related to human activities along streams, which influence the rate of MP release through surface runoff, industrial discharges, and agricultural practices. These discharges also contribute to the gradual accumulation of MPs as they move downstream toward the sea (Kukkola et al. 2023; Lebreton et al. 2017). The study by Wagner et al. (2019) reported an increasing concentration of 79 MP/1000 m^3 from the rural to urban sub-catchment. However, research on MPs' spatial and temporal distribution along streams and rivers and their variation by management practices remains limited (Baldwin et al. 2016; Strokal et al. 2023). The study by Kukkola et al. (2023) showed that water flow diversions result in over 500 MP removals per second from the river. However, this process resulted in MPs' dispersion into a larger catchment area and suggested that models developed for the estimation of MPs' flux should consider the role of management practices.

Unlike nitrogen and phosphates, which are soluble contaminants and thoroughly mixed in the water column, MPs behave as colloidal particles, floating, sinking, or remaining suspended. This results in their heterogeneous and variable spatial and temporal distribution. Heterogeneous distribution can cause substantial errors in sampling and analysis, which are often overlooked in most research (Ali et al. 2024; Lebreton et al. 2017; Niu et al. 2023; Stanton et al. 2020; Talbot and Chang 2022). Additionally, there are limited investigations on the influence of factors such as flow variability (e.g., changes from rainfall or drought), seasonal dynamics (e.g., wet and dry season discharge), and localized pollution sources (e.g., runoff and industrial discharges) on the spatial and temporal distribution of MPs (Ali et al. 2024; Lebreton et al. 2017; Niu et al. 2023; Stanton et al. 2020; Talbot and Chang 2022). Moses et al. (2023) reported that 80% of all quantified MPs were found in samples collected during high-discharge events that occurred from December to March in a German River. However, a study by Huang et al. (2023) reported a greater concentration of MPs in the Houjin River during the dry season (183 ± 129 items/m^3) compared with the wet season (102 ± 46 items/m^3). Currently, there are only a limited number of studies that examine the variations in MP concentrations, as they relate to the flow and transport dynamics of specific river systems (Campanale et al. 2020; Wagner et al. 2019). For instance, Campanale et al. (2020) reported that the MPs' abundance ranges from 0.9 ± 04 MP/m^3 to 13 ± 5 MP/m^3 and indicated an increased abundance of MPs with higher water levels in the Ofanto River in Italy. Wagner et al. (2019) suggested the combined sewer overflow as a potential source that contributed to additional plastic load during high-flow conditions.

2.4 Stream Parameters Influencing MP Transport

Understanding the transport dynamics of MPs in riverine systems requires analyzing multiple river-specific parameters that directly or indirectly influence MPs' transport behavior. Table 2.3 provides an overview of the key riverine parameters that influence the transport, suspension, and deposition of MPs in freshwater systems. The parameters, including channel morphology, flow velocity, water depth, discharge, hydraulic indices, and stream power, influence the transport characteristics of MPs as they move through streams. Channel morphology, including features such as channel width, slope, bed roughness, substrate composition, and clast size, significantly influences flow dynamics and depositional environments, thereby affecting the MP trapping or mobilization (Kumar et al. 2021; Newbould et al. 2021). Findings by Wang et al. (2021) indicate that wider river channels are positively correlated with larger MPs (>2 mm) but negatively associated with smaller MPs. Hydraulic parameters, specifically flow velocity and water depth, are fundamental in determining MPs' distribution within the water column. High flow velocities favor MPs' suspension and downstream transport, whereas low velocities promote their deposition (Miller and Orbock Miller 2020). Moreover, deeper river sections can enhance the suspension of low-density MPs through increased vertical mixing (Owowenu et al. 2023).

Discharge, the volume of water moving past a river cross section, is another important factor. It was reported that higher discharge levels can increase the energy available for MPs' mobilization (Owowenu et al. 2023). Hydraulic conditions are generally quantified using indices such as Reynolds number (Re), Froude number (Fr), shear velocity (U*), and roughness Reynolds number (Re*) that can be related to MP mobility in river systems (Wadeson 1994; Wadeson and Rowntree 1998). For example, it was reported that laminar flows ($Re < 1$) can promote the settling of high-density MPs, whereas turbulent flows ($Re > 100$) maintain low-density MPs in suspension (Kumar et al. 2021). The flow regime, distinguished as subcritical ($Fr < 1$) or supercritical ($Fr > 1$), also substantially affects MPs' transport behavior. Subcritical flows favor MPs' deposition, while supercritical flows, often associated with floods, enhance MPs' transport and aggregation (Owowenu et al. 2023). Finally, stream power (Ω), the energy available for geomorphic work, strongly affects MPs' transport dynamics. Rivers with higher stream power can mobilize high-density and biomass-laden MPs more effectively, while lower stream power conditions favor MPs' sedimentation (Biswas and Chandra Das 2016; Owowenu et al. 2023).

Table 2.3 A summary of key riverine parameters influencing the transport, suspension, and deposition of MPs

River parameter	Definition	Influence on MP transport	Reference(s)
Channel morphology	Physical characteristics of the river (width, slope, bed roughness, substrate)	Controls depositional zones; wide channels favor larger MPs; rough beds enhance the trapping of MPs	Kumar et al. (2021), Newbould et al. (2021), and Owowenu et al. (2023)
Flow velocity	Speed of water movement	High velocity maintains MPs in suspension; low velocity promotes MPs' deposition	Miller and Orbock Miller (2020) and Owowenu et al. (2023)
Water depth	Vertical distance from surface to riverbed	Greater depth increases MPs' suspension potential due to enhanced turbulence	Owowenu et al. (2023)
Discharge	Volume of water flowing per unit time	High discharge increases transport capacity and MPs' mobility	Owowenu et al. (2023)
Hydraulic indices (Re, Fr, U*, Re*)	Quantitative measures of flow patterns (laminar vs. turbulent, subcritical vs. supercritical)	Laminar flow promotes the settling of high-density MPs; turbulent flow suspends low-density MPs	Kumar et al. (2021), Wadeson (1994), and Wadeson and Rowntree (1998)
Stream power (Ω)	The river's energy to perform geomorphic work	High stream power mobilizes heavier MPs, while low power favors sedimentation	Biswas and Chandra Das (2016) and Owowenu et al. (2023)

2.4.1 Parameters Impacting Microplastic Transport in Soil and Porous Media

Microplastics entering riverine systems can become trapped within vegetation or buried under sediments and subsequently remain there for an extended period. These processes increase the MPs' residence time in inland waters as they move toward the oceans (Mancini et al. 2023). It was reported that the greatest abundance of MPs may be found in the vegetation and sediment parts of streams, such as hyporheic, riparian, and even floodplain zones, rather than water in the main channel (Lechthaler et al. 2021). The transport and fate mechanism of MPs in hyporheic, riparian, and floodplain zones can be more complex than water flow movement due to the presence of dense vegetation and slower or stagnant water movement. For instance, complex vertical transport processes into soil and sediment layers and biological and chemical activities in these areas have become a significant challenge (Ding et al. 2019; Roebroek et al. 2021; Scheurer and Bigalke 2018; Woodward et al. 2021).

Although research into the vertical transport of MPs through sedimentary porous systems is still limited, emerging studies have identified key parameters that control this process, as summarized in Table 2.4. The transport behavior of MPs is strongly

Table 2.4 A summary of key environmental and physicochemical parameters influencing MP transport in soil porous media

Factor	Effect on MP transport	Reference(s)
Grain size	Larger grain size increases porosity and MP mobility	Ye et al. (2022)
Particle size of medium	Smaller particle size enhances MP retention via increased adsorption sites; smaller MPs can penetrate deeper layers	Dong et al. (2022), Quevedo and Tufenkji (2012), Ashiq et al. (2023), Boos et al. (2024), and Jazaei et al. (2022)
Porosity	Smaller porosity favors MP retention	Ryu et al. (2021)
Surface roughness of medium	Rougher surfaces promote MP adsorption and reduce mobility	Zhou et al. (2021a, b) and Tan et al. (2021)
Flow velocity	Higher flow velocity increases MP mobility by reducing interaction time with the media	Ye et al. (2022) and Ling et al. (2022)
Solution pH	Higher pH increases MP mobility by making the surface charge more negative	Dong et al. (2021b)
Ionic strength	Higher ionic strength promotes MP aggregation and retention	Geng et al. (2024)
Cation valence (Ca^{2+} vs. Na^{+})	Divalent cations (Ca^{2+}, Mg^{2+}) inhibit MP migration more than monovalent cations (Na^{+}, K^{+})	Zhao et al. (2021)
Ca:Na ratio	Increasing Ca:Na ratio enhances particle retention	Zhang et al. (2020)
Dissolved organic matter (humic acid)	Humic acid introduces negative charges and steric hindrance, enhancing MP mobility	Wu et al. (2019)

influenced by sediment characteristics. Grain size plays a significant role, where larger grain sizes lead to increased porosity and higher MP mobility, as observed by Ye et al. (2022). In contrast, smaller particle sizes of the porous medium enhance MP retention by providing more adsorption sites and reducing permeability (Dong et al. 2022; Quevedo and Tufenkji 2012). Porosity also directly impacts transport, with smaller porosity favoring greater retention of MPs (Ryu et al. 2021). Surface roughness of the medium has been shown to promote MP adsorption and reduce mobility, as reported by Zhou et al. (2021a, b) and Tan et al. (2021). Furthermore, flow velocity is an important hydrodynamic factor; higher flow velocities were found to enhance MP mobility by reducing the interaction time between MPs and soil surfaces (Ye et al. 2022; Ling et al. 2022). The solution pH also affects MP transport, with higher pH levels increasing the negative surface charge and thus enhancing MP mobility (Dong et al. 2021b). Regarding ionic effects, ionic strength generally promotes MP aggregation and retention (Geng et al. 2024), and cation valence plays a key role, where divalent cations like Ca^{2+} and Mg^{2+} are more effective at inhibiting MP migration compared to monovalent cations like Na^{+} and K^{+} (Zhao et al. 2021). Increasing the Ca:Na ionic concentration ratio further enhances particle retention in the soil (Zhang et al. 2020). Additionally, dissolved organic matter, such as humic acid, can promote MP mobility by introducing more negative charges and creating steric hindrance that reduces retention (Wu et al. 2019; Bonyadinejad and Salehi 2024). A study by Owowenu et al. (2023) suggested that root development created a preferential path and resulted in a more even distribution of MPs along the depth of an urban stormwater detention pond. Moreover, Luo et al. (2020) reported that the sorption of polystyrene MPs on soil was significantly influenced by soil organic carbon, Fe_2O_3, and clay contents.

2.5 Characteristics of Microplastics Influencing Their Transport in Water and Soil Matrices

Understanding the transport and fate of MPs in environmental systems, particularly in riverine waters and porous soils, requires investigating both environmental factors and intrinsic properties of the MPs. Key characteristics such as density, shape, size, surface morphology, and surface chemistry critically influence MP transport dynamics (Owowenu et al. 2023; Kumar et al. 2021). Density is a fundamental property influencing MPs' vertical and horizontal movement. MPs with densities lower than freshwater typically float or remain suspended, while denser MPs tend to settle into riverbeds and soil matrices (Kumar et al. 2021). However, MP density is dynamic; environmental processes like biofouling, aggregation, and photochemical degradation can increase MPs' effective density over time, altering their transport behavior (Zhang 2017). Aggregation with organic and inorganic matter, for instance, can substantially increase MP mass, leading to faster sedimentation (Zhang 2017). Moreover, once deposited, denser fragments may resist remobilization, depending on their embedded depth within sediments (Hoellein et al. 2019). Shape is also a significant factor influencing the transport of MPs. Common shapes include fibers, pellets, fragments, films, foams, and beads (Kumar et al. 2021). Generally, MPs with shapes like fibers and films show greater buoyancy and prolonged suspension, while compact shapes like fragments and beads tend to be deposited more quickly (Hoellein et al. 2019). In soil environments, nonspherical particles demonstrate higher retention compared to spherical ones, as shown by Rillig et al. (2017a, b). Experimental results by Waldschläger and Schüttrumpf (2020) confirmed that in the same experimental conditions, spherical MPs could penetrate up to 30 cm into sand columns, whereas fibrous MPs reached only 12 cm. Size significantly impacts MPs' transport behavior, both in rivers and soils. Smaller MPs are more mobile and prone to aggregation and can penetrate deeper layers (Wang et al. 2022a, b). In soils, studies show that smaller MPs, particularly those in the 10–20 μm range, exhibit enhanced vertical migration through porous matrices (Ashiq et al. 2023; Boos et al. 2024; Jazaei et al. 2022). Larger MPs are more influenced by their shapes and typically settle more quickly, unless aggregation occurs.

In addition to physical characteristics, surface chemistry parameters, including surface functional groups, surface charge, and hydrophilicity, play an essential role in MP transport. Dong et al. (2019a, b) demonstrated that MPs modified with carboxyl, sulfonic acid, and amino groups exhibited different retention patterns in saturated sandy media. Carboxylated MPs showed higher mobility compared to sulfonic acid– or amino-modified MPs. Similarly, Ren et al. (2022) found that MPs with more oxygen-containing functional groups were more effective at trapping heavy metals like cadmium in soil environments. Surface charge also influences electrostatic interactions that control MP aggregation and sedimentation (Dong et al. 2019a, b; Li et al. 2020a, b), while hydrophilic MPs are more readily transported through aqueous environments (Dong et al. 2019a, b; Jiang et al. 2021). Environmental exposure could result in the degradation of MPs, which alters their surface properties and behavior. Processes like oxidative degradation introduce new surface functional groups, such as carbonyls (Geng et al. 2024). Aging increases surface roughness, enhances oxygen-containing groups, and shifts MPs' hydrophobicity toward hydrophilicity. These changes enhance MPs' adsorption capacity and alter their mobility in soil and water matrices (Liu et al. 2020).

2.6 Mitigation Strategies and Management Practices

Protection of freshwater resources from MPs released by land-based sources should be done through a multifaceted approach that focuses on source reduction, improved removal practices, and public awareness. Source reduction can involve using more biodegradable and nonplastic alternatives to synthetic plastic products for everyday items; banning disposable plastic bags; utilizing organic, biodegradable, or reusable materials for agricultural practices; and limiting the usage of MPs for personal care products. On the other hand, advancing wastewater treatment practices for the improved removal of MPs from municipal and industrial wastewater can prevent the release of a considerable amount of MPs into freshwater systems. Moreover, developing standard methodologies for the quantification of the flux of MPs via various pathways to water resources and conducting regular, reliable monitoring for atmospheric, aquatic, and terrestrial systems can inform the decision-makers regarding their essential role in developing effective policies for better management of MP pollution (Mani et al. 2016; Wang et al. 2022c). Moreover, creating open-access platforms and comprehensive databases can assist decision-makers, researchers, and community coalitions to better understand their associated risks (Jenkins et al. 2022). Additionally, knowledge advancement regarding MP transport by stormwater runoff can lead to the development of more effective removal strategies (Krause et al. 2021; Zhang and Choi 2022). Although policy initiatives, public outreach, and education programs are critical in terms of the mitigation of land-based plastic pollution sources (da Costa et al. 2020; Sinha et al. 2024; Walther et al. 2021), in this chapter, we solely focus on best practices for stormwater management and sustainable agricultural practices to minimize MP release from terrestrial sources to freshwater systems.

2.6.1 Best Practices for MP Management in Stormwater Runoff

Decentralized runoff management approaches have been recognized as effective in mitigating MP pollution (Moore et al. 2017; Roy et al. 2008). In urban areas, planning can be integrated with stormwater management practices through water-sensitive design approaches. Green roofs that decrease runoff volumes, bioretention systems (e.g., rain gardens) that filter stormwater through vegetated basins, vegetated swales that slow and filter runoff, and permeable pavements that promote water infiltration are some of the major examples of sensitive design approaches (Coutts et al. 2012; Liu et al. 2022a; Nichols and Lucke 2015; Ristianti et al. 2022; Wanitchayapaisit et al. 2023). These engineered solutions, as well as constructed wetlands and infiltration trenches, emphasize source control and natural treatment processes (Rodak et al. 2020; Roy et al. 2008). The implementation of these measures maintains predevelopment hydrologic conditions while reducing contaminated runoff reaching surface waters (Walsh et al. 2016, 2022). Figure 2.2 illustrates the major examples of runoff management practices that could mitigate MP. Example practices are green infrastructure (GI, aka low-impact development, stormwater control measures, and sustainable urban drainage systems) and nonstructural practices. These practices can work together to achieve effective stormwater management through structural and nonstructural interventions (Borah et al. 2024). Each practice can play a role in MP management. Bioretention systems have demonstrated removal efficiencies of 60–95% for MPs, particularly those larger than 100 μm (Sharma and Malaviya 2021; Werbowski et al. 2021), while stormwater retention ponds can achieve 50–80% removal through sedimentation processes (Adeyemo et al. 2010; Snodgrass et al. 2017). A combination of these systems can enhance their effectiveness, as each one targets specific sizes and types of MPs (Braga Moruzzi et al. 2020; Lange et al. 2022). The MP concentration has been reported to be reduced from 3.8–59 MPs/L in untreated runoff to 1.8–32 MPs/L after stormwater treatment (Ziajahromi et al. 2023).

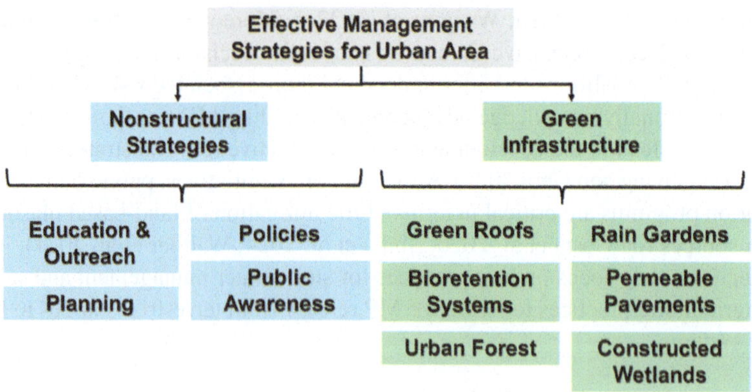

Fig. 2.2 Effective management strategies for controlling MPs in surface runoff

Integrated location-specific GI (e.g., wetlands) can be utilized as an effective tool for better MP management in urban stormwater (Moore et al. 2018; Rodak et al. 2020). However, its effectiveness can be influenced by its incorporation into broader urban planning strategies and the availability of funding resources for its long-term maintenance (Liu et al. 2016; Walsh et al. 2016; Yoo and Park 2022; Zhao et al. 2019). Conducting active community outreach programs can enhance the sustainability of developed integrated stormwater management programs by promoting public support and engagement (Brown et al. 2014; Rodak et al. 2020). Community outreach and education programs can present multiple benefits of sustainable runoff management practices (Brown et al. 2014; Rodak et al. 2020). Moreover, by implementing integrated monitoring plans, we can ensure that green infrastructures are performing well in removing MPs and other runoff pollutants (Liu et al. 2016; McIntyre et al. 2016). Runoff control practices across different spatial scales and jurisdictions can be coordinated via integrated watershed management strategies Walsh et al. 2016). However, ongoing coordination among researchers, community stakeholders, and local and regional decision-makers can enhance the chance of success for these strategies.

2.6.2 Sustainable Agricultural Practices to Minimize MP Release

Replacing regular synthetic plastic mulch with more sustainable alternatives is essential due to its significant contribution to plastic pollution worldwide. Cover crops, biodegradable and paper mulch, and organic mulching materials (e.g., straw, woodchips, and compost) can be used as alternatives to plastic mulch. By using these sustainable alternatives, we can promote soil health and food safety by limiting the sources of MP release to the farmlands and, subsequently, to the adjacent freshwater systems. For instance, straw (e.g., rice and wheat straw) can reduce fertilizer leaching, suppress weeds, retain moisture, and support beneficial insect communities, promoting integrated pest management (Azimi et al. 2023). Deep compost mulching, which involves applying a thick layer of compost to bury weed seeds, can be expensive and unsuitable for erosion-prone areas due to nutrients in stormwater runoff. Woodchips are another effective organic option for moisture retention, though they should not be placed near root zones to avoid nitrogen depletion. Organic mulches also increase soil organic matter, provide nutrients through biodegradation, enhance soil biological activity, and improve soil structure (Uwah and Iwo 2011). Additionally, paper mulch, made from cellulose-based materials, can be a sustainable alternative to plastic mulch, though it is less durable throughout the season. Moreover, instead of using plastic mulch, cash crops can be directly planted into a terminated cover crop. By this practice, soil fertility is enhanced, and soil erosion resistance is promoted, while weeds are also suppressed (Vincent-Caboud et al. 2019).

Biodegradable plastic mulches can be an alternative to the conventional synthetic disposable plastic mulch. They are primarily designed to be tilled into the soil and reduce the cost associated with their removal from farmlands. However, they are not allowed to be tilled in the soil in organic-certified farms everywhere. An example is the United States that does not allow biodegradable plastic. Thus, they should be fully retrieved from the soil. However, their susceptibility to environmental weathering makes their retrieval from organic farms very challenging. Using the residual biodegradable plastic mulch as carbon and energy sources for microorganisms results in their final mineralization into water and carbon dioxide (Kyrikou and Briassoulis 2007). Although biodegradable plastic provides multiple benefits compared to common synthetic plastic mulch, its widespread adoption is limited due to financial considerations (Shah and Wu 2020). Using woven plastic fabric is advantageous compared to conventional disposable plastics, as it can be reused for several years, thus reducing the extent of plastic waste generation. Moreover, the efficient removal of disposable plastic mulch after harvesting can limit the long-term accumulation of its residual plastic particles within the soil and consequently limit its subsequent transport to adjacent freshwater systems. Future research is needed to examine how various operational factors, material characteristics, and soil properties influence the retrieval efficiency of plastic mulch from farmlands. Moreover, recycling is generally recommended as an efficient practice in decreasing the plastic pollutants caused by thermoplastic disposable products; however, recycling synthetic plastic mulch is not a feasible process due to its contamination with soil and plant residuals.

2.7 Conclusions and Future Research Needs

The potential negative impacts of MPs on human health and natural ecosystems have created significant global concerns. Understanding the MP emission sources, accumulation dynamics, and transport pathways across different environmental systems can promote the development of reliable and accurate predictions for their flux and subsequent exposure rates for human, aquatic, and terrestrial organisms. The significant contribution of the terrestrial sources to MP pollution in freshwater systems is evident by the huge number of MPs released by municipal and industrial wastewater treatment plants and urban and agricultural surface runoff. The major transport pathways for releasing MPs from terrestrial sources to freshwater consist of wastewater effluent, stormwater runoff, atmospheric deposition, and inland streams. Despite many case studies investigating the contribution of certain terrestrial sources in MP release and tracking their transport dynamics by quantifying them in runoff, sediments, streams, and atmospheric precipitates, the knowledge of the physicochemical variations of MPs as they transport through this pathway is limited. For instance, limited information is available about the physicochemical alterations of MPs during tertiary wastewater treatment processes such as disinfection. Although the vertical distribution of MPs in the water column and penetration into soil depth have attracted a lot of attention from researchers, future investigations are needed to better understand the interrelation among MP surface characteristics, dimensions, shapes, and their vertical distribution in vegetated systems. Moreover, the inherent variability in MP dispersion in riverine systems resulted in significant challenges in developing standard methodologies for water sampling and quantification of MP distribution in this system. Future studies should examine the drivers of MP distribution in riverine systems as they are influenced by flow and hydrologic conditions. Moreover, future research should be conducted to investigate the variations of MPs' atmospheric deposition rates as influenced by climate conditions (tropical, arid, and semiarid), topographies (e.g., mountains, valleys, and coastal areas), and land cover/land uses (e.g., industrial, residential, and commercial) to develop effective mitigation plans and control measures. Future studies are needed to examine how the interaction of MPs with other air pollutants affects their long-term degradation and settling characteristics. Advancing our understanding of these processes is essential to developing integrated strategies for MP mitigation and ensuring both environmental safety and human health.

Acknowledgments Funding for this work was provided by the US Department of Agriculture's National Resources Conservation Service (USDA/NRCS) under awards NR236424XXXXG003 NR234209XXXXG004; US National Science Foundation (NSF) under grants CBET-2305189, CBET-2347006, CBET-2446126, and EAR-2340588; and USDA National Institute of Food and Agriculture, grant 67019-31166-2020.

References

Adeyemo J, Adeyemo F, Otieno F (2010) Assessment of pollutant loads of runoff in Pretoria, South Africa. Int J Soc Ecol Sustain Dev 1(3):1–12. https://doi.org/10.4018/jsesd.2010070101

Afkarina KII, Moersidik S, Utomo SW (2020) Distribution and environmental risk of microplastics pollution in freshwater of Citarum watershed. E3S Web Conf 211:03012. https://doi.org/10.1051/e3sconf/202021103012

Aghilinasrollahabadi K, Salehi M, Fujiwara T (2021) Investigate the influence of microplastics weathering on their heavy metals uptake in stormwater. J Hazard Mater 408:124439. https://doi.org/10.1016/j.jhazmat.2020.124439

Ali N, Khan MH, Ali M, Sidra, Ahmad S, Khan A, Nabi G, Ali F, Bououdina M, Kyzas GZ (2024) Insight into microplastics in the aquatic ecosystem: properties, sources, threats and mitigation strategies. Sci Total Environ 913:169489. https://doi.org/10.1016/j.scitotenv.2023.169489

Anggraini RR, Risjani Y, Yanuhar U (2020) Plastic litter as pollutant in the aquatic environment: a mini-review. Jurnal Ilmiah Perikanan Dan Kelautan 12(1):167. https://doi.org/10.20473/jipk.v12i1.17963

Ashiq MM, Jazaei F, Bell K, Ali ASA, Bakhshaee A, Babakhani P (2023) Abundance, spatial distribution, and physical characteristics of microplastics in stormwater detention ponds. Front Environ Sci Eng 17(10):124. https://doi.org/10.1007/s11783-023-1724-y

Aves AR, Revell LE, Gaw S, Ruffell H, Schuddeboom A, Wotherspoon NE, LaRue M, McDonald AJ (2022) First evidence of microplastics in Antarctic snow. Cryosphere 16(6):2127–2145. https://doi.org/10.5194/tc-16-2127-2022

Azimi S, Amini R, Hosseingolizadeh M (2023) Suppression of weed and insect populations by living and straw mulches in sesame (*Sesamum indicum* L.). Sci Rep 13:21586. https://doi.org/10.1038/s41598-023-48978-6

Baensch-Baltruschat B, Kocher B, Stock F, Reifferscheid G (2020) Tyre and road wear particles (TRWP)—a review of generation, properties, emissions, human health risk, ecotoxicity, and fate in the environment. Sci Total Environ 733:137823. https://doi.org/10.1016/j.scitotenv.2020.137823

Baldwin AK, Corsi SR, Mason SA (2016) Plastic debris in 29 great lakes tributaries: relations to watershed attributes and hydrology. Environ Sci Technol 50(19):10377–10385. https://doi.org/10.1021/acs.est.6b02917

Barkmann-Metaj L, Weber F, Bitter H, Wolff S, Lackner S, Kerpen J, Engelhart M (2023) Quantification of microplastics in wastewater systems of German industrial parks and their wastewater treatment plants. Sci Total Environ 881:163349. https://doi.org/10.1016/j.scitotenv.2023.163349

Battacharjee L, Jazaei F, Salehi M (2023) Insights into the mechanism of plastics' fragmentation under abrasive mechanical forces: an implication for agricultural soil health. CLEAN Soil Air Water 51(8):2200395. https://doi.org/10.1002/clen.202200395

Beheshtimaal A, Alamdari N, Wang B, Kamali M, Salehi M (2024) Understanding the dynamics of microplastics transport in urban stormwater runoff: implications for pollution control and management. Environ Pollut 356:124302. https://doi.org/10.1016/j.envpol.2024.124302

Bennett SJ, Simon A (2004) Riparian vegetation and fluvial geomorphology: problems and opportunities. In: Riparian vegetation and fluvial geomorphology, vol 8. https://doi.org/10.1029/WS008

Bergmann M, Mützel S, Primpke S, Tekman MB, Trachsel J, Gerdts G (2019) White and wonderful? Microplastics prevail in snow from the Alps to the Arctic. Sci Adv 5(8):eaax1157. https://doi.org/10.1126/sciadv.aax1157

Bharath KM, Natesan U, Vaikunth R, Kumar PR, Ruthra R, Srinivasalu S (2021) Spatial distribution of microplastic concentration around landfill sites and its potential risk on groundwater. Chemosphere 277:130263. https://doi.org/10.1016/j.chemosphere.2021.130263

Bhattacharjee L, Gopakumar AN, Beheshtimaal A, Jazaei F, Ccanccapa-Cartagena A, Salehi M (2025) Mechanisms of microplastic generation from polymer-coated controlled-release fertilizers (PC-CRFs). J Hazard Mater 486:137082. https://doi.org/10.1016/j.jhazmat.2024.137082

Biswas B, Chandra Das B (2016) Hydraulic parameters and morphometric variables interactions in bedrock channel. Quaest Geogr 35(3):75–88. https://doi.org/10.1515/quageo-2016-0028

Bocker R, Silva EK (2025) Microplastics in our diet: a growing concern for human health. Sci Total Environ 968:178882. https://doi.org/10.1016/j.scitotenv.2025.178882

Bonyadinejad G, Salehi M (2024) A simple methodology for in situ study of microplastics' aggregation. CLEAN Soil Air Water 52:2300378. https://doi.org/10.1002/clen.202300378

Bonyadinejad G, Salehi M, Herath A (2022) Investigating the sustainability of agricultural plastic products, combined influence of polymer characteristics and environmental conditions on microplastics aging. Sci Total Environ 839:156385. https://doi.org/10.1016/j.scitotenv.2022.156385

Boon PJ, Raven PJ (eds) (2012) River conservation and management. Wiley. https://doi.org/10.1002/9781119961819

Boos J-P, Dichgans F, Fleckenstein JH, Gilfedder BS, Frei S (2024) Assessing the behavior of microplastics in fluvial systems: infiltration and retention dynamics in streambed sediments. Water Resour Res 60(2):e2023WR035532. https://doi.org/10.1029/2023WR035532

Braga Moruzzi R, Galileu Speranza L, Tomazini da Conceição F, Teodoro de Souza Martins S, Busquets R, Cintra Campos L (2020) Stormwater detention reservoirs: an opportunity for monitoring and a potential site to prevent the spread of urban microplastics. Water 12(7):1994. https://doi.org/10.3390/w12071994

Brahney J, Hallerud M, Heim E, Hahnenberger M, Sukumaran S (2020) Plastic rain in protected areas of the United States. Science 368(6496):1257–1260. https://doi.org/10.1126/science.aaz5819

Bretas Alvim C, Mendoza-Roca JA, Bes-Piá A (2020) Wastewater treatment plant as microplastics release source—quantification and identification techniques. J Environ Manag 255:109739. https://doi.org/10.1016/j.jenvman.2019.109739

Brown H, Bos DG, Walsh CJ, Fletcher TD, Rossrakesh S (2014) More than money: how multiple factors influence householder participation in at-source stormwater management. J Environ Plan Manag 59(1):79–97. https://doi.org/10.1080/09640568.2014.984017

Cakaj A, Lisiak-Zielińska M, Drzewiecka K, Budka A, Borowiak K, Drapikowska M, Cakaj A, Qorri E, Szkudlarz P (2023) Potential impact of urban land use on microplastic atmospheric deposition: a case study in Pristina City, Kosovo. Sustainability 15(23):16464. https://doi.org/10.3390/su152316464

Campanale C, Stock F, Massarelli C, Kochleus C, Bagnuolo G, Reifferscheid G, Uricchio VF (2020) Microplastics and their possible sources: the example of the Ofanto River in Southeast Italy. Environ Pollut 258:113284. https://doi.org/10.1016/j.envpol.2019.113284

Canchari F, Iannacone J (2023) Microplastics in sediments deposited by rainwater runoff in a populated center in the Peruvian Andes. RBRH 28:e7. https://doi.org/10.1590/2318-0331.282320220108

Carr SA (2017) Sources and dispersive modes of micro-fibers in the environment. Integr Environ Assess Manag 13(3):466–469. https://doi.org/10.1002/ieam.1916

Chen Y, Niu J, Xu D, Zhang M, Sun K, Gao B (2023) Wet deposition of globally transportable microplastics (<25 μm) hovering over the megacity of Beijing. Environ Sci Technol 57(30):11152–11162. https://doi.org/10.1021/acs.est.3c03474

Chicharo L, Müller F, Fohrer N (2015) Introduction. In: Chicharo L, Müller F, Fohrer N (eds) Ecosystem services and River Basin Ecohydrology. Springer, Dordrecht. https://doi.org/10.1007/978-94-017-9846-4_1

Cohen QM, Glaese M, Meng K, Geissen V, Huerta-Lwanga E (2022) Parks and recreational areas as sinks of plastic debris in urban sites: the case of light-density microplastics in the City of Amsterdam, The Netherlands. Environments 9(1):5. https://doi.org/10.3390/environments9010005

Conard WM, O'Reilly KE, Hartlage C, Lamberti GA (2023) Widespread microplastic pollution in Indiana, USA. River Res Appl 39(10):2092–2101. https://doi.org/10.1002/rra.4204

Coutts AM, Tapper NJ, Beringer J, Loughnan M, Demuzere M (2012) Watering our cities: the capacity for water sensitive urban design to support urban cooling and improve human thermal comfort in the Australian context. Prog Phys Geogr 37(1):2–28. https://doi.org/10.1177/0309133312461032

Cristaldi A, Fiore M, Zuccarello P, Oliveri Conti G, Grasso A, Nicolosi I, Copat C, Ferrante M (2020) Efficiency of wastewater treatment plants (WWTPs) for microplastic removal: a systematic review. Int J Environ Res Public Health 17(21):8014. https://doi.org/10.3390/ijerph17218014

da Costa JP, Mouneyrac C, Costa M, Duarte AC, Rocha-Santos T (2020) The role of legislation, regulatory initiatives and guidelines on the control of plastic pollution. Front Environ Sci 8. https://www.frontiersin.org/journals/environmental-science/articles/10.3389/fenvs.2020.00104

Deva K., Borah Harry X., Zhang Moira, Zellner Ebrahim, Ahmadisharaf Meghna, Babbar-Sebens Nigel W. T., Quinn Saurav, Kumar Vamsi Krishna, Sridharan Navaratnam, Leelaruban Craig, Lott (2024) Advances in Total Maximum Daily Load Implementation Planning by Modeling Best Management Practices and Green Infrastructures Journal of Environmental Engineering 150(7) 10.1061/JOEEDU.EEENG-7578

Ding L, fan Mao R, Guo X, Yang X, Zhang Q, Yang C (2019) Microplastics in surface waters and sediments of the Wei River, in the northwest of China. Sci Total Environ 667:427–434. https://doi.org/10.1016/j.scitotenv.2019.02.332

Dong Z, Zhang W, Qiu Y, Yang Z, Wang J, Zhang Y (2019a) Cotransport of nanoplastics (NPs) with fullerene (C_{60}) in saturated sand: effect of NPs/C_{60} ratio and seawater salinity. Water Res 148:469–478. https://doi.org/10.1016/j.watres.2018.10.071

Dong Z, Zhu L, Zhang W, Huang R, Lv X, Jing X, Yang Z, Wang J, Qiu Y (2019b) Role of surface functionalities of nanoplastics on their transport in seawater-saturated sea sand. Environ Pollut (Barking, Essex: 1987) 255(Pt 1):113177. https://doi.org/10.1016/j.envpol.2019.113177

Dong H, Wang L, Wang X, Xu L, Chen M, Gong P, Wang C (2021a) Microplastics in a remote lake basin of the Tibetan Plateau: impacts of atmospheric transport and glacial melting. Environ Sci Technol 55(19):12951–12960. https://doi.org/10.1021/acs.est.1c03227

Dong S, Xia J, Sheng L, Wang W, Liu H, Gao B (2021b) Transport characteristics of fragmental polyethylene glycol terephthalate (PET) microplastics in porous media under various chemical conditions. Chemosphere 276:130214. https://doi.org/10.1016/j.chemosphere.2021.130214

Dong S, Zhou M, Su X, Xia J, Wang L, Wu H, Suakollie EB, Wang D (2022) Transport and retention patterns of fragmental microplastics in saturated and unsaturated porous media: a real-time pore-scale visualization. Water Res 214:118195. https://doi.org/10.1016/j.watres.2022.118195

Dris R, Gasperi J, Saad M, Mirande C, Tassin B (2016) Synthetic fibers in atmospheric fallout: a source of microplastics in the environment? Mar Pollut Bull 104(1–2):290–293. https://doi.org/10.1016/j.marpolbul.2016.01.006

Drummond JD, Nel HA, Packman AI, Krause S (2020) Significance of hyporheic exchange for predicting microplastic fate in rivers. Environ Sci Technol Lett 7(10):727–732. https://doi.org/10.1021/acs.estlett.0c00595

Du J, Xu S, Zhou Q, Li H, Fu L, Tang J, Wang Y, Peng X, Xu Y, Du X (2020) A review of microplastics in the aquatic environmental: distribution, transport, ecotoxicology, and toxicological mechanisms. Environ Sci Pollut Res 27(11):11494–11505. https://doi.org/10.1007/s11356-020-08104-9

Esquinas GGMS, Mantala AP, Atilano MG, Apugan RP, Galarpe VRKR (2020) Physical characterization of litter and microplastic along the urban coast of Cagayan de Oro in Macajalar Bay, Philippines. Mar Pollut Bull 154:111083. https://doi.org/10.1016/j.marpolbul.2020.111083

Fayshal MA (2024) Current practices of plastic waste management, environmental impacts, and potential alternatives for reducing pollution and improving management. Heliyon 10(23):e40838. https://doi.org/10.1016/j.heliyon.2024.e40838

Food and Agriculture Organization of the United Nations (2021) Assessment of agricultural plastics and their sustainability: a call for action. FAO. https://doi.org/10.4060/cb7856en

Forster NA, Wilson SC, Tighe MK (2023) Microplastic surface retention and mobility on hiking trails. Environ Sci Pollut Res 30(16):46368–46382. https://doi.org/10.1007/s11356-023-25635-z

Gambino I, Terzaghi E, Baldini E, Bergna G, Palmisano G, Guardo AD (2025) Microcontaminants and microplastics in water from the textile sector: a review and a database of physicochemical properties, use in the textile process, and ecotoxicity data for detected chemicals. Environ Sci: Processes Impacts 27(2):297–319. https://doi.org/10.1039/D4EM00639A

Geng C, Gao Y, Zhang H et al (2024) Microplastic migration in porous media at various scales: a review. Environ Chem Lett 22:691–713. https://doi.org/10.1007/s10311-023-01688-x

Giechaskiel B, Grigoratos T, Mathissen M, Quik J, Tromp P, Gustafsson M, Franco V, Dilara P (2024) Contribution of road vehicle tyre wear to microplastics and ambient air pollution. Sustainability 16(2):522. https://doi.org/10.3390/su16020522

Hadiuzzaman M, Salehi M, Fujiwara T (2022) Plastic litter fate and contaminant transport within the urban environment, photodegradation, fragmentation, and heavy metal uptake from storm runoff. Environ Res 212:113183. https://doi.org/10.1016/J.ENVRES.2022.113183

Herath A, Salehi M (2022) Studying the combined influence of microplastics' intrinsic and extrinsic characteristics on their weathering behavior and heavy metal transport in storm runoff. Environ Pollut 308:119628. https://doi.org/10.1016/j.envpol.2022.119628

Hidayaturrahman H, Lee TG (2019) A study on characteristics of microplastic in wastewater of South Korea: identification, quantification, and fate of microplastics during treatment process. Mar Pollut Bull 146:696–702. https://doi.org/10.1016/j.marpolbul.2019.06.071

Hoellein TJ, Shogren AJ, Tank JL et al (2019) Microplastic deposition velocity in streams follows patterns for naturally occurring allochthonous particles. Sci Rep 9:3740. https://doi.org/10.1038/s41598-019-40126-3

Horton AA, Walton A, Spurgeon DJ, Lahive E, Svendsen C (2017) Microplastics in freshwater and terrestrial environments: evaluating the current understanding to identify the knowledge gaps and future research priorities. Sci Total Environ 586:127–141. https://doi.org/10.1016/j.scitotenv.2017.01.190

Huang Y, Liu Q, Jia W, Yan C, Wang J (2020) Agricultural plastic mulching as a source of microplastics in the terrestrial environment. Environ Pollut 260:114096. https://doi.org/10.1016/j.envpol.2020.114096

Huang CW, Li Y, Lin C, Bui X, Vo T, Ngo HH (2023) Seasonal influence on pollution index and risk of multiple compositions of microplastics in an urban river. Sci Total Environ 859(Part 1):160021. https://doi.org/10.1016/j.scitotenv.2022.160021

Hurley RR, Nizzetto L (2018) Fate and occurrence of micro(nano)plastics in soils: knowledge gaps and possible risks. Curr Opin Environ Sci Health 1:6–11. https://doi.org/10.1016/J.COESH.2017.10.006

Ibna Hafiz AM, Ahmadisharaf E, Salehi M, Farner J, White JC, Zeng EY, Nazari B (2025) A review of processes and models for the export of microplastics from terrestrial to aquatic systems. WIREs Water 12(1):e70004. https://doi.org/10.1002/wat2.70004

Jayalakshmamma M, Na Nagara V, Borgaonkar A, Sarkar D, Sadik O, Boufadel M (2023) Characterizing microplastics in urban runoff: a multi-land use assessment with a focus on 1–125 μm size particles. Sci Total Environ 904:166685. https://doi.org/10.1016/j.scitotenv.2023.166685

Jazaei F, Chy TJ, Salehi M (2022) Can microplastic pollution change soil-water dynamics? Results from controlled laboratory experiments. Water 14(21):Article 21. https://doi.org/10.3390/w14213430

Jenkins T, Persaud BD, Cowger W, Szigeti K, Roche DG, Clary E, Slowinski S, Lei B, Abeynayaka A, Nyadjro ES, Maes T, Thornton Hampton L, Bergmann M, Aherne J, Mason SA, Honek JF, Rezanezhad F, Lusher AL, Booth AM et al (2022) Current state of microplastic pollution research data: trends in availability and sources of open data. Front Environ Sci 10. https://www.frontiersin.org/journals/environmental-science/articles/10.3389/fenvs.2022.912107

Jiang Y, Yin X, Xi X, Guan D, Sun H, Wang N (2021) Effect of surfactants on the transport of polyethylene and polypropylene microplastics in porous media. Water Res 196:117016. https://doi.org/10.1016/j.watres.2021.117016

Kabir SMA, Bhuiyan MA, Zhang G, Pramanik BK (2024) Microplastic distribution and ecological risks: investigating road dust and stormwater runoff across land uses. Environ Sci Adv 3(1):62–75. https://doi.org/10.1039/D3VA00128H

Kanhai LDK, Gardfeldt K, Krumpen T et al (2020) Microplastics in sea ice and seawater beneath ice floes from the Arctic Ocean. Sci Rep 10:5004. https://doi.org/10.1038/s41598-020-61948-6

Karakurt O, Altuntaş O, Şimşek İ, Hatinoğlu D, Sanin FD (2025) Microplastics from industrial sources: a known but overlooked problem. J Water Process Eng 72:107487. https://doi.org/10.1016/j.jwpe.2025.107487

Kasirajan S, Ngouajio M (2012) Polyethylene and biodegradable mulches for agricultural applications: a review. Agron Sustain Dev 32(2):501–529. https://doi.org/10.1007/s13593-011-0068-3

Keep America Beautiful (2021) 2020 national litter study: summary report, May. https://kab.org/wp-content/uploads/2021/05/Litter-Study-Summary-Report-May-2021_final_05172021.pdf

Kole PJ, Löhr AJ, Van Belleghem FGAJ, Ragas AMJ (2017) Wear and tear of tyres: a stealthy source of microplastics in the environment. Int J Environ Res Public Health 14(10):1265. https://doi.org/10.3390/ijerph14101265

Krause S, Baranov V, Nel HA, Drummond JD, Kukkola A, Hoellein T, Sambrook Smith GH, Lewandowski J, Bonet B, Packman AI, Sadler J, Inshyna V, Allen S, Allen D, Simon L, Mermillod-Blondin F, Lynch I (2021) Gathering at the top? Environmental controls of microplastic uptake and biomagnification in freshwater food webs. Environ Pollut 268:115750. https://doi.org/10.1016/j.envpol.2020.115750

Kukkola A, Runkel RL, Schneidewind U, Murphy SF, Kelleher L, Sambrook Smith GH, Nel HA, Lynch I, Krause S (2023) Prevailing impacts of river management on microplastic transport in contrasting U.S. streams: rethinking global microplastic flux estimations. Water Res 240:120112. https://doi.org/10.1016/j.watres.2023.120112

Kumar R, Sharma P, Verma A, Jha PK, Singh P, Gupta PK, Chandra R, Prasad PVV (2021) Effect of physical characteristics and hydrodynamic conditions on transport and deposition of microplastics in riverine ecosystem. Water 13(19):2710. https://doi.org/10.3390/w13192710

Kumar A, Mishra S, Pandey R, Yu ZG, Kumar M, Khoo KS, Thakur TK, Show PL (2023) Microplastics in terrestrial ecosystems: un-ignorable impacts on soil characterises, nutrient storage and its cycling. TrAC Trends Anal Chem 158:Article 116869. https://doi.org/10.1016/j.trac.2022.116869

Kyrikou I, Briassoulis D (2007) Biodegradation of agricultural plastic films: a critical review. J Polym Environ 15:125–150. https://doi.org/10.1007/s10924-007-0053-8

Lambert S, Wagner M (2018) Microplastics are contaminants of emerging concern in freshwater environments: an overview. In: Wagner M, Lambert S (eds) Freshwater microplastics. The handbook of Environmental chemistry, vol 58. Springer, Cham. https://doi.org/10.1007/978-3-319-61615-5_1

Lange K, Österlund H, Viklander M, Blecken G-T (2022) Occurrence and concentration of 20–100 μm sized microplastic in highway runoff and its removal in a gross pollutant trap—bioretention and sand filter stormwater treatment train. Sci Total Environ 809:151151. https://doi.org/10.1016/j.scitotenv.2021.151151

Leads RR, Weinstein JE (2019) Occurrence of tire wear particles and other microplastics within the tributaries of the Charleston Harbor Estuary, South Carolina, USA. Mar Pollut Bull 145:569–582. https://doi.org/10.1016/j.marpolbul.2019.06.061

Lebreton LCM, van der Zwet J, Damsteeg J-W, Slat B, Andrady A, Reisser J (2017) River plastic emissions to the world's oceans. Nat Commun 8(1):15611. https://doi.org/10.1038/ncomms15611

Lechthaler S, Esser V, Schüttrumpf H, Stauch G (2021) Why analysing microplastics in floodplains matters: application in a sedimentary context. Environ Sci: Processes Impacts 23(1):117–131. https://doi.org/10.1039/D0EM00431F

Li L, Luo Y, Li R et al (2020a) Effective uptake of submicrometre plastics by crop plants via a crack-entry mode. Nat Sustain 3:929–937. https://doi.org/10.1038/s41893-020-0567-9

Li M, He L, Zhang X, Rong H, Tong M (2020b) Different surface charged plastic particles have different cotransport behaviors with kaolinite *particles in porous media. Environ Pollut (Barking, Essex: 1987) 267:115534. https://doi.org/10.1016/j.envpol.2020.115534

Ling X, Yan Z, Lu G (2022) Vertical transport and retention behavior of polystyrene nanoplastics in simulated hyporheic zone. Water Res 219:118609. https://doi.org/10.1016/j.watres.2022.118609

Liu Y, Cibin R, Bralts VF, Chaubey I, Bowling LC, Engel BA (2016) Optimal selection and placement of BMPs and LID practices with a rainfall-runoff model. Environ Model Softw 80:281–296. https://doi.org/10.1016/j.envsoft.2016.03.005

Liu F, Olesen KB, Borregaard AR, Vollertsen J (2019) Microplastics in urban and highway stormwater retention ponds. Sci Total Environ 671:992–1000. https://doi.org/10.1016/j.scitotenv.2019.03.416

Liu P, Zhan X, Wu X, Li J, Wang H, Gao S (2020) Effect of weathering on environmental behavior of microplastics: properties, sorption and potential risks. Chemosphere 242:125193. https://doi.org/10.1016/j.chemosphere.2019.125193

Liu A, Egodawatta P, Goonetilleke A (2022a) Ranking three water sensitive urban design (WSUD) practices based on hydraulic and water quality treatment performance: implications for effective stormwater treatment design. Water 14(8):1296. https://doi.org/10.3390/w14081296

Liu Y, Chen H, Wu S, Gao J, Li Y, An Z, Mao B, Tu R, Li T (2022b) Impact of vehicle type, tyre feature and driving behaviour on tyre wear under real-world driving conditions. Sci Total Environ 842:156950. https://doi.org/10.1016/j.scitotenv.2022.156950

Lopatta K, Buchholz F, Kaspereit T (2016) Asymmetric information and corporate social responsibility. Bus Soc 55(3):458–488. https://doi.org/10.1177/0007650315575488

Luo Y, Zhang Y, Xu Y, Guo X, Zhu L (2020) Distribution characteristics and mechanism of microplastics mediated by soil physicochemical properties. Sci Total Environ 726:138389. https://doi.org/10.1016/j.scitotenv.2020.138389

Lv X, Dong Q, Zuo Z, Liu Y, Huang X, Wu WM (2019) Microplastics in a municipal wastewater treatment plant: fate, dynamic distribution, removal efficiencies, and control strategies. J Clean Prod 225:579–586. https://doi.org/10.1016/j.jclepro.2019.03.321

Magalhães S, Alves L, Romano A, Medronho B, da Graça Rasteiro M (2022) Extraction and characterization of microplastics from Portuguese industrial effluents. Polymers 14(14):2902. https://doi.org/10.3390/polym14142902

Magalhães S, Paciência D, Rodrigues JMM, Lindman B, Alves L, Medronho B, da Graça Rasteiro M (2024) Insights on microplastic contamination from municipal and textile industry effluents and their removal using a cellulose-based approach. Polymers 16(19):2803. https://doi.org/10.3390/polym16192803

Mancini M, Solari L, Colomer J, Serra T (2023) Retention of microplastics by interspersed lagoons in both natural and constructed wetlands. J Water Process Eng 56:104559. https://doi.org/10.1016/j.jwpe.2023.104559

Mandal M, Roy A, Binha SK, Popek R, Przybysz A, Koczoń P, Prasad D, Sarkar A (2025) Waste dumps as microplastic hotspots: a comparative investigation at urban, suburban, and rural areas of eastern India and associated risk assessment. Environ Toxicol Chem:vgaf086. https://doi.org/10.1093/etojnl/vgaf086

Mani T, Hauk A, Walter U et al (2016) Microplastics profile along the Rhine River. Sci Rep 5:17988. https://doi.org/10.1038/srep17988

Mason SA, Garneau D, Sutton R, Chu Y, Ehmann K, Barnes J, Fink P, Papazissimos D, Rogers DL (2016) Microplastic pollution is widely detected in U.S. municipal wastewater treatment plant effluent. Environ Pollut 218:1045–1054. https://doi.org/10.1016/j.envpol.2016.08.056

McIntyre JK, Edmunds RC, Redig MG, Mudrock EM, Davis JW, Incardona JP, Stark JD, Scholz NL (2016) Confirmation of stormwater bioretention treatment effectiveness using molecular indicators of cardiovascular toxicity in developing fish. Environ Sci Technol 50(3):1561–1569. https://doi.org/10.1021/acs.est.5b04786

Miller J, Orbock Miller S (2020) A geomorphic framework for the analysis of microplastics in riverine sediments. E3S Web Conf 202:01002. https://doi.org/10.1051/e3sconf/202020201002

Moore TL, Rodak CM, Vogel JR (2017) Urban stormwater characterization, control, and treatment. Water Environ Res 89(10):1876–1927. https://doi.org/10.2175/106143017X15023776270692

Moore TL, Rodak CM, Ahmed F, Vogel JR (2018) Urban stormwater characterization, control and treatment. Water Environ Res 90(10):1821–1871. https://doi.org/10.2175/106143018X15289915807452

Moses SR, Löder MGJ, Herrmann F, Laforsch C (2023) Seasonal variations of microplastic pollution in the German River Weser. Sci Total Environ 902:166463. https://doi.org/10.1016/j.scitotenv.2023.166463

Murphy F, Quinn B (2018) The effects of microplastic on freshwater *Hydra attenuata* feeding, morphology & reproduction. Environ Pollut 234:487–494. https://doi.org/10.1016/j.envpol.2017.11.029

National Research Council (2002) Riparian areas: functions and strategies for management. The National Academies Press. https://doi.org/10.17226/10327

Newbould RA, Powell DM, Whelan MJ (2021) Macroplastic debris transfer in rivers: a travel distance approach. Front Water 3. https://www.frontiersin.org/journals/water/articles/10.3389/frwa.2021.724596

Nichols PWB, Lucke T (2015) Local level stormwater harvesting and reuse: a practical solution to the water security challenges faced by urban trees. Sustainability 7(7):8635–8648. https://doi.org/10.3390/su7078635

Niu L, Wang Y, Li Y, Lin L, Chen Y, Shen J (2023) Occurrence, degradation pathways, and potential synergistic degradation mechanism of microplastics in surface water: a review. Curr Pollut Rep 9(2):312–326. https://doi.org/10.1007/s40726-023-00262-x

Niyobuhungiro RV, Schenck CJ (2021) The dynamics of indiscriminate/illegal dumping of waste in Fisantekraal, Cape Town, South Africa. J Environ Manag 293:112954. https://doi.org/10.1016/j.jenvman.2021.112954

Nizzetto L, Futter M, Langaas S (2016) Are agricultural soils dumps for microplastics of urban origin? Environ Sci Technol 50(20):10777–10779. https://doi.org/10.1021/acs.est.6b04140

Olesen KB, Stephansen DA, Van Alst N, Vollertsen J (2019) Microplastics in a stormwater pond. Water 11(7):1466. https://doi.org/10.3390/w11071466

Olubusoye BS, Cizdziel JV, Wontor K, Heinen E, Grandberry T, Bennett ER, Moore MT (2024) Removal of microplastics from agricultural runoff using biochar: a column feasibility study. Front Environ Sci 12:1388606. https://doi.org/10.3389/fenvs.2024.1388606

Ortega DE, Cortés-Arriagada D (2023) Atmospheric microplastics and nanoplastics as vectors of primary air pollutants: a theoretical study on the polyethylene terephthalate (PET) case. Environ Pollut 318:120860. https://doi.org/10.1016/j.envpol.2022.120860

Owowenu EK, Nnadozie CF, Akamagwuna F et al (2023) A critical review of environmental factors influencing the transport dynamics of microplastics in riverine systems: implications for ecological studies. Aquat Ecol 57:557–570. https://doi.org/10.1007/s10452-023-10029-7

Piehl S, Leibner A, Löder MGJ et al (2018) Identification and quantification of macro- and microplastics on an agricultural farmland. Sci Rep 8:17950. https://doi.org/10.1038/s41598-018-36172-y

Prata JC (2018) Microplastics in wastewater: state of the knowledge on sources, fate and solutions. Mar Pollut Bull 129(1):262–265. https://doi.org/10.1016/j.marpolbul.2018.02.046

Prata JC, Silva ALP, da Costa JP, Mouneyrac C, Walker TR, Duarte AC, Rocha-Santos T (2019) Solutions and integrated strategies for the control and mitigation of plastic and microplastic pollution. Int J Environ Res Public Health 16(13):2411. https://doi.org/10.3390/ijerph16132411

Priya KL, Iqbal S, Archana AR, Gopika B, Mina M, Haddout S, Madhu AM (2023) Implications of solid waste dumps on the microplastic abundance in groundwater in Kollam, India. J Environ Manag 348:119224. https://doi.org/10.1016/j.jenvman.2023.119224

Quevedo IR, Tufenkji N (2012) Mobility of functionalized quantum dots and a model polystyrene nanoparticle in saturated quartz sand and loamy sand. Environ Sci Technol 46(8):4449–4457. https://doi.org/10.1021/es2045458

Radford F, Horton A, Hudson M, Shaw P, Williams I (2023) Agricultural soils and microplastics: are biosolids the problem? Front Soil Sci 2. https://www.frontiersin.org/journals/soil-science/articles/10.3389/fsoil.2022.941837

Rehm R, Zeyer T, Schmidt A, Fiener P (2021) Soil erosion as a transport pathway of microplastic from agricultural soils to aquatic ecosystems. Sci Total Environ 795:148774. https://doi.org/10.1016/j.scitotenv.2021.148774

Ren Z, Gui X, Xu X, Zhao L, Qiu H, Wang X, Cao X (2022) Weathering of microplastics and their enhancement on the retention of cadmium in coastal soil saturated with seawater. J Hazard Mater 440:129850. https://doi.org/10.1016/j.jhazmat.2022.129850

Ren S-Y, Sun Q, Xia S-Y, Tong D, Ni H-G (2023) Microplastics in wastewater treatment plants and their contributions to surface water and farmland pollution in China. Chemosphere 343:140239. https://doi.org/10.1016/j.chemosphere.2023.140239

Research and Markets (2021) Mulch films market—forecast (2021–2026). Research and Markets, Dublin. Retrieved from https://www.researchandmarkets.com/reports/5566618/mulch-films-market-forecast-2021-2026. Accessed 30 Oct 2022

Rillig MC, Ingraffia R, de Souza Machado AA (2017a) Microplastic incorporation into soil in agroecosystems. Front Plant Sci 8. https://www.frontiersin.org/journals/plant-science/articles/10.3389/fpls.2017.01805

Rillig MC, Ziersch L, Hempel S (2017b) Microplastic transport in soil by earthworms. Sci Rep 7(1):1362. https://doi.org/10.1038/s41598-017-01594-7

Ristianti NS, Bashit N, Ulfiana D, Windarto YE (2022) Bioretention Basin, rain garden, and swales track concepts through vegetated-WSUD: sustainable rural stormwater management in klaten regency. IOP Conf Ser Earth Environ Sci 1082(1):012029. https://doi.org/10.1088/1755-1315/1082/1/012029

Rodak CM, Jayakaran AD, Moore TL, David R, Rhodes ER, Vogel JR (2020) Urban stormwater characterization, control, and treatment. Water Environ Res 92(10):1552–1586. https://doi.org/10.1002/wer.1403

Roebroek CTJ, Harrigan S, van Emmerik THM, Baugh C, Eilander D, Prudhomme C, Pappenberger F (2021) Plastic in global rivers: are floods making it worse? Environ Res Lett 16(2):025003. https://doi.org/10.1088/1748-9326/abd5df

Ross MS, Loutan A, Groeneveld T, Molenaar D, Kroetch K, Bujaczek T, Kolter S, Moon S, Huynh A, Khayam R, Franczak BC, Camm E, Arnold VI, Ruecker NJ (2023) Estimated discharge of microplastics via urban stormwater during individual rain events. Front Environ Sci 11. https://www.frontiersin.org/journals/environmental-science/articles/10.3389/fenvs.2023.1090267

Roy AH, Wenger SJ, Fletcher TD, Walsh CJ, Ladson AR, Shuster WD, Thurston HW, Brown RR (2008) Impediments and solutions to sustainable, watershed-scale urban stormwater management: lessons from Australia and the United States. Environ Manag 42(2):344–359. https://doi.org/10.1007/s00267-008-9119-1

Ryu H, Moon J, Kim H, Lee J (2021) Modeling and parametric simulation of microplastic transport in groundwater environments. Appl Sci 11(16):7189. https://doi.org/10.3390/app11167189

Sakti AD, Sembiring E, Rohayani P, Fauzan KN, Anggraini TS, Santoso C, Patricia VA, Ihsan KTN, Ramadan AH, Arjasakusuma S, Candra DS (2023) Identification of illegally dumped plastic waste in a highly polluted river in Indonesia using Sentinel-2 satellite imagery. Sci Rep 13(1):5039. https://doi.org/10.1038/s41598-023-32087-5

Salehi M (2022) Global water shortage and potable water safety; today's concern and tomorrow's crisis. Environ Int 158:106936. https://doi.org/10.1016/j.envint.2021.106936

Salehi M, Pincus LN, Deng B, Peters CA (2024) Microplastics: from intrinsic properties to Environmental fate. Environ Eng Sci 41(11):425–435. https://doi.org/10.1089/ees.2024.0232

Scheurer M, Bigalke M (2018) Microplastics in Swiss floodplain soils. Environ Sci Technol 52(6):3591–3598. https://doi.org/10.1021/acs.est.7b06003

Šedová B (2016) On causes of illegal waste dumping in Slovakia. J Environ Plan Manag 59(7):1277–1303. https://doi.org/10.1080/09640568.2015.1072505

Shah F, Wu W (2020) Chapter five—use of plastic mulch in agriculture and strategies to mitigate the associated environmental concerns. In: Sparks DL (ed) Advances in agronomy, vol 164. Academic Press, pp 231–287. https://doi.org/10.1016/bs.agron.2020.06.005

Shamskhany A, Li Z, Patel P, Karimpour S (2021) Evidence of microplastic size impact on mobility and transport in the marine environment: a review and synthesis of recent research. Front Mar Sci 8. https://www.frontiersin.org/journals/marine-science/articles/10.3389/fmars.2021.760649

Sharma R, Malaviya P (2021) Management of stormwater pollution using green infrastructure: the role of rain gardens. WIREs Water 8(2):e1507. https://doi.org/10.1002/wat2.1507

Siegfried M, Koelmans AA, Besseling E, Kroeze C (2017) Export of microplastics from land to sea. A modelling approach. Water Res 127:249–257. https://doi.org/10.1016/j.watres.2017.10.011

Sinha RK, Kumar R, Phartyal SS, Sharma P (2024) Interventions of citizen science for mitigation and management of plastic pollution: understanding sustainable development goals, policies, and regulations. Sci Total Environ 955:176621. https://doi.org/10.1016/j.scitotenv.2024.176621

Smyth K, Tan S, Van Seters T, Gasperi J, Dris R, Drake J, Passeport E (2024) Small-size microplastics in urban Stormwater runoff are efficiently trapped in a bioretention cell. ACS ES&T Water 4(6):2522–2531. https://doi.org/10.1021/acsestwater.4c00037

Snodgrass JW, Moore J, Lev SM, Casey RE, Ownby DR, Flora RF, Izzo G (2017) Influence of modern stormwater management practices on transport of road salt to surface waters. Environ Sci Technol 51(8):4165–4172. https://doi.org/10.1021/acs.est.6b03107

Soltani N, Keshavarzi B, Moore F, Busquets R, Nematollahi MJ, Javid R, Gobert S (2022) Effect of land use on microplastic pollution in a major boundary waterway: the Arvand River. Sci Total Environ 830:154728. https://doi.org/10.1016/j.scitotenv.2022.154728

Sotamenou J, De Jaeger S, Rousseau S (2019) Drivers of legal and illegal solid waste disposal in the global south—the case of households in Yaoundé (Cameroon). J Environ Manag 240:321–330. https://doi.org/10.1016/j.jenvman.2019.03.098

Sridharan S, Kumar M, Bolan NS, Singh L, Kumar S, Kumar R, You S (2021) Are microplastics destabilizing the global network of terrestrial and aquatic ecosystem services? Environ Res 198:111243. https://doi.org/10.1016/j.envres.2021.111243

Stanton T, Johnson M, Nathanail P, MacNaughtan W, Gomes RL (2020) Freshwater microplastic concentrations vary through both space and time. Environ Pollut 263:114481. https://doi.org/10.1016/j.envpol.2020.114481

Strokal M, Vriend P, Bak MP, Kroeze C, van Wijnen J, van Emmerik T (2023) River export of macro- and microplastics to seas by sources worldwide. Nat Commun 14(1):4842. https://doi.org/10.1038/s41467-023-40501-9

Sugiura M, Takada H, Takada N, Mizukawa K, Tsuyuki S, Furuma H (2021) Microplastics in urban wastewater and estuarine water: importance of street runoff. Environ Monit Contam Res 1(1):54–65. https://doi.org/10.5985/emcr.20200006

Szewc K, Graca B, Dołęga A (2021) Atmospheric deposition of microplastics in the coastal zone: characteristics and relationship with meteorological factors. Sci Total Environ 761:143272. https://doi.org/10.1016/j.scitotenv.2020.143272

Talbot R, Chang H (2022) Microplastics in freshwater: a global review of factors affecting spatial and temporal variations. Environ Pollut 292:118393. https://doi.org/10.1016/j.envpol.2021.118393

Tamis JE, Koelmans AA, Dröge R, Kaag NHBM, Keur MC, Tromp PC, Jongbloed RH (2021) Environmental risks of car tire microplastic particles and other road runoff pollutants. Micropl Nanopl 1(1):10. https://doi.org/10.1186/s43591-021-00008-w

Tamminga M, Hengstmann E, Deuke AK, Fischer EK (2022) Microplastic concentrations, characteristics, and fluxes in water bodies of the Tollense catchment, Germany, with regard to different sampling systems. Environ Sci Pollut Res Int 29(8):11345–11358. https://doi.org/10.1007/s11356-021-16106-4

Tan M, Liu L, Zhang M, Liu Y, Li C (2021) Effects of solution chemistry and humic acid on the transport of polystyrene microplastics in manganese oxides coated sand. J Hazard Mater 413:125410. https://doi.org/10.1016/j.jhazmat.2021.125410

Tanentzap AJ, Cottingham S, Fonvielle J, Riley I, Walker LM et al (2021) Microplastics and anthropogenic fibre concentrations in lakes reflect surrounding land use. PLoS Biol 19(9):e3001389. https://doi.org/10.1371/journal.pbio.3001389

Tatsii D, Bucci S, Bhowmick T, Guettler J, Bakels L, Bagheri G, Stohl A (2024) Shape matters: long-range transport of microplastic fibers in the atmosphere. Environ Sci Technol 58(1):671–682. https://doi.org/10.1021/acs.est.3c08209

Townsend K, Lu H-C, Sharley D, Pettigrove V (2019) Associations between microplastic pollution and land use in urban wetland sediments. Environ Sci Pollut Res 26:22551. https://doi.org/10.1007/s11356-019-04885-w

Turner A, Bridgwater A, Marshall E (2024) Environmental transport and sorting of glass retroreflective microbeads and their potential as proxies for road marking paints. Sci Total Environ 954:176057. https://doi.org/10.1016/j.scitotenv.2024.176057

U.S. Environmental Protection Agency (2021) Basic information about biosolids. U.S. Environmental Protection Agency. https://19january2021snapshot.epa.gov/biosolids/basic-information-about-biosolids_.html

Uwah DF, Iwo GA (2011) Effectiveness of organic mulch on the productivity of maize (Zea mays L.) and weed growth. J Anim Plant Sci 21(3):525–530

Vincent-Caboud L, Casagrande M, David C et al (2019) Using mulch from cover crops to facilitate organic no-till soybean and maize production. A review. Agron Sustain Dev 39:45. https://doi.org/10.1007/s13593-019-0590-2

Wadeson RA (1994) A geomorphological approach to the identification and characterization of instream flow environments. S Afr J Aquat Sci 20(1–2):38–61. https://doi.org/10.1080/10183469.1994.9631326

Wadeson RA, Rowntree KM (1998) Application of the hydraulic biotope concept to the description of channel morphology and aquatic habitat: a case study on the Sabie River. S Afr J Aquat Sci 24(1–2):124–136

Wagner S, Klöckner P, Stier B, Römer M, Seiwert B, Reemtsma T, Schmidt C (2019) Relationship between discharge and river plastic concentrations in a rural and an urban catchment. Environ Sci Technol 53(17):10082–10091. https://doi.org/10.1021/acs.est.9b03048

Waldschläger K, Schüttrumpf H (2020) Infiltration behavior of microplastic particles with different densities, sizes, and shapes-from glass spheres to natural sediments. Environ Sci Technol 54(15):9366–9373. https://doi.org/10.1021/acs.est.0c01722

Walsh CJ, Booth DB, Burns MJ, Fletcher TD, Hale RL, Hoang LN, Livingston G, Rippy MA, Roy AH, Scoggins M, Wallace A (2016) Principles for urban stormwater management to protect stream ecosystems. Freshw Sci 35(1):398–411. https://doi.org/10.1086/685284

Walsh CJ, Burns MJ, Fletcher TD, Bos DG, Poelsma P, Kunapo J, Imberger M (2022) Linking stormwater control performance to stream ecosystem outcomes: incorporating a performance metric into effective imperviousness. PLOS Water 1(2):e0000004. https://doi.org/10.1371/journal.pwat.0000004

Walther BA, Yen N, Hu C-S (2021) Strategies, actions, and policies by Taiwan's ENGOs, media, and government to reduce plastic use and marine plastic pollution. Mar Policy 126:104391. https://doi.org/10.1016/j.marpol.2021.104391

Wang F, Wang B, Duan L, Zhang Y, Zhou Y, Sui Q, Xu D, Qu H, Yu G (2020) Occurrence and distribution of microplastics in domestic, industrial, agricultural and aquacultural wastewater sources: a case study in Changzhou, China. Water Res 182:115956. https://doi.org/10.1016/j.watres.2020.115956

Wang W, Ndungu AW, Li Z, Wang J (2021) Microplastics pollution in inland freshwaters: a case study of river microplastic contamination and dynamics. Water Res 197:117093. https://doi.org/10.1016/j.watres.2021.117093

Wang Y, Xie Y, Fan W, Yang Z, Tan W, Huo M, Huo Y (2022a) Mechanism comparisons of transport-deposition-reentrainment between microplastics and natural mineral particles in porous media: a theoretical and experimental study. Sci Total Environ 850:157998. https://doi.org/10.1016/j.scitotenv.2022.157998

Wang Y, Xu L, Chen H, Zhang M (2022b) Retention and transport behavior of microplastic particles in water-saturated porous media. Sci Total Environ 808:152154. https://doi.org/10.1016/j.scitotenv.2021.152154

Wang C, Tang J, Yu H, Wang Y, Li H, Xu S, Li G, Zhou Q (2022c) Microplastic pollution in the soil environment: characteristics, influencing factors, and risks. Sustainability 14(20):13405. https://doi.org/10.3390/su142013405

Wanitchayapaisit C, Charoenlertthanakit N, Surinseng V, Yaipimol E, Rinchumphu D, Suppakittpaisarn P (2023) Enhancing water-sensitive urban design in Chiang Mai through a research–design collaboration. Sustainability 15(22):16127. https://doi.org/10.3390/su152216127

Werbowski LM, Gilbreath AN, Munno K, Zhu X, Grbic J, Wu T, Sutton R, Sedlak MD, Deshpande AD, Rochman CM (2021) Urban stormwater runoff: a major pathway for anthropogenic particles, black rubbery fragments, and other types of microplastics to urban receiving waters. ACS ES&T Water 1(6):1420–1428. https://doi.org/10.1021/acsestwater.1c00017

Woodward J, Li J, Rothwell J, Hurley R (2021) Acute riverine microplastic contamination due to avoidable releases of untreated wastewater. Nat Sustain 4(9):793–802. https://doi.org/10.1038/s41893-021-00718-2

Wright SL, Ulke J, Font A, Chan KLA, Kelly FJ (2020) Atmospheric microplastic deposition in an urban environment and an evaluation of transport. Environ Int 136:105411. https://doi.org/10.1016/j.envint.2019.105411

Wu J, Jiang R, Lin W, Ouyang G (2019) Effect of salinity and humic acid on the aggregation and toxicity of polystyrene nanoplastics with different functional groups and charges. Environ Pollut (Barking, Essex: 1987) 245:836–843. https://doi.org/10.1016/j.envpol.2018.11.055

Xiao S, Cui Y, Brahney J, Mahowald NM, Li Q (2023) Long-distance atmospheric transport of microplastic fibres influenced by their shapes. Nat Geosci 16(10):863–870. https://doi.org/10.1038/s41561-023-01264-6

Yan H, Zhang L, Liu L, Wen S (2021) Investigation of the external conditions and material compositions affecting the formation mechanism and size distribution of tire wear particles. Atmos Environ 244:118018. https://doi.org/10.1016/j.atmosenv.2020.118018

Yang S, Brasseur G, Walters S et al (2025) Global atmospheric distribution of microplastics with evidence of low oceanic emissions. npj Clim Atmos Sci 8:81. https://doi.org/10.1038/s41612-025-00914-3

Ye X, Cheng Z, Wu M, Hao Y, Hu BX, Mo C, Li Q, Xiang L, Zhao H, Wu J, Wu J, Lu G (2022) Investigating transport kinetics of polystyrene nanoplastics in saturated porous media. Ecotoxicol Environ Saf 241:113820. https://doi.org/10.1016/j.ecoenv.2022.113820

Yoo J, Park K (2022) Stormwater utility fee estimation method for individual land use areas. Sustainability 14(16):10211. https://doi.org/10.3390/su141610211

Yu Y, Velandia M, Hayes DG, DeVetter LW, Miles CA, Flury M (2024) Chapter Three—Biodegradable plastics as alternatives for polyethylene mulch films. In: Sparks DL (ed) , vol 183. Academic Press, pp 121–192. https://doi.org/10.1016/bs.agron.2023.10.003

Yücel N, Tutsak E, Kılıç E (2023) First evidence of microplastic deposition in snow from Turkey. J Anatol Environ Anim Sci 8(1):95–102. https://doi.org/10.35229/jaes.1233006

Zhang H (2017) Transport of microplastics in coastal seas. Estuar Coast Shelf Sci 199:74–86. https://doi.org/10.1016/j.ecss.2017.09.032

Zhang J, Choi CE (2022) Improved settling velocity for microplastic fibers: a new shape-dependent drag model. Environ Sci Technol 56(2):962–973. https://doi.org/10.1021/acs.est.1c06188

Zhang Y, Gao T, Kang S, Sillanpää M (2019) Importance of atmospheric transport for microplastics deposited in remote areas. Environ Pollut 254:112953. https://doi.org/10.1016/j.envpol.2019.07.121

Zhang B, Chen L, Chao J, Yang X, Wang Q (2020) Research progress of microplastics in freshwater sediments in China. Environ Sci Pollut Res 27:31046. https://doi.org/10.1007/s11356-020-09473-x

Zhao JZ, Fonseca C, Zeerak R (2019) Stormwater utility fees and credits: a funding strategy for sustainability. Sustainability 11(7):1913. https://doi.org/10.3390/su11071913

Zhao H, Zhou Y, Han Y, Sun Y, Ren X, Zhang Z, Wang Q (2022) Pollution status of microplastics in the freshwater environment of China: a mini review. Water Emerg Contam Nanopl 1:5. https://doi.org/10.20517/wecn.2021.05

Zhou J, Luo D (2024) The global governance of marine plastic pollution: rethinking the extended producer responsibility system. Front Mar Sci 11. https://doi.org/10.3389/fmars.2024.1363269

Zhou H, Zhou L, Ma K (2020) Microfiber from textile dyeing and printing wastewater of a typical industrial park in China: occurrence, removal and release. Sci Total Environ 739:140329. https://doi.org/10.1016/j.scitotenv.2020.140329

Zhao P, Cui L, Zhao W et al (2021) Cotransport and deposition of colloidal polystyrene microplastic particles and tetracycline in porous media: the impact of ionic strength and cationic types. Sci Total Environ 753:142064.

Zhou M, Wang R, Cheng S et al (2021a) Bibliometrics and visualization analysis regarding research on the development of microplastics. Environ Sci Pollut Res 28:8953–8967. https://doi.org/10.1007/s11356-021-12366-2

Zhou Y, He G, Jiang X, Yao L, Ouyang L, Liu X, Liu W, Liu Y (2021b) Microplastic contamination is ubiquitous in riparian soils and strongly related to elevation, precipitation and population density. J Hazard Mater 411:125178. https://doi.org/10.1016/j.jhazmat.2021.125178

Zhu X, Munno K, Grbic J, Werbowski LM, Bikker J, Ho A, Guo E, Sedlak M, Sutton R, Box C, Lin D, Gilbreath A, Holleman RC, Fortin M-J, Rochman C (2021) Holistic assessment of microplastics and other anthropogenic microdebris in an urban bay sheds light on their sources and fate. ACS ES&T Water 1(6):1401–1410. https://doi.org/10.1021/acsestwater.0c00292

Ziajahromi S, Lu H-C, Drapper D, Hornbuckle A, Leusch FDL (2023) Microplastics and tire wear particles in urban stormwater: abundance, characteristics, and potential mitigation strategies. Environ Sci Technol 57(34):12829–12837. https://doi.org/10.1021/acs.est.3c03949

Zubris KAV, Richards BK (2005) Synthetic fibers as an indicator of land application of sludge. Environ Pollut 138(2):201–211. https://doi.org/10.1016/j.envpol.2005.04.013

Understanding the Various Pathways of Microplastic Pollution in Freshwater Systems

3

Mozim Shafi, Ayan Lodh, Khalid Muzamil Gani, and Sudha Goel

Abstract

Microplastics (MPs) have emerged as an important scientific issue of global concern and have been extensively reported in each environmental compartment, with aquatic environments regarded as the ultimate receptors and sinks. Identifying relevant pathways through which MPs reach freshwater systems is difficult due to the complex and interconnected nature of the activities involved in generating plastic waste. Nevertheless, most studies have consistently reported wastewater treatment plants (WWTPs) as an important pathway through which MPs reach surface waters. However, recent studies have exposed other important routes, particularly stormwater runoff, atmospheric deposition, agricultural inputs, and littering. Despite the efforts to highlight the key routes of MP contamination, the relative importance of these pathways is likely to vary geographically and temporally, influenced by various factors that collectively hinder our

M. Shafi
Environmental Engineering and Management, Department of Civil Engineering, Indian Institute of Technology Kharagpur, Kharagpur, West Bengal, India

A. Lodh
School of Environmental Science and Engineering, Institute of Technology Kharagpur, Kharagpur, West Bengal, India

K. M. Gani
Department of Civil Engineering, National Institute of Technology, Srinagar, Jammu, and Kashmir, India

S. Goel (✉)
Environmental Engineering and Management, Department of Civil Engineering, Indian Institute of Technology Kharagpur, Kharagpur, West Bengal, India

School of Environmental Science and Engineering, Institute of Technology Kharagpur, Kharagpur, West Bengal, India
e-mail: sudhagoel@civil.iitkgp.ac.in

K. Bala et al. (eds.), *Occurrence, Detection, and Fate of Microplastics in Freshwater Ecosystems*, https://doi.org/10.1007/978-981-95-0225-7_3

ability to develop practical and effective management policies. Therefore, to address the pervasive issue of MP pollution in freshwater systems, the first step is to comprehensively summarize the critical pathways to fully understand how MPs enter and disperse in these systems. Future research should explore these questions, which could help allocate resources more effectively by first targeting the most significant pathway. In addition, we need standardized methods to quantify and analyze MPs across different pathways. To this end, this chapter addresses these questions specifically in the context of freshwater ecosystems, providing key insights into sustainable MP management.

Keywords

Microplastics · Freshwater · Pathways · Sources

3.1 Introduction

The menace of plastic pollution is universal and accepted as an emerging environmental and human health crisis. This crisis is projected to escalate in the upcoming years due to the increased usage and mismanagement of plastic waste (Bank and Hansson 2019). Recently, plastic waste management has been addressed in the United Nations' Sustainable Development Goal 14, which focuses on protecting marine ecosystems. Most of the studies so far have highlighted an enormous rise in the amount of plastic produced worldwide, from 2 to 380 megatons, between 1950 and 2015. More than 70% of this remained mismanaged in terrestrial settings, and only 10% was recycled (Geyer et al. 2017). The lower recycling rate is linked to the rising usage of single-use plastics, alongside insufficient public awareness and inadequate infrastructure (Vanapalli et al. 2021). Only 9.4% of plastic is recycled in the United States because of the high collection cost, lack of awareness, and lack of necessary infrastructure (Oudejans 2017). In addition, a report published by the United Nations Environment Programme in 2015 highlighted that globally, nearly two billion individuals lack access to reliable waste management infrastructure, leading to improper plastic waste handling.

As mismanaged waste remains in the terrestrial environment for a longer time, in response to various environmental stresses over time, the plastic disintegrates into smaller particles of size <5 mm and are called secondary microplastics (MPs) (Kiran et al. 2022), whereas for some applications (in cosmetics and pharmaceutical industries), MPs are purposely produced and referred to as primary MPs (Sharma et al. 2021). Not only does the pervasiveness of MPs in every environmental compartment represent our limitations in finding solutions for managing plastic waste, but it also highlights the intricate network of channels through which MPs accumulate in aquatic ecosystems. Most of the research on MPs so far has focused on the oceans. Before they reach the oceans, freshwater systems play a significant role because they serve as reservoirs and conduits, allowing them to influence MPs movement and fate (Wagner et al. 2014; Wu et al. 2024).

MPs in freshwater ecosystems originate from various point and nonpoint sources, with more than 80% contribution from the terrestrial environment through littering, agricultural activities, and poor waste management infrastructure (Jambeck et al. 2023). This complexity is further amplified due to the intricate network of pathways associated with MP generation, thus making it difficult to understand the fate and transport of MPs in these systems. Moreover, MPs in freshwater systems have shown resistance to degradation, which increases their chances of interacting with surrounding harmful contaminants, hence further amplifying their environmental repercussions (Mammo et al. 2020). At the same time, as the MPs move through the water column, they are ingested by aquatic organisms, resulting in biomagnification to higher trophic levels of the food chain (Liu et al. 2023). This is because aquatic species perceive MPs as food due to their color resemblance to natural prey. Furthermore, MPs can enhance the spread and growth of antibiotic resistance by providing a favorable niche for microbial growth. Increasing surface area enables higher rates of biofilm growth due to the adsorption of organics and nutrients on its surface. These biofilms can contain antibiotic-resistant bacteria as well. Microorganisms favor the microplastic surface because attachment can help them survive external environmental stresses. The attachment of antibiotic resistance has implications for the regulation and management of the spread of diseases (Zhang et al. 2020a). Investigations by Song et al. (2020) in Germany confirmed the attachment of *Escherichia coli* on microplastics after incubation in water bodies, and 64% of the attached *Escherichia coli* conferred resistance against three antibiotic groups.

Therefore, it can be inferred that MPs in freshwater systems are more diverse and dynamic, and thus, the elucidation of the mechanisms involved in their transport, transformation, and accumulation needs a comprehensive understanding. This chapter is expected to give detailed information on the different pathways through which MPs affect freshwater systems, thereby enhancing the understanding of their sources, transport mechanisms, and the factors that influence their distribution. This chapter will provide a foundation for informed mitigation strategies and future research directions.

3.2 Microplastic Sources

Microplastics are classified into two broad categories that can be distinguished from one another based on how they originate: primary and secondary. Figure. 3.1 provides the detailed characteristics of microplastics in freshwater systems.

3.2.1 Primary Microplastics

By definition, primary MPs are manufactured as pellets, nurdles, industrial plastic powders, microfibers, and microbeads that are to be used to produce cosmetics, toothpaste, pharmaceutical drugs, etc. According to Zhao et al. (2022), primary MPs reach freshwater systems in their original form, either through domestic

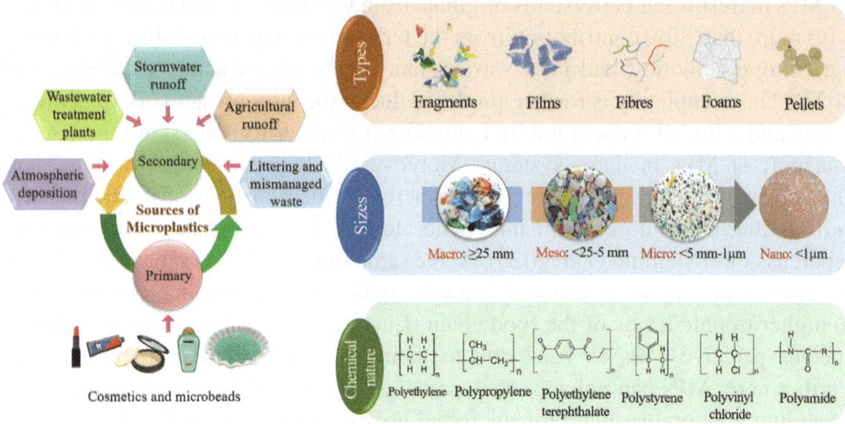

Fig. 3.1 Characteristics of MPs in freshwater systems

effluents via wastewater treatment plants (WWTPs) or as an inadvertent leak during the production and transportation processes, both of which are notoriously difficult to manage. From the share of primary MPs, plastic microbeads, widely used in personal care products, are the most dominant type of MPs observed in the freshwater systems (Ranjan et al. 2024). Habib et al. (2020a) suggested that a single application of a personal care product could roughly discharge around 45,000–90,000 microbeads. Likewise, Kukkola et al. (2024) reported that the use of 1 ml of face scrub will potentially result in the release of approximately 19,000 PE microbeads. Furthermore, every year, 880 microbeads per capita are released into Vietnam's aquatic ecosystem (Dung et al. 2024). Plastic-based paint industries use binder agents composed of polymers, including acrylic, vinyl, and polyester, to ensure the durability and adhesive nature of the paint. Effluents from such industries and their products have the potential to release MPs directly (Fang et al. 2024). An investigation estimates that the annual emission of MPs from paint varies between approximately 5 and 10 million tons (Hurtado Jorge 2023). This indicates that paint particles are an important type of MP in natural water systems (Tamburri et al. 2022). In addition, synthetic textile–based industries and domestic laundry are considered to be significant primary sources of MP pollution in freshwater systems (Cai et al. 2020). Other sources for the primary MPs reported on a percentage basis comprise resins containing road-marking chemicals (7%) and automobile tire wear particles (38%), which are transported to nearby water bodies through surface runoff (Burghardt et al. 2022; Järlskog et al. 2022).

3.2.2 Secondary Microplastics

Secondary MPs are formed as plastic waste disintegrates over time. Many plastic products are discarded into the environment rather than recycled. As a result of numerous environmental pressures, such as ultraviolet (UV) radiation, fluctuating temperatures, and bacterial activity, these unmanaged plastics deteriorate and eventually degrade into MPs (Li et al. 2020). After formation, there are multiple routes for MPs to enter the freshwater ecosystems, including stormwater runoff, WWTPs, atmospheric deposition, aquaculture, and agriculture. In aquatic systems, an additional level of weathering occurs due to abrasive forces from interactions between polymers and sediments, pebbles, waves, and tides (Duan et al. 2021). This mechanical breakdown further speeds up MP formation within the system, which wind can transport from uncontrolled dumping sites into adjoining water bodies. Kannankai and Devipriya (2024) showed that a fallout rate of 37.29 MPs/m^2/day was observed in atmospheric deposition close to a municipal landfill site. Another significant contributor to secondary MPs is the frequent application of plastic mulch films in agricultural fields (Long et al. 2023). As a result of mechanical wear and long-term exposure to ultraviolet light, these thin films break down into tiny particles, which are then carried by surface runoff into nearby irrigation canals during periods of heavy rainfall and eventually end up in downstream bodies of water. In addition, aquaculture-related equipment, like fishing nets, ropes, and traps, are also important and underestimated sources of secondary MPs (Sharma et al. 2024). Chandrakanthan et al. (2024) pointed out that recreational parks and activities like boating and camping produce plastic trash that decomposes into MPs. Water bodies like reservoirs, rivers, and lakes that attract many people are especially vulnerable to this kind of pollution.

3.3 Point and Nonpoint Sources of MPs

Primary and secondary MPs, derived from both point and nonpoint sources, have become pollutants of emerging concern, affecting the concentration and impact of MPs in freshwater systems (Bailey et al. 2021). MPs originating from point sources are identifiable and are released into the surrounding surface waters directly from specific sites. WWTP effluents are considered the most critical point source of MP pollution in freshwater systems (Jessieleena et al. 2023). Dishwashing, laundry, and bathing are examples of routine domestic activities involving the use of plastic items containing primary MPs (Lin et al. 2023). Wastewater from laundry-based industries can contain MP concentrations as high as 6900 P/L, significantly higher than levels found in domestic sewage (Iordachescu et al. 2024). Around 48% of global wastewater remains untreated; therefore, a higher load of MPs with untreated wastewater will directly reach downstream water bodies (Jones et al. 2021). Industrial facilities, especially those involved in plastic manufacturing, processing, or recycling, also directly contribute MPs to nearby water bodies (Suzuki et al. 2022). Additionally, separate or combined stormwater drainage systems are another vital point source for transporting MPs to aquatic systems. Pilgrim centers and tourism activities near lakes and rivers are other significant point sources of MP pollution. Studies have also reported a rise in MP pollution linked to anthropogenic activities at pilgrim centers in southern India (Amrutha and Warrier 2020).

Conversely, nonpoint sources are challenging to trace and manage because regulating intermittent pollution inputs and identifying specific sources are difficult. These MPs can enter freshwater environments through complex routes, including surface runoff, wind storms, and atmospheric depositions. Studies have shown that following a precipitation event, the annual MP load released via runoff is nearly sixfold more than that released from WWTP effluents (Chen et al. 2020). The probable sources of MPs in River Yulin, China, were analyzed from the perspective of both point and nonpoint pollution sources (Cho et al. 2023; Mao et al. 2020). The authors concluded that surface runoff is the predominant nonpoint source of MP pollution. Plastic products such as mulching films, plant protectors, silage, bale nets, twines, pipes, coated fertilizers, and pesticides are growing with the development of the agricultural industries. All of these products contribute to on-field MP formation through their degradation over time (Lwanga et al. 2022). The movement of MPs through vertical and lateral seepage flow from farmland to surrounding water bodies has not yet been studied (Schell et al. 2020). Unfavorable conditions, such as flooding or heavy storms, can elevate MP's occurrence from nonpoint sources in freshwater systems (Chen et al. 2021; Lahon and Handique 2023). Polymerized resin is used in conjunction with other types of chemicals in manufacturing rubber tires, road paint, varnishes, and construction paints, and it releases MPs with time (Wright et al. 2020). Moreover, due to the fact that this small particle has the potential to go airborne and eventually settle in nearby freshwater bodies, regions that are often located in close proximity to municipal landfills frequently experience a large influx of atmospheric deposition of MPs (Hu et al. 2022).

3.4 Pathways of Microplastic Pollution

To reduce MP leakage, it is necessary to have a thorough understanding of the complex network of pathways through which MPs reach freshwater systems. This knowledge will help us spotlight the most important pathway first, thereby developing targeted waste management strategies and policies to prevent this pervasive environmental issue. Figure 3.2 illustrates important routes of MP pollution in freshwater systems.

3.4.1 Atmospheric Deposition

Recently, atmospheric deposition has become an important pathway through which MPs reach freshwater systems (Dris et al. 2016; Liu et al. 2019a). With the help of intricate meteorological processes, the atmosphere acts as a vector for MPs, hence making it possible for them to travel over large distances (Allen et al. 2019; Nafea et al. 2024). Sources of airborne MPs are complex and originate from a combination of sources, such as MPs derived from tire and road abrasion, building materials, landfills, textile fibers, and urban dust (Shao et al. 2022). In any case, once in the atmosphere, MPs can be transported to locations significantly far from their source

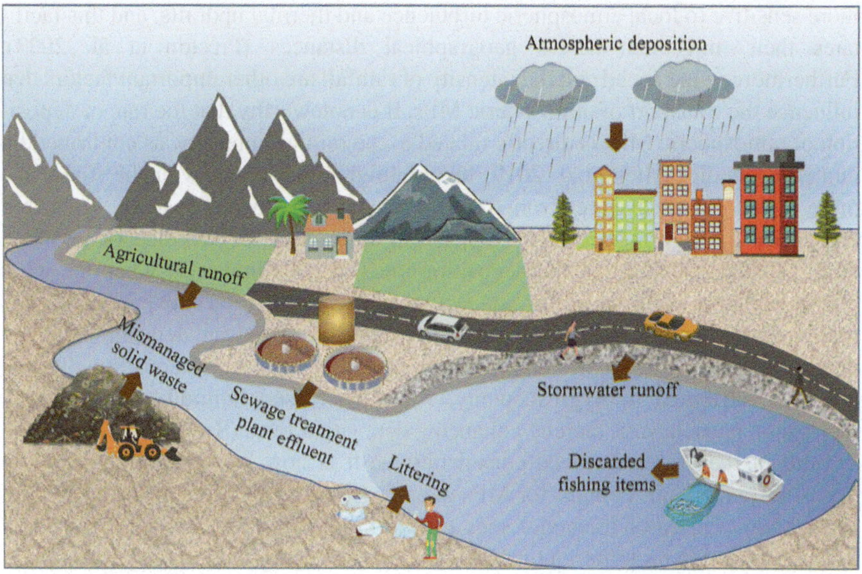

Fig. 3.2 A conceptual picture showing the important pathways of MP pollution

and, as a result, affect even the most isolated environmental spheres (Evangeliou et al. 2020). The process of atmospheric deposition occurs through both dry and wet depositions, thereby introducing MPs directly into lakes, rivers, and streams, bypassing traditional terrestrial sources (Zhang et al. 2020b; Shafi et al. 2024). So far, few studies have elucidated the transport mechanisms of airborne MPs. Nonetheless, meteorological processes such as wind and rainfall play a key role in the migration of atmospheric MPs into surface water bodies (Munyaneza et al. 2022; Osterlund et al. 2023). Most of the research has recently elucidated that atmospheric deposition can contribute substantial loads of MPs to freshwater environments. Around 1.2 tons of MPs are carried from terrestrial sources to freshwater bodies every year via atmospheric depositions (Liu et al. 2019a). A contrasting difference in MP deposition rates was observed between the urban and suburban areas of Paris, with average values of 110 ± 96 MP/m^2/day and 53 ± 38 MP/m^2/day, respectively (Dris et al. 2016). Likewise, Bergmann et al. (2019) reported MPs in remote and urban areas with yearly depositions of approximately 1.4 and 66 MP/m^2, respectively. In London, the United Kingdom, Wright et al. (2020) observed 712 ± 162 MPs/m^2/day in atmospheric samples collected from a highly urbanized megacity. So far, studies across geographical regions have consistently reported fibers, especially those originating from textiles, to be the dominant source of airborne MPs (Zhang et al. 2020b). Moreover, the way in which airborne MPs are transported during dry and wet deposition processes varies regionally and is influenced by both natural and anthropogenic factors. In general, airborne MPs have low settling velocities as they are in the smaller size range and, as a result, can stay in suspension for a longer time (Batool et al. 2022). Hence, these smaller MPs are more sensitive to local atmospheric turbulence and thermal updrafts, and this facilitates their movement across geographical distances (Preston et al. 2023). Furthermore, wind speed and the intensity of rainfall are other important factors that influence the transport of atmospheric MPs. It is noteworthy that the rate of deposition of atmospheric MPs is directly related to the rainfall intensity, as confirmed by numerous studies Allen et al. 2019 this highlights that the heavier the storm, the more MPs will be scavenged from the atmosphere to freshwater systems. To further emphasize the significance of wind speed in the process of MP dispersion, Allen et al. (2019) observed that the deposition rates of MPs increased when the wind speed was more than one meter per second. Szewc et al. (2021) expanded on this finding by noting that atmospheric MPs' average wind speed and deposition rate are directly related, with greater humidity rates resulting in larger deposition. This is probably attributed to hygroscopicity, where increased humidity results in an increase in particle size, thereby enhancing deposition rates (Szewc et al. 2021). As a result of the aforementioned considerations, it is now known that atmospheric deposition is a major conduit for MPs to enter freshwater systems. However, standardization of research methodologies and working out the mechanisms of airborne MP transport remain important priorities in this emerging field to ensure comparable and reliable findings.

3.4.2 Wastewater Treatment Plants

Many studies have focused on wastewater treatment systems as a potential entry point for MPs into aquatic ecosystems (Enfrin et al. 2019; Wong et al. 2020). MPs in wastewater treatment plants (WWTPs) may originate from various sources, including but not limited to industrial effluents, domestic effluents, and laundry effluents (Andrady 2011; Habib et al. 2020b). Furthermore, MPs in these systems also result from poor waste management systems, residues from the plastic recycling and processing sectors, and abrasion from common home products, such as synthetic fabrics and plastic materials (Prata 2018; Falco et al. 2019; Liang et al. 2022). Abrasives, drilling fluids, and engine-cleaning chemicals all include MPs, which may lead to a substantial amount of MPs entering wastewater (Galafassi et al. 2019; Naji et al. 2021).

In Canada, Grbic et al. (2020) studied various pathways of MP pollution in an urban lake, Lake Ontario, and observed that WWTPs are the dominant pathway, with a mean abundance of 13.3 MP/L. From the effluents of 22 WWTPs globally, Sun et al. (2019) reported that MPs' abundance varied between 0 and 447 MPs/L. Polyester (28–89%) and Polyethylene (4–51%) were the most observed types of MPs. Fibers and fragments were highly dominant across 22 studies, contributing approximately 53% and 29% of the total. Similarly, Yaseen et al. (2022), in their systematic review of 121 WWTPs in 17 countries, found that PE and fibers were the dominant types of MPs observed. This can be attributed to the increased influx of sewage containing high amounts of fibers through domestic laundry washing (Napper and Thompson 2016) and the lower efficiency of fiber removal compared to fragments in the treatment process (Kim et al. 2022). In the United States, the load of three WWTPs per capita for MPs for one year was approximately 10^3 million particles (Conley et al. 2019). Likewise, from 17 WWTPs in the United States, on average, 4.4 million particles are discharged daily into surface waters (Mason et al. 2016). Notably, most MPs identified in the effluents of WWTPs are not formed within the system itself but are also funneled from other external sources, which highlight the function of WWTPs as receiving systems for transporting MPs to receiving water bodies (Schell et al. 2021). These studies emphasize the information that WWTPs are not only the potential sink of MPs but also act as a main pathway for transporting MPs via effluents to freshwater systems.

Although WWTPs have great promise and are capable of removing a high percentage of MPs through various physical and biological treatment processes, there are various limitations that hinder their ability to remove the entire spectrum of MP sizes and types (Talukdar et al. 2024). According to numerous studies, WWTPs have shown good efficiency, thereby eliminating approximately 97–99% of MPs from the influent (Sun et al. 2019; Parashar and Hait 2023). Even though MPs in the final effluent of WWTPs may have low concentration, the overall discharge rate of MPs will still be high due to large quantities of wastewater discharged daily into receiving waters (Enfrin et al. 2019). Moreover, MPs are also retained in sludge during treatment processes and become an additional source of secondary contamination. This sludge is usually used as fertilizer on agricultural lands, potentially leading to the release of MPs into soils, from where they are subsequently mobilized into the downstream water bodies through surface runoff. Thus, effectively reducing the release of MPs into freshwater systems first requires the improvement of treatment systems, along with innovative upstream plastic management strategies aimed at reducing their emission at the source.

3.4.3 Stormwater Runoff

Stormwater runoff is defined as the surface flow resulting from rainfall or snowmelt. This runoff, which does not soak into the soil or pavement due to the impervious surface, is a relatively new pathway for MPs to enter freshwater systems (Shruti et al. 2021; Wang et al. 2022). When the water in the form of runoff flows across a surface, it sweeps accumulated MPs, such as packaging fragments, tire wear particles, and synthetic fibers, eventually transporting them into rivers, lakes, and streams (Shafi et al. 2024). Unlike WWTPs, as previously discussed, which often undergo different levels of treatment, stormwater generally flows untreated into natural water bodies, thereby increasing the risk to downstream water bodies (Yano et al. 2021).

Research has affirmed that stormwater runoff has high levels of MPs, especially after heavy rainfall or prolonged dry spells, when MPs accumulate on urban surfaces (de Jesus Pinon-Colin et al. 2020; Ziajahromi et al. 2020). The first flush, which is the initial period of the storm, is particularly important in MP transport, as it can carry disproportionately high concentrations of MPs that may have accumulated during a dry spell. Within the Korean city of Gumi, the first flush phenomenon mobilized 90% of the total MP load during rainstorm events (Cho et al. (2023)). Similarly, the analysis of rainfall-runoff dynamics showed that MP loads decreased by 91% within the initial two-hour period of rainfall (Cheung et al. 2019). Likewise, Lin et al. (2024) observed that about 80% of the total mass of MPs was carried by the initial 60–70% of runoff volume in these rainfall events, which typically occurred within the initial 20 min of runoff generation. The spatial distribution of MPs in stormwater research represents the influence of urbanization, with higher levels of MPs reported from intensively developed regions (Liu et al. 2019b; Shafi et al. 2024). A further factor that contributes to the complexity of the process as a whole includes the characteristics of the urban infrastructure. Impervious surfaces, for instance, generate faster flow patterns as a result of their lack of permeability, which makes it easier for MPs that have accumulated on these surfaces to readily mobilize and transfer themselves (Cho et al. 2023).

However, there is a significant knowledge gap in relation to quantifying what amount of MPs reach surface waters via stormwater runoff. The complexity arises due to the dynamic nature of the stormwater runoff, influenced by various factors, including the characteristics of MPs (Shafi et al. 2024). Additionally, offsetting how much plastic waste ends up in the open environment can be difficult because various areas have varying standards for waste management and disposal practices. However, few studies have tried to quantify MP loads transported through stormwater. Sutton et al. (2019) applied concentration data to quantify the discharge of 7200 billion MP particles annually into San Francisco Bay, California. Further, it was observed that individual rainstorm events had the potential to release roughly 6–9 million MPs per hectare via stormwater (de Jesus Pinon-Colin et al. 2020). Likewise, based on the combination of MPs and the hydrograph data in Canada, Ross et al. (2023) observed that each storm event carried an average of 6.8 billion MP particles. However, sample representativeness, variability in sampling methodology, and analytical methods make comparisons of these results difficult.

Consequently, future studies should prioritize improving these areas to facilitate comparisons and develop a deeper understanding of the movement and fate of stormwater MPs in order to draw better conclusions from the findings.

Recent research on green infrastructure, such as rain gardens and permeable pavements, has demonstrated that these systems may effectively retain MP (Johansson et al. 2024; Azad et al. 2024), but little is known about how well these systems will work in the long run. Many studies have explained how stormwater treatment systems such as retention ponds, bioswales, and infiltration basins play an important role in trapping MPs from stormwater runoff. However, there are limitations in these systems, which majorly depend on various factors such as MP properties and the design of the treatment systems (Liu et al. 2019b; Stang et al. 2022). It is important to understand that these treatment units can also function as temporary reservoirs, accumulating MPs that may be remobilized during subsequent severe storm occurrences, further exacerbating the MP pollution issue (Koutnik et al. 2022).

3.4.4 Agricultural Runoff

It has been suggested that agricultural soils serve as a significant sink for terrestrial MPs (Rochman 2018; Jin et al. 2022). The application of plastics in modern agriculture is quite broad due to its low cost, easy processing, and at the same time, its ability to tolerate pressure and abrasion for a long time (Briassoulis 2023). The type of plastic used in agricultural practices is mostly single-use, which leads to its fragmentation into MPs, which are carried to watercourses with precipitation or through irrigation runoff. This is a nontrivial system through which MPs from different upstream agronomic activities find their way down to freshwater systems. In line with the 2021 report of the Food and Agriculture Organization of the United Nations, the global consumption of agricultural plastics has reached significant levels. The total plastic usage is estimated at 12.5 million metric tons (MMT), with the crop and animal sectors responsible for ten million metric tons. A primary source of this pathway is the widespread use of agricultural plastic products, including greenhouses, plastic tunnels, irrigation pipes, and mulching films. These materials degrade with time as a result of weathering, which, in turn, generates MPs that are further mobilized by runoff (Zhang et al. 2022). Additionally, biosolids or sewage sludge used as a fertilizer represents another important route of MPs found in agricultural soils. WWTPs remove the majority of the MPs, leading to their accumulation in sludge. The accumulation of MPs in sludge occurs because WWTPs remove most of them. Soil amendment applications of this sludge expose the resultant MPs to weathering and runoff, and they ultimately reach nearby water bodies (Naderi Beni et al. 2023). Researchers have indicated that wastewater treatment plants eliminate 95% of incoming microplastics, concentrating them in sewage sludge at concentrations varying from 1000 to 186,700 particles per kilogram of dry weight (Gao et al. 2020). Approximately 35% of the European sewage sludge is used for agricultural practices, and some countries like Australia and Spain use even up to 70% and 80%, respectively (Schell et al. 2022; Ziajahromi et al. 2024). While the majority of the studies have reported the transport of MPs to freshwater systems through stormwater runoff, only a few studies so far have provided evidence that agricultural runoff is also an important yet overlooked pathway of MP transport. In field-scale experiments, a higher abundance of MPs was reported in the runoff samples collected from the field plots where biosolids were applied compared to the control plots (Naderi Beni et al. 2023). MPs from the biosolid-amended plots in this study ranged from 16 to 31 P/L, which was higher compared to the MP concentration reported in stormwater runoff (Naderi Beni et al. 2023). In another study, the analysis of rainfall-induced runoff in field plots showed a positive correlation between MP characteristics and their transport potential, where smaller MPs (<1 mm), less dense MPs (Polypropylene), and fibers showed higher transport potential (Han et al. 2022). Zhang et al. (2022) also confirmed similar findings, highlighting that MP characteristics (size, density, and shape) influence both the vertical and horizontal movement of MPs in soils. Recently, Liu et al. (2024) reported the occurrence of MPs in agricultural runoff and highlighted that fertilizers and MPs co-migrated within agricultural runoff. The authors further revealed that MPs with a low density and smaller size, specifically those falling within the range of 20–50 μm, exhibited greater mobility within agricultural runoff. The size-dependent mobility of MPs not only poses risks to agricultural environments but also negatively impacts freshwater ecosystems. Smaller MPs can easily bypass conventional treatment systems, leading to their higher abundance in nearby freshwater habitats. Research has shown that these tiny MPs are often ingested by various aquatic organisms, such as plankton and fish, due to their bioavailability and resemblance to natural food items (Lehtiniemi et al. 2018; Sang et al. 2021). In addition, these MPs have the potential to act as carriers for toxic chemicals that have been absorbed from agricultural fields, such as pesticides and heavy metals, and then transported into freshwater systems, where they might potentially contaminate the food chain (Kinigopoulou et al. 2022; Jin et al. 2022). As a result, agricultural runoff has serious implications for the downstream water bodies, highlighting the need for ecologically sustainable land management practices. Implementing strategies such as buffer strips, reducing plastic usage, and establishing regulated drainage systems (Naderi Beni et al. 2023) can help prevent the transfer of MPs from agricultural areas to nearby waterways. Therefore, it is essential to raise awareness and manage the agricultural sources of MPs locally by adopting sustainable waste disposal methods.

3.4.5 Littering and Mismanaged Waste

Plastic materials have become such an essential part of many applications that their use cannot be avoided in many ways. We agree that plastic has enormous applications and, as a result of its widespread use, has led to a huge environmental burden on the existing waste management infrastructure. As per the available statistics, roughly 70% of all plastic produced ends up in the environment as solid waste (Geyer et al. 2017; Osterlund et al. 2023; Teo et al. 2022). The significant dependence on plastic materials in many applications eventually leads to their considerable accumulation as discarded litter (Osterlund et al. 2023). Therefore, littering and mismanaged plastic waste are also important pathways through which MPs can enter global water bodies (Magnusson et al. 2016; Waldschlager et al. 2020). Commonly littered plastic items, such as plastic food containers, plastic cups, plastic straws, plastic bottles, face masks, and cigarette butts, accumulate due to both intentional and unintentional disposal (Youngblood et al. 2022). A study reported that every year, more than 0.3 million metric tons of microfibers derived from cigarette butts reach freshwater systems (Belzagui et al. 2021). However, accurately estimating how much plastic litter reaches freshwater systems is difficult. First, the weathering of plastic debris is a complex phenomenon, and we still do not know the rate at which plastic litter degrades once thrown into the open environment. Second, measuring plastic waste flows is challenging because there are no sufficient up-to-date, detailed records of how much waste is generated and managed locally (Cottom et al. 2024). Jambeck et al. (2015) quantified the tonnage of mismanaged plastic waste by applying a conversion factor to estimate the amount of waste entering the ocean. The study projected that in 2010, approximately between 5 and 13 million metric tons (MMT) of plastic waste originated from terrestrial sources and entered the ocean. The estimates were derived using a robust methodology that incorporated country-specific statistics on waste generation, waste management procedures, and the population residing within 50 km of the coastline. In the Ganges River Basin, Youngblood et al. (2022) observed that low-population areas had higher litter counts per 1000 people, with more than 75% of the items recorded being plastic, primarily tobacco products and food wrappers. These littered plastic trash fragments from various environmental stresses like UV radiation, physical abrasion, and microbial activity result in the formation of MPs that disperse widely and subsequently reach freshwater systems through wind and surface runoff. In a suburban catchment in France, Treilles et al. (2021) observed that surface runoff had a significant amount of plastic litter, making up 62% of anthropogenic waste. The most prevalent types of litter included cigarette butts, plastic bags, food wrappers, and medical waste. The abundance of plastic litter in the runoff ranged from 7 to 134 mg/m^3. They further highlighted that in Greater Paris, surface runoff could discharge approximately 8–33 tons of plastic litter into the downstream water bodies annually. In addition, a report published by the United Nations Environment Programme in 2015 highlighted that globally, nearly two billion individuals lack access to reliable waste management infrastructure, which leads to improper practices of handling plastic waste. Mismanaged plastic waste refers to waste that is not recycled and remains in landfills or is directly dumped into open waters (Ferronato and Torretta 2019). In most of the low- and middle-income developing countries, open dumping is a common waste disposal practice, often due to limited access to proper waste management facilities (Meijer et al. 2021; Shafi et al. 2023). Furthermore, these dumping sites are not equipped with systems to contain leachates, and therefore, they have a high potential of reaching downstream water bodies. Lodh et al. (2025) did a comparative study on leachates from controlled and uncontrolled landfills and highlighted that leachates from uncontrolled landfills have a greater potential to contaminate the nearby environment. The Cisadane River in Indonesia receives approximately 80,640 MP/day through leachate discharge from the upstream Galuga landfill (Cordova and Riani 2021). Similarly, Sulistyowati et al. (2022) noted that in the Cisadane River, MP concentration levels were fourfold higher at a point 1 kilometer downstream of the Galuga landfill site in comparison to upstream. Some researchers have also observed that landfill leachate is a major source of MP pollution in nearby water bodies (Imhof et al. 2013; Kazour et al. 2019). In the future, tackling the issues of littering and improper waste management practices will require a comprehensive approach that combines innovation in waste management infrastructure, the raising of public awareness, and the implementation of specific governmental interventions. The implementation of extended producer responsibility schemes and deposit-return systems has shown promise in reducing plastic leakage into the environment. Additionally, the development of strict interventions for identifying and addressing illegal dumping sites close to water bodies could help prevent MPs from leaking.

3.5 Conclusion and Future Outlook

This chapter comprehensively summarizes MP pathways in freshwater systems, highlighting the complexity of the problem involved and the urgent need to prioritize important pathways. The pathways through which MPs enter freshwater systems are diverse and include atmospheric deposition, WWTPs, stormwater runoff, agricultural runoff, and mismanaged waste. The complexity of these pathways shows that prioritizing pathways is important to finding solutions and controlling MP pollution. Each pathway varies, depending on location, season, and local human activities, which means that targeted, site-specific management strategies are essential. To the best of our knowledge, no study has yet compared the MP pollution from each of these pathways in a single ecosystem, highlighting a significant knowledge gap. The majority of the studies have focused on each pathway independently. These studies have helped identify the roles each pathway plays in MP pollution; they have not explored how these pathways interact within a single ecosystem. However, due to limited comparative data on MP pollution from each of these pathways within a single ecosystem, it becomes challenging to prioritize interventions. Are MPs from stormwater runoff more critical in contributing to the freshwater system's total MP load than those from atmospheric sources or agricultural runoff? Does wastewater contribute the largest amount, or does stormwater deliver more MPs to freshwater systems during wet weather? Future research should explore these questions, which could help allocate resources more effectively by first targeting the most significant pathway. In addition, we need standardized methods to quantify and analyze MPs across different pathways. Current variations in sampling and analysis make it difficult to compare studies and understand the true scale of the problem. Also, the role of climate change in affecting MP transport remains poorly understood. Changes in rainfall patterns and extreme weather events could alter how MPs move from land-based sources to freshwater systems. In terms of practical applications, this chapter has important implications for urban planning, water resource management, and environmental policy. Targeting specific pathways, such as enhancing treatment processes at WWTPs or using green infrastructure for stormwater management, can reduce MP loads entering freshwater systems. However, these efforts must be complemented by upstream actions, such as promoting alternatives to single-use plastics, improving waste collection systems, and enforcing policies that support sustainable plastic use. Additionally, educating the public on the environmental impacts of improper waste disposal can significantly reduce littering and the subsequent introduction of MPs into freshwater systems.

References

Allen S, Allen D, Phoenix VR, Le Roux G, Durántez Jiménez P, Simonneau A et al (2019) Atmospheric transport and deposition of microplastics in a remote mountain catchment. Nat Geosci 12(5):339–344

Amrutha K, Warrier AK (2020) The first report on the source-to-sink characterization of microplastic pollution from a riverine environment in tropical India. Sci Total Environ 739:140377

Andrady, A. L. (2011). Microplastics in the marine environment. Marine pollution bulletin, 62(8), 1596–1605.

Azad A, Sheikh MN, Hai FI (2024) A critical review of the mechanisms, factors, and performance of pervious concrete to remove contaminants from stormwater runoff. Water Res:121101

Bailey K, Sipps K, Saba GK, Arbuckle-Keil G, Chant RJ, Fahrenfeld NL (2021) Quantification and composition of microplastics in the Raritan Hudson Estuary: comparison to pathways of entry and implications for fate. Chemosphere 272:129886

Bank MS, Hansson SV (2019) The plastic cycle: a novel and holistic paradigm for the Anthropocene

Batool I, Qadir A, Levermore JM, Kelly FJ (2022) Dynamics of airborne microplastics, appraisal and distributional behaviour in atmosphere; a review. Sci Total Environ 806:150745

Belzagui F, Buscio V, Gutiérrez-Bouzán C, Vilaseca M (2021) Cigarette butts as a microfiber source with a microplastic level of concern. Sci Total Environ 762:144165

Bergmann M, Mützel S, Primpke S, Tekman MB, Trachsel J, Gerdts G (2019) White and wonderful? Microplastics prevail in snow from the Alps to the Arctic. Sci Adv 5(8):eaax1157

Briassoulis D (2023) Agricultural plastics as a potential threat to food security, health, and environment through soil pollution by microplastics: problem definition. Sci Total Environ 892:164533

Burghardt TE, Pashkevich A, Babić D, Mosböck H, Babić D, Żakowska L (2022) Microplastics and road markings: the role of glass beads and loss estimation. Transp Res Part D Transp Environ 102:103123

Cai Y, Yang T, Mitrano DM, Heuberger M, Hufenus R, Nowack B (2020) Systematic study of microplastic fiber release from 12 different polyester textiles during washing. Environ Sci Technol 54(8):4847–4855

Chandrakanthan K, Fraser MP, Herckes P (2024) Recreational activities as a major source of microplastics in aquatic environments. Water Emerg Contam Nanoplastics 3(3)

Chen H, Jia Q, Zhao X, Li L, Nie Y, Liu H, Ye J (2020) The occurrence of microplastics in water bodies in urban agglomerations: impacts of drainage system overflow in wet weather, catchment land-uses, and environmental management practices. Water Res 183:116073

Chen L, Li J, Tang Y, Wang S, Lu X, Cheng Z et al (2021) Typhoon-induced turbulence redistributed microplastics in coastal areas and reformed plastisphere community. Water Res 204:117580

Cheung PK, Hung PL, Fok L (2019) River microplastic contamination and dynamics upon a rainfall event in Hong Kong, China. Environ Process 6:253–264

Cho Y, Shim WJ, Ha SY, Han GM, Jang M, Hong SH (2023) Microplastic emission characteristics of stormwater runoff in an urban area: intra-event variability and influencing factors. Sci Total Environ 866:161318

Conley K, Clum A, Deepe J, Lane H, Beckingham B (2019) Wastewater treatment plants as a source of microplastics to an urban estuary: removal efficiencies and loading per capita over one year. Water Res X 3:100030. https://doi.org/10.1016/j.wroa.2019.100030

Cordova MR, Riani E (2021) Micro-and mesoplastics release from the Indonesian municipal solid waste landfill leachate to the aquatic environment: case study in Galuga Landfill Area, Indonesia. Mar Pollut Bull 163:111986

Cottom JW, Cook E, Velis CA (2024) A local-to-global emissions inventory of macroplastic pollution. Nature 633(8028):101–108

de Jesus Piñon-Colin T, Rodriguez-Jimenez R, Rogel-Hernandez E, Alvarez-Andrade A, Wakida FT (2020) Microplastics in stormwater runoff in a semiarid region, Tijuana, Mexico. Sci Total Environ 704:135411

Dong H, Wang L, Wang X, Xu L, Chen M, Gong P, Wang C (2021) Microplastics in a remote lake basin of the Tibetan Plateau: Impacts of atmospheric transport and glacial melting. Environ Sci Technol 55(19):12951–12960

Dris R, Gasperi J, Saad M, Mirande C, Tassin B (2016) Synthetic fibers in atmospheric fallout: a source of microplastics in the environment? Mar Pollut Bull 104(1–2):290–293

Duan J, Bolan N, Li Y, Ding S, Atugoda T, Vithanage M et al (2021) Weathering of microplastics and interaction with other coexisting constituents in terrestrial and aquatic environments. Water Res 196:117011

Dung TTT, Ngoc HN, Trinh NN (2024) Microbeads in exfoliating products: occurrence, abundance, and potential for water contamination in Ho Chi Minh City, Vietnam. Discov Environ 2(1):92

Enfrin M, Dumée LF, Lee J (2019) Nano/microplastics in water and wastewater treatment processes–origin, impact and potential solutions. Water Res 161:621–638. https://doi.org/10.1016/j.watres.2019.06.049

Evangeliou N, Grythe H, Klimont Z, Heyes C, Eckhardt S, Lopez-Aparicio S, Stohl A (2020) Atmospheric transport is a major pathway of microplastics to remote regions. Nat Commun 11(1):3381

Falco F, Di Pace E, Cocca M, Avella M (2019) The contribution of washing processes of synthetic clothes to microplastic pollution. Sci Rep 9(1):6633

Fang C, Zhou W, Hu J, Wu C, Niu J, Naidu R (2024) Paint has the potential to release microplastics, nanoplastics, inorganic nanoparticles, and hybrid materials. Environ Sci Eur 36(1):17

Ferronato N, Torretta V (2019) Waste mismanagement in developing countries: a review of global issues. Int J Environ Res Public Health 16(6):1060

Galafassi S, Nizzetto L, Volta P (2019) Plastic sources: a survey across scientific and grey literature for their inventory and relative contribution to microplastics pollution in natural environments, with an emphasis on surface water. Sci Total Environ 693:133499. https://doi.org/10.1016/j.scitotenv.2019.07.305

Gao D, Li XY, Liu HT (2020) Source, occurrence, migration and potential environmental risk of microplastics in sewage sludge and during sludge amendment to soil. Sci Total Environ 742:140355

Geyer R, Jambeck JR, Law KL (2017) Production, use, and fate of all plastics ever made. Sci Adv 3(7):e1700782

Grbić J, Helm P, Athey S, Rochman CM (2020) Microplastics entering northwestern Lake Ontario are diverse and linked to urban sources. Water Res 174:115623

Habib RZ, Abdoon MMS, Al Meqbaali RM, Ghebremedhin F, Elkashlan M, Kittaneh WF et al (2020a) Analysis of microbeads in cosmetic products in the United Arab Emirates. Environ Pollut 258:113831

Habib RZ, Thiemann T, Al Kendi R (2020b) Microplastics and wastewater treatment plants – a review. J Water Resour Prot 12(01):1

Han N, Zhao Q, Ao H, Hu H, Wu C (2022) Horizontal transport of macro-and microplastics on soil surface by rainfall induced surface runoff as affected by vegetations. Sci Total Environ 831:154989

Hu T, He P, Yang Z, Wang W, Zhang H, Shao L, Lü F (2022) Emission of airborne microplastics from municipal solid waste transfer stations in downtown. Sci Total Environ 828:154400

Hurtado Jorge (2023) Empowering paint manufacturers amidst the challenges of microplastic pollution - Polymers Paint Colour Journal. https://www.polymerspaintcolourjournal.com/empowering-paint-manufacturers-amidst-the-challenges-of-microplastic-pollution/

Imhof HK, Ivleva NP, Schmid J, Niessner R, Laforsch C (2013) Contamination of beach sediments of a subalpine lake with microplastic particles. Curr Biol 23(19):R867–R868

Iordachescu L, Nielsen RV, Papacharalampos K, Barritaud L, Denieul MP, Plessis E et al (2024) Point-source tracking of microplastics in sewerage systems. Finding the culprit. Water Res 257:121696

Jambeck JR, Geyer R, Wilcox C, Siegler TR, Perryman M, Andrady A et al (2015) Plastic waste inputs from land into the ocean. Science 347(6223):768–771

Jambeck J, Moss E, Dubey B, Arifin Z, Godfrey L, Hardesty BD et al (2023) Leveraging multi-target strategies to address plastic pollution in the context of an already stressed ocean. In: The blue compendium: from knowledge to action for a sustainable ocean economy. Springer, Cham, pp 141–184

Järlskog I, Jaramillo-Vogel D, Rausch J, Gustafsson M, Strömvall AM, Andersson-Sköld Y (2022) Concentrations of tire wear microplastics and other traffic-derived non-exhaust particles in the road environment. Environ Int 170:107618

Jessieleena A, Rathinavelu S, Velmaiel KE, John AA, Nambi IM (2023) Residential houses – a major point source of microplastic pollution: insights on the various sources, their transport, transformation, and toxicity behaviour. Environ Sci Pollut Res 30(26):67919–67940

Jin T, Tang J, Lyu H, Wang L, Gillmore AB, Schaeffer SM (2022) Activities of microplastics (MPs) in agricultural soil: a review of MPs pollution from the perspective of agricultural ecosystems. J Agric Food Chem 70(14):4182–4201

Johansson G, Fedje KK, Modin O, Haeger-Eugensson M, Uhl W, Andersson-Sköld Y, Strömvall AM (2024) Removal and release of microplastics and other environmental pollutants during the start-up of bioretention filters treating stormwater. J Hazard Mater 468:133532

Jones ER, Van Vliet MT, Qadir M, Bierkens MF (2021) Country-level and gridded estimates of wastewater production, collection, treatment and reuse. Earth Syst Sci Data 13(2):237–254

Kannankai MP, Devipriya SP (2024) Atmospheric microplastic deposition in a coastal city of India: the influence of a landfill source on monsoon winds. Sci Total Environ 908:168235

Kazour M, Terki S, Rabhi K, Jemaa S, Khalaf G, Amara R (2019) Sources of microplastics pollution in the marine environment: Importance of wastewater treatment plant and coastal landfill. Mar Pollut Bull 146:608–618

Kim MJ, Na SH, Batool R, Byun IS, Kim EJ (2022) Seasonal variation and spatial distribution of microplastics in tertiary wastewater treatment plant in South Korea. J Hazard Mater 438:129474. https://doi.org/10.1016/j.jhazmat.2022.129474

Kinigopoulou V, Pashalidis I, Kalderis D, Anastopoulos I (2022) Microplastics as carriers of inorganic and organic contaminants in the environment: a review of recent progress. J Mol Liq 350:118580

Kiran BR, Kopperi H, Venkata Mohan S (2022) Micro/nano-plastics occurrence, identification, risk analysis and mitigation: challenges and perspectives. Rev Environ Sci Biotechnol 21(1):169–203

Koutnik VS, Leonard J, Glasman JB, Brar J, Koydemir HC, Novoselov A et al (2022) Microplastics retained in stormwater control measures: where do they come from and where do they go? Water Res 210:118008

Kukkola A, Chetwynd AJ, Krause S, Lynch I (2024) Beyond microbeads: Examining the role of cosmetics in microplastic pollution and spotlighting unanswered questions. J Hazard Mater:135053

Lahon J, Handique S (2023) Impact of flooding on microplastic abundance and distribution in freshwater environment: a review. Environ Sci Pollut Res 30(56):118175–118191

Lehtiniemi M, Hartikainen S, Näkki P, Engström-Öst J, Koistinen A, Setälä O (2018) Size matters more than shape: Ingestion of primary and secondary microplastics by small predators. Food Webs 17:e00097

Li C, Busquets R, Campos LC (2020) Assessment of microplastics in freshwater systems: a review. Sci Total Environ 707:135578

Liang H, Ji Y, Ge W, Wu J, Song N, Yin Z, Chai C (2022) Release kinetics of microplastics from disposable face masks into the aqueous environment. Sci Total Environ 816:151650

Lin Q, Pang L, Ngo HH, Guo W, Zhao S, Liu L et al (2023) Occurrence of microplastics in three types of household cleaning products and their estimated emissions into the aquatic environment. Sci Total Environ 902:165903

Lin Y, Wang Y, Ho YW, Fang JKH, Li Y (2024) Characterization and ecological risks of microplastics in urban road runoff. Sci Total Environ 954:176590

Liu F, Olesen KB, Borregaard AR, Vollertsen J (2019b) Microplastics in urban and highway stormwater retention ponds. Sci Total Environ 671:992–1000

Liu K, Wu T, Wang X, Song Z, Zong C, Wei N, Li D (2019a) Consistent transport of terrestrial microplastics to the ocean through atmosphere. Environ Sci Technol 53(18):10612–10619

Liu W, Liao H, Wei M, Junaid M, Chen G, Wang J (2023) Biological uptake, distribution and toxicity of micro (nano) plastics in the aquatic biota: a special emphasis on size-dependent impacts. TrAC Trends Anal Chem 117477

Liu D, Yang Z, Gong Y, Song D, Chen Y (2024) Occurrence and emission characteristics of microplastics in agricultural surface runoff under different natural rainfall and short-term fertilizer application. J Hazard Mater 477:135254

Lodh A, Shafi M, Goel S (2025) Microplastics in municipal solid waste landfill leachate and their removal in treatment units: a perspective of controlled and uncontrolled landfills. Environ Pollut:125853

Long B, Li F, Wang K, Huang Y, Yang Y, Xie D (2023) Impact of plastic film mulching on microplastic in farmland soils in Guangdong province, China. Heliyon 9(6)

Lwanga EH, Beriot N, Corradini F, Silva V, Yang X, Baartman J et al (2022) Review of microplastic sources, transport pathways and correlations with other soil stressors: a journey from agricultural sites into the environment. Chem Biol Technol Agric 9(1):20

Magnusson K, Eliaeson K, Fråne A, Haikonen K, Olshammar M, Stadmark J, Hultén J (2016) Swedish sources and pathways for microplastics to the marine environment

Mammo FK, Amoah ID, Gani KM, Pillay L, Ratha SK, Bux F, Kumari S (2020) Microplastics in the environment: Interactions with microbes and chemical contaminants. Sci Total Environ 743:140518

Mao Y, Li H, Gu W, Yang G, Liu Y, He Q (2020) Distribution and characteristics of microplastics in the Yulin River, China: role of environmental and spatial factors. Environ Pollut 265:115033

Mason SA, Garneau D, Sutton R, Chu Y, Ehmann K, Barnes J et al (2016) Microplastic pollution is widely detected in US municipal wastewater treatment plant effluent. Environ Pollut 218:1045–1054

Meijer LJ, Van Emmerik T, Van Der Ent R, Schmidt C, Lebreton L (2021) More than 1000 rivers account for 80% of global riverine plastic emissions into the ocean. Sci Adv 7(18):eaaz5803

Munyaneza J, Jia Q, Qaraah FA, Hossain MF, Wu C, Zhen H, Xiu G (2022) A review of atmospheric microplastics pollution: in-depth sighting of sources, analytical methods, physiognomies, transport and risks. Sci Total Environ 822:153339

Naderi Beni N, Karimifard S, Gilley J, Messer T, Schmidt A, Bartelt-Hunt S (2023) Higher concentrations of microplastics in runoff from biosolid-amended croplands than manure-amended croplands. Commun Earth & Environ 4(1):42

Nafea TH, Chan FKS, Xu Y, Wang C, Wang X, Zhao W et al (2024) Microplastics Aloft: a comprehensive exploration of sources, transport, variations, interactions and their implications on human health in the atmospheric realm. Earth Sci Rev:104864

Naji A, Azadkhah S, Farahani H, Uddin S, Khan FR (2021) Microplastics in wastewater outlets of Bandar Abbas city (Iran): a potential point source of microplastics into the Persian Gulf. Chemosphere 262:128039

Napper IE, Thompson RC (2016) Release of synthetic microplastic plastic fibres from domestic washing machines: effects of fabric type and washing conditions. Mar Pollut Bull 112(1–2):39–45

Österlund H, Blecken G, Lange K, Marsalek J, Gopinath K, Viklander M (2023) Microplastics in urban catchments: review of sources, pathways, and entry into stormwater. Sci Total Environ 858:159781

Oudejans L (2017) Report on the 2016 US Environmental Protection Agency (EPA) International decontamination research and development conference. US Environmental Protection Agency (Agence de protection de l'environnement des Etats-Unis)

Parashar N, Hait S (2023) Abundance, characterization, and removal of microplastics in different technology-based sewage treatment plants discharging into the middle stretch of the Ganga River, India. Sci Total Environ 905:167099

Prata JC (2018) Microplastics in wastewater: State of the knowledge on sources, fate and solutions. Mar Pollut Bull 129(1):262–265. https://doi.org/10.1016/j.marpolbul.2018.02.046

Preston CA, McKenna Neuman CL, Aherne J (2023) Effects of shape and size on microplastic atmospheric settling velocity. Environ Sci Technol 57(32):11937–11947

Ranjan VP, Joseph A, Srivastava S, Sharma HB, Biswas B, Goel S, Kumar S (2024) From cosmetics to contamination: microplastics in personal care products as vectors for chromium in aquatic environments. Waste Manage Bull 2(3):229–240

Rochman CM (2018) Microplastics research – from sink to source. Science 360(6384):28–29

Ross, M. S., Loutan, A., Groeneveld, T., Molenaar, D., Kroetch, K., Bujaczek, T., ... & Ruecker, N. J. (2023). Estimated discharge of microplastics via urban stormwater during individual rain events. Frontiers in Environmental Science, 11, 1090267.

Sang W, Chen Z, Mei L, Hao S, Zhan C, Bin Zhang W et al (2021) The abundance and characteristics of microplastics in rainwater pipelines in Wuhan, China. Sci Total Environ 755:142606

Schell T, Rico A, Vighi M (2020) Occurrence, fate and fluxes of plastics and microplastics in terrestrial and freshwater ecosystems. Rev Environ Contam Toxicol 250:1–43

Schell T, Hurley R, Nizzetto L, Rico A, Vighi M (2021) Spatio-temporal distribution of microplastics in a Mediterranean river catchment: the importance of wastewater as an environmental pathway. J Hazard Mater 420:126481

Schell T, Hurley R, Buenaventura NT, Mauri PV, Nizzetto L, Rico A, Vighi M (2022) Fate of microplastics in agricultural soils amended with sewage sludge: is surface water runoff a relevant environmental pathway? Environ Pollut 293:118520

Shafi M, Jan R, Gani KM (2023) Selection of priority emerging contaminants in surface waters of India, Pakistan, Bangladesh, and Sri Lanka. Chemosphere:139976

Shafi M, Lodh A, Khajuria M, Ranjan VP, Gani KM, Chowdhury S, Goel S (2024) Are we underestimating stormwater? Stormwater as a significant source of microplastics in surface waters. J Hazard Mater:133445

Shao L, Li Y, Jones T, Santosh M, Liu P, Zhang M et al (2022) Airborne microplastics: a review of current perspectives and environmental implications. J Clean Prod 347:131048

Sharma S, Basu S, Shetti NP, Nadagouda MN, Aminabhavi TM (2021) Microplastics in the environment: occurrence, perils, and eradication. Chem Eng J 408:127317

Sharma D, Dhanker R, Bhawna, Tomar A, Raza S, Sharma A (2024) Fishing gears and nets as a source of microplastic. In: Microplastic pollution. Singapore, Springer, pp 127–140

Shruti VC, Pérez-Guevara F, Elizalde-Martínez I, Kutralam-Muniasamy G (2021) Current trends and analytical methods for evaluation of microplastics in stormwater. Trends Environ Anal Chem 30:e00123

Song J, Jongmans-Hochschulz E, Mauder N, Imirzalioglu C, Wichels A, Gerdts G (2020) The travelling particles: investigating microplastics as possible transport vectors for multidrug resistant E. coli in the Weser estuary (Germany). Sci Total Environ 720:137603

Stang C, Mohamed BA, Li LY (2022) Microplastic removal from urban stormwater: Current treatments and research gaps. J Environ Manag 317:115510

Sulistyowati L, Riani E, Cordova MR (2022) The occurrence and abundance of microplastics in surface water of the midstream and downstream of the Cisadane River, Indonesia. Chemosphere 291:133071

Sun J, Dai X, Wang Q, Van Loosdrecht MC, Ni BJ (2019) Microplastics in wastewater treatment plants: detection, occurrence and removal. Water Res 152:21–37. https://doi.org/10.1016/j.watres.2018.12.050

Sutton R, Franz A, Gilbreath A, Lin D, Miller L, Box C et al (2019) Understanding microplastic levels, pathways, and transport in the San Francisco Bay region

Suzuki G, Uchida N, Tanaka K, Matsukami H, Kunisue T, Takahashi S et al (2022) Mechanical recycling of plastic waste as a point source of microplastic pollution. Environ Pollut 303:119114

Szewc K, Graca B, Dołęga A (2021) Atmospheric deposition of microplastics in the coastal zone: characteristics and relationship with meteorological factors. Sci Total Environ 761:143272

Talukdar A, Kundu P, Bhattacharya S, Dutta N (2024) Microplastic contamination in wastewater: sources, distribution, detection and remediation through physical and chemical-biological methods. Sci Total Environ:170254

Tamburri MN, Soon ZY, Scianni C, Øpstad CL, Oxtoby NS, Doran S, Drake LA (2022) Understanding the potential release of microplastics from coatings used on commercial ships. Front Mar Sci 9:1074654

Teo JY, Ong A, Tan TT, Li X, Loh XJ, Lim JY (2022) Materials from waste plastics for CO 2 capture and utilisation. Green Chem 24(16):6086–6099

Treilles R, Gasperi J, Gallard A, Saad M, Dris R, Partibane C et al (2021) Microplastics and microfibers in urban runoff from a suburban catchment of Greater Paris. Environ Pollut 287:117352

Vanapalli KR, Sharma HB, Ranjan VP, Samal B, Bhattacharya J, Dubey BK, Goel S (2021) Challenges and strategies for effective plastic waste management during and post COVID-19 pandemic. Sci Total Environ 750:141514

Wagner M, Scherer C, Alvarez-Muñoz D, Brennholt N, Bourrain X, Buchinger S et al (2014) Microplastics in freshwater ecosystems: what we know and what we need to know. Environ Sci Eur 26:1–9

Waldschläger K, Lechthaler S, Stauch G, Schüttrumpf H (2020) The way of microplastic through the environment–application of the source-pathway-receptor model. Sci Total Environ 713:136584

Wang C, O'Connor D, Wang L, Wu WM, Luo J, Hou D (2022) Microplastics in urban runoff: global occurrence and fate. Water Res 225:119129

Wong JKH, Lee KK, Tang KHD, Yap PS (2020) Microplastics in the freshwater and terrestrial environments: prevalence, fates, impacts and sustainable solutions. Sci Total Environ 719:137512. https://doi.org/10.1016/j.scitotenv.2020.137512

Wright SL, Ulke J, Font A, Chan KLA, Kelly FJ (2020) Atmospheric microplastic deposition in an urban environment and an evaluation of transport. Environ Int 136:105411

Wu F, Ha X, Wang S, Li J, Gao Y (2024) Microplastic occurrences, transport, and quantification and associated effects on primary productivity and carbon cycling processes in freshwater ecosystems. TrAC Trends Anal Chem 117611

Yano KA, Geronimo FK, Reyes NJ, Kim LH (2021) Characterization and comparison of microplastic occurrence in point and non-point pollution sources. Sci Total Environ 797:148939

Yaseen A, Assad I, Sofi MS, Hashmi MZ, Bhat SU (2022) A global review of microplastics in wastewater treatment plants: understanding their occurrence, fate and impact. Environ Res 212:113258

Youngblood K, Brooks A, Das N, Singh A, Sultana M, Verma G et al (2022) Rapid characterization of macroplastic input and leakage in the Ganges river basin. Environ Sci Technol 56(7):4029–4038

Zhang Y, Kang S, Allen S, Allen D, Gao T, Sillanpää M (2020b) Atmospheric microplastics: a review on current status and perspectives. Earth Sci Rev 203:103118

Zhang S, Abbas M, Rehman MU, Huang Y, Zhou R, Gong S et al (2020a) Dissemination of antibiotic resistance genes (ARGs) via integrons in Escherichia coli: a risk to human health. Environ Pollut 266:115260

Zhang X, Chen Y, Li X, Zhang Y, Gao W, Jiang J et al (2022) Size/shape-dependent migration of microplastics in agricultural soil under simulative and natural rainfall. Sci Total Environ 815:152507

Zhao M, Cao Y, Chen T, Li H, Tong Y, Fan W et al (2022) Characteristics and source-pathway of microplastics in freshwater system of China: a review. Chemosphere 297:134192

Ziajahromi S, Drapper D, Hornbuckle A, Rintoul L, Leusch FD (2020) Microplastic pollution in a stormwater floating treatment wetland: Detection of tyre particles in sediment. Sci Total Environ 713:136356

Ziajahromi S, Lu HC, Dwyer J, Fernandes M, Griffith M, Leusch FD (2024) Transport and accumulation of microplastics from biosolids to Australian agricultural soils: detection of microplastics down to 1 μm. Environ Sci Technol

Distribution Pattern of Microplastics in Aquatic Sources

4

Akshay Botle, Sayli Salgaonkar, Divyanka Thakur, Mihir Herlekar, and Gayatri Barabde

Abstract

Plastic is constantly finding its way into aquatic environments due to its use in everyday consumer goods, many of which are single-use. Plastic waste can break down into smaller pieces (<5 mm), leading to increasing ecotoxicological concerns in the aquatic environment. Despite microplastic (MP) pollution affecting air, water, and soil widely, these ecosystems are often viewed distinctly while being interlinked. Because it is ubiquitously present in the aquatic environment, MP pollution has become a significant global issue. Thus, this chapter focuses on the gaps limited to the understanding of the occurrence and characterization of MPs, their distribution in various segments of aquatic ecosystems, their transfer and fate at different trophic levels, and the implementation of policies for the mitigation of MPs in freshwater. Last, the chapter highlights future scope, focusing on long-term fate and transport studies, as well as interconnectivity and toxicity studies, followed by early detection technologies for MPs in aquatic ecosystems.

A. Botle · S. Salgaonkar (✉) · D. Thakur
Department of Environmental Science, The Institute of Science, Dr. Homi Bhabha State University, Mumbai, India
e-mail: 23res.evs.sayli@iscm.ac.in

M. Herlekar
Symbiosis Centre for Management Studies, Pune, Symbiosis International (Deemed University), Pune, India

G. Barabde
Department of Environmental Science, The Institute of Science, Dr. Homi Bhabha State University, Mumbai, India

Department of Analytical Chemistry, The Institute of Science, Dr. Homi Bhabha State University, Mumbai, India

K. Bala et al. (eds.), *Occurrence, Detection, and Fate of Microplastics in Freshwater Ecosystems*, https://doi.org/10.1007/978-981-95-0225-7_4

Keywords

Microplastics · Freshwater · Distribution and characterization · Ecological impacts · Impacts on organisms

4.1 Introduction

Plastic is an essential commodity in day-to-day life (Ahmad et al. 2020). Plastic usage has expanded along with the global population, yet academicians are still concerned about how to manage plastic garbage (Rezania et al. 2018). Plastic materials have a relatively brief history, spanning approximately 60 years. Human activities significantly influence and contribute to the aquatic ecosystem. A study by Vivekanand et al. (2021) indicated that between 0.0048 and 0.0127 billion metric tons of plastic waste is introduced into the oceans each year from coastline nations. Even though plastic's special qualities—such as its electrical insulation, corrosion resistance, tensile strength, and lightweight nature—make it more valuable in industrial settings, it hardly ever biodegrades when discharged into the environment. Its disintegration into smaller particles known as microplastics (MPs) is the outcome of its interaction with biological, physical, and chemical processes (Alimba and Faggio 2019). MPs are defined by the European Chemicals Agency as plastic particles less than 5 mm. Since 2004, when Thompson et al. conducted research on ocean plastics at Plymouth University in the United Kingdom, there has been an increase in contemporary concern regarding MPs. The reduced speed of degradation of MPs, biological ingestion by aquatic life, and their role in accumulating and migrating various types of organic pollutants from the ecosystem to aquatic fauna make MP pollution in freshwater and marine environments a growing environmental concern (Jiang 2018). MPs are classified into various groups: films (thin and 2D plastic films), micro-pellets (hard, rounded particles), fibers (thin, uniform plastic strands), fiber bundles, fragments (gagged-edge particles), spheres, and foam (Styrofoam-type material, and polyethylene (PE), polyvinyl chloride (PVC), polyamide (PA), polypropylene (PP), polyethylene terephthalate (PET), polystyrene (PS), and polyurethane (PUR) are the six forms into which they can be separated (Fig. 4.1) (Rezania et al. 2018).

Numerous investigations have recorded the occurrence and dispersion of MP pollution in freshwater environments, encompassing the sediments and water column. Poor waste management practices, domestic waste, and uncontrolled littering are the major contributors of plastic waste in riverine systems. According to Nan et al. (2020), rivers and estuaries act as dumping grounds for MP waste. A few studies that have detected MPs in surface waters, seawater, and sediments are as follows: Monterey Bay—1.32 ± 0.70 particles/m^3 (Kashiwabara et al. 2021), Brazilian coast—30.4 particles/m^2 (Maynard et al. 2021), sediments of the Black Sea—106.7 particles/kg (Cincinelli et al. 2021), Gulf of California—1.7–2.0 particles/m^3 (Rios-Mendoza et al. 2021), and Bohai Sea—0.35–0.13 particles/m^3 (Zhang et al. 2020b). This suggests the contamination of aquatic ecosystems with MPs. Furthermore, MP

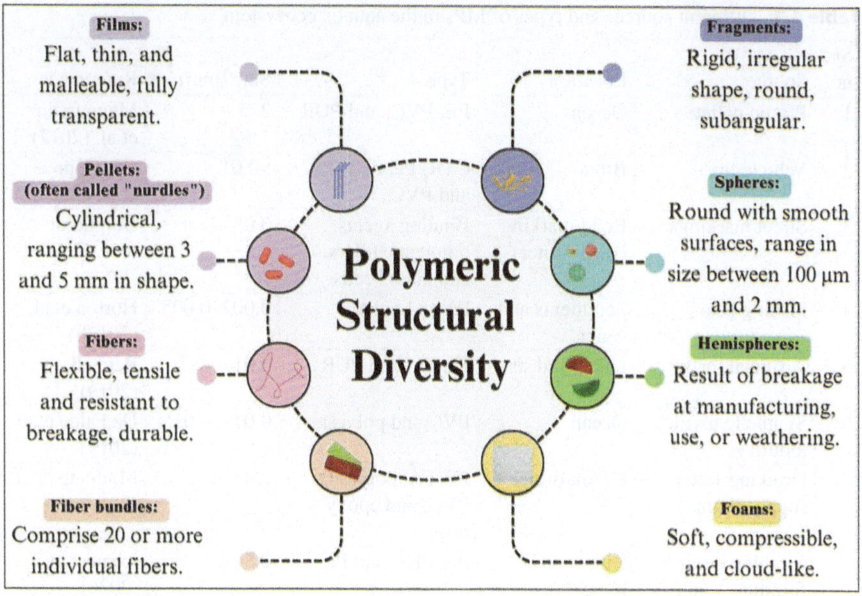

Fig. 4.1 Classification and polymeric structural diversity of MPs

contamination can have more damaging impacts on humanity and the ecosystem. Recent research studies have detailed the undesirable impacts and ecological consequences of MP pollution. For instance, aquatic species may consume MPs as food particles because of their microscopic size, allowing them to enter the food chain.

This chapter aims to provide information about the emerging microplastic pollution in water, its sources, occurrence, and characterization, along with its distribution in water, sediments, groundwater, estuaries, and oceans. It also discusses the transport and fate of MPs and emphasizes uptake by organisms and the impacts of MPs on water, biota, and sediments. The chapter concludes with policies and regulations for microplastic pollution in aquatic sources.

4.2 MPs: Occurrence, Sources, and Characterization

Industrial activities, including the fabric industry, manufacturing processes, and ports, along with human activities like tourism, urban runoff, and sewage treatment facilities, are major contributors of MPs (Rezania et al. 2018). Additionally, thermoplastic composite paints, textile fibers, small pieces of packaged materials, and plastic bottles (Horton et al. 2017a), followed by land-based MPs dumped straight into aquatic bodies as industrial discharge or wastewater treatment plant (WWTP) effluent, contribute to MP pollution. Table 4.1 presents the types and sources of MPs in the aquatic ecosystem. The presence of MPs in man-made and natural water cycles can be attributed to their prevalence. MPs can be found in urban sewage as well as

Table 4.1 Different sources and types of MPs in the aquatic ecosystem

Sr. no.	Source	Location	Type	Size (mm)	References
1.	Plastic pellets	Ocean	PE, PVC, and PUR	2–5	Magnusson et al. (2017)
2.	Vehicle tires	Biota	CTR, PP, PET, PS, and PVC	>0.01	Capolupo et al. (2020)
3.	Street markings	Road marking (wash water)	Binding agents, pigments, fillers, and glass beads	0.05–4	Burghardt et al. (2022)
4.	Fishing gear	Sediments and water	PA and nylon	0.002–0.005	Horton et al. (2017b)
5.	Artificial turfs	Lakes and sea	PP, PA, and PUR	0.01–0.2	Regnell (2019)
6.	Synthetic textile industry	Ocean	PVC and polyester	0.012–0.016	De Falco et al. (2019)
7.	Drinking water supply chain	Groundwater	PE, PA, polyester, PVC, and epoxy resin	0.05–0.15	Mintenig et al. (2019)
8.	Clothing items, furniture, and fishing gear	River and biota	PE, PET, and PP	0.5–1	Khan et al. (2020)

PE polyethylene, *PVC* polyvinyl chloride, *PA* polyamide, *CTR* car tire rubber, *PP* polypropylene, *PET*—polyethylene terephthalate, *PS* polystyrene, *PUR* polyurethane

WWTPs. MPs also consist of microfibers and microbeads originating from household washing operations and personal care products. It has been discovered that MPs predominate in the WWTP's inlet (Vivekanand et al. 2021). Since the 1990s, microbeads have been discovered in seawater and are a frequent source of MPs in WWTPs. Moreover, studies indicate that after each wash, over 94,500 microbeads are liberated from cosmetics and personal hygiene items (Napper et al. 2015). WWTPs are the main cause for the release of MPs into the aquatic ecosystem, even though various processes efficiently eliminate plastics of larger sizes. One thing to keep in mind is that the majority of WWTPs are situated close to bodies of water, either the sea or the ocean, which causes a larger percentage of MPs to be released (Murphy et al. 2016). The following is the order of the several sources of surface water MP pollution: business, public, recreational, farmland activities, industrialization, and urban areas (Chen et al. 2020). As per Zhang et al. (2020a), aquaculture facilities are a significant source of MP sheets in the soil samples and MP fibers in the river water, in addition to industrial structure and population density. The rise in tourism activities, such as the dumping of waste from hotels and cruise ships, is an additional contributor to the elevated levels of MPs in water bodies (Retama et al. 2016).

Depending on their origin or source in marine ecosystems, MPs are classified into primary and secondary (Fig. 4.2). Primary MPs are discharged into the marine ecosystem straight from land-based sources. Primary MPs are tiny particles that appear as pellets in plastic production or as raw materials to make consumer goods

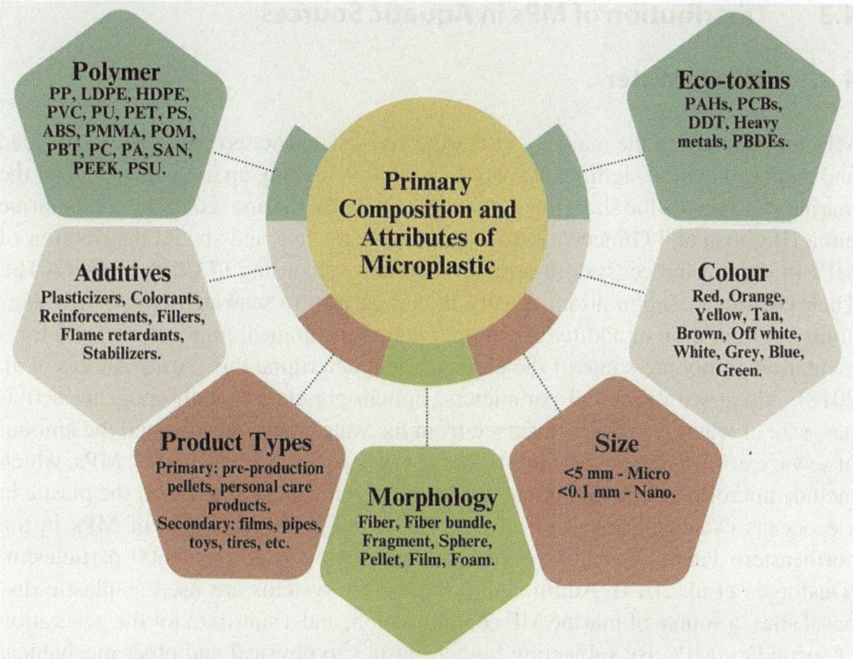

Fig. 4.2 Sources and morphological characteristics of MPs found in the aquatic ecosystem

for personal care and hygiene products like scrubs, beauty products, and abrasives used in cleaning supplies, sunscreen, insect repellents, air-blasting media, synthetic clothing, and drilling fluids (Alimba and Faggio 2019). In this context, primary MPs that are <0.5 mm and around 5 mm are employed as raw materials in personal care and hygiene products (Thompson 2015). The in situ fragmentation of larger plastic materials into smaller particles gives rise to secondary MPs in the marine ecosystem. The breakdown of the cross-links and bonds in the plastic polymers is accelerated by a single or a mixture of chemical, biological, and physical processes. Plastic fibers and pieces are continuously created during secondary MP production and discharged into the marine ecosystem by oxidation, mechanical abrasion, or photo-thermal breakdown (Wagner et al. 2014). According to prior investigations, the majority of MPs that are created by primary or secondary practices are predicted to either continuously degrade over unknown timescales till the particle is fully disintegrated into water, biomass, and carbon dioxide or continue to fragment until they reach the nanoparticle size (Dawson et al. 2018).

4.3 Distribution of MPs in Aquatic Sources

4.3.1 MPs in Water

MPs are reported in the majority of marine ecosystems across the globe. Their load and dispersal exhibit significant spatial variability, making up more than 95% of the marine trash across the shorelines (beaches), seabeds, marine sediments, and marine biota (Barboza and Gimenez 2015). The long-term fate and spatial distribution of MPs in the marine ecosystem remain uncertain, according to Clark et al. (2016). Their chemical configuration, density in comparison to seawater, ease of weatherability, formulation of additives, polymer nature, ecological implications, and fragmentation ability are some of the elements that determine this (Arias-Andres et al. 2018). Along with physical parameters, human population, anthropogenic activities, size of water reservoirs, distance from the water body, density, and the amount of sewage effluent are significant factors (Eerkes-Medrano et al. 2015). MPs, which include microfibers, are thought to make up a significant proportion of the plastic in the oceans (Veerasingam et al. 2016). The average concentration of MPs in the northeastern Pacific Ocean's subsurface seawaters was around 4600 particles/m^3 (Desforges et al. 2014). Additionally, freshwater systems are used as plastic disposal sites, a source of marine MP contamination, and a substrate for the generation of secondary MPs. By subjecting bigger plastics to physical and other mechanical stressors present in rivers, rivers can also serve as MP producers by facilitating the breakdown of these materials through mechanical and/or UV weathering (Maes et al. 2017). MPs are not evenly dispersed across rivers; instead, their distribution and quantity are influenced by wind, waves, surface circulation, and proximity to urban areas (Schrank et al. 2022). There have been investigations suggesting elevated MP concentrations in the St. Lawrence River, Pearl River, and the Antua River that can be attributed to urbanization and anthropogenic activities. WWTP effluent is another significant source of MPs in waterways (Yan et al. 2019). Different sources and types of MPs are highlighted in Table 4.1.

4.3.2 MPs in Sediments

Research reports on MPs from areas like lakes, reservoirs, and rivers and freshwater sediments have been made worldwide. River deltas are thought to be hotspots for the buildup of MPs. Rivers gather local MPs from plastic production facilities, plastic waste, and sewage discharge, which is a significant means of moving MPs from inland to the ocean (Yang et al. 2021). The MPs settle with the sinking silt because the rivers carry a large amount of plastic particles across longer distances. This settling is seen in areas where energy flow is reduced, such as when riverbed growth or advancement is delayed. Lakes are bodies of water with relatively high sedimentation rates and the least amount of water flow, which leads to a large buildup of MPs (Vivekanand et al. 2021). Long-distance migration of MPs in various strata allows them to enter the ocean sink (Yu et al. 2018). Small MPs that could be released into the sea and ocean may be stored in beach habitats. The greatest concentration of MPs (851 ± 177 items/kg) was found in the sediments of the Pearl River Estuary (Zuo et al. 2020). MP contamination was discovered in the sediments from beaches, harbors, and sublittoral ecosystems along the coast of Belgium (Claessens et al. 2011). MPs and other debris have been found in coastal sediments in the Indian Ocean, according to reports from Singapore and India. On India's western coast, intertidal zones were found to contain polyester, nylon, PS, and polyurethane. MPs, primarily from secondary sources, were common on Singapore's east coast tourist beaches (Ng and Obbard 2006; Reddy et al. 2006). The River Thames in the United Kingdom had a mean concentration of 660 items/kg of MPs in its riverside silt (Horton et al. 2017a).

4.3.3 MPs in Groundwater

The possibility exists that MPs may be carried vertically, into soil sub-layers, and ultimately into deeper soil layers once they are deposited on the soil's surface (Li et al. 2021). Additionally to being a possible sink for MPs, soil is also a practical pathway to groundwater or aquifers (O'Connor et al. 2019). MPs are found in soils or groundwater as fibers, which suggest that they may originate from carpets, furniture, or clothing (Dalla Fontana et al. 2021). Similarly, human activities like agrarian activities, fishing, and wastewater treatment are connected to groundwater pollution by MPs (Wainkwa Chia et al. 2021). MPs in groundwater in wealthy and industrialized nations may be the result of leaching, particularly in landfills (Bellasi et al. 2020), followed by estuary and karst. Sources of pellet- and film-shaped MPs can be linked to personal care items like cleaning supplies, cosmetics, plastic bags, plastic containers, and plastic covers (Yu et al. 2021).

4.3.4 MPs in Biota

Freshwater species have been discovered to contain MPs. Numerous freshwater species, such as fish, turtles, whales, and shellfish, absorb plastic fragments, which can build up in their digestive tracts (Matsuguma et al. 2017). The feeding habits of various species of fish determine how quickly they consume MPs. In a lab setting, Rosenkranz et al. (2009) show that *Daphnia magna* quickly consumes MPs. Because organisms are more likely to come into contact with MPs, an increase in the quantity of MPs in the marine ecosystem will also have an impact on their bioavailability. Accordingly, the ecological implications of the harmful pollutants' bioaccumulation and biomagnification in species are uncertain. Thus, it is essential to look into how MPs impact the health of various aquatic fauna. MPs are consumed by marine species, which allows them to enter the food chain (Rezania et al. 2018). According to estimates, marine plastic pollution impacted about 690 species in 2015, and ~ 10% of those species consumed MPs (Yu et al. 2020). Additionally, MPs have been found in the tissues or stomachs of numerous aquatic creatures, such as fish and zooplankton. Through tangling and ingestion, plastic or MP trash directly affects aquatic organisms mechanically (Yu et al. 2020). Aquatic species consume less food as a result of the digestive tract's mechanical wear and tear, which ultimately causes famine and death.

4.4 Transport and Fate of MPs in Freshwater

The ocean is considered a sink for MPs, where freshwater and terrestrial ecosystems serve as prominent suppliers and dominant pathways for transporting the MPs to the oceans (Horton and Dixon 2018). Thus, to evaluate the influence of MPs on humans and habitats, it is significant to comprehend the fate and possibility of MPs in these ecosystems. MPs remain dormant in the ecosystems due to their nondegradable abilities and long-term durability (Rios Mendoza et al. 2021). In the past decades, a shift in investigations on MPs has been observed from the marine to inland ecosystems, where freshwater ecosystems have been considered vital, considering the availability of water, ecosystem health, pollutant control and management, and various other resources they provide (Li et al. 2024). Once the MPs are subjected to the natural settings, they are subjected to a variety of transportation and modification processes (Li et al. 2024).

4.4.1 Transport of MPs in Freshwater

To verify the process of transport of MPs, it is vital to understand how MP fragments are settled and transported in freshwater systems (Kumar et al. 2021). Thus, to lower the impacts of MPs on these ecosystems, it is vital to comprehend how these MPs are transported and deposited, as environmental conditions, such as charge, area, currents, hydrophobicity, particle density, wind, size, and shape, affect the transport of MPs in freshwater (Wang et al. 2021). For example, it was determined that flow velocity was the important factor for the concentrations of MP in the River Tame, United Kingdom, where elevated MP concentrations were observed in locations with reduced flow rates (Tibbetts et al. 2018). Tibbetts et al. (2018) determined elevated concentrations when the flow velocities reduced as the water reached the lake areas, leading to high deposition rates of MPs. Furthermore, it was found that concentrations were elevated in floodplains, dams, lakes, and habitats with low velocities than in rivers and streams (Petersen and Hubbart 2021). Additionally, for particles with diameters <2 m, the influence of gravity and buoyancy is minimal, while particle mobility is influenced by interfacial tension (Kumar et al. 2021). Urbanization, water retention duration, waste management style, proximity to dense populations, and sewage spills also influence the stress of MPs in aquatic systems (Wang et al. 2021).

Additionally, because most MPs have a density that is lower than fresh- or seawater, they are frequently buoyant at the water's surface and are carried by rivers and seas in a process called surface transport (Wu et al. 2019). In general, plastics tend to drift more easily in the topmost layer if they are larger. MPs are, therefore, more likely to be transported offshore because they are less impacted by stokes drift caused by wind waves in surficial water (Isobe et al. 2014). MPs of various origins are anticipated to have distinct transit characteristics, varying likelihood of burial in riverbed sediments, and consequently, varying forms of association with aquatic species at various trophic levels, depending on their buoyancy and density (Krause et al. 2021). For example, it was discovered that low-density synthetic polymers were present in numerous Canadian Great Lakes rivers, compared to the high-density polymers observed in benthic and hyporheic accumulation zones (Ballent et al. 2016). A study by Zylstra (2013) demonstrated that MPs with higher densities were more likely to infiltrate terrestrial and aquatic ecosystems by virtue of surface runoffs and wind speeds.

The density of MPs also plays a complex role as it is linked with various interactions of biotic and abiotic components (Wong et al. 2020). To understand this, a term "plastic biofouling" is considered, which refers to the mechanism by which microorganisms adhere to the outer surface of the MPs, thus modifying their density, buoyancy, and geometry (Nava and Leoni 2021). These mechanisms incorporated complex interactions that go beyond the straightforward idea of creating a homogeneous molecular layer on the particle surface (Junaid and Wang 2021). Meanwhile, the extracellular matrix of polymeric compounds reacts with MPs to become sticky in the process of biofilm formation, thus promoting the development of hetero-aggregates (Wu et al. 2019). Thus, biological activities like biofilm adhesion, particle decay and aging, entrapment, and biomolecule adsorption influence the MPs' fate and transportation (Krause et al. 2021). On the contrary, biofilms that tend to develop on the surface of aging MPs have the tendency to modify the hydrodynamic and buoyancy properties, which directly impact their fate, accumulation, transport, and adsorption properties (Corcoran et al. 2015). Furthermore, rivers that have elevated suspended particulate concentrations encourage associations with MPs, particularly following biofouling, to create flocs, which have a significant impact on MP dynamics by promoting sediment deposition and buildup (Guo et al. 2024).

Moreover, the movement of MPs is studied using a variety of models. The amount of MPs in the marine ecosystem has been measured using the Lagrangian numerical simulation model (Guo et al. 2024) that determines the spatial and temporal motion in a turbulent flow and integrates with the ocean circulation models to trace the fate, paths, and origins under multiple environmental conditions. As per the study by Atugoda et al. (2020), the model generates assumptions considering that MPs are biochemically neutral and buoyant. Another model used to determine the transport mechanisms is the "DUFLOW model" that considers the areas of accumulation and sedimentation of MPs in aquatic ecosystems by considering the hydrological parameters and fate processes. The model considers the hydrological factors and advection-diffusion equations, in addition to the river's mean velocity, discharge, and water level, to compute the particle transfer (Quik et al. 2015). Similarly, the "Global NEWS" model considers how large-scale anthropogenic activities affect the nutrient inputs in the river (Mayorga et al. 2010). Siegfried et al.

(2017) investigated the transport of MPs from European rivers to the seas using the "Global NEWS" model to understand the trends till 2050. The MPs' transport output was estimated as the gap between the input from sources in the river and the proportion that is eliminated after river retention (Atugoda et al. 2020). Thus, to determine the behavior and fate of nanoplastics and MPs in the riverine system, the "NanoDUFLOW model" was recently modified (Besseling et al. 2017). The first water quality and hydrological model to be created for the determination of fate and distribution of pollutants and nutrients in riverine and terrestrial ecosystems is the "Integrated Catchment (INCA) Contaminant Pollutant model" (Bussi et al. 2016). The model is horizontally and vertically integrated and based on process variables and operations that determine dynamic and flow inputs from diffuse and discrete sources. The model also tracks spatial variations consisting of properties of soil, land use, river network, catchment boundaries, and length and slope of rivers (Nizzetto et al. 2016).

4.4.2 Fate of MPs in Freshwater

MPs end up in the oceans once they reach aquatic bodies. MPs and litter of microbeads originate from land-based sources, which contribute to 80% of the remnants found in the aquatic environments (Sharma and Chatterjee 2017). As soon as MPs are disintegrated from the bigger plastics and are subjected to freshwater ecosystems, they are transported by numerous spatial scales (Meng et al. 2020; Salgaonkar et al. 2024). Nonetheless, MPs are also trapped in sediments for a prolonged time and get resuspended by agents such as waves and tides (Hasan Anik et al. 2021). A study by Zain et al. (2022) reported that MPs trapped in sediments are translocated to aquatic ecosystems or in the food chain, where they impose detrimental impacts and disrupt the functioning of the ecosystem. Similarly, MPs are engulfed by biofilms consisting of algae, fungi, and bacteria, which change the chemical and physical properties of MPs (Raza et al. 2022). Thus, the the fate of MPs, ingested by fish in the form of biofilms, will be changed (Wang et al. 2021). MPs once consumed are then either expelled or moved into tissues, and then they are moved up in the trophic levels to another food chain (Hasan Anik et al. 2021). Currently, MPs do not have a defined fate, and if they enter natural ecosystems, they pose a danger to the flora and fauna of that ecosystem (Raza et al. 2022).

4.5 Impacts of MP Pollution

The literature has found MPs in surface water, sediments, beaches, soil and other terrestrial niches, and even in our diet (Hasan Anik et al. 2021). Furthermore, a variety of faunal taxa, from tiny plankton to huge fish, birds, and mammals, have been found to have negative effects on an ecosystem due to MPs (Ašmonaitė and Almroth 2018) (Table 4.2). Fibers or fragments make up the bulk of MPs discovered in organisms, which have detrimental impacts on the flora and fauna (Burns and

Table 4.2 Impacts of MPs on human health and the aquatic ecosystem

Sources	Impacts	References
Humans	Inflammation in the liver	Deng et al. (2017)
	Adverse effects on neurotransmission	Deng et al. (2017)
	Altered expression of scavenger receptors	Fuchs et al. (2016)
	Apoptosis in all intestinal epithelial cells	Inkielewicz-Stepniak et al. (2018)
	Reduced cell viability with a reduction of adenosine triphosphate (ATP) and an increase in reactive oxygen species (ROS) concentrations	Mahadevan and Valiyaveettil (2021)
	Changes in amino acid and bile acid metabolism	Jin et al. (2019)
	Induced gut microbiota dysbiosis and intestinal barrier dysfunction	Luo et al. (2019)
	Disrupted intestinal iron transport and cellular uptake	Mahler et al. (2012)
Algae	Growth inhibition, induction of ROS generation, reduced growth rate, and decreased chlorophyll content	Besseling et al. (2014, 2017) and Lagarde et al. (2016)
Crustaceans	Lower feeding and reproduction	Ogonowski et al. (2016)
Fish	Altered immune response, changes in metabolic function, and tumor formation	Rochman et al. (2013b)
Mussel	Decreased enzymatic activity and induced effects on neurotransmission	Brandts et al. (2018)

Boxall 2018). MP consumption by any biota can cause blockage and problems in the gastrointestinal tract, which can result in false satiety, hunger, and mortality (da Costa et al. 2017). A wide array of lipophilic additives can infiltrate cell membranes and affect biochemical processes within the cells, leading to reproductive and behavioral problems (Lambert et al. 2014). Moreover, monomers of potential toxicity are found in PS, PC, and PVC, leading to cancers and reproductive issues (Boyle and Örmeci 2020).

Considering their minute size and their proliferation in aquatic media, MPs are ingested by several benthic, pelagic, and top predator species, such as polychaete worms, amphipods, tubifex worms, zooplankton, phytoplankton, algae, and fishes (Hasan Anik et al. 2021). Field observations in the English Channel have reported that 36.5% of ten fish species contain MPs in their gastrointestinal systems (Lusher et al. 2013). MPs build up in the reproductive and digestive systems of several trophic freshwater species, including *Lumbricus variegatus*, *Oryzias latipes*, and *Hyalella azteca* (Au et al. 2015; Rochman et al. 2013a). On the other hand, when the toxicity of tiny MPs was examined in animals, including mice, rats, and dogs, significant consequences were discovered (Carr et al. 2012; Deng et al. 2017). Impacts of MPs on humans and various aquatic flora and fauna are highlighted in Table 4.2.

Worldwide, there have been several reports of MPs in sediments (Chaukura et al. 2021). MPs in soil are accumulated by various processes, such as erosion, street runoff, air deposition, plastic mulching, and littering (Bläsing and Amelung 2018; Corradini et al. 2019). MPs are also associated with various fauna present in the soil, thus causing detrimental consequences on the health of the organisms and inhibiting various soil processes (Huerta Lwanga et al. 2017). A study conducted by Huerta Lwanga et al. (2017) stressed the structural alteration in earthworm burrows caused by MPs, which impacted the functioning and aggregation of soil. Additionally, many investigations reported the presence of MPs in the digestive systems of deceased birds, goats, and sheep in China (Omidi et al. 2012; Zhu et al. 2018). Consequently, elevated concentrations were reported in sediments from the Changjiang Estuary in China (Peng et al. 2017) and Edgbaston Pool in the United Kingdom (Vaughan et al. 2017). MPs' concentration ranged from 20% to 30% (w/w) in the Edgbaston Pool, United Kingdom, while mean concentrations of 121 ± 9 per kg of dry weight were recorded in the Changjiang Estuary in China. Browne et al. (2013) investigated the effects of sediments containing 5% w/w PVC particles pre-sorbed with phenanthrene and nonylphenol, triclosan, and diphenyl ether on lugworms (*Arenicola marina*). It was discovered that these pollutants and additives entered the lugworms' tissues, increasing oxidative stress and vulnerability to infections, with an increase in mortality rate by 55%. Conversely, Rodriguez-Seijo et al. (2017) investigated the effects of low-density polyethylene pellets on earthworms in sediments and discovered no negative effects on growth, reproduction, or survival, but inflammation and damage to the gastrointestinal tract's epithelium were observed, triggering an immune reaction.

4.6 Policy Development

Several treaties were established to address plastic pollution seven decades after synthetic plastic was invented (Fig. 4.3) (Munhoz et al. 2022). In order to track and test the consequences of MP contamination in wastewater, freshwater, and marine environments, indigenous test methods are too costly and tedious, and current international standards are not unified (Li et al. 2020). This is due to the relatively young nature of this field of study, the lack of pertinent reference materials, and the dearth of more comprehensive studies on the biological degradation of plastic materials in these settings (Harrison et al. 2018). Numerous national governments and international organizations have implemented specific steps to combat the issue of plastic pollution in response to the acknowledged risks associated with it (Shi et al. 2024).

China, in the year 2001, stopped the manufacturing of disposable foam plastic dinnerware, while, in order to combat plastic pollution, the "Fifth session of the United Nations Environment Assembly" put forth the demand of establishing policy guidelines (Shi et al. 2024). In 2011, Canada developed the "Honolulu Strategy" to mitigate the marine debris and minimize contamination from sources like fishing cargo and fishing gear. The "Barcelona Convention" in 2012 also focused on marine litter by establishing monitoring programs for protecting the oceans. The World Bank's "Pollution Management and Environmental Health (PMEH) Program" in 2016, focused on promoting awareness and emphasizing on the reduction of MPs (Vivekanand et al. 2021). The European Union (EU) has also undertaken initiatives for eliminating plastic waste and promoting sustainability

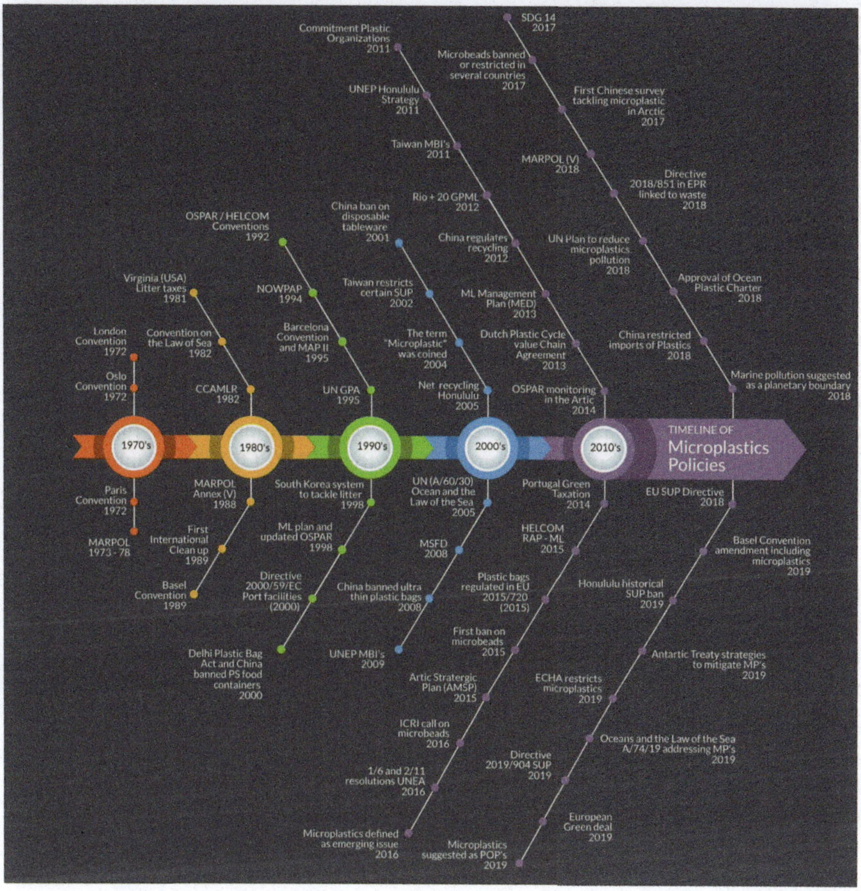

Fig. 4.3 Various policies implemented for the abatement of MPs in the aquatic ecosystem

(Boyle and Örmeci 2020). Moreover, the EU has implemented an array of policies for the management of plastics, addressing plastic trash, disposable plastics, packaging waste, and other associated areas. Thus, minimizing the usage of plastic will tend to mitigate the production of MPs at their source (Shi et al. 2024). The EU has also adopted a circular model in 2018, which implements regulations on all packaging materials that are sold in the markets of the EU should be reusable and recyclable by 2030 having extended durability, efficient recycling, and reuse. According to Boyle and Örmeci (2020), the European Commission has implemented regulations on industries who generate or recycle plastics that they must showcase a "concrete and ambitious set of voluntary commitments to back the strategy and its vision for 2030."

In the past 5 years, various European nations, along with the USA, South Korea, New Zealand, and Canada, have applied stringent restrictions on the usage of MPs in personal hygiene products and cosmetics. Moreover, numerous international organizations, including the G20, G7, United Nations, and the Arctic Council, are developing policies that will mitigate the issue of plastic pollution in the environment. The "G7 Ocean Plastics Charter," the "United Nations Global Partnership on Marine Litter," the "G20 Implementation Framework for Actions on Marine Plastic Litter," the "Arctic Council Desktop Study on Marine Litter Including Microplastics in the Arctic," and the "United Nations Sustainable Development Goals" are a few of these initiatives (Boyle and Örmeci 2020). Despite the fact that the majority of legislation initially dealt with plastic pollution in general, it had a direct impact on regulations that targeted MPs from the mid-2000s to 2020. Globally, legislation and activities to address plastic pollution or even outright prohibit plastics increased dramatically between 2013 and 2020 (Munhoz et al. 2022). The following are noteworthy because they serve as a foundation for the existing policies: national laws; directives (EU)2015/720, (EU)2018/851, and (EU)2019/904; tax reforms, such as the Portuguese Law n_ 82-D/2014; restrictions on the importation of plastic waste; national, regional, and collective agreements; the MARPOL 73/78 annex amendment; UN resolutions promoting stakeholder involvement and monitoring programs; regional plans; and bans on single-use plastics (Munhoz et al. 2022).

Thus, there are significant issues with MP contamination in the freshwater environment, and current regulations have not yet produced a thorough management system. Few studies have emphasized how important it is to concentrate on creating and enforcing regulations to control freshwater MPs. The macro-policy establishment stage of MP management is now underway, but the overall management framework is still insufficient. To improve the governance and management system for MPs, additional specialists must be hired. Even though nations and areas have made some strides in recent years to reduce MP pollution, freshwater MPs receive comparatively less attention and specific policies than marine MPs, which are the main focus of current laws and regulations (Shi et al. 2024).

4.7 Conclusion and Future Directions

MP contamination has caused enormous challenges around the globe, primarily in aquatic ecosystems. MPs emerge from various anthropogenic activities and persist in nature for prolonged periods. This chapter significantly explored their occurrence, distribution, and transport, followed by their impacts on the aquatic ecosystem. The findings from the chapter indicate that there is an urgent need to monitor, assess mitigation strategies, and implement polices to curb the growing MP pollution. Additionally, there is a lack of comprehensive research on MPs' fate, transformation, and long-term ecological consequences. Thus, to bridge the gap, future research should focus on monitoring long-term fate and transport studies, the interconnectedness of MPs in ecosystems' toxicity, and health risk assessment studies. Upcoming research should also focus on developing advanced methodologies for the early detection and characterization of MPs in freshwater ecosystems. Last, to curb the risks of MPs in freshwater, strict environmental regulations and plastic waste management systems that are environmentally sustainable, economically sound, and socially acceptable should be developed.

References

Ahmad M, Li J-L, Wang P-D, Hozzein W, Li W-J (2020) Environmental perspectives of microplastic pollution in the aquatic environment: a review. Mar Life Sci Technol 2:414. https://doi.org/10.1007/s42995-020-00056-w

Alimba CG, Faggio C (2019) Microplastics in the marine environment: current trends in environmental pollution and mechanisms of toxicological profile. Environ Toxicol Pharmacol 68:61–74. https://doi.org/10.1016/j.etap.2019.03.001

Arias-Andres M, Klümper U, Rojas-Jimenez K, Grossart H-P (2018) Microplastic pollution increases gene exchange in aquatic ecosystems. Environ Pollut 237:253–261. https://doi.org/10.1016/j.envpol.2018.02.058

Ašmonaitė G, Almroth BC (2018) Effects of microplastics on organisms and impacts on the environment. Goteborgs Universitet, Goteborgs, pp 1–70

Atugoda T, Piyumali H, Liyanage S, Mahatantila K, Vithanage M (2020) Fate and behavior of microplastics in freshwater systems. In: Rocha-Santos T, Costa M, Mouneyrac C (eds) Handbook of microplastics in the environment. Springer, pp 1–31. https://doi.org/10.1007/978-3-030-10618-8_42-1

Au SY, Bruce TF, Bridges WC, Klaine SJ (2015) Responses of *Hyalella azteca* to acute and chronic microplastic exposures. Environ Toxicol Chem 34(11):2564–2572. https://doi.org/10.1002/etc.3093

Ballent A, Corcoran PL, Madden O, Helm PA, Longstaffe FJ (2016) Sources and sinks of microplastics in Canadian Lake Ontario nearshore, tributary and beach sediments. Mar Pollut Bull 110(1):383–395. https://doi.org/10.1016/j.marpolbul.2016.06.037

Barboza LGA, Gimenez BCG (2015) Microplastics in the marine environment: current trends and future perspectives. Mar Pollut Bull 97(1–2):5–12. https://doi.org/10.1016/j.marpolbul.2015.06.008

Bellasi A, Binda G, Pozzi A, Galafassi S, Volta P, Bettinetti R (2020) Microplastic contamination in freshwater environments: a review, focusing on interactions with sediments and benthic organisms. Environments 7(4):30. https://doi.org/10.3390/environments7040030

Besseling E, Wang B, Lürling M, Koelmans AA (2014) Nanoplastic affects growth of *S. obliquus* and reproduction of *D. magna*. Environ Sci Technol 48(20):12336–12343. https://doi.org/10.1021/es503001d

Besseling E, Quik JTK, Sun M, Koelmans AA (2017) Fate of nano- and microplastic in freshwater systems: a modeling study. Environ Pollut 220:540–548. https://doi.org/10.1016/j.envpol.2016.10.001

Bläsing M, Amelung W (2018) Plastics in soil: analytical methods and possible sources. Sci Total Environ 612:422–435. https://doi.org/10.1016/j.scitotenv.2017.08.086

Boyle K, Örmeci B (2020) Microplastics and nanoplastics in the freshwater and terrestrial environment: a review. Water 12(9):2633. https://doi.org/10.3390/w12092633

Brandts I, Teles M, Gonçalves AP, Barreto A, Franco-Martinez L, Tvarijonaviciute A, Martins MA, Soares AMVM, Tort L, Oliveira M (2018) Effects of nanoplastics on Mytilus galloprovincialis after individual and combined exposure with carbamazepine. Sci Total Environ 643:775–784. https://doi.org/10.1016/j.scitotenv.2018.06.257

Browne MA, Niven SJ, Galloway TS, Rowland SJ, Thompson RC (2013) Microplastic moves pollutants and additives to worms, reducing functions linked to health and biodiversity. Curr Biol 23(23):2388–2392

Burghardt TE, Pashkevich A, Babić D, Mosböck H, Babić D, Żakowska L (2022) Microplastics and road markings: the role of glass beads and loss estimation. Transp Res Part D: Transp Environ 102:103123. https://doi.org/10.1016/j.trd.2021.103123

Burns EE, Boxall ABA (2018) Microplastics in the aquatic environment: evidence for or against adverse impacts and major knowledge gaps. Environ Toxicol Chem 37(11):2776–2796. https://doi.org/10.1002/etc.4268

Bussi G, Dadson SJ, Prudhomme C, Whitehead PG (2016) Modelling the future impacts of climate and land-use change on suspended sediment transport in the River Thames (UK). J Hydrol 542:357–372. https://doi.org/10.1016/j.jhydrol.2016.09.010

Capolupo M, Sørensen L, Jayasena KDR, Booth AM, Fabbri E (2020) Chemical composition and ecotoxicity of plastic and car tire rubber leachates to aquatic organisms. Water Res 169:115270. https://doi.org/10.1016/j.watres.2019.115270

Carr KE, Smyth SH, McCullough MT, Morris JF, Moyes SM (2012) Morphological aspects of interactions between microparticles and mammalian cells: intestinal uptake and onward movement. Prog Histochem Cytochem 46(4):185–252. https://doi.org/10.1016/j.proghi.2011.11.001

Chaukura N, Kefeni KK, Chikurunhe I, Nyambiya I, Gwenzi W, Moyo W, Nkambule TTI, Mamba BB, Abulude FO (2021) Microplastics in the aquatic environment—the occurrence, sources, ecological impacts, fate, and remediation challenges. Pollutants 1(2):95–118. https://doi.org/10.3390/pollutants1020009

Chen H, Jia Q, Zhao X, Li L, Nie Y, Liu H, Ye J (2020) The occurrence of microplastics in water bodies in urban agglomerations: impacts of drainage system overflow in wet weather, catchment land-uses, and environmental management practices. Water Res 183:116073. https://doi.org/10.1016/j.watres.2020.116073

Cincinelli A, Scopetani C, Chelazzi D, Martellini T, Pogojeva M, Slobodnik J (2021) Microplastics in the Black Sea sediments. Sci Total Environ 760:143898. https://doi.org/10.1016/j.scitotenv.2020.143898

Claessens M, De Meester S, Van Landuyt L, De Clerck K, Janssen CR (2011) Occurrence and distribution of microplastics in marine sediments along the Belgian coast. Mar Pollut Bull 62(10):2199–2204. https://doi.org/10.1016/j.marpolbul.2011.06.030

Clark JR, Cole M, Lindeque PK, Fileman E, Blackford J, Lewis C, Lenton TM, Galloway TS (2016) Marine microplastic debris: a targeted plan for understanding and quantifying interactions with marine life. Front Ecol Environ 14(6):317–324. https://doi.org/10.1002/fee.1297

Corcoran PL, Norris T, Ceccanese T, Walzak MJ, Helm PA, Marvin CH (2015) Hidden plastics of Lake Ontario, Canada and their potential preservation in the sediment record. Environ Pollut 204:17–25. https://doi.org/10.1016/j.envpol.2015.04.009

Corradini F, Meza P, Eguiluz R, Casado F, Huerta-Lwanga E, Geissen V (2019) Evidence of microplastic accumulation in agricultural soils from sewage sludge disposal. Sci Total Environ 671:411–420. https://doi.org/10.1016/j.scitotenv.2019.03.368

da Costa JP, Duarte AC, Rocha-Santos TA (2017) Microplastics–occurrence, fate and behaviour in the environment. In: Comprehensive analytical chemistry, vol 75. Elsevier, pp 1–24. https://www.sciencedirect.com/science/article/pii/S0166526X16301532

Dalla Fontana G, Mossotti R, Montarsolo A (2021) Influence of sewing on microplastic release from textiles during washing. Water Air Soil Pollut 232(2):50. https://doi.org/10.1007/s11270-021-04995-7

Dawson AL, Kawaguchi S, King CK, Townsend KA, King R, Huston WM, Bengtson Nash SM (2018) Turning microplastics into nanoplastics through digestive fragmentation by Antarctic krill. Nat Commun 9(1):1001. https://doi.org/10.1038/s41467-018-03465-9

De Falco F, Di Pace E, Cocca M, Avella M (2019) The contribution of washing processes of synthetic clothes to microplastic pollution. Sci Rep 9(1):6633. https://doi.org/10.1038/s41598-019-43023-x

Deng Y, Zhang Y, Lemos B, Ren H (2017) Tissue accumulation of microplastics in mice and biomarker responses suggest widespread health risks of exposure. Sci Rep 7(1):46687. https://doi.org/10.1038/srep46687

Desforges J-PW, Galbraith M, Dangerfield N, Ross PS (2014) Widespread distribution of microplastics in subsurface seawater in the NE Pacific Ocean. Mar Pollut Bull 79(1–2):94–99. https://doi.org/10.1016/j.marpolbul.2013.12.035

Eerkes-Medrano D, Thompson RC, Aldridge DC (2015) Microplastics in freshwater systems: a review of the emerging threats, identification of knowledge gaps and prioritisation of research needs. Water Res 75:63–82. https://doi.org/10.1016/j.watres.2015.02.012

Fuchs A-K, Syrovets T, Haas KA, Loos C, Musyanovych A, Mailänder V, Landfester K, Simmet T (2016) Carboxyl- and amino-functionalized polystyrene nanoparticles differentially affect the polarization profile of M1 and M2 macrophage subsets. Biomaterials 85:78–87. https://doi.org/10.1016/j.biomaterials.2016.01.064

Guo M, Noori R, Abolfathi S (2024) Microplastics in freshwater systems: dynamic behaviour and transport processes. Resour Conserv Recycl 205:107578. https://doi.org/10.1016/j.resconrec.2024.107578

Harrison JP, Boardman C, O'Callaghan K, Delort A-M, Song J (2018) Biodegradability standards for carrier bags and plastic films in aquatic environments: a critical review. R Soc Open Sci 5(5):171792. https://doi.org/10.1098/rsos.171792

Hasan Anik A, Hossain S, Alam M, Binte Sultan M, Hasnine MT, Rahman MM (2021) Microplastics pollution: a comprehensive review on the sources, fates, effects, and potential remediation. Environ Nanotechnol Monit Manag 16:100530. https://doi.org/10.1016/j.enmm.2021.100530

Horton AA, Dixon SJ (2018) Microplastics: an introduction to environmental transport processes. WIREs Water 5(2):e1268. https://doi.org/10.1002/wat2.1268

Horton AA, Svendsen C, Williams RJ, Spurgeon DJ, Lahive E (2017a) Large microplastic particles in sediments of tributaries of the River Thames, UK–abundance, sources and methods for effective quantification. Mar Pollut Bull 114(1):218–226. https://doi.org/10.1016/j.marpolbul.2016.09.004

Horton AA, Walton A, Spurgeon DJ, Lahive E, Svendsen C (2017b) Microplastics in freshwater and terrestrial environments: evaluating the current understanding to identify the knowledge gaps and future research priorities. Sci Total Environ 586:127–141. https://doi.org/10.1016/j.scitotenv.2017.01.190

Huerta Lwanga E, Gertsen H, Gooren H, Peters P, Salánki T, van der Ploeg M, Besseling E, Koelmans AA, Geissen V (2017) Incorporation of microplastics from litter into burrows of *Lumbricus terrestris*. Environ Pollut 220:523–531. https://doi.org/10.1016/j.envpol.2016.09.096

Inkielewicz-Stepniak I, Tajber L, Behan G, Zhang H, Radomski MW, Medina C, Santos-Martinez MJ (2018) The role of mucin in the toxicological impact of polystyrene nanoparticles. Materials 11(5):724. https://doi.org/10.3390/ma11050724

Isobe A, Kubo K, Tamura Y, Nakashima E, Fujii N (2014) Selective transport of microplastics and mesoplastics by drifting in coastal waters. Mar Pollut Bull 89(1–2):324–330. https://doi.org/10.1016/j.marpolbul.2014.09.041

Jiang J-Q (2018) Occurrence of microplastics and its pollution in the environment: a review. Sustain Prod Consum 13:16–23. https://doi.org/10.1016/j.spc.2017.11.003

Jin Y, Lu L, Tu W, Luo T, Fu Z (2019) Impacts of polystyrene microplastic on the gut barrier, microbiota and metabolism of mice. Sci Total Environ 649:308–317. https://doi.org/10.1016/j.scitotenv.2018.08.353

Junaid M, Wang J (2021) Interaction of nanoplastics with extracellular polymeric substances (EPS) in the aquatic environment: a special reference to eco-corona formation and associated impacts. Water Res 201:117319. https://doi.org/10.1016/j.watres.2021.117319

Kashiwabara LM, Kahane-Rapport SR, King C, DeVogelaere M, Goldbogen JA, Savoca MS (2021) Microplastics and microfibers in surface waters of Monterey Bay National Marine Sanctuary, California. Mar Pollut Bull 165:112148. https://doi.org/10.1016/j.marpolbul.2021.112148

Khan FR, Shashoua Y, Crawford A, Drury A, Sheppard K, Stewart K, Sculthorp T (2020) 'The plastic Nile': first evidence of microplastic contamination in fish from the Nile River (Cairo, Egypt). Toxics 8(2):22. https://doi.org/10.3390/toxics8020022

Krause S, Baranov V, Nel HA, Drummond JD, Kukkola A, Hoellein T, Sambrook Smith GH, Lewandowski J, Bonet B, Packman AI, Sadler J, Inshyna V, Allen S, Allen D, Simon L, Mermillod-Blondin F, Lynch I (2021) Gathering at the top? Environmental controls of microplastic uptake and biomagnification in freshwater food webs. Environ Pollut 268:115750. https://doi.org/10.1016/j.envpol.2020.115750

Kumar R, Sharma P, Verma A, Jha PK, Singh P, Gupta PK, Chandra R, Prasad PVV (2021) Effect of physical characteristics and hydrodynamic conditions on transport and deposition of microplastics in riverine ecosystem. Water 13(19):2710. https://doi.org/10.3390/w13192710

Lagarde F, Olivier O, Zanella M, Daniel P, Hiard S, Caruso A (2016) Microplastic interactions with freshwater microalgae: hetero-aggregation and changes in plastic density appear strongly dependent on polymer type. Environ Pollut 215:331–339. https://doi.org/10.1016/j.envpol.2016.05.006

Lambert S, Sinclair C, Boxall A (2014) Occurrence, degradation, and effect of polymer-based materials in the environment. In: Whitacre DM (ed) Reviews of environmental contamination and toxicology, volume 227, vol 227. Springer, pp 1–53. https://doi.org/10.1007/978-3-319-01327-5_1

Li C, Busquets R, Campos LC (2020) Assessment of microplastics in freshwater systems: a review. Sci Total Environ 707:135578. https://doi.org/10.1016/j.scitotenv.2019.135578

Li H, Lu X, Wang S, Zheng B, Xu Y (2021) Vertical migration of microplastics along soil profile under different crop root systems. Environ Pollut 278:116833. https://doi.org/10.1016/j.envpol.2021.116833

Li C, Luo D, Shi Y, Bai X, Krause S, Wang R, Li K (2024) Advances in quantifying the drivers of the occurrence, transport, and fate of freshwater microplastics. Crit Rev Environ Sci Technol 55:1–23. https://doi.org/10.1080/10643389.2024.2380537

Luo T, Zhang Y, Wang C, Wang X, Zhou J, Shen M, Zhao Y, Fu Z, Jin Y (2019) Maternal exposure to different sizes of polystyrene microplastics during gestation causes metabolic disorders in their offspring. Environ Pollut 255:113122. https://doi.org/10.1016/j.envpol.2019.113122

Lusher AL, Mchugh M, Thompson RC (2013) Occurrence of microplastics in the gastrointestinal tract of pelagic and demersal fish from the English Channel. Mar Pollut Bull 67(1–2):94–99

Maes T, Van der Meulen MD, Devriese LI, Leslie HA, Huvet A, Frère L, Robbens J, Vethaak AD (2017) Microplastics baseline surveys at the water surface and in sediments of the North-East Atlantic. Front Mar Sci 4:135. https://doi.org/10.3389/fmars.2017.00135

Magnusson K, Eliasson K, Fråne A, Haikonen K, Hultén J, Olshammar M, Stadmark J, Voisin A (2017) Swedish sources and pathways for microplastics to the marine environment. Swedish Environmental Research Institute, Stockholm

Mahadevan G, Valiyaveettil S (2021) Understanding the interactions of poly(methyl methacrylate) and poly(vinyl chloride) nanoparticles with BHK-21 cell line. Sci Rep 11(1):2089. https://doi.org/10.1038/s41598-020-80708-0

Mahler GJ, Esch MB, Tako E, Southard TL, Archer SD, Glahn RP, Shuler ML (2012) Oral exposure to polystyrene nanoparticles affects iron absorption. Nat Nanotechnol 7(4):264–271. https://doi.org/10.1038/nnano.2012.3

Matsuguma Y, Takada H, Kumata H, Kanke H, Sakurai S, Suzuki T, Itoh M, Okazaki Y, Boonyatumanond R, Zakaria MP, Weerts S, Newman B (2017) Microplastics in sediment cores from Asia and Africa as indicators of temporal trends in plastic pollution. Arch Environ Contam Toxicol 73(2):230–239. https://doi.org/10.1007/s00244-017-0414-9

Maynard IFN, Bortoluzzi PC, Nascimento LM, Madi RR, Cavalcanti EB, Lima ÁS, Jeraldo VDLS, Marques MN (2021) Analysis of the occurrence of microplastics in beach sand on the Brazilian coast. Sci Total Environ 771:144777. https://doi.org/10.1016/j.scitotenv.2020.144777

Mayorga E, Seitzinger SP, Harrison JA, Dumont E, Beusen AHW, Bouwman AF, Fekete BM, Kroeze C, Van Drecht G (2010) Global nutrient export from WaterSheds 2 (NEWS 2): model development and implementation. Environ Model Softw 25(7):837–853. https://doi.org/10.1016/j.envsoft.2010.01.007

Meng Y, Kelly FJ, Wright SL (2020) Advances and challenges of microplastic pollution in freshwater ecosystems: a UK perspective. Environ Pollut 256:113445. https://doi.org/10.1016/j.envpol.2019.113445

Mintenig SM, Löder MGJ, Primpke S, Gerdts G (2019) Low numbers of microplastics detected in drinking water from ground water sources. Sci Total Environ 648:631–635. https://doi.org/10.1016/j.scitotenv.2018.08.178

Munhoz DR, Harkes P, Beriot N, Larreta J, Basurko OC (2022) Microplastics: a review of policies and responses. Microplastics 2(1):1–26. https://doi.org/10.3390/microplastics2010001

Murphy F, Ewins C, Carbonnier F, Quinn B (2016) Wastewater treatment works (WwTW) as a source of microplastics in the aquatic environment. Environ Sci Technol 50(11):5800–5808. https://doi.org/10.1021/acs.est.5b05416

Nan B, Su L, Kellar C, Craig NJ, Keough MJ, Pettigrove V (2020) Identification of microplastics in surface water and Australian freshwater shrimp Paratya australiensis in Victoria, Australia. Environ Pollut 259:113865. https://doi.org/10.1016/j.envpol.2019.113865

Napper IE, Bakir A, Rowland SJ, Thompson RC (2015) Characterisation, quantity and sorptive properties of microplastics extracted from cosmetics. Mar Pollut Bull 99(1–2):178–185. https://doi.org/10.1016/j.marpolbul.2015.07.029

Nava V, Leoni B (2021) A critical review of interactions between microplastics, microalgae and aquatic ecosystem function. Water Res 188:116476. https://doi.org/10.1016/j.watres.2020.116476

Ng KL, Obbard JP (2006) Prevalence of microplastics in Singapore's coastal marine environment. Mar Pollut Bull 52(7):761–767. https://doi.org/10.1016/j.marpolbul.2005.11.017

Nizzetto L, Bussi G, Futter MN, Butterfield D, Whitehead PG (2016) A theoretical assessment of microplastic transport in river catchments and their retention by soils and river sediments. Environ Sci: Processes Impacts 18(8):1050–1059. https://doi.org/10.1039/C6EM00206D

O'Connor D, Pan S, Shen Z, Song Y, Jin Y, Wu W-M, Hou D (2019) Microplastics undergo accelerated vertical migration in sand soil due to small size and wet-dry cycles. Environ Pollut 249:527–534. https://doi.org/10.1016/j.envpol.2019.03.092

Ogonowski M, Schür C, Jarsén Å, Gorokhova E (2016) The effects of natural and anthropogenic microparticles on individual fitness in Daphnia magna. PLoS One 11(5):e0155063. https://doi.org/10.1371/journal.pone.0155063

Omidi A, Naeemipoor H, Hosseini M (2012) Plastic debris in the digestive tract of sheep and goats: an increasing environmental contamination in Birjand, Iran. Bull Environ Contam Toxicol 88(5):691–694. https://doi.org/10.1007/s00128-012-0587-x

Peng G, Zhu B, Yang D, Su L, Shi H, Li D (2017) Microplastics in sediments of the Changjiang Estuary, China. Environ Pollut 225:283–290. https://doi.org/10.1016/j.envpol.2016.12.064

Petersen F, Hubbart JA (2021) The occurrence and transport of microplastics: the state of the science. Sci Total Environ 758:143936. https://doi.org/10.1016/j.scitotenv.2020.143936

Quik JTK, De Klein JJM, Koelmans AA (2015) Spatially explicit fate modelling of nanomaterials in natural waters. Water Res 80:200–208. https://doi.org/10.1016/j.watres.2015.05.025

Raza M, Lee J-Y, Cha J (2022) Microplastics in soil and freshwater: understanding sources, distribution, potential impacts, and regulations for management. Sci Prog 105(3):00368504221126676. https://doi.org/10.1177/00368504221126676

Reddy MS, Basha S, Adimurthy S, Ramachandraiah G (2006) Description of the small plastics fragments in marine sediments along the Alang-Sosiya ship-breaking yard, India. Estuar Coast Shelf Sci 68(3–4):656–660. https://doi.org/10.1016/j.ecss.2006.03.018

Regnell F (2019) Dispersal of microplastic from a modern artificial turf pitch with preventive measures. https://www.genan.eu/wp-content/uploads/2020/02/MP-dispersal-from-Bergavik-IP-Kalmar-Report.pdf. Last accessed date: 24 July 2025

Retama I, Jonathan MP, Shruti VC, Velumani S, Sarkar SK, Roy PD, Rodríguez-Espinosa PF (2016) Microplastics in tourist beaches of Huatulco Bay, Pacific coast of southern Mexico. Mar Pollut Bull 113(1–2):530–535. https://doi.org/10.1016/j.marpolbul.2016.08.053

Rezania S, Park J, Din MFM, Taib SM, Talaiekhozani A, Yadav KK, Kamyab H (2018) Microplastics pollution in different aquatic environments and biota: a review of recent studies. Mar Pollut Bull 133:191–208. https://doi.org/10.1016/j.marpolbul.2018.05.022

Rios Mendoza LM, Leon Vargas D, Balcer M (2021) Microplastics occurrence and fate in the environment. Curr Opin Green Sustain Chem 32:100523. https://doi.org/10.1016/j.cogsc.2021.100523

Rios-Mendoza LM, Ontiveros-Cuadras JF, Leon-Vargas D, Ruiz-Fernández AC, Rangel-García M, Pérez-Bernal LH, Sanchez-Cabeza J-A (2021) Microplastic contamination and fluxes in a touristic area at the SE Gulf of California. Mar Pollut Bull 170:112638. https://doi.org/10.1016/j.marpolbul.2021.112638

Rochman CM, Hoh E, Hentschel BT, Kaye S (2013a) Long-term field measurement of sorption of organic contaminants to five types of plastic pellets: implications for plastic marine debris. Environ Sci Technol 130109073312009:1646. https://doi.org/10.1021/es303700s

Rochman CM, Hoh E, Kurobe T, Teh SJ (2013b) Ingested plastic transfers hazardous chemicals to fish and induces hepatic stress. Sci Rep 3(1):3263. https://doi.org/10.1038/srep03263

Rodriguez-Seijo A, Lourenço J, Rocha-Santos TAP, Da Costa J, Duarte AC, Vala H, Pereira R (2017) Histopathological and molecular effects of microplastics in Eisenia andrei Bouché. Environ Pollut 220:495–503. https://doi.org/10.1016/j.envpol.2016.09.092

Rosenkranz P, Chaudhry Q, Stone V, Fernandes TF (2009) A comparison of nanoparticle and fine particle uptake by *Daphnia magna*. Environ Toxicol Chem 28(10):2142–2149. https://doi.org/10.1897/08-559.1

Salgaonkar S, Botle A, Barabde G, Herlekar M (2024) Micro-nanoplastics from stormwater runoffs to water bodies: an in-depth investigation. In: Bhat SA, Kumar V, Li F, Kumar S (eds) Management of micro and nano-plastics in soil and biosolids. Springer Nature, Cham, pp 95–117. https://doi.org/10.1007/978-3-031-51967-3_4

Schrank I, Löder MGJ, Imhof HK, Moses SR, Heß M, Schwaiger J, Laforsch C (2022) Riverine microplastic contamination in Southwest Germany: a large-scale survey. Front Earth Sci 10:794250. https://doi.org/10.3389/feart.2022.794250

Sharma S, Chatterjee S (2017) Microplastic pollution, a threat to marine ecosystem and human health: a short review. Environ Sci Pollut Res 24(27):21530–21547. https://doi.org/10.1007/s11356-017-9910-8

Shi Y, Chen W, Yang S, Fan Y, Lu L (2024) Freshwater microplastics governance and sustainable development: pollution status, interactions, policies, and prospective studies. Desalin Water Treat 320:100704. https://doi.org/10.1016/j.dwt.2024.100704

Siegfried M, Koelmans AA, Besseling E, Kroeze C (2017) Export of microplastics from land to sea. A modelling approach. Water Res 127:249–257. https://doi.org/10.1016/j.watres.2017.10.011

Thompson RC (2015) Microplastics in the marine environment: sources, consequences and solutions. In: Bergmann M, Gutow L, Klages M (eds) Marine anthropogenic litter. Springer, pp 185–200. https://doi.org/10.1007/978-3-319-16510-3_7

Thompson RC, Olsen Y, Mitchell RP, Davis A, Rowland SJ, John AWG, McGonigle D, Russell AE (2004) Lost at sea: where is all the plastic? Science 304(5672):838–838. https://doi.org/10.1126/science.1094559

Tibbetts J, Krause S, Lynch I, Sambrook Smith GH (2018) Abundance, distribution, and drivers of microplastic contamination in urban river environments. Water 10(11):1597. https://doi.org/10.3390/w10111597

Vaughan R, Turner SD, Rose NL (2017) Microplastics in the sediments of a UK urban lake. Environ Pollut 229:10–18. https://doi.org/10.1016/j.envpol.2017.05.057

Veerasingam S, Saha M, Suneel V, Vethamony P, Rodrigues AC, Bhattacharyya S, Naik BG (2016) Characteristics, seasonal distribution and surface degradation features of microplastic pellets along the Goa coast, India. Chemosphere 159:496–505. https://doi.org/10.1016/j.chemosphere.2016.06.056

Vivekanand AC, Mohapatra S, Tyagi VK (2021) Microplastics in aquatic environment: challenges and perspectives. Chemosphere 282:131151. https://doi.org/10.1016/j.chemosphere.2021.131151

Wagner M, Scherer C, Alvarez-Muñoz D, Brennholt N, Bourrain X, Buchinger S, Fries E, Grosbois C, Klasmeier J, Marti T, Rodriguez-Mozaz S, Urbatzka R, Vethaak AD, Winther-Nielsen M, Reifferscheid G (2014) Microplastics in freshwater ecosystems: what we know and what we need to know. Environ Sci Eur 26(1):12. https://doi.org/10.1186/s12302-014-0012-7

Wainkwa Chia R, Lee J-Y, Kim H, Jang J (2021) Microplastic pollution in soil and groundwater: a review. Environ Chem Lett 19:4211. https://doi.org/10.1007/s10311-021-01297-6

Wang C, Zhao J, Xing B (2021) Environmental source, fate, and toxicity of microplastics. J Hazard Mater 407:124357. https://doi.org/10.1016/j.jhazmat.2020.124357

Wong JKH, Lee KK, Tang KHD, Yap P-S (2020) Microplastics in the freshwater and terrestrial environments: prevalence, fates, impacts and sustainable solutions. Sci Total Environ 719:137512. https://doi.org/10.1016/j.scitotenv.2020.137512

Wu P, Huang J, Zheng Y, Yang Y, Zhang Y, He F, Chen H, Quan G, Yan J, Li T, Gao B (2019) Environmental occurrences, fate, and impacts of microplastics. Ecotoxicol Environ Saf 184:109612. https://doi.org/10.1016/j.ecoenv.2019.109612

Yan M, Nie H, Xu K, He Y, Hu Y, Huang Y, Wang J (2019) Microplastic abundance, distribution and composition in the Pearl River along Guangzhou city and Pearl River estuary, China. Chemosphere 217:879–886. https://doi.org/10.1016/j.chemosphere.2018.11.093

Yang L, Zhang Y, Kang S, Wang Z, Wu C (2021) Microplastics in freshwater sediment: a review on methods, occurrence, and sources. Sci Total Environ 754:141948. https://doi.org/10.1016/j.scitotenv.2020.141948

Yu Y, Zhou D, Li Z, Zhu C (2018) Advancement and challenges of microplastic pollution in the aquatic environment: a review. Water Air Soil Pollut 229(5):140. https://doi.org/10.1007/s11270-018-3788-z

Yu Q, Hu X, Yang B, Zhang G, Wang J, Ling W (2020) Distribution, abundance and risks of microplastics in the environment. Chemosphere 249:126059. https://doi.org/10.1016/j.chemosphere.2020.126059

Yu L, Zhang J, Liu Y, Chen L, Tao S, Liu W (2021) Distribution characteristics of microplastics in agricultural soils from the largest vegetable production base in China. Sci Total Environ 756:143860. https://doi.org/10.1016/j.scitotenv.2020.143860

Zain NM, Fauzi N, Subki NS, Ghazali ZZ (2022) Occurrence of microplastics in immature aquatic insects of Gua Musang tributaries in Kelantan. IOP Conf Ser Earth Environ Sci 1102(1):012047. https://doi.org/10.1088/1755-1315/1102/1/012047

Zhang L, Liu J, Xie Y, Zhong S, Yang B, Lu D, Zhong Q (2020a) Distribution of microplastics in surface water and sediments of Qin River in Beibu Gulf, China. Sci Total Environ 708:135176. https://doi.org/10.1016/j.scitotenv.2019.135176

Zhang W, Zhang S, Zhao Q, Qu L, Ma D, Wang J (2020b) Spatio-temporal distribution of plastic and microplastic debris in the surface water of the Bohai Sea, China. Mar Pollut Bull 158:111343. https://doi.org/10.1016/j.marpolbul.2020.111343

Zhu L, Bai H, Chen B, Sun X, Qu K, Xia B (2018) Microplastic pollution in North Yellow Sea, China: observations on occurrence, distribution and identification. Sci Total Environ 636:20–29. https://doi.org/10.1016/j.scitotenv.2018.04.182

Zuo L, Sun Y, Li H, Hu Y, Lin L, Peng J, Xu X (2020) Microplastics in mangrove sediments of the Pearl River Estuary, South China: correlation with halogenated flame retardants' levels. Sci Total Environ 725:138344. https://doi.org/10.1016/j.scitotenv.2020.138344

Zylstra ER (2013) Accumulation of wind-dispersed trash in desert environments. J Arid Environ 89:13–15. https://doi.org/10.1016/j.jaridenv.2012.10.004

Wong JKH, Lee KK, Tang KHD, Yap PS (2020) Microplastics in the freshwater and terrestrial environments: prevalence, fates, impacts and sustainable solutions. Sci Total Environ 719:137512. https://doi.org/10.1016/j.scitotenv.2020.137512
Wu P, Huang J, Zheng Y, Yang Y, Zhang Y, He F, Chen H, Quan G, Yan J, Li T, Gao B (2019) Environmental occurrences, fate, and impacts of microplastics. Ecotoxicol Environ Saf 184:109612. https://doi.org/10.1016/j.ecoenv.2019.109612
[illegible]
[illegible]
[illegible]
[illegible]
[illegible]
[illegible]
[illegible]
[illegible]
[illegible]
[illegible]

Tracing the Path: Fate and Transport of Microplastics from Aquatic Systems to Soils

5

Sheha Shaji and Baranidharan Sundaram

Abstract

Microplastics (MPs), which are tiny plastic particles (size<5 mm), have permeated every corner of the environment, emerging as contaminants in aquatic and soil ecosystems. This article explores the fate and transport of MPs from water systems into soils. MPs enter aquatic environments through various sources, such as wastewater, run-off, and industrial discharges. In water bodies, their movement is influenced by factors such as size, shape, and density. Over time, these particles can sink and become embedded in sediments. Sediments, through processes like flooding and irrigation, serve as conduits, transferring MPs into terrestrial soils. Once in soils, MPs can alter soil structure, hinder nutrient cycles, and affect microbial communities. They may also act as carriers for several pollutants, exacerbating environmental contamination. Understanding these transport mechanisms is crucial for developing predictive models and mitigation strategies to address the growing problem of MP pollution in interconnected ecosystems. Improved wastewater treatment, policies to reduce plastic use, and ecosystem restoration techniques are vital for managing MP contamination in water and soil environments.

Keywords

Microplastics · Aquatic systems · Fate · Co-pollutants · Soil

---

S. Shaji
Research Scholar, Department of Civil Engineering, National Institute of Technology, Tadepalligudem, Andhra Pradesh, India

B. Sundaram (✉)
Department of Civil Engineering, National Institute of Technology, Tadepalligudem, Andhra Pradesh, India
e-mail: baranis@nitandhra.ac.in

K. Bala et al. (eds.), *Occurrence, Detection, and Fate of Microplastics in Freshwater Ecosystems*, https://doi.org/10.1007/978-981-95-0225-7_5

5.1 Introduction

The proliferation of plastic waste has become a global environmental crisis, with plastic pollution now recognised as one of the most pervasive threats to ecosystems. Annually, 30 million metric tonnes of plastic waste are produced (Plastics Europe 2024), of which nine million metric tonnes end up in natural habitats, especially aquatic environments. According to the Organisation for Economic Co-operation and Development (OECD 2023), without substantial improvements, plastic consumption would rise, resulting in a 50% escalation in macro-plastic leakage by 2040. Over time, large plastic debris breaks down into smaller particles, known as microplastics (MPs), which are defined as plastic particles less than 5 millimetres in size (Burrows et al. 2024; Mghili et al. 2024; Zhang et al. 2024e). MPs have now been detected in almost all major water bodies, from oceans to rivers and lakes (Afzal et al. 2024; Jain et al. 2024; Rajan et al. 2023; Sun et al. 2024b), and even in remote locations (Neelavannan et al. 2022). This reflects the extent of MPs' spread across various water bodies. MPs are considered unique and complex pollutants owing to their persistent nature and ability to carry other pollutants, thereby introducing significant risks to both terrestrial and aquatic ecosystems. MPs can be classified broadly into primary and secondary MPs. Primary MPs include manufactured small particles that are often used in most of the personal care products, including microbeads in exfoliants, and industrial abrasives (Bian et al. 2022). Meanwhile, secondary MPs are being developed as a result of the disintegration of larger plastic items that fragment into smaller pieces due to environmental weathering. This includes UV radiation and mechanical abrasion (Baek et al. 2024; Rajan et al. 2023; Sarin and Klomjek 2022). The major portion of MPs present in the aquatic environments is often classified as the secondary type. These MPs are formed due to the degradation of plastic waste, fishing nets, and other products.

MPs that contribute to the intricate contamination observed in aquatic ecosystems have diverse and numerous sources. One of the major pathways through which the MPs enter the freshwater ecosystem has been wastewater treatment plants (WWTPs) (Sukmono et al. 2024). The effluents from these WWTPs often contain MPs generated by various activities or processes, including household activities, industrial activities, and storm water run-off. It is worth noting that urban run-off is found to be another major source that contributes towards the accumulation of MPs. The run-off, especially from roads, where the wearing of tyres occurs, contributes a significant amount of MPs. Moreover, agricultural run-off also plays a significant role in introducing MPs into water bodies. This includes the addition of MPs through the use of various fertilisers that contain plastic particles and plastic mulch films. Additionally, the reused treated sewage sludge added to increase the soil fertility also adds to the MP contamination (Rahman et al. 2024; Rajan et al. 2023; Sukmono et al. 2024). Various industrial processes and activities, including the manufacturing of plastics and improper waste disposal, further increase the contamination.

MPs in water bodies often sink and get accumulated in sediments. Therefore, they are not confined to the water column alone. MPs are subjected to different transport mechanisms when they enter the aquatic systems, and these mechanisms determine the movement of MPs within water bodies. Several factors, including shape, environmental conditions, size, and density of particles, influence these mechanisms. It may be noted that these MPs can remain suspended in the water column by floating on the surface or get settled at the bottom along with the sediments. This phenomenon is based on the density of the particles. For instance, less dense particles like polyethylene (PE) and polypropylene (PP) tend to float or accumulate near the surface, owing to their low density. Meanwhile, the denser particles such as polyvinyl chloride (PVC) tend to sink comparatively faster and get accumulated in sediments (Mercy et al. 2023; Shao et al. 2024; Sukmono et al. 2024; Zhang et al. 2024a). The behaviour of MPs is also influenced by the bio-fouling process in which the microorganisms colonise plastic surfaces, thereby increasing the density of particles. This, in turn, leads to sinking of these particles. Some of the other factors that play a crucial role in the distribution of MPs include tides, water currents, and waves. MPs are carried by currents in rivers and coastal regions. This, in turn, allows them to get dispersed widely across aquatic systems. For example, river currents transport MPs over a long distance and eventually deliver these particles into deltas, estuaries, oceans, etc. These ocean currents and gyres are considered major drivers of MP transport in marine environments. This results in large-scale MP accumulation zones like the Great Pacific Garbage Patch (Reineccius and Waniek 2024; Sambandam et al. 2022). Such oceanic "garbage patches" are high-density regions in which the MPs get trapped by circulating currents, thereby resulting in long-term accumulation. Table 5.1 represents the occurrence of MPs at different locations across the world in the aquatic ecosystem.

It may be worth noting that sedimentation is another process that majorly contributes towards the transport of MPs in aquatic ecosystems (Godoy et al. 2020; Pervez and Wang 2022). It is mainly due to the reason that these MPs get embedded in sediments present on the lake bottoms, river beds, ocean floors, etc. when they sink. This sediment-bound storage can be temporary or long-lasting, depending on environmental disturbances. For example, storm events, dredging, and changes in water flow can re-suspend sediments, re-releasing MPs into the water column. Once in sediments, MPs can interact with benthic organisms, leading to ingestion by bottom-dwelling species and subsequent transfer through food webs. Sediments act as a reservoir, holding MPs that can later be transported to terrestrial systems, especially during events such as flooding and through agricultural irrigation with contaminated water. Floodwaters carrying sediment-bound MPs can deposit them on floodplains, leading to their incorporation into soils. Irrigation using water contaminated with MPs is another mechanism through which these particles reach agricultural soils, impacting soil structure and potentially influencing crop health. MPs in soils and sediments have far-reaching implications for environmental health, biodiversity, and potentially, human well-being. This disrupts the soil structure and also hinders nutrient cycling. In this way, they pose risks to organisms and transport the toxic pollutants across the ecosystems. Table 5.2 illustrates the occurrence of MPs in the soil/sediments at various locations in the world.

The interconnected nature of these ecosystems is evident from the presence of MPs in both aquatic as well as terrestrial environments. Once the MPs enter the soil,

Table 5.1 MPs' occurrence at various locations across the world in aquatic systems

Sl. No.	Location/ country	Sampling site	MP concentration	Reference
1	Amin Bazar landfill, Dhaka	Landfill leachate	270 ± 13 MPs/L	Rahman et al. (2024)
		Surface water	2090 ± 100MPs/L	
		Groundwater	647 ± 22 MPs/L	
2	Mahakam River, Indonesia	Mahakam River	9.2 ± 1.8–58.5 ± 3.5 particles/l	Sukmono et al. (2024)
3	Da Nang, Vietnam	Seawater	111–304 MPs/L	Phan et al. (2024)
		Marine sediment	2267–4600 MPs/kg	
4	Andhra Pradesh, India	Bottled drinking water	2.89 ± 0.48 items/L	Phan et al. (2024)
5	Chennai, India	Cooum River	0–26 particles/L	Afzal et al. (2024)
6	Western Cape, South Africa	River water	5.13 ± 6.62 MPs/L	Apetogbor et al. (2023)
		River sediment	1587.50 ± 599.32 MPs/kg	
7	Dhaka, Bangladesh	Lake water	36 items/L	Mercy et al. (2023)
		Lake sediment	67 items/kg	
8	India	Ganga River	92.85 ± 50.69 particles/m^3	Rajan et al. (2023)
9	Italy	Po River	1.89–8.22 particles/m^3	Fiore et al. (2022)
10	China	Chin Ling-Wei River	2.30–21.05 items/L	Bian et al. (2022)
11	Thailand	Bueng Boraphet wetland	1.44 ± 3.4 particles/m^3	Sarin and Klomjek (2022)
12	Thailand	Surface water from the Inner Gulf	Wet season: 34.59 ± 46.02 pieces/L	Vibhatabandhu and Srithongouthai (2022)
			Dry season: 8.70 ± 15.34 pieces/L	
13	China	Surface water of Alpine rivers	0.48 ± 0.28 items/L	Liu et al. (2023a, b)
14	China	Shenzhen Bay watershed	Wet season: 6.53 ± 0.96 items/L Dry season: 10.87 ± 0.57 items/L	Shao et al. (2024))
15	Western Indian Ocean		0.23 ± 0.04 particles/m^3	Baek et al. (2024))
16	India	Muttukadu backwater estuary		Ramakrishnan and Sathiyamoorthy (2024)
		Water	Wet season: 2.75 items/m^3 Dry season: 1.95 items/m^3	
		Sediment	Wet season: 3.15 items/kg Dry season: 2.55 items/kg	
17	Hong Kong	Marine water	2.07 ± 4.00 particles/m^3	Zhang et al. (2024a, b, c, d, e)

(continued)

Table 5.1 (continued)

Sl. No.	Location/ country	Sampling site	MP concentration	Reference
18	China	South Yellow Sea	1.36–51.68 items/L	Sun et al. (2024b)
19	Denmark	Marine water	103 ± 86 items/ m^3	Liu et al. (2023a, b)
20	Japan	River water	0–273.3 particles/m^3	Nihei et al. (2024))
21	China	Dongguan city, urban rivers	Wet season: 1646.22 ± 154.73 items/m^3	Liu et al. (2024a, b)
			Dry season: 351.09 ± 34.20 items/m^3	
22	South Africa	Orange-Senqu River basin		Graham et al. (2024))
		River water	1.00 ± 0.71–69.75 ± 68.55 items/L	
		River sediment	0–4000 particles/kg	
23	Jamshedpur, India	River water	9.55 ± 4.87 items/L	Patidar et al. (2024))
		River sediment	270.88 ± 105.89 items/kg	
24	Egypt	Nile River	2.82 ± 0.95 particles/L	Khedre et al. (2024))
25	India	Nainital Lake	8.6–56.0 particles/L	Jain et al. (2024))

they can interfere with soil aeration, nutrient cycling, etc. They also affect the water retention capacity of the soil and the microbial communities that are essential for maintaining the health of the soil. Moreover, MPs act as vectors for various other environmental contaminants, including heavy metals and persistent organic pollutants (POPs), which amplify their potential to harm both ecosystems and human health (Sun et al. 2024b; Wang et al. 2023a, b; Xu et al. 2023b). Considering the global scale of plastic pollution and its pervasive nature, it is very much critical to understand the fate and transport mechanisms of MPs, especially between aquatic and terrestrial systems. This would be beneficial in developing effective mitigation strategies.

This study explores the fate and transport of MPs from the water bodies to the soil. It may be noted that for developing predictive models and mitigation strategies to address the growing problem of MP pollution in interconnected ecosystems, it is crucial to understand their transport mechanisms. In view of the current state of research, it is absolutely necessary to implement more stringent laws, more advanced treatment technologies, and improved waste management practices in order to lessen the amount of MPs that are released into the environment.

5.2 MPs' Transport Pathways in the Aquatic System

Various release pathways originating from different sources are mostly responsible for the entry of MPs into the aquatic environment (Afzal et al. 2024; Sukmono et al. 2024). These sources include industrial practices or processes, WWTP discharge, consumer products, and in situ breakdown of larger plastic debris. Once these

Table 5.2 Occurrence of MPs in soil/sediments at various locations across the world

Sl. No.	Location/country	Sampling site	MP concentration	Reference
1	Mediterranean coast, Spain	Beach	Summer: 339 ± 92 MPs/kg Winter: 71 ± 92 MPs/kg	Soursou et al. (2024)
2	Southwest England	Beach	160–450 MPs/kg	Burrows et al. (2024)
3	Shilaoren beach, China	Beach	Spring: 452.2 ± 176.5 n/kg Autumn: 748.5 ± 395.5 n/kg in	Luan and Wang (2024)
4	Bangladesh	Beach	242.86 particles/kg dw	Tasnim et al. (2023)
5	Singapore	Beach	13.1 particles/kg	Curren and Yew Leong (2023)
6	Vietnam	Beach sediment samples near refineries	1582 ± 660 MPs/kg	Le et al. (2023)
7	Gulf of Khambhat, India	Muddy shore site (21 sites)	0.32 ± 0.18 particles/g–2.81 ± 0.50 particles/g	Rabari et al. (2023)
8	Eastern coast, India	Beach	4 ± 2 MPs/kg	Mandal et al. (2023)
9	Baltic Sea coast, Germany	Beach	2 particles/500 g	Lenz et al. (2023)
10	China	Coal mining area		Zhang et al. (2024b)
		Soil	593.9 MPs/kg–4949.0 MPs/kg	
		River sediment	125.8 MPs/kg–821.2 MPs/kg	
		River water	4123.0 MPs/m^3–22,030.5 MPs/m^3	
11	China	Yellow River delta		Ding et al. (2024)
		Soil	100–1400 items/kg	
		Sediments	200–4200 items/kg	
		River water	0.50–7.83 items/L	
12	Ximen Island, China	Mangrove	673.3 ± 102.6 items/kg	Li et al. (2022)
		Saltmarsh	693.3 ± 140.5 items/kg	
		Bare beach	406.7 ± 110.2 items/kg	
13	Zhenjiang City, China	River sediment	Summer: 2049.09 ± 883.78 items/kg	Wang et al. (2024a)
			Winter: 2216.36 ± 826.21 items/kg	

(continued)

Table 5.2 (continued)

Sl. No.	Location/country	Sampling site	MP concentration	Reference
14	Andhra Pradesh, India	Beach sediment	564.5 particles/kg	Shaji et al. (2024)
15	Black Sea coast, Bulgaria	Beach sediment	464.44 ± 272.50 MPs/kg	Bobchev et al. (2024)
		Seabed sediment	825.93 ± 396.30 MPs/kg	
16	Hatia Island, Bangladesh	Surface water	139 ± 44 items/m^3	Fardullah et al. (2024)
		Sediments	493 ± 80 items/kg	
17	Aqua Lake, Italy	Surface water	0.13 ± 0.04 MPs/L	Pierdomenico et al. (2024)
		Sediments	0–540 MPs/kg	
18	United Kingdom	Estuary sediments	170.80 particles/ kg	Trusler et al. (2024)
19	Madura Island, Indonesia	Seawater	0.17 and 0.59 items/L	Efendy et al. (2024)
		Sediments	35.7 and 123.3 items/kg	
20	Brazil (Amazonian mangroves)	Tidal channel sediments	433 ± 261.6 items/ kg	Mendes et al. (2024)
21	North-west Himalayas, India	Freshwater lake sediments	606 ± 360 MPs/kg	Neelavannan et al. (2022)
22	Tibetan plateau, China	Surface water	2.26 ± 2.38 items/L	Liang et al. (2024)
		Sediments	0–1.24 items/ g	

particles enter the environment, they are subjected to intricate processes that govern their transformation, movement, and ultimate fate. The physical, biological, and chemical factors act together to dictate the transport and fate of MPs. It may be noted that these factors also significantly affect the interaction and distribution of MPs within organisms, sediments, and water bodies (Kumar et al. 2024; Phan et al. 2024; Shao et al. 2024). Owing to their minute size and buoyancy, MPs have been influenced by unique transport mechanisms in aquatic ecosystems. This would enable them to traverse longer distances and eventually penetrate diverse strata in various water bodies.

5.2.1 Physical Mechanism

MPs are redistributed and transported through various physical mechanisms in aquatic ecosystems. This redistribution and transportation are based on the characteristics of MPs, such as their density, form, particle size, and local water conditions. The main physical mechanisms through which MPs are transported in aquatic environments encompass sediment movement and its deposition, hydrodynamic forces, and air deposition. These factors collectively facilitate the extensive distribution of MPs from the water surface to deep-sea sediments. Moreover, these factors also dictate the transmission of MPs from freshwater sources to ocean systems, which would eventually result in crucial implications for the quality of water and marine organisms.

Hydrodynamic Forces

The transportation of MPs by means of hydrodynamic forces has been majorly governed by three driving forces, which include turbulence, waves, and currents within the water bodies. The particles that are less dense or buoyant MPs, including PE and PP, tend to remain at the surface of water, and their transportation is significantly governed by surface currents (Sukmono et al. 2024) across large distances. The MPs from river mouths are often distributed into the ocean by these currents and would eventually develop concentrated areas known as oceanic gyres (e.g. the Great Pacific Garbage Patch), and it may be noted that these areas would have large amounts of plastic debris accumulation (Atlantic et al. 2024). Meanwhile, the MPs having higher density would either sink or remain suspended, depending on the mixing and movement of water.

Moreover, wind-driven waves also have a significant role to play, especially in shallow coastal areas. In such coastal areas, these waves push the MPs floating on the surface onto the shore, which eventually cause accumulation along the coastlines. Nevertheless, these waves continually expose the MPs to mixing forces and thereby affect their residing time and movement on the water surface. Additionally, the turbulence prevailing either near the water surface or in estuaries mix the MPs vertically throughout the water column. This results in keeping the particles suspended for a longer period of time and aids their transportation between different depths across the water bodies (Sarin and Klomjek 2022; Trusler et al. 2024). The significant difference in the density of particles and their dimensions also influences their transit. For example, the lighter particles remain at the surface, and the denser particles either sink or move with sediments along the seafloor. The dynamic distribution of MPs is facilitated by the interplay of these hydrodynamic forces. All these factors collectively affect their spread across rivers, lakes, and oceans. Moreover, these factors influence their eventual accumulation in certain areas, including the seafloor and shorelines.

Atmospheric Deposition, Run-Off, and Effluents from Industries and WWTPs

Atmospheric deposition and transmission are often considered critical factors that contribute towards MP pollution in aquatic ecosystems. These MPs are transported over a long distance through the atmosphere and subsequently get either deposited into terrestrial surfaces or directly deposited into aquatic ecosystems by means of precipitation. Dust storms, wind erosion from urban locales, and other atmospheric phenomena disseminate MP particles, especially in proximity to populated or industrial zones. The atmospheric deposition of MPs can be categorised into dry deposition and wet deposition, based on varying climatic circumstances. Sun et al. (2022) reported that MPs' atmospheric deposition fluxes ranged from 0.91×10^3 to 3.5×10^3 MP/m^2/day. It has been further reported that the deposition flux of MPs exhibited moderate-to-strong correlations with atmospheric particulate matter concentrations, particularly PM2.5 levels, suggesting that the routinely monitored PM2.5 concentration may serve as an indicator for estimating MPs' deposition flux. The analysis of MPs gathered under varying meteorological conditions indicated that precipitation events may enhance MPs' deposition fluxes when air quality is consistent (Sun et al. 2022). Studies conducted in the South China Sea by

Ding et al. (2021) reported that the dry deposition of atmospheric MPs into the ocean amounted to 1.4×10^3 t, surpassing the contribution from riverine inputs of MPs in this region, indicating that atmospheric MPs represent a significant source of marine plastic litter with considerable environmental risk. As reported by various studies, wet weather conditions facilitated MPs' deposition, while high wind speed and air humidity increased MPs' deposition during dry periods (Szewc et al. 2021).

Rainwater and air deposition contribute MPs to rivers, lakes, and even isolated regions of the ocean, augmenting the intricate network of transport channels. Agricultural run-off is also a significant pathway to MP pollution in aquatic environments, primarily from plastic mulching films, sludge-based fertilisers, synthetic pesticides, and farming equipment (Bigalke et al. 2022). Mulching films fragment over time, while sludge-based fertilisers introduce MPs directly from wastewater treatment byproducts. Microencapsulated pesticides, plastic containers, and farm machinery also degrade, releasing MPs that run-off can carry into rivers and lakes (Bigalke et al. 2022). Additionally, surface run-off from metropolitan regions following precipitation can transport MPs from roadways, soils, and various land surfaces into adjacent water bodies, creating a direct pathway for MPs' infiltration into aquatic ecosystems.

The release of MPs from industrial effluents and WWTPs impacts aquatic ecosystems in multiple ways. WWTPs, untreated wastewater, sewer overflow systems, storm water, and urban run-off are some of the most common entry points for MPs into aquatic systems (Kumar et al. 2024). Treated and untreated effluents enter rivers, lakes, and oceans directly, increasing pollution levels and allowing MPs to spread throughout the environment. Rivers and streams transport MPs across watersheds and eventually discharge them into larger bodies of water, where they become widely distributed. In coastal areas, WWTPs and industrial discharges often create "hotspots" of MP pollution, affecting local marine organisms and habitats (Bobchev et al. 2024; Kumar et al. 2024). Once in the environment, MPs are subject to hydrodynamic forces, which transport them across long distances. Further, Bobchev et al. (2024) reported that the prolonged discharge of untreated sewage into the Black Sea coast has led to a pronounced gradient of escalating MP concentrations in sediments, which has endured for six years after the commissioning of the local WWTP.

5.2.2 Biological and Chemical Influences

The transport of MPs in aquatic systems is also governed by both biological and chemical factors, which can significantly influence their fate, distribution, and persistence. These factors, in conjunction with the physical factors such as hydrodynamic forces, influence how MPs move through water bodies and ultimately interact with various environmental compartments, including sediments and biota (Efendy et al. 2024; Graham et al. 2024; Leiser et al. 2020). The transport of MPs influenced by the biological and chemical factors would eventually alter their properties, enhance or retard their movement, and affect their interactions with various aquatic organisms. It is worth noting that the aquatic creatures, ranging from plankton to fish consume MPs, and these MPs are subsequently transferred via trophic levels in the food web. Moreover, these ingested MPs might be expelled at a considerable distance away from their original location of ingestion. This results in modifying the dispersion of these particles. Meanwhile, the other crucial factor that influences the transport mechanism of MPs is bio-fouling. It results in increasing the concentration of MPs that remain floating on the surface of water. Additionally, in the aquatic systems, this bio-fouling also impacts the interactions with various other particles present there. The physical characteristics of MPs get modified due to this bio-fouling process. Moreover, due to different aggregation and buoyancy properties of contaminated particles in comparison with those of untainted MPs, the interactions with the hydrodynamic as well as the sediment transport forces are also affected.

5.2.2.1 Bio-fouling and Aggregation

Microorganisms, including fungi, bacteria, and algae, colonise on the surface of MPs, and this process is referred to as bio-fouling (Leiser et al. 2020). It is worth noting that the bio-fouling often starts immediately after the MPs enter the aquatic systems. Once the MPs enter the aquatic systems, the microbial communities start to eventually create biofilms on the plastic surfaces (Leiser et al. 2020; Nguyen et al. 2023a). Based on the various studies conducted, it has been found that the biofilm developed has the ability to alter the physical characteristics of MPs. It may be noted that on adding the biological layers on the particle by the bio-fouling process, the weight of MPs increases. Therefore, the buoyancy of the particle gets decreased subsequently. Moreover, it is found that the density of MPs also increases owing to the increase in weight due to the bio-fouling process (Leiser et al. 2020; Priya et al. 2022), especially in the case of MPs having lower density initially, such as polyethylene and polypropylene. This potentially causes them to settle and sink into benthic zones and eventually interact with benthic organisms and sediment as well. Research studies, such as those by Leiser et al. (2020) and Nguyen et al. (2023a), illustrated that bio-fouling not only influences MPs' transportation but also affects their degradation. Nevertheless, the microorganisms present within the biofilms might sometimes degrade certain components of the plastic polymers, but the complete biodegradation process would remain slow and rare. A stable habitat on the MPs' surface is created by this microbial colonisation. This results in facilitating the interaction of MPs with the aquatic food webs. Studies have shown that MPs covered with biofilm attract planktonic organisms and also the higher trophic levels, which results in ingestion by a wide range of marine species present in the ecosystem (Nguyen et al. 2023a; Shi et al. 2023; Wang et al. 2021). Therefore, it is evident that this bio- fouling-based transport enhances the MPs' entry into the aquatic food chains and eventually results in potential biomagnification of the pollutants that are associated with MPs (Fred-Ahmadu et al. 2022b; Liu et al. 2022; Rafa et al. 2024).

Moreover, it has been reported that bio-fouling may also promote the aggregation of MPs with inorganic and organic particles. This results in clusters that descend more easily. It may be noted that the aggregation of MPs frequently transpires as a result of hydrophobic processes and electrostatic interactions and also due to bridging by organic or inorganic colloids. It has been reported by Long et al. (2017) that in saline circumstances, including those prevalent in marine or estuary settings, the aggregation behaviour of MPs is intensified. This behaviour is significantly due to reduced electrostatic repulsion created between the particles. Furthermore, the microscopic plastic particles having lower density (particles that would never sink) ascend to the surface sediment owing to their inclusion into organic aggregates (Leiser et al. 2020). Moreover, the MPs aggregated are deposited in the seabed, and as a result of this, sedimentary build-up of plastic particles gets facilitated. Additionally, these aggregated MPs either persist there for a longer period of time or are re-suspended by currents or waves (hydrodynamic forces). It may be noted that the studies conducted by Long et al. (2017) have depicted that the addition of 2-μm polystyrene (PS) spheres (1050 kg m^{-3}) into diatom aggregates would result in a significant change in the density of these aggregates and would slow down or accelerate sinking in two species.

Owing to the combined influence of aggregation and bio-fouling frequently, a dynamic mechanism is introduced in the transportation of MPs. This dynamic interaction is more prominent in environments, especially where organic matter is present in such higher levels. Under such scenarios, the contaminated MPs readily adhere to the mineral deposits. Moreover, these contaminated particles also adhere to both organic particles and phytoplankton and eventually lead to the development of "hetero-aggregates," which get carried away along the water column as bigger entities (Long et al. 2017). The combined synergistic effect of bio-fouling and aggregation influences the spatial distribution of MPs and also the potential environmental consequences introduced by them. Mghili et al. (2024) illustrated that the cumulative mass of aggregates and biofilms significantly promote prolonged retention of MPs in the sediment layers. This accumulation of MPs in benthic habitats eventually pose a danger to sediment-dwelling animals that may either consume the MPs directly or may encounter modified habitat conditions.

5.2.2.2 Chemical Degradation and Sorption

In aquatic environments, MPs are exposed to chemical degradation processes driven by sunlight (photodegradation), temperature fluctuations, and reactions with waterborne chemicals. Over time, these processes cause plastics to become brittle and fragment, further creating even smaller particles called nanoplastics (size <1 μm) (De Souza MacHado et al. 2018), wherein the smaller MPs may be more mobile in the water column due to their reduced size and increased buoyancy, and they may be more readily ingested by aquatic organisms. Degraded MPs exhibit altered physical and chemical properties, such as increased surface roughness and modified functional groups on their surfaces. This alteration increases their reactivity and their likelihood of adsorbing pollutants (Wang et al. 2023b). It was also reported that PVC and PS exhibited greater physicochemical changes under photodegradation than PE and PP, which could be attributed to the dissolution of chlorine and benzene ring of PVC and PS, respectively. MPs can also adsorb contaminants such as heavy metals, pesticides, and persistent organic pollutants (POPs) from the surrounding water (Bao et al. 2024; Su et al. 2023; Xu et al. 2023a). By acting as carriers for these toxic substances, MPs introduce pollutants into food webs, posing a potential risk to aquatic organisms and, ultimately, to human consumers (Ateia et al. 2020).

5.3 Transfer and Fate of MPs in the Soil Ecosystem

MPs do not remain confined to water bodies; various processes can transfer these particles from aquatic systems to terrestrial environments, particularly soils. Reportedly, rivers act as carriers of MPs to other environmental zones. Moreover, MPs that settle in aquatic sediments can be transported to soil environments through a variety of pathways, especially in coastal and riverine systems, where water bodies and land are closely connected. MPs have intricate interactions with sediments that affect their persistence and mobility (Fig. 5.1). PVC and other dense MPs have a tendency to sink in sediments and may eventually be assimilated into sedimentary layers. Furthermore, bio-fouling increases the density of buoyant particles, causing them to sink. Over time, sediment-settling MPs can build up and form permanent reservoirs in lake bottoms, river beds, and ocean floors. Strong currents, storms, or human operations like dredging, however, frequently re-suspend these settled particles, bringing them back into the water column and prolonging their environmental residence period.

Fig. 5.1 MPs' transfer from the aquatic to the soil environment and their potential impacts

5.3.1 Transfer of MPs from Sediments to Soil Environments

In many coastal and urban areas, sediments are dredged to maintain waterways and prevent flooding. The dredged material, frequently containing MPs, is occasionally utilised as a soil amendment for agricultural fields or land restoration initiatives. MPs from these sediments are consequently immediately delivered into terrestrial habitats, where they accumulate in soils. Dredged river sediments collected in south-east China (Ji et al. 2021) reportedly contained very high MP concentrations (ranging from 6000 to 37,600 particles/kg). The MPs from river sediments were conveyed to dredged sediment storage piles, where they eventually underwent degradation and were disseminated into adjacent soils and surface waterways while awaiting a permanent disposal solution (Ji et al. 2021). Flooding can convey sediment-laden MPs to neighbouring floodplains, agricultural lands, and urban regions (Liang et al. 2024). In a flood occurrence, water can transport suspended MPs from river beds and lake beds to adjacent regions, where they are deposited into the soil when floodwaters diminish. This technique is especially pertinent in regions experiencing seasonal flooding because MPs are consistently deposited onto land. Tasnim et al. (2023) reported that sewage irrigation and sludge usage contribute to MPs in cultivated soil. During intense precipitation or snow-melt, MPs from aquatic sediments can be mobilised and conveyed to adjacent soil ecosystems through run-off. Erosion along riverbanks or lake coasts can also result in the deposition of sediment particles containing MPs onto land (Fardullah et al. 2024; Mendes et al. 2024). When water from rivers, lakes, or wastewater effluents containing MPs is utilised for irrigation, MPs are directly injected into the soil. These particles collect over time, particularly in agricultural soils with frequent watering, resulting in a steady accumulation of MPs.

5.3.2 MPs' Deposition, Retention, and Impacts on the Soil Environment

MPs get embedded in the matrix of the soil. Once embedded in the soil, they interact with the soil particles, plants, and other organisms as well. The permanence and behaviour of MPs in the soil have been influenced by both their physical and chemical qualities, along with different soil factors, such as moisture content, texture, and organic matter. It may be noted that owing to the water penetration, these MPs ascend vertically across the soil strata. This ascending of MPs could also be due to the actions of soil organisms, including earthworms, which would accelerate the transport of these particles within the soil profile. The soil properties, such as its pH, soil bulk density, and water-holding capacity, are affected due to the ingestion and transport of MPs in the soil (Wang et al. 2019). Moreover, the presence of MPs also affect nutrient cycling (Rong et al. 2021), microbial activity (Zhang et al. 2022), and enzymatic reactions (Wang et al. 2024a, b). Table 5.3 represents the various research studies conducted on the fate of MPs in the soil ecosystem. Additionally, the results

Table 5.3 Fate of MPs in the soil system

Sl. no.	Polymer type	Soil properties/ organism studied	Potential impacts	Reference
1	RY, PE, PES	Physicochemical properties of soil	PE reduced the soil silt content, while it increased the nitrate content PE increased MPs, degrading bacterial abundance	Zhou et al. (2024a)
2	PAN, PE	Physicochemical properties of soil	0.1% PAN increased total porosity and reduced bulk density PE and PAN decreased cation exchange capacity More than 0.1% dosage of PE and PAN increased TN and TOC and increased microbial activity but decreased soil pH	Oliveira de Miranda et al. (2024)
3	PU	Maize (*Zea mays* L.)	PU impacted maize seedling root development and reduced lignin levels	Zhang et al. (2024c)
4	LDPE, HDPE, PVC	Sandy soil index and engineering properties	Increased concentration of MPs reduced the shear strength of sandy soil but increased the cohesion It also reduced the coefficient of permeability	Dike and Apte (2023)
5	PP, PE, PS, PVC	Physicochemical properties of soil	MPs do not significantly impact soil pH at concentrations $\leq$7% MPs' concentration $\geq$ 7% altered soil water constants and capacity Cr bioavailability in soil increased due to MPs No effect of MPs on Cr vertical migration in soil	Li et al. (2024)
6	LDPE, HDPE, PVC	Sandy, silty, and clayey soil index and engineering properties	Sandy soil is mostly affected due to MPs Shear strength is reduced for all soil types	Dike and Apte (2024)
7	PP, PLA	Soil geochemical properties and microbial activity	MPs increased pH and altered carbon and nitrogen nutrients MPs affected cellulose breakdown and ureolysis The effect of MPs on soil bacteria depended on polymer type and dosage	Zhang et al. (2024d)
8	PE	Soil nitrogen cycling	Unlike PE, PE resin reduced DON and increased microbial biomass	Zhou et al. (2024b)
9	PE (aged)	Earthworm (*Eisenia fetida*)	Combined exposure to azoxystrobin and aged PE reduced body weight and caused oxidative stress and intestine damage	Bao et al. (2024)

(continued)

Table 5.3 (continued)

Sl. no.	Polymer type	Soil properties/ organism studied	Potential impacts	Reference
10	PE, PLA	Earthworm	Short-term exposure negatively impacted and increased the mortality rate Prolonged exposure rendered earthworms adapted to the MP environment	Lai et al. (2024)
11	PS, PE, PVC	Cucumber (*Cucumis sativus L.*)	Combined exposure to MPs and DOP leads to intercellular CO_2 concentration, increased transpiration rate, and decreased WUE	Zhuang et al. (2024)
12	PLA	Rice seedling	Induced oxidative stress and reduced photosynthetic activity	Irshad et al. (2024)
13	LDPE	Green gram (*Vigna radiata L. Wilczek*)	Retarded growth and yield and influenced nutrient uptake	Sravya et al. (2024)
14	HDPE	Soil springtail	MPs damaged the gut barrier and mitochondria and increased pathogenic fungi	Kwak et al. (2024)

RY rayon, *PE* polyethylene, *PES* polyester, *PAN* polyacrylonitrile, *PU* polyurethane, *LDPE* low-density polyethylene, *HDPE* high-density polyethylene, *PVC* polyvinyl chloride, *PP* polypropylene, *PS* polystyrene, *PLA* polylactic acid, *TN* total nitrogen, *TOC* total organic carbon, *Cr* chromium, *DON* dissolved organic nitrogen, *DOP* di-n-octyl phthalate, *WUE* water use efficiency

of the research are reported based on the investigation of the effects of MPs on soil properties, soil organisms, and plants.

5.3.2.1 Soil Structure and Porosity

The physical structure of the soil is disrupted by the presence of MPs. Also, the soil porosity, along with the aggregate stability, gets altered. Research studies conducted have illustrated that fibrous or filamentous types of MPs significantly increase the porosity of soil by creating spaces within the matrix of the soil. Even though this improves the aeration in certain cases, it could also lead to reduced soil cohesion. Moreover, the aggregate stability could also get lowered, which would make the soil more prone to compaction and erosion. In contrast, smaller and irregularly shaped MPs may fill soil pores, reducing porosity and negatively impacting root growth and water infiltration. De Souza Machado et al. (2019) reported that the addition of PET (polyethylene terephthalate), PP, PS, and polyester (PES) MPs significantly decreased the soil bulk density. It also reportedly impacted the soil structure and water-holding capacity. Additionally, Wen et al. (2022) found a linear relation between the low-density polyethylene (LDPE) concentration and the change in field capacity.

Soil's Water-Holding Capacity

MPs can affect the soil's water retention and hydraulic properties. For instance, MPs with high hydrophobicity (e.g. PE and PP) repel water, potentially reducing the capacity of the soil to retain moisture (De Souza MacHado et al. 2018). This effect is particularly pronounced in sandy soils having relatively poor water-holding capacity. Conversely, specific MPs can enhance soil water retention by capturing water in micro-pockets surrounding the particles (Li et al. 2024). However, this impact is generally more significant at elevated MP concentrations. As reported by various recent studies, the higher concentration of MPs can obstruct water infiltration and drainage, hence influencing moisture availability for plants and soil organisms (Li et al. 2019;. Junhao et al. 2022). These alterations result in waterlogging, accelerated surface run-off, and erosion.

Soil Nutrient Cycling

MPs influence soil nutrient cycling by altering microbial communities and enzyme activities. They undergo degradation in the soil that results in the dispersal of various chemical additives, such as plasticisers, flame retardants, and antioxidants, which can either promote or inhibit microbial growth (Wang et al. 2022a, b, 2024a, b). Research indicates that MPs can alter microbial diversity and abundance, creating imbalances that favour certain bacterial species over others. This microbial imbalance can impact organic matter decomposition rates and nutrient cycling, particularly of nitrogen, phosphorus, and carbon (Wang et al. 2022a, b, 2024a, b). Additionally, MPs adsorb nutrients and heavy metals from the surrounding soil, functioning as carriers for these substances. While this adsorption can temporarily remove nutrients from the soil solution, reducing their bioavailability, it may also result in the gradual release of nutrients or pollutants under favourable conditions (Fajardo et al. 2022; Fred-Ahmadu et al. 2022a). This potentially leads to localised nutrient or contaminant hotspots (Nguyen et al. 2023b).

Soil-Dwelling Organisms

The presence of MPs in the soil environment can disrupt life processes, impacting the health and behaviour of soil-dwelling organisms, including earthworms, arthropods, microbes, and other invertebrates (Table 5.3). These organisms may ingest the MP particles, erroneously identifying them as food, which can result in numerous adverse effects (Bao et al. 2024; Lai et al. 2024; Rolf et al. 2022). MP ingestion can cause blockages, injury, or inflammation in the digestive systems of soil organisms; reduce feeding efficiency that can hinder growth; and eventually lead to the death of the species(Fajardo et al. 2022; Liu et al. 2024a, b; Wang et al. 2022a, b). Lai et al. (2024) investigated the effects of polyethylene and polylactic acid (PLA) MPs on earthworms subjected to an exposure period of 14–28 days. The study observed increased worm mortality rates, which could possibly have occurred due to intestinal skin damage and an eventual lack of nutrition. However, prolonged exposure rendered the earthworm adapted to the exposure conditions, which resulted in fewer negative effects. It can be inferred that the toxicity depends on exposure duration as well. The decline in soil fauna populations caused by pollution stressors can destabilise the complex relationships within soil food webs (Chen et al. 2020; Roscher et al. 2021). Since soil organisms serve as food sources for higher trophic levels, their decline can cause cascading effects throughout the ecosystem. Additionally,

MPs are also reported to impact the soil microbial community, wherein the effects can be either positive, negative, or sometimes insignificant. Studies reported that soils polluted with MPs showed higher abundance of a few fungi genera, including *Aspergillus*, *Fusarium*, and *Penicillium*, which utilised the carbon sources obtained from MPs (Zhang et al. 2021).

5.4 Conclusion

In conclusion, the transfer of MPs from aquatic systems to the soil represents a significant and complex environmental issue with far-reaching consequences for terrestrial ecosystems. The main ways that MPs infiltrate soil are by flooding, contaminated water irrigation, and wastewater treatment sludge application, often carrying additional pollutants that compound their impact. Once in the soil, MPs interact with various soil components and organisms, leading to alterations in soil structure and physicochemical properties, and interfere with water retention and nutrient cycling while also impacting the microbial communities essential for soil health. These effects can cascade through the ecosystem, eventually affecting plant health and, ultimately, food safety. The enduring presence of MPs in the soil, facilitated by their resistance to breakdown, poses significant long-term difficulties for soil health, especially as MPs build up in regions of elevated human activity. Furthermore, their function in the transport and concentration of contaminants, including heavy metals and POPs, exacerbates their ecological influence, endangering not only soil organisms but also higher trophic levels through bioaccumulation and biomagnification. Given the global scale of plastic pollution and its pervasive nature, understanding the fate and transport mechanisms of MPs between aquatic and terrestrial systems is crucial for developing effective mitigation strategies. The current state of research highlights the urgency for improved waste management practices, stricter regulations, and advanced treatment technologies to reduce MPs' inputs into the environment.

References

Afzal JM, Selvam S, Saravanan P, Roy PD, Sanju P, Muthukumar P (2024) Microplastics in water from the Cooum River, Chennai, India: An assessment of their distribution, composition, and environmental impact. Groundw Sustain Dev 27:101362. https://doi.org/10.1016/j.gsd.2024.101362

Apetogbor K, Pereao O, Sparks C, Opeolu B (2023) Spatio-temporal distribution of microplastics in water and sediment samples of the Plankenburg river, Western cape. South Africa Environ Pollut 323:121303. https://doi.org/10.1016/J.ENVPOL.2023.121303

Ateia M, Zheng T, Calace S, Tharayil N, Pilla S, Karanfil T (2020) Sorption behavior of real microplastics (MPs): insights for organic micropollutants adsorption on a large set of well-characterized MPs. Sci Total Environ 720:137634. https://doi.org/10.1016/j.scitotenv.2020.137634

Atlantic N, Rodrigues C, Rodríguez Y, Frias J, Carriço R, Sobral P, Duncan EM, Pham CK (2024) Microplastics in beach sediments of the Azores. Mar Pollut Bull 201:1–8

Baek ER, Kim M, Kang DJ, Kang JH (2024) Distribution characteristics of microplastics in the surface mixed layer of the western Indian Ocean. Deep Sea Res Part II Top Stud Oceanogr 218:105424. https://doi.org/10.1016/J.DSR2.2024.105424

Bao X, Wang Z, Liu L, Wang D, Gu Y, Chen L, Chen X, Meng Z (2024) The combined effects of azoxystrobin and different aged polyethylene microplastics on earthworms (Eisenia fetida): a systematic evaluation based on oxidative damage and intestinal function. Sci Total Environ 923:171494. https://doi.org/10.1016/J.SCITOTENV.2024.171494

Bian P, Liu Y, Zhao K, Hu Y, Zhang J, Kang L, Shen W (2022) Spatial variability of microplastic pollution on surface of rivers in a mountain-plain transitional area: a case study in the Chin Ling-Wei River plain. China Ecotoxicol Environ Saf 232:113298. https://doi.org/10.1016/J.ECOENV.2022.113298

Bigalke M, Fieber M, Foetisch A, Reynes J, Tollan P (2022) And aquatic microplastic pollution. Sci Total Environ 806:150709

Bobchev N, Berov D, Klayn S, Karamfilov V (2024) High microplastic pollution in marine sediments associated with urbanised areas along the SW Bulgarian Black Sea coast. Mar Pollut Bull 209:117150. https://doi.org/10.1016/J.MARPOLBUL.2024.117150

Burrows S, Anderson RJ, Turner A (2024) Microplastic transport and deposition in a beach-dune system (Saunton Sands-Braunton Burrows, Southwest England), vol 909, p 168535

Chen Y, Liu X, Leng Y, Wang J (2020) Defense responses in earthworms (Eisenia fetida) exposed to low-density polyethylene microplastics in soils. Ecotoxicol Environ Saf 187:109788. https://doi.org/10.1016/J.ECOENV.2019.109788

Curren E, Yew Leong SC (2023) Spatiotemporal characterisation of microplastics in the coastal regions of Singapore. Heliyon 9:e12961. https://doi.org/10.1016/j.heliyon.2023.e12961

De Souza MacHado AA, Lau CW, Till J, Kloas W, Lehmann A, Becker R, Rillig MC (2018) Impacts of microplastics on the soil biophysical environment. Environ Sci Technol 52:9656–9665. https://doi.org/10.1021/ACS.EST.8B02212/ASSET/IMAGES/LARGE/ES-2018-02212S_0007.JPEG

De Souza Machado AA, Lau CW, Kloas W, Bergmann J, Bachelier JB, Faltin E, Becker R, Görlich AS, Rillig MC (2019) Microplastics can change soil properties and affect plant performance. Environ Sci Technol 53:6044–6052. https://doi.org/10.1021/acs.est.9b01339

Dike S, Apte S (2023) Impact of microplastic pollution in terrestrial ecosystem on index and engineering properties of sandy soil: An experimental investigation. Sci Total Environ 887:164049. https://doi.org/10.1016/j.scitotenv.2023.164049

Dike S, Apte SD (2024) Impact of microplastics on strength parameters of clayey, Sandy, silty soil: a comparative assessment. Sci Total Environ 946:174161. https://doi.org/10.1016/j.scitotenv.2024.174161

Ding Y, Zou X, Wang C, Feng Z, Wang Y, Fan Q, Chen H (2021) The abundance and characteristics of atmospheric microplastic deposition in the northwestern South China Sea in the fall. Atmos Environ 253:118389. https://doi.org/10.1016/J.ATMOSENV.2021.118389

Ding R, Li Q, Wang K, Tian J, Lu L, Li W, Xu L (2024) Occurrence and distribution of microplastics in the adjacent environment of Yellow River Delta. China Mar Pollut Bull 199:116019. https://doi.org/10.1016/J.MARPOLBUL.2023.116019

Efendy M, Amir N, Mahasin MZ, Wahyu FM, Aroua MK, Gozan M (2024) Distribution and characteristics of microplastics in seawater, sediment, and crude salt from Madura Island, the largest solar salt production Island in Indonesia. Reg Stud Mar Sci 78:103803. https://doi.org/10.1016/j.rsma.2024.103803

Fajardo C, Martín C, Costa G, Sánchez-Fortún S, Rodríguez C, de Lucas Burneo JJ, Nande M, Mengs G, Martín M (2022) Assessing the role of polyethylene microplastics as a vector for organic pollutants in soil: Ecotoxicological and molecular approaches. Chemosphere 288:132460. https://doi.org/10.1016/J.CHEMOSPHERE.2021.132460

Fardullah M, Hossain MT, Islam MS, Islam MR, Rahman MR, Akther K, Uddin A, Morshed S, Sultana N, Alam MA, Bahadur NM, Robel FN (2024) Occurrence and spatial distribution of microplastics in water and sediments of Hatiya Island, Bangladesh and their risk assessment. J Environ Manag 370:122697. https://doi.org/10.1016/j.jenvman.2024.122697

Fiore L, Serranti S, Mazziotti C, Riccardi E, Benzi M, Bonifazi G (2022) Classification and distribution of freshwater microplastics along the Italian Po river by hyperspectral imaging. Environ Sci Pollut Res 29:48588–48606. https://doi.org/10.1007/S11356-022-18501-X/TABLES/1

Fred-Ahmadu OH, Ayejuyo OO, Tenebe IT, Benson NU (2022a) Occurrence and distribution of micro(meso)plastic-sorbed heavy metals and metalloids in sediments, gulf of Guinea coast (SE Atlantic). Sci Total Environ 813:152650. https://doi.org/10.1016/j.scitotenv.2021.152650

Fred-Ahmadu OH, Tenebe IT, Ayejuyo OO, Benson NU (2022b) Microplastics and associated organic pollutants in beach sediments from the Gulf of Guinea (SE Atlantic) coastal ecosystems. Chemosphere 298:134193. https://doi.org/10.1016/j.chemosphere.2022.134193

Godoy V, Prata JC, Blázquez G, Almendros AI, Duarte AC, Rocha-Santos T, Calero M, Martín-Lara MÁ (2020) Effects of distance to the sea and geomorphological characteristics on the quantity and distribution of microplastics in beach sediments of Granada (Spain). Sci Total Environ 746:142023. https://doi.org/10.1016/j.scitotenv.2020.142023

Graham PM, Pattinson NB, Bakir A, McGoran AR, Nel HA (2024) Determination of microplastics in sediment, water, and fish across the Orange-Senqu river basin. Water Res 266:122394. https://doi.org/10.1016/J.WATRES.2024.122394

Irshad MK, Kang MW, Aqeel M, Javed W, Noman A, Khalid N, Lee SS (2024) Unveiling the detrimental effects of polylactic acid microplastics on rice seedlings and soil health. Chemosphere 355:141771. https://doi.org/10.1016/j.chemosphere.2024.141771

Jain Y, Govindasamy H, Kaur G, Ajith N, Ramasamy K, Robin RS, Ramachandran P (2024) Microplastic pollution in high-altitude Nainital lake, Uttarakhand. India Environ Pollut 346:123598. https://doi.org/10.1016/J.ENVPOL.2024.123598

Ji X, Ma Y, Zeng G, Xu X, Mei K, Wang Z, Chen Z, Dahlgren R, Zhang M, Shang X (2021) Transport and fate of microplastics from riverine sediment dredge piles: implications for disposal. J Hazard Mater 404:124132. https://doi.org/10.1016/j.jhazmat.2020.124132

Junhao C, Pengpeng C, Xiaodong G, Qifang Z, Yunjie F, Xiaobo G, Xining Z, Yuannong L (2022) Effects of plastic film residue and emitter flow rate on soil water infiltration and redistribution under different initial moisture content and dry bulk density. Sci Total Environ 807:151381. https://doi.org/10.1016/J.SCITOTENV.2021.151381

Khedre AM, Ramadan SA, Ashry A, Alaraby M (2024) Abundance and risk assessment of microplastics in water, sediment, and aquatic insects of the Nile River. Chemosphere 353:141557. https://doi.org/10.1016/J.CHEMOSPHERE.2024.141557

Kumar V, Umesh M, Chakraborty P, Sharma P, Sarojini S, Basheer T, Kaur K, Pasrija R, Barcelo D (2024) Origin, ecotoxicity, and analytical methods for microplastic detection in aquatic systems. TrAC - Trends Anal Chem 170:117392. https://doi.org/10.1016/j.trac.2023.117392

Kwak JI, Kim L, An YJ (2024) Microplastics promote the accumulation of negative fungal groups and cause multigenerational effects in springtails. J Hazard Mater 466:133574. https://doi.org/10.1016/J.JHAZMAT.2024.133574

Lai H, Han S, Sun J, Fang Y, Liu P, Zhao H (2024) The comparison effect on earthworms between conventional and biodegradable microplastics. Heliyon 10:e37308. https://doi.org/10.1016/j.heliyon.2024.e37308

Le XTT, Nguyen DT, Pham MT, Trinh MV, Le PC, Do VM (2023) Risk assessment of microplastic exposure: a case study near a refinery factory at the central coast of Vietnam. Mar Pollut Bull 196:115636. https://doi.org/10.1016/j.marpolbul.2023.115636

Leiser R, Wu GM, Neu TR, Wendt-Potthoff K (2020) Biofouling, metal sorption and aggregation are related to sinking of microplastics in a stratified reservoir. Water Res 176:115748. https://doi.org/10.1016/j.watres.2020.115748

Lenz M, Brennecke D, Haeckel M, Knickmeier K, Kossel E (2023) Spatio-temporal variability in the abundance and composition of beach litter and microplastics along the Baltic Sea coast of Schleswig-Holstein. Germany Mar Pollut Bull 190:114830. https://doi.org/10.1016/j.marpolbul.2023.114830

Li, J., Song, Y., Cai, Y., 2019. Focus topics on microplastics in soil: analytical methods, occurrence, transport, and ecological risks. Environmental Pollution, 257, 113570. https://doi.org/10.1016/J.ENVPOL.2019.113570

Li Y, Huang R, Hu L, Zhang C, Xu X, Song L, Wang Z, Pan X, Christakos G, Wu J (2022) Microplastics distribution in different habitats of Ximen Island and the trapping effect of blue carbon habitats on microplastics. Mar Pollut Bull 181:113912. https://doi.org/10.1016/J.MARPOLBUL.2022.113912

Li Jl, Li N-y, Ni J, Wan R-l, Yang J-y (2024) Effects of microplastics on soil physicochemical properties and vertical migration and transformation of chromium. J Environ Chem Eng 12:114577. https://doi.org/10.1016/j.jece.2024.114577

Liang T, Ho YW, Wang Q, Wang P, Sun S, Fang JKH, Liu X (2024) Distribution and risk assessment of microplastics in water, sediment and brine shrimps in a remote salt lake on the Tibetan plateau. China J Hazard. Mater. 476:134959. https://doi.org/10.1016/J.JHAZMAT.2024.134959

Liu Q, Wu H, Chen J, Guo B, Zhao X, Lin H, Li W, Zhao X, Lv S, Huang C (2022) Adsorption mechanism of trace heavy metals on microplastics and simulating their effect on microalgae in river. Environ Res 214:113777. https://doi.org/10.1016/j.envres.2022.113777

Liu Y, Lorenz C, Vianello A, Syberg K, Nielsen AH, Nielsen TG, Vollertsen J (2023a) Exploration of occurrence and sources of microplastics (>10 μm) in Danish marine waters. Sci Total Environ 865:161255. https://doi.org/10.1016/J.SCITOTENV.2022.161255

Liu Q, Xiong X, Wang K, Wang H, Ling Y, Li Q, Xu F, Wu C (2023b) Homogenization of microplastics in alpine rivers: analysis of microplastic abundance and characteristics in rivers of Qilian Mountain. China J Environ Manage 340:118011. https://doi.org/10.1016/J.JENVMAN.2023.118011

Liu W, Li S, Zhou Y, Cai Y, Liu C, Yang Z (2024a) Characteristics, drivers and ecological risk assessment of microplastics in the surface water of urban rivers in Guangdong-Hong Kong-Macao Greater Bay Area cities – a case study of Dongguan city. Environ Pollut 362:125024. https://doi.org/10.1016/J.ENVPOL.2024.125024

Liu B, Zhao S, Qiu T, Cui Q, Yang Y, Li L, Chen J, Huang M, Zhan A, Fang L (2024b) Interaction of microplastics with heavy metals in soil: mechanisms, influencing factors and biological effects. Sci Total Environ 918:170281. https://doi.org/10.1016/j.scitotenv.2024.170281

Long M, Paul-Pont I, Hégaret H, Moriceau B, Lambert C, Huvet A, Soudant P (2017) Interactions between polystyrene microplastics and marine phytoplankton lead to species-specific hetero-aggregation. Environ Pollut 228:454–463. https://doi.org/10.1016/J.ENVPOL.2017.05.047

Luan J, Wang YH (2024) Abundance, morphology, spatial-temporal distribution and sources of microplastics in Shilaoren Beach, Qingdao. China Reg Stud Mar Sci 71:103422. https://doi.org/10.1016/j.rsma.2024.103422

Mandal A, Singh N, Mondal A, Talib M, Basu R, Biswas MK, Darbha GK (2023) The extent of microplastic pollution along the eastern coast of India: Focussing on marine waters, beach sand, and fish. Mar Pollut Bull 194:115265. https://doi.org/10.1016/j.marpolbul.2023.115265

Mendes DS, Silva DNN, Santiago LEP, Gomes VJC, Beasley CR, Fernandes MEB (2024) Comprehensive risk assessment of microplastics in tidal channel sediments in amazonian mangroves (northern Brazil). J Environ Manag 366:121826. https://doi.org/10.1016/J.JENVMAN.2024.121826

Mercy FT, Alam AKMR, Akbor MA (2023) Abundance and characteristics of microplastics in major urban lakes of Dhaka, Bangladesh. Heliyon 9:2405–8440. https://doi.org/10.1016/J.HELIYON.2023.E14587/ASSET/07CE7F16-E0CD-4DBC-8EA6-A5BA16FB2FD1/MAIN.ASSETS/GR4_LRG.JPG

Mghili B, Hasni S, Ben-haddad M, Rangel-buitrago N, Ahannach Y, Hassou N, Cherradi S, Aksissou M (2024) Plastic pollution on Moroccan beaches : Toward baselines for large-scale assessment. Mar Pollut Bull 201:116288

Neelavannan K, Sen IS, Lone AM, Gopinath K (2022) Microplastics in the high-altitude Himalayas: assessment of microplastic contamination in freshwater lake sediments, northwest Himalaya (India). Chemosphere 290:133354. https://doi.org/10.1016/J.CHEMOSPHERE.2021.133354

Nguyen HT, Lee YK, Kwon JH, Hur J (2023a) Microplastic biofilms in water treatment systems: fate and risks of pathogenic bacteria, antibiotic-resistant bacteria, and antibiotic resistance genes. Sci Total Environ 892:164523. https://doi.org/10.1016/j.scitotenv.2023.164523

Nguyen M-K, Rakib MRJ, Lin C, Hung NTQ, Le V-G, Nguyen H-L, Malafaia G, Idris AM (2023b) A comprehensive review on ecological effects of microplastic pollution: An interaction with pollutants in the ecosystems and future perspectives. TrAC Trends Anal Chem 168:117294. https://doi.org/10.1016/j.trac.2023.117294

Nihei Y, Ota H, Tanaka M, Kataoka T, Kashiwada J (2024) Comparison of concentration, shape, and polymer composition between microplastics and mesoplastics in Japanese river waters. Water Res 249:120979. https://doi.org/10.1016/j.watres.2023.120979

OECD 2023 Towards eliminating plastic pollution by 2040 a policy scenario analysis

Oliveira de Miranda C, Leal L, de Souza JJ, Gonçalves Reynaud Schaefer CE, Huerta Lwanga E, Nadal Junqueira Villela F (2024) Short-term impacts of polyethylene and polyacrylonitrile microplastics on soil physicochemical properties and microbial activity of a marine terrace environment in maritime Antarctica. Environ Pollut 347:123791. https://doi.org/10.1016/j.envpol.2024.123791

Patidar K, Ambade B, Younis AM, Alluhayb AH (2024) Characteristics, fate, and sources of microplastics contaminant in surface water and sediments of river water. Phys Chem Earth, Parts A/B/C 134:103596. https://doi.org/10.1016/J.PCE.2024.103596

Pervez R, Wang YH (2022) Microplastic distribution within core sediments of beach and its responses to anthropogenic activities. Mar Pollut Bull 174:113256. https://doi.org/10.1016/j.marpolbul.2021.113256

Phan NT, Thanh Thao LX, Do VM, Nguyen DD (2024) Assessment of microplastic presence in coastal environments and organisms of Da Nang. Vietnam Mar Pollut Bull 204:116516. https://doi.org/10.1016/j.marpolbul.2024.116516

Pierdomenico M, Morgana S, Chiocci FL (2024) First report of microplastics in water and sediments of the alkaline Bagno dell'Acqua Lake (Pantelleria Island, southern Italy). Environ Pollut 362:124962. https://doi.org/10.1016/j.envpol.2024.124962

Plastics Europe (2024) The circular economy for plastics—a European analysis 2024 • plastics Europe [WWW document]. URL. https://plasticseurope.org/knowledge-hub/the-circular-economy-for-plastics-a-european-analysis-2024/. Accessed 7 Nov 2024

Priya KL, Renjith KR, Joseph CJ, Indu MS, Srinivas R, Haddout S (2022) Fate, transport and degradation pathway of microplastics in aquatic environment — a critical review. Reg Stud Mar Sci 56:102647. https://doi.org/10.1016/j.rsma.2022.102647

Rabari V, Patel H, Patel K, Patel A, Bagtharia S, Trivedi J (2023) Quantitative assessment of microplastic contamination in muddy shores of gulf of Khambhat. India Mar Pollut Bull 192:115131. https://doi.org/10.1016/j.marpolbul.2023.115131

Rafa N, Ahmed B, Zohora F, Bakya J, Ahmed S, Ahmed SF, Mofijur M, Chowdhury AA, Almomani F (2024) Microplastics as carriers of toxic pollutants: source, transport, and toxicological effects. Environ Pollut 343:123190. https://doi.org/10.1016/j.envpol.2023.123190

Rahman MA, Haque MM, Tareq SM (2024) Abundance and characteristics of microplastics in the landfill leachate of Amin bazar, Dhaka: a potential risk to aquatic environments. Phys Chem Earth 134:103573. https://doi.org/10.1016/j.pce.2024.103573

Rajan K, Khudsar FA, Kumar R (2023) Urbanization and population resources affect microplastic concentration in surface water of the river ganga. J Hazard Mater Adv 11:100342. https://doi.org/10.1016/J.HAZADV.2023.100342

Ramakrishnan D, Sathiyamoorthy M (2024) Seasonal distribution, source apportionment and risk exposure of microplastic contaminants along the Muttukadu backwater estuary, Tamil Nadu. India Results Eng 23:102776. https://doi.org/10.1016/J.RINENG.2024.102776

Reineccius J, Waniek JJ (2024) Critical reassessment of microplastic abundances in the marine environment. Sci Total Environ 954:176449. https://doi.org/10.1016/j.scitotenv.2024.176449

Rolf M, Laermanns H, Kienzler L, Pohl C, Möller JN, Laforsch C, Löder MGJ, Bogner C (2022) Flooding frequency and floodplain topography determine abundance of microplastics in an alluvial Rhine soil. Sci Total Environ 836:155141. https://doi.org/10.1016/J.SCITOTENV.2022.155141

Rong L, Zhao L, Zhao L, Cheng Z, Yao Y, Yuan C, Wang L, Sun H (2021) LDPE microplastics affect soil microbial communities and nitrogen cycling. Sci Total Environ 773:145640. https://doi.org/10.1016/j.scitotenv.2021.145640

Roscher L, Fehres A, Reisel L, Halbach M, Scholz-Böttcher B, Gerriets M, Badewien TH, Shiravani G, Wurpts A, Primpke S, Gerdts G (2021) Microplastic pollution in the Weser estuary and the German North Sea. Environ Pollut 288:117681. https://doi.org/10.1016/j.envpol.2021.117681

Sambandam M, Dhineka K, Sivadas SK, Kaviarasan T, Begum M, Hoehn D, Sivyer D, Mishra P, Murthy MVR (2022) Occurrence, characterization, and source delineation of microplastics in the coastal waters and shelf sediments of the central east coast of India. Bay of Bengal Chemosphere 303:135135. https://doi.org/10.1016/j.chemosphere.2022.135135

Sarin C, Klomjek P (2022) Spatial and seasonal distribution of microplastic in surface water of Bueng Boraphet wetland—a Ramsar wetland in Thailand. Environ Monit Assess 194:1–13. https://doi.org/10.1007/S10661-022-10578-8/FIGURES/5

Shaji S, Vannarath A, Rao YRS, Sundaram B (2024) Distribution, characteristics and ecological risk of microplastics in beach sediments along the northern coast of Andhra Pradesh. India Reg Stud Mar Sci 77:103716. https://doi.org/10.1016/j.rsma.2024.103716

Shao H, Wang Q, Wang L, Lei X, Dai S, Li T, Zhu X, Mao X, Mao X-z (2024) Source identification of microplastics in highly urbanized river environments and its implications for watershed management. Sci Total Environ 950:175308. https://doi.org/10.1016/J.SCITOTENV.2024.175308

Shi J, Lv B, Wang B, Xie B (2023) Insight into the responses of antibiotic resistance genes in microplastic biofilms to zinc oxide nanoparticles and zinc ions pressures in landfill leachate. J Hazard Mater 459:132096. https://doi.org/10.1016/j.jhazmat.2023.132096

Soursou V, Campo J, Picó Y (2024) Spatio-temporal variation and ecological risk assessment of microplastics along the touristic beaches of a mediterranean coast transect (Valencia province, East Spain). J Environ Manag 354:120315. https://doi.org/10.1016/j.jenvman.2024.120315

Sravya GC, Gautam S, Kumar KU, Poonguzhali RS, Manuel RI (2024) Assessing LDPE microplastics' impact on green gram (Vigna radiata L. Wilczek) cultivation: a greenhouse pot experiment. Phys Chem Earth 136:103710. https://doi.org/10.1016/j.pce.2024.103710

Su X, Liu M, Yuan J, Huang X, Lu Z, Xu J, He Y (2023) Potential risk of co-occurrence of microplastics and chlorinated persistent organic pollutants to coastal wetlands: evidence from a case study. Environ Pollut 320:121087. https://doi.org/10.1016/J.ENVPOL.2023.121087

Sukmono Y, Hadibarata T, Kristanti RA, Singh A, Al Farraj DA, Elshikh MS (2024) Occurrence and visual characterization of microplastics from Mahakam River at Tenggarong City. Indonesia J Contam Hydrol 267:104440. https://doi.org/10.1016/j.jconhyd.2024.104440

Sun J, Peng Z, Zhu ZR, Fu W, Dai X, Ni BJ (2022) The atmospheric microplastics deposition contributes to microplastic pollution in urban waters. Water Res 225:119116. https://doi.org/10.1016/J.WATRES.2022.119116

Sun X, Yang R, Ji J, Zhu Z, White JC, Shen Y (2024b) An evaluation of microplastic contamination in the marine waters and species in the coastal region of the South Yellow Sea. China J Hazard Mater 469:134018. https://doi.org/10.1016/J.JHAZMAT.2024.134018

Szewc K, Graca B, Dołęga A (2021) Atmospheric deposition of microplastics in the coastal zone: characteristics and relationship with meteorological factors. Sci Total Environ 761:143272. https://doi.org/10.1016/J.SCITOTENV.2020.143272

Tasnim J, Ahmed MK, Hossain KB, Islam MS (2023) Spatiotemporal distribution of microplastic debris in the surface beach sediment of the southeastern coast of Bangladesh. Heliyon 9:e21864. https://doi.org/10.1016/j.heliyon.2023.e21864

Trusler MM, Moss-Hayes VL, Cook S, Lomax BH, Vane CH (2024) Microplastics pollution in sediments of the Thames and Medway estuaries, UK: organic matter associations and predominance of polyethylene. Mar Pollut Bull 208:116971. https://doi.org/10.1016/j.marpolbul.2024.116971

Vibhatabandhu P, Srithongouthai S (2022) Influence of seasonal variations on the distribution characteristics of microplastics in the surface water of the inner gulf of Thailand. Mar Pollut Bull 180:113747. https://doi.org/10.1016/J.MARPOLBUL.2022.113747

Wang J, Liu X, Li Y, Powell T, Wang X, Wang G, Zhang P (2019) Microplastics as contaminants in the soil environment: a mini-review. Sci Total Environ 691:848–857. https://doi.org/10.1016/j.scitotenv.2019.07.209

Wang J, Guo X, Xue J (2021) Biofilm-developed microplastics as vectors of pollutants in aquatic environments. Environ Sci Technol. https://doi.org/10.1021/acs.est.1c04466

Wang Y, Wang X, Li Y, Liu Y, Sun Y, Hansen HCB, Xia S, Zhao J (2022a) Effects of struvite-loaded zeolite amendment on the fate of copper, tetracycline and antibiotic resistance genes in microplastic-contaminated soil. Chem Eng J 430:130478. https://doi.org/10.1016/j.cej.2021.130478

Wang X, Xing Y, Lv M, Zhang T, Ya H, Jiang B (2022b) Recent advances on the effects of microplastics on elements cycling in the environment. Sci Total Environ 849:157884. https://doi.org/10.1016/j.scitotenv.2022.157884

Wang Y, Munir U, Huang Q (2023a) Soil & Environmental Health Occurrence of per- and poly fl uoroalkyl substances (PFAS) in soil : sources, fate, and remediation 1, vol 1, p 100004

Wang H, Zhu J, He Y, Wang J, Zeng N, Zhan X (2023b) Photoaging process and mechanism of four commonly commercial microplastics. J Hazard Mater 451:131151. https://doi.org/10.1016/j.jhazmat.2023.131151

Wang C, Liu X, Ma Q, Xing S, Yuan L, Ma Y (2024a) Distribution and effects of microplastics as carriers of heavy metals in river surface sediments. J Contam Hydrol 266:104396. https://doi.org/10.1016/J.JCONHYD.2024.104396

Wang W, Zhang Z, Gao J, Wu H (2024b) The impacts of microplastics on the cycling of carbon and nitrogen in terrestrial soil ecosystems: Progress and prospects. Sci Total Environ 915:169977. https://doi.org/10.1016/j.scitotenv.2024.169977

Wen X, Yin L, Zhou Z, Kang Z, Sun Q, Zhang Y, Long Y, Nie X, Wu Z, Jiang C (2022) Microplastics can affect soil properties and chemical speciation of metals in yellow-brown soil. Ecotoxicol Environ Saf 243:113958. https://doi.org/10.1016/j.ecoenv.2022.113958

Xu Z, Bai X, Li Y, Weng Y, Li F (2023a) New insights into the decrease in Cd2+ bioavailability in sediments by microplastics: role of geochemical properties. J Hazard Mater 442:1–10. https://doi.org/10.1016/j.jhazmat.2022.130103

Xu B, Yang G, Lehmann A, Riedel S, Rillig MC (2023b) Effects of perfluoroalkyl and polyfluoroalkyl substances (PFAS) on soil structure and function. Soil Ecol Lett 5:108–117. https://doi.org/10.1007/s42832-022-0143-5

Zhang X, Li Y, Ouyang D, Lei J, Tan Q, Xie L, Li Z, Liu T, Xiao Y, Farooq TH, Wu X, Chen L, Yan W (2021) Systematical review of interactions between microplastics and microorganisms in the soil environment. J Hazard Mater 418:126288. https://doi.org/10.1016/j.jhazmat.2021.126288

Zhang W, Liu X, Liu L, Lu H, Wang L, Tang J (2022) Effects of microplastics on greenhouse gas emissions and microbial communities in sediment of freshwater systems. J Hazard Mater 435:129030. https://doi.org/10.1016/j.jhazmat.2022.129030

Zhang K, Cheng MCY, Liu M, Xu S, Ma Y, Chau HS, Chen L, Cao Y, Yan M, Xu X, Thoe W, Sun SWC, Yang RR, Leung KMY, Lam PKS (2024a) Microplastics in Hong Kong's marine waters: impact of rainfall and Pearl River discharge. Mar Pollut Bull 205:116635. https://doi.org/10.1016/J.MARPOLBUL.2024.116635

Zhang W, Hu J, Yin C, Ling L, Gu L, Zheng L (2024b) Characterization of microplastic contamination and influencing factors in different environmental compartments of a coal mining subsidence area in northern Anhui. China Sci Total Environ 955:176881. https://doi.org/10.1016/j.scitotenv.2024.176881

Zhang Y, Song M, Zhu Y, Li H, Zhang Y, Wang G, Chen X, Zhang W, Wang H, Wang Y, Shao R, Guo J, Yang Q (2024c) Impact of microplastic particle size on physiological and biochemical properties and rhizosphere metabolism of Zea mays L.: comparison in different soil types. Sci Total Environ 908:168219. https://doi.org/10.1016/J.SCITOTENV.2023.168219

Zhang Z, Wang W, Liu J, Wu H (2024d) Discrepant responses of bacterial community and enzyme activities to conventional and biodegradable microplastics in paddy soil. Sci Total Environ 909:168513. https://doi.org/10.1016/J.SCITOTENV.2023.168513

Zhang T, Yang W, Pang S, Cao X, Chen Y, Seif M, Ren P, Liu Y, Feng Z (2024e) Accumulation of microplastics in the marine sediments of the Chukchi Sea. Arctic Ocean Reg Stud Mar Sci 70:103363. https://doi.org/10.1016/j.rsma.2023.103363
Zhou Y, Dou M, Zhang Y, Ning K, Li Y (2024a) Distribution characteristics of soil microplastics and their impact on soil physicochemical properties in agricultural areas of the North China plain. Environ Sci Process Impacts 471:1556. https://doi.org/10.1039/d4em00242c
Zhou Z, Hua J, Xue J, Yu C (2024b) Differential impacts of polyethylene microplastic and additives on soil nitrogen cycling: a deeper dive into microbial interactions and transformation mechanisms. Sci Total Environ 942:173771. https://doi.org/10.1016/J.SCITOTENV.2024.173771
Zhuang H, Li Z, Wang M, Liu B, Chu Y, Lin Z (2024) Effects of microplastics and combined pollution of polystyrene and di-n-octyl phthalate on photosynthesis of cucumber (Cucumis sativus L.). Sci Total Environ 947:174426. https://doi.org/10.1016/j.scitotenv.2024.174426

Techniques for Sampling, Detection, and Characterization of Microplastics in Water, Sediment, and Tissue Samples

6

Pratyusha Nayak, Nishigandha Muduli, Smruti Prajna Pradhan, Surya Narayan Dash, and Lipika Patnaik

Abstract

Water systems are closely coupled with the terrestrial environment; they provide a key pathway for the infiltration of pollutants. Microplastics (MPs) are widely acknowledged as a dangerous threat to all forms of life. Overuse, improper release, and mismanagement of microplastics lead to their ingestion, accumulation, and transfer through different trophic levels and gradually pose a health risk to all the organisms at different levels. Various methods have been standardized for microplastic sampling, extraction, separation, visualization, identification, and characterization. The most widely accepted method of microplastics processing is the digestion of the samples using hydrogen peroxide as an oxidizing agent, characterization by Fourier Transformed Infrared Spectroscopy (FTIR), scanning electron microscope (SEM), fluorescence microscopy, Raman spectrometry, and Pyr-GC-MS. Most of the methods and techniques have shown promising results. Microplastics persist and accumulate in the environment for a long period of time as they undergo a very slow degradation process. The degradation process may depend on multiple variables existing in the surrounding environment and the composition of different types of polymers. This chapter provides a comprehensive discussion about sample collection using specific equipment, chemical extraction of MPs, the Nile red staining method, and light and dark contrast microscopy to isolate, quantify, and characterize the MPs from water, soil, and tissue samples. It also highlights the challenges that have been faced throughout the MP's research spectrum and offers brief recommendations

P. Nayak · N. Muduli · S. P. Pradhan · L. Patnaik (✉)
Environmental Science Laboratory, Department of Zoology, Ravenshaw University, Cuttack, Odisha, India

S. N. Dash
Institute of Biotechnology, Viikinkaari, University of Helsinki, Helsinki, Finland

K. Bala et al. (eds.), *Occurrence, Detection, and Fate of Microplastics in Freshwater Ecosystems*, https://doi.org/10.1007/978-981-95-0225-7_6

and suitable countermeasures to improve our understanding of MP extraction and isolation.

Keywords

Microplastic extraction · Nets · FTIR

6.1 Section Overview

With development in science and technology and population explosion, plastics have become an integral and indispensable part of our lives. The plastic industry is growing every year due to its availability, low-cost production, long-term durability, and application in various sectors. The plastic production rate is around 330 million tons per year (Jiang et al. 2019). Microplastics are everywhere in our daily life, beginning from drinking water to tea bags, food, and personal care products, and the intake of plastic particles in the human body is about 39,000–52,000 particles/person/year (Cox et al. 2019). The microplastics study is very crucial to know the impact of microplastics on human health and the environment. They can affect human health in various ways and show impact on immune system response, onset of allergies, lung disorders, an increase in oxidative stress, microbial infection, and resistance to microbes (Chen et al. 2023). Marine organisms can ingest microplastics, which may affect their growth, population, diversity, and ecology. Plastic wastes from various sources are transferred to garbage dump yard sites or released directly into water (Stoett et al. 2024). These plastics break down into smaller particles as a result of physical, chemical, and biological processes taking place in the environment. The concern regarding microplastic contamination is due to their chemically stable nature, large surface area, long persistence time period, binding capacity with other substances, movement through different ecosystems along with other contaminants, and accumulation capacity by penetrating biological barriers (Wright and Kelly 2017). Microplastics are small plastics added to the environment from different sources (from the manufacturing stage to environmental degradation). Microplastics from the direct manufacturing stage are called primary microplastics, whereas those that result from the environmental degradation process are called secondary microplastics. Primary includes microbeads from the cosmetic industry, synthetic fibers from the cloth and net producing industry, and pellets from the plastic production industry. Secondary microplastics are degraded from plastic materials by UV radiation, wave action, and saltwater degradation. Humans are exposed to microplastics through inhalation, the oral route, and through skin (Li et al. 2023). Food containers widely used for different purposes are composed of polymers like polypropylene (PP), polystyrene (PS), polyethylene (PE), and polyethylene terephthalate (PET) (He et al. 2021; Du et al. 2020). These small microplastics leach out into the food and drink that is consumed and can pose a risk inside the body. Microplastics are toxic in their own form and can act as carriers for other toxic compounds in the environment, making the process of pollutant accumulation very complex. The distribution of microplastics in water is regulated and influenced by various anthropogenic factors. Hence, it is essential to monitor microplastic pollution on a long-term basis. Microplastics have been detected in both invertebrates and vertebrates, indicating their bioavailability. The most common forms of available microplastics in water are polyethylene, polystyrene, and polypropylene.

This chapter attempts to provide a systematic compilation of the methods standardized by the authors for microplastic extraction from water, sediment, and biota samples, along with an overview of other extraction methods being followed worldwide from different sources.

6.2 Sampling Methods

Microplastics are present far and wide in the water column, on the water surface, in sediments, on beaches, and in organisms. Hence, the selection of sampling method depends upon the choice of sources to be sampled, the targeted size of microplastics, and the equipment available. As the distribution of microplastics is greatly influenced by climatic, temporal, and physical factors, it may affect the efficacy of the results.

6.2.1 Water

For the surface water, manta nets and neuston nets are mainly used. They can filter a large amount of water in less time and give a concentrated sample for analysis. Zooplankton nets, bottom trawl nets used for benthic organisms, submersible pumps, and continuous plankton recorder (CPR) are generally used for water column sampling. Besides this, the Niskin water sampler can also be used (Zobkov and Esiukova 2018).

Trawls/Nets:

Microplastics can be collected from the surface water using different types of nets such as (Campanale et al. 2020).

(a) Manta Net/Trawl:

The name is derived from the Manta ray, a filter feeder having paired extensions on either side of the mouth. It consists of a stainless steel or aluminum frame whose width can be from 30 to 120 cm and height from 10 to 75 cm with a rectangular opening (Wang et al. 2020; Isobe et al. 2015). The mouth part is followed by a net, having a length of 3 m and a mesh size of 300–350 μm with a detachable collection bag (Fig. 6.1a). The hauling speed and time of hauling are important factors for the collection of microplastics. The preferred trawling speed is 2–3 knots, and the duration is <60mins. Higher speed and duration above 90 minutes should be avoided to prevent clogging and organic matter contamination (Tamminga et al. 2018; Pasquier et al. 2022). Manta net should be deployed 3–4 m away from the wake zone to prevent collection from turbu-

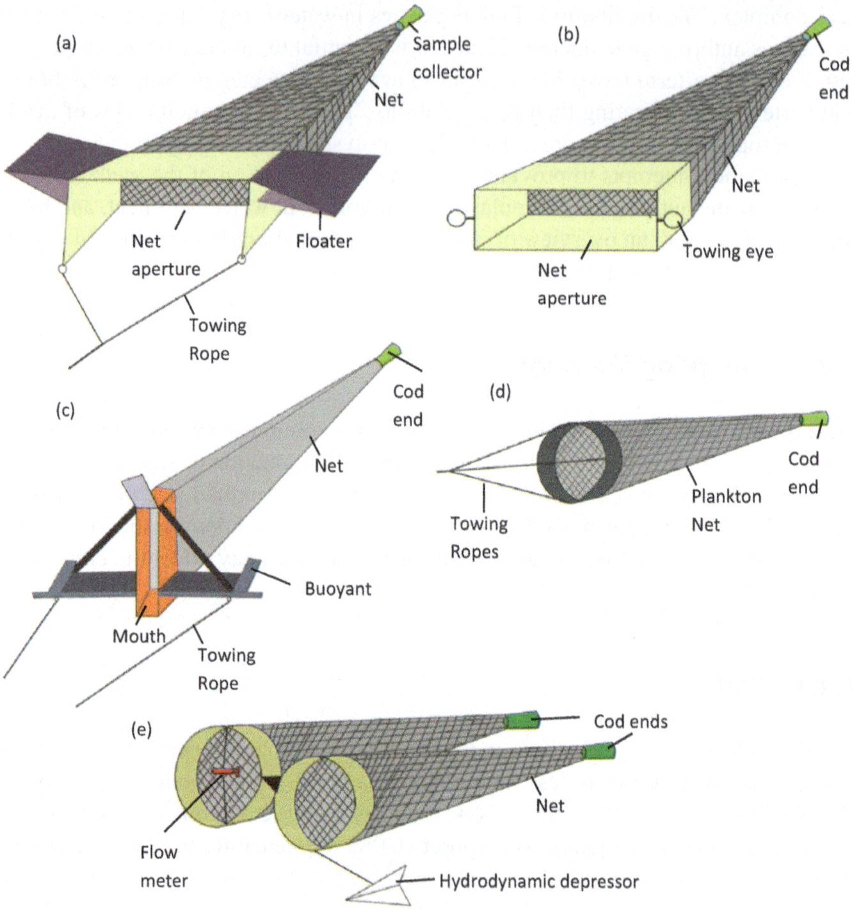

Fig. 6.1 Types of nets used for sampling of microplastics: (**a**) Manta net, (**b**) Neuston net, (**c**) AVANI net, (**d**) Plankton net, and (**e**) Bongo net

lent water (Li et al. 2020a; Pasquier et al. 2022). The volume of water filtered can be calculated through a flow meter or using the following formula:

$$\text{Volume of water} = \text{The sampling distance} \times \left(\text{Width of the trawl} \times \frac{1}{2} \text{Height of the trawl} \right)$$

The sampling distance can be calculated based on GPS coordinates (Schönlau et al. 2020). The collected samples from the collecting bag are then washed and sieved through a 300 μm steel sieve and stored in glass bottles in a dark condition.

(b) DiSalvo Neuston Net:

It was first designed by DiSalvo (1988) for the collection of zooplankton, fish eggs, and larvae but is now used for microplastic sampling. It has a rectangular aperture of height 40 cm and width 80 cm, with an attached PVC pipe on both sides, which helps in floating. The length of the net is 2.2 m, and the mesh size is 300 μm. The net ends with a 30×15cm^2 collection bag (Fig. 6.1b). The neuston net is also trawled at a speed of 23 knots (Eriksen et al. 2018; Gusmao et al. 2022).

(c) AVANI Trawl:

The All-purpose Velocity Accelerated Net Instrument (AVANI) has a rectangular mouth with a height of 60 cm and a 14 cm width, which is divided by an aluminum plate and placed at the same plane as two skis (left and right). The bottom compartment is 20 cm high and 14 cm wide and remains beneath the surface water. The attached net is 4 m long with a mesh size of 335 μm, with a collection bag (30×10cm^2). This design is mostly used along long transects, as in the open sea (Fig. 6.1c). Trawls designed by Eriksen et al. (2018) can work at a high speed of eight knots to reduce the time of sampling and are used while sailing on a cruise (Pasquier et al. 2022). It can scoop off the surface water or can submerge at a high speed (Eriksen et al. 2018; Gago et al. 2019).

(d) Plankton Net:

Plankton nets consist of towing lines, nylon mesh with different mesh sizes, and a cod end (Fig. 6.1d). The targeted microplastics can be obtained using different mesh sizes, such as Kataoka et al. (2019) used 335 μm, 300 μm by Lahens et al. (2018), 100 μm by Kapp and Yeatman (2018), 112 μm by Tan et al. (2019), and 64 μm by Xiong et al. (2019). These nets can be towed, or buckets can be used to collect microplastics (Razeghi et al. 2021).

(e) Bongo Net:

Water column sampling can be done using bongo nets. The bongo net is made up of nylon mesh of different sizes. Di Mauro et al. (2017) used 335 μm mesh size at 15 m depth for 3.6 min, Green et al. (2018) 500 μm mesh size at 1 m depth for 5 min at five knots, and Cai et al. (2018) carried out their experiment using 333 μm at 200 m deep, whereas Rist et al. (2020) used 300 μm pore-sized mesh at 0.5 m depth for 20 min. The volume of water passed can be determined using a flow meter. After towing, the contents of the cod end are washed into a clean glass vial using distilled water (Fig. 6.1e).

6.2.1.1 Pumps (Teflon Pump)

The smallest microplastic particles can be collected using pumps. The pump has a metal hose protected by 5 mm steel mesh to prevent clogging, through which water is pumped, an inverter for energy, and filtering devices. The outlet is provided with a flow meter to record the volume of water. Rist et al. (2020) used pumps at 5 m depth and filtered 1 m^3 water at each sampling station, and filter meshes were 300 μm, 10 μm, and 10 μm. Cai et al. (2018) used a pump equipped with 5 mm, 154 μm, and 44 μm filter meshes for their experiment. They discarded the 5 mm debris and stored the product of other meshes at 4 °C for further analysis.

6.2.1.2 Collection of Water: Niskin Bottles/Jars/Bottles/Buckets/Rosette/Integrated Water Sampler (IWS)/Ruttner Bottles/Friedinger Bottles/Bernatowicz Bottles

Niskin bottles with a CTD-Rosette sampler are commonly used in surface water and subsurface water sampling. Water collected through a Niskin sampler can be stored in glass jars in the dark for further filtration of microplastics. The used Niskin bottles are cleaned using Milli-Q water to avoid further contamination (Gago et al. 2019; Di Mauro et al. 2017).

Sediment

The microplastics tend to remain in sediment due to their buoyancy and durability. The distribution of microplastics in the sediment is irregular considering their nature and environmental factors such as water current, wind, and temperature. Microplastic concentrations will be high on tide lines. Hence, the water current, type of sediments, and proximity to the urban location must be recorded before selecting the sites (Hanvey et al. 2017). After site selection, the following methods can be used to collect sediment samples.

1. Manual Grab Sampler:
   This method uses hand spades, stainless steel spoons, spatulas, and scoops for the collection of sediments on the riverbanks. The transect method helps to obtain sediment samples using quadrates of different sizes and depths (Prata et al. 2019a, 2019b; Hanvey et al. 2017). Collection of sediment sample from beaches ideally should cover the entire stretch and not limited to only tidal zones (Hidalgo-Ruz et al. 2012; Lv et al. 2019).
2. Deep Sediment Sampler (Grabs and Corers): For the riverbed or deep-water regions, different types of grabs and corers are used. The commonly used samplers are the Ekman and Van Veen grab, box corers, and multicorer (Merga et al. 2020; Vianello et al. 2013; Campanale et al. 2020).

The sediment sampling also depends on the depth of sampling. As reported by many authors, around 95% of microplastics are present in the surface layer of beach sampling (Carson et al. 2011), while it was 10% in underwater sediments (Turra et al. 2014).

6.2.2 Biota

Multiple elements play a major role during sampling of biota, for instance, spatial distribution of microplastics along the water column, density, sinking rate, feeding habits of organisms, type of species, and their habitat (Hermsen et al. 2018). The feeding approach of organisms in the aquatic system has an impact on the type and concentration of microplastics ingested. Plankton eaters and suspension feeders can possibly ingest microplastic particles (low density) mistakenly (Setälä et al. 2016). Microplastics tend to settle down; hence, benthic animals can easily accumulate more microplastics compared to other organisms due to constant exposure. In addition, migratory fish also encounter microplastics in their changing environment. Sampling strategies can also affect the results, as large mesh size nets can influence the number of organisms caught, whereas small mesh sizes can result in cod-end feeding (Hermsen et al. 2018; Davison and Asch 2011). Biota sampling can be carried out using mesh nets, traps, trawls, and grabs (for benthic organisms). Frequently reported organisms, like plankton, bivalves, marine worms, fish, turtles, and seabirds, have shown the presence of microplastics (Peters and Bratton 2016; Wedemeyer-Strombel et al. 2015; Van Franeker et al. 2011). However, a large number of samples are required to produce accurate results, which is challenging in the case of top predators (Mai et al. 2018). Benthic invertebrates, such as crustaceans, can be collected using otter trawls, beam trawls, grabs, and traps. Similarly, for plankton species, manta and bongo nets are used. Gill nets are used for collecting fish samples, as demonstrated by Lusher et al. 2020b and Stock et al. 2019 for quantifying microplastics in various tissues. Macroinvertebrates and fish samples can be manually handpicked from the fish catch landing at a fishing port or harbor (Bour et al. 2018). Sampling methods and strategies can greatly influence the outcome of the experiment; hence, repetitive and reproducible studies should be carried out to ensure accuracy.

For the collection of biota samples, different types of nets are used according to the size, habitat, and nature.

1. Neuston Net: These types of nets are used for the collection of debris and biota from the surface water (Frias et al. 2014).
2. Bongo Net: This net is used for the collection of small fishes and zooplankton.
3. Manta Net: This wing-shaped net is used for the collection of surface-dwelling organisms (neuston) as well as microplastics.
4. WP 2 Net: It is a rectangular net used for zooplankton and small fish (Frias et al. 2014).
5. Gill Nets: Gill nets are used for capturing fish samples (Lusher et al. 2020a, 2020b) (Stock et al. 2019).
6. Plankton Net: Phytoplankton and zooplankton are collected using a plankton net of mesh size 20–300 microns. A conical plankton net is used for the collection of both surface and benthic samples by towing at an average speed of 1–4 knots for 15–30 meters with a digital flow meter (Lima et al. 2014).
7. Longhurst Hardy Plankton Recorder (Pro-LHPR): It is used for zooplankton sampling (Frias et al. 2014).

6.3 Extraction Protocol

Extraction is mandatory for proper visualization and characterization.

6.3.1 Water

Water samples are categorized into two subtypes as volume-reduced samples (collected from nets and trawls) and bulk samples (collected through a Niskin sampler). The large volume of water samples is reduced in volume to remove unwanted materials.

6.3.1.1 Sieving

Sieving is the most commonly used method for sorting microplastics. A sieve is made up of steel, copper, or nylon of different pore sizes. Multitier sieving is used to categorize and segregate microplastic particles according to their size. Each sieve can entrap particles larger than the mesh size. Bulk samples after sieving may contain unwanted organic matter, which then undergoes the digestion process.

6.3.1.2 Physical Separation

Through this process, the microplastics can be isolated during sample collection on the site by sieving and handpicking. The microplastics >1 mm can be successfully collected through this process. However, particles <1 mm cannot be analyzed, which need further processing (Hanvey et al. 2017).

6.3.1.3 Digestion

This process aims at the removal of natural organic matter like plankton, biofilms, or remnants of aquatic organisms present within the sample for better identification and characterization. Various techniques used to digest organic matter are:

6.3.1.4 Acid Digestion

HNO_3, H_2SO_4, and HCl are used along with high temperature and pressure to digest the organic matter present in the sample. Naidoo et al. (2017) reported that a 55% nitric acid (HNO_3) solution can digest fish tissue 26 times faster, accompanied by heating at 80 °C. Although nitric acid is popular for decomposition of organic material, it may cause melting of nylon, polystyrene (PS), low-density polyethylene (LDPE), PET, and HDPE; yellowing of PP (polypropylene), PVC (polyvinyl chloride), and PET; whitening of PVC; and oily residue patches (Claessens et al. 2013; Catarino et al. 2017). Use of hydrochloric acid (HCl) seems to have incomplete tissue digestion with a risk of polymer degradation. The treatment with 1 M HCl and 2 M HCl can digest 82.6% and 72.1% of organic matter, respectively (Campanale et al. 2020). Karami et al. (2017) suggested that reaction with 37% HCl at 25 °C can digest up to 95% of tissue but can also melt PET. The alternative acids that can also be used for digestion are sulfuric acid (H_2SO_4) with hydrogen peroxide (H_2O_2) (Huppertsberg and Knepper 2018) and 68% perchloric acid ($HClO_4$), with 65% nitric acid in the ratio of 1:4 (Vandermeersch et al. 2015). Although acids are efficient in digesting organic matter, they might be detrimental to many polymers present in the sample.

6.3.1.5 Alkali Digestion

Digestion with sodium hydroxide (NaOH) and potassium hydroxide (KOH) is remarkable and an alternative to acid digestion. Treatment with 10 M NaOH at 60 °C for 24 hr. and 1 M KOH at 18–21 °C for 48 hr. can result in 90% degradation of the organic matter. However, NaOH is equipped with the degradation of PA, PET, and PVC, while KOH can degrade cellulose and biodegradable plastics (Campanale et al. 2020). The other protocols have been developed using 1 M NaOH, 10% KOH at 60 °C overnight, and 10% KOH at 60 °C for 24 hr. (Hurley et al. 2018; Maes et al. 2017; Dehaut et al. 2016).

6.3.1.6 Oxidizing Agents

Hydrogen peroxide (H_2O_2) is an efficient oxidizing agent and capable of digesting organic matter. 30% of H_2O_2 is commonly used with little effect on polymer physiology. The treatment with H_2O_2 might show some discoloration to polymers like PVC, nylon, acrylonitrile butadiene styrene (ABS), PC, PET, Polyurethane (PU), PP, LDPE, and HPDE. While 35% H_2O_2 can degrade nylon and change color in PET at 50 °C for 96 hr. (Nuelle et al. 2014; Karami et al. 2017). Hence, optimum temperature and concentration of H_2O_2 are required for extracting microplastics from the sample. One alternative can be wet peroxide oxidation (WPO), which uses Fenton's reagent, a mixture of 30% H_2O_2 and 0.05 M $FeSO_4$ has been proven as an effective solution for degrading organic matter (Klein et al. 2015; Masura et al. 2015; Wen et al. 2018). This method has been recommended by NOAA and is one of the most common methods for the digestion of organic matter, followed by alkaline (sodium hydroxide, potassium hydroxide), acidic (nitric acid, hydrochloric acid), and enzymatic (protease K) methods. It involves heating Fenton's reagent along with the sample at 75 °C.

6.3.1.7 Enzymatic Digestion

This process of organic matter digestion is less hazardous to polymers and the environment. The first experiment was performed by Cole et al. (2014) using proteinase K (500 μg ml^{-1}), which digested more than 97% of the organic matter without affecting polymer physiology. Other alternative enzymes like cellulase (1 ml + 25 ml NaOAc buffer), lipase (1 ml + 25 ml Tris-HCl buffer), chitinase (1 ml + 25 ml NaOAc buffer), protease (5 ml + 25 ml Tris-HCl), and amylase (5 ml + 25 ml NaOAc buffer) have also been used to remove organic matter (Campanale et al. 2020). A combination of enzyme and hydrogen peroxide is also an efficient alternative for digestion (Karlsson et al. 2017). Although enzymatic treatment is a reliable and less hazardous method, the primary disadvantages are cost-ineffectiveness, time-consuming, temperature, and pH dependence, due to which only a small amount of sample can be processed (Lusher et al. 2020b).

Hence, the most suitable digestive method would be the WPO method, as experimentally proven by our team. The addition of 20 ml of Fenton's reagent to the beaker containing the filtrate is followed by mixing at room temperature. Then the mixture can be heated on the magnetic stirrer-hot plate at 75 °C and 350 rpm for 30 mins.

6.3.1.8 Density Separation: Flotation and Elutriation

This method is based on the density difference between microplastics and unwanted materials. The specific density of most of the plastic particles ranges between 0.8 and 1.7 g cm^{-3}; hence, they can be separated using sodium chloride (NaCl). The separation through NaCl is a widely used method due to its low cost and eco-friendly nature. However, high-density microplastics such as PVC (1.3–1.7 g cm^{-3}), PET (1.4–1.6 g cm^{-3}), and HDPE (0.94–0.98 g cm^{-3}) cannot be efficiently separated by NaCl and may remain underestimated. Therefore, other salts like sodium iodide (NaI), sodium bromide (NaBr), zinc chloride ($ZnCl_2$), zinc bromide ($ZnBr_2$), sodium tungstate dehydrate, and sodium polytungstate have proven useful in extracting high-density microplastics (Campanale et al. 2020; Prata et al. 2019a, 2019b; Wang and Wang 2018). These high-density salts are expensive and some of them can be hazardous to the environment (Wang and Wang 2018).

For the water sample, 6 g of NaCl can be added to the solution after the WPO reaction to extract microplastic particles as experimentally standardized by our team.

6.3.1.9 Filtration

After digestion and density separation, the supernatant containing microplastics should be filtered through multisieves arranged from 5 mm to 1 μm. The filter papers, such as Whatman microfiber filters, polycarbonate filters (15 μm), nylon filters (20 μm), can also be used for the filtration process (Poli et al. 2024; Ghanadi et al. 2024; Liu et al. 2022a). The pore size of filter papers can range from 0.45 to 20 μm. Some microplastics can remain attached to the glassware, which can be avoided by rinsing thoroughly with distilled water (Wang and Wang 2018). The filtered microplastics can be stored in clean, covered beakers and stored in a dry place (Table 6.1).

6.3.2 Sediment

6.3.2.1 Drying

The sediment samples should be completely dried to remove the water content. Oven drying is a widely accepted method as it provides a closed space and fast drying without contamination. Air drying for a longer period makes samples susceptible to airborne contaminants. Freeze-drying leads to the formation of porous structures created by ice crystals, which helps in further processing (Rao et al. 2020; He et al. 2020; Adomat and Grischek 2021).

6.3.2.2 Physical Separation

The on-site sieving technique is used for dry sediment samples to extract large-sized microplastics, which are stored in collection vials for further analysis. This is also known as dry sieving. The limitation of this technique is the non-retention of small microplastics (<1 mm).

Table 6.1 Separation of MPs from water samples

Sl no	Types of samples	Filter used	Digestion of organic matter	Density separation	Reference
1	Sea surface water (southwest coast of Kerala)	44 μm sieve	30% H_2O_2, 0.5 M $FeSO_4$	$ZnCl_2$	Nikhil et al. (2023)
2	Sea water sample (west coast of Tamil Nadu and Kerala)		30% H_2O_2		Patterson et al. (2021)
3	Marine water sample (Auckland)	Polycarbonate filter (15 μm)	30% H_2O_2	$CaCl_2$	Ghanadi et al. (2024)
4	Surface seawater (Zhongsha Atoll, South China Sea)	Nylon filters (20 μm)	30% H_2O_2	NaCl	Liu et al. (2022a)
5	Mineral water stored in different plastic bottles	Gold-coated polycarbonate filters	No digestion	No density separation	Schymanski et al. (2018)
6	Freshwater sample (St. Lawrence River)	Borosilicate filter (1 μm)	Extraction through oil followed by digestion by 30% H_2O_2	Density independent	Crew et al. (2020)
7	Wastewater treatment plant		Plastic preserving enzymatic maceration	$ZnCl_2$	Mintenig et al. (2017)

6.3.2.3 Density Separation: Flotation and Elutriation

The density separation involves the following four steps:

(a) Introduction of water solvents of specific density so that the higher-density polymer will float. Mostly, NaCl solution is used to create a density difference.
(b) Mixing of the solvent for a period of time to ensure the detachment of particles from the sediment.
(c) Resting time to settle down the sediment particles.
(d) Filtration to obtain different sizes of microplastics.

The microplastics can be separated from sediment samples by mixing sediment (2.7 g cm^{-3}) with salt solution, followed by filtration. Although NaCl has a lower efficacy (<90%) than NaI and $ZnCl_2$ (99%), it is widely used because of its cost-effectiveness and nonhazardous nature (Quinn et al. 2017). Three subsequent washes with NaCl solution are needed to extract 93% of the microplastics from sediment (Wang and Wang 2018). An apparatus, Sediment-Microplastic-Isolater, uses $ZnCl_2$ for the extraction of 95.8% of microplastics from sediment samples (Coppock et al. 2017). The Munich Plastic Sediment Separator (MPSS) uses the elutriation technique in which a high-density ZnCl solution is injected into the bottom of the column; hence, microplastics can diffuse into the solution and be extracted out through the supernatant (Imhof et al. 2012). It requires 1–2 hr. for separation.

Oils like pine oil, canola oil, corn oil, olive oil, sesame oil, and peanut oil can also be an alternative for density separation (Crichton et al. 2017; Scopetani et al. 2020; Yuan et al. 2022). Oils are added to the salt solution to increase the recovery rate of microplastics (Prata et al. 2019a, 2019b).

6.3.2.4 Wet Sieving

It is used to separate polymers floating on the solvent after density separation, which is carried out by vacuum filtration. The commonly used filters are glass fiber filters and nitrocellulose filters. Sieves of different sizes (less than 5 mm) can also be used (Hanvey et al. 2017).

6.3.2.5 Digestion of Organic Matter

Digestion of biological matter decreases the overestimation of microplastics, as they can interfere with identification. As explained in detail in the water extraction process, the digestion can be of the following types: such as a) acid digestion, b) alkali digestion, c) oxidising agents, and d) enzymatic digestion.

6.3.2.6 Acid Digestion

Nitric acid and hydrochloric acid are extensively used methods; however, some polymers can show low resistance and be susceptible to digestion at high concentrations and high temperatures. Hence, an optimum concentration and temperature should be maintained to achieve maximum results.

6.3.2.7 Alkali Digestion

Alkali digestion is an alternative to acid digestion, but it also has its limitations in the form of discoloration and deposit of tissue fragments on the upper layer of microplastics, which in turn can interfere with the characterization processes. Ten percent potassium hydroxide and 10% sodium hydroxide are commonly used for alkali digestion with promising results.

6.3.2.8 Oxidizing Agents

As mentioned previously, H_2O_2 is an effective oxidizing agent to digest organic matter. H_2O_2 coupled with high temperature and other chemicals like Fenton's reagent, is widely used and an excellent oxidizing agent.

6.3.2.9 Enzymatic Digestion

It is a less hazardous and reliable method, but at high temperatures, it cannot be used due to the denaturation process. For large-scale samples, the enzymatic degradation method is not preferred, as temperature and pH regulate enzyme efficacy and, in turn, can impact maximum microplastic extraction.

6.3.2.10 Filtration

After digestion, the solvent is filtered, and purified polymers are stored for further analysis (Table 6.2).

6.3.3 Biota

Extraction of microplastics from freshwater/marine biota is of concern to assess the impact on the food chain and marine ecosystem. MPs can enter the food chain, accumulate, and harm the higher trophic organisms. Aquatic biota consume pieces of microplastic, mistaking them for food. Extraction, identification, and quantification are useful to know the accurate result. The extraction process includes tissue dissection, purification, homogenization, and digestion.

6.3.3.1 Dissection

The required tissue is dissected from the sampled organism of interest. Mostly, the digestive tract is used for microplastic extraction and analysis. Post-dissection microplastic sorting through visual inspection is a preferable method in freshly acquired samples in which dietary remains can be visualized clearly. In other cases, GI tracts also can be rinsed with water to observe the presence of microplastic, but it can lead to underrepresentation of microplastic as it can adhere to dietary remains and the GI tract (Silva-Cavalcanti et al. 2017; Vendel et al. 2017; Faure et al. 2015; Holland et al. 2016).

6.3.3.2 Digestion

Digestion of organic substances is essential to extract the pure form of microplastics. Several methods for digestion can be done through the application of alkaline, acidic, oxidizing agents, and enzymes. The method and efficacy of extraction will vary in plankton species, bivalves, crustaceans, fish, and other marine species (mammals, birds, and turtles).

6.3.3.3 Acid Digestion

Acid digestion of the tissue sample is required for the efficient breakdown of organic components, which separates microplastics easily from the tissue sample. Common acids such as HCl, HNO_3, H_2SO_4, and perchloric acid ($HClO_4$) are used for acid digestion of tissue samples.

6.3.3.4 Alkali Digestion

The alkaline method of digestion is commonly used for the fish gastrointestinal tract. The alkaline digestion method involves the addition of KOH, followed by 10% NaOH for a period of 24 hrs at 60 °C using a magnetic stirrer at 125 rpm (Lusher and Hernandez-Milian 2018; Amponsah et al. 2024). The concentration of KOH influences the breakdown of natural particles and can increase it by 10–20% (Milne et al.2024). A mixture of a strong acid (KOH) and a weak base (NaClO) has been used for the digestion of Nile tilapia (*Oreochromis niloticus*) filters.

Table 6.2 Separation of MPs from soil and sediment samples

Sl. no.	Type of sample	Pretreatment	Density separation	Digestion	Device	Reference
1	Field soil (Japan)	Air-dried and oven-dried	Oil extraction (canola oil, saturated NaCl)	30% H_2O_2	Separating funnel and glass filter	Kononov et al. (2022)
2	Sea bottom sediment (Baltic Sea—Russian part)		$ZnCl_2$ treatment (specific density 1.6 g/ml)	Wet peroxide oxidation, calcite fraction digestion (30% H_2O_2, Fe(II) catalyst solution, 4.5% HCl solution)	Density separator, polyamide filter nets	Zobkov and Esiukova (2017)
3	Low organic content soil samples	Homogenization	Density separation by sodium polytungstate (SPT) and sodium iodide (density 1.8 g cm^{-1})		Centrifuge rotor	Pfohl et al. (2024)
4	Sediments from the Western English Channel	Dried at 50 °C	Density flotation ($ZnCl_2$, NaCl, NaI)		Sediment–microplastic isolation (SMI) unit	Coppock et al. (2017)
5	Sediments from Lake Idro (Northern Italy)		Density separation by NaCl	10% KOH digestion	Magnetic stirrer, centrifuge, Whatman glass microfiber filters (grade GF/C, 0.45 µ)	Nava and Leoni (2021)
6	Sediment from Bekkelaget WWTP, Oslo municipality	500 g homogenized samples fed to Bauta microplastic-sediment separator (BMSS)	$ZnCl_2$, $CaCl_2$	30% H_2O_2, 10 M NaOH	Bauta microplastic-sediment separator (BMSS), magnetic stirrer	Mahat (2017)

7	Intertidal sediment samples (Frei Antarctic region-Fildes Bay)		NaCl		Vacuum filtration unit with glass fiber prefilters of 1.6 μ pore size	Perfetti-Bolaño et al. (2022)
8	Local soil samples		$ZnCl_2$	Fenton's reagent (30% H_2O_2)	Nylon filters	Rowe et al. (2018)
9	Municipality waste sample (Sydney)			Methanol, hexane, dichloromethane	Pressurized fluid extractor (Dionex ASE-350, USA), glass fiber filters (Dionex#056781, USA)	Fuller and Gautam (2016)
10	Soil samples from the Ecological Environment Research Centre, China		$ZnCl_2$	30% H_2O_2	Plastic flotation and separator system (PFSS)	Li et al. (2024)
11	Agricultural field, Tome, Japan	Sieved with a 2 m sieve	$CaCl_2$ solution	Fenton oxidation	Centrifuge rotor	Grause et al. (2022)
12	Agricultural soil, Paotai town, Xinjiang, China	Soil dissolved in pressurized distilled water	Flotation	98% H_2SO_4	Ultra Sonicator	Li et al. (2020c)
13	Landfill-mined soil		Density separation by NaI and $ZnCl_2$	Fenton's reagent	Centrifuge and ultrasonic bath Sonicator (UBS-Bandelin RK 106)	Goli and Singh (2023)

(continued)

Table 6.2 (continued)

Sl. no.	Type of sample	Pretreatment	Density separation	Digestion	Device	Reference
14	Beach sediment, Grape bathing beach, Weihai, China	Sieved by 5 mm and 2 mm	Density flotation ($NaH_2PO_4 \cdot 2H_2O$, NaCl)	H_2O_2	Magnetic stirrer	Zhang et al. (2020)
15	Local beach sediment samples, Vancouver, British Columbia, Canada		Oil extraction and density separation (canola oil, NaI, and $CaCl_2$)		Glass filter separating funnel equipped with polycarbonate membrane (47 mm diameter, 1 μ pore size, Sterlitech, Kent, USA)	Crichton et al. (2017)
16	Soil sample from campus of East China Normal University and the suburb of Shanghai		Circulation extraction ($CaCl_2$, NaCl, NaBr)	30% H_2O_2	Automatic cycling device with vacuum filtration system	Liu et al. (2019)
17	Soil samples, campus of Shaanxi University of Science and Technology (Xi'an, China)		Density separation by NaBr		Separation and extraction device with a combined system of circulation and recovery	Li et al. (2021)
18	Soil from oat fields of Finland and compost (Labio Ltd.)		Oil extraction (olive oil)	30% H_2O_2, $FeSO_4 \cdot 7H_2O$	Separating funnel with Whatman glass microfiber filters (grade GF/A, 90 mm diameter)	Scopetani et al. (2020)

19	Alluvial soil (banks of Sava River), compost (JP VOKA Snaga Ltd., Slovenia)		Oil extraction by olive oil, and $ZnCl_2$	Oxidation (5 M KOH, Fenton's reagent)	Separating funnel and centrifuge (Hettich Universal centrifuge, Germany)	Prosenc et al. (2021)
20	Sea floor sediment (2.7 miles off the city of Barcelona)		Density separation (NaCl)		Buchi speed extractor E-916	Raguso et al. (2021)
21	Farm soil, Glensaugh Farm Research Facility (Laurencekirk, Aberdeenshire)		High gradient magnetic separation (ethanol, hexadecyltrimethoxysaline)		High gradient magnetic separation (HGMS) system, Whatman glass microfiber filters (grade GF/C, 1.2 μ)	Ramage et al. (2022)
22	Agricultural soil (Jiangsu, China), sewage sludge (wastewater treatment plants, Nanjing, China)	Dried	Gravity flotation separation (NaCl, $ZnCl_2$, NaI)	Oxidation (30% H_2O_2)	Gravity flotation and vacuum filtration apparatus	Li et al. (2019)

(continued)

Table 6.2 (continued)

Sl. no.	Type of sample	Pretreatment	Density separation	Digestion	Device	Reference
23	Silt-loam agricultural soil, Stuttgart, Southwest Germany	Freeze-dried	$ZnCl_2$ (specific density 1.8 g cm^{-3})	Enzymatic oxidative digestion with SDS (48 hrs), Fenton's reagent (1 hr), protease (12 hrs), pectinase (48 hrs), Viscozyme(48 hrs), Cellulase (24 hrs) and again with Fenton's reagent (1 hr)	Separating funnels and filters	Möller et al. (2022)
24	Organic matter removed soil and diatomaceous earth		Dichloromethane		Dionex ASE 350 fast solvent extraction instrument (Thermo Fisher, Waltham, MA)	Wen et al. (2021)
25	Soil from creeks (Oklahoma), Rivers (Tulsa, Oklahoma), and agricultural fields	Dried	Elutriation, flotation, and filtration (7.5% NaOCl, 6 M $ZnCl_2$)	Digestion	Ultra Sonicator, centrifuge	Forsythe et al. (2024)
26	Sediment samples	Sieving	Density separation (NaI)	Oxidation (Fenton's reagent)	Separating funnels	Lusher et al. (2020a)
27	Sand+ commercial compost (John Innes Manufacturers Association)		Density separation ($ZnCl_2$, NaCl), oil separation (canola oil)	Oxidation (Fenton's reagent, sodium hexametaphosphate)	Separating apparatus, crystallizing dish, Whatman glass microfiber filters (grade GF/A, 1.6 μ)	Radford et al. (2021)

28	Soil samples (University of Macau, China)		Hydrophobic separation (NaCl, corn oil, olive, peanut oil, and sesame oil)		Microplastic extractor device with peristaltic pump, recovery tank, magnetic stirrer, separating column, and screen sieve (1 mm)	Yuan et al. (2022)
29	Sediment samples (VNF, Navigable Waterways of France) and along Aa River (St-Omer, France)		Hand stirring, centrifugation, and aeration (NaCl, canola oil)		Centrifuge, Whatman glass microfiber filters (grade GF/F, 0.7 μ)	Constant et al. (2021)
30	Sediment samples from tidal flats of the western coast of Bohai Bay (Dashentang beach, Yongdingxin, Haihe, Dagupaiwu, and Duliujian River estuary)		Density flotation (NaCl, NaI)		Optimized extraction setup with peristaltic pump, flotation unit, air flow meter, water circulation vacuum pump, and vacuum filtration unit	Han et al. (2019)

To remove any form of undigested organic matter, the tissue samples can be subjected to peroxide oxidation after alkaline digestion. Peroxide oxidation includes sulfuric acid, iron heptahydrate, and 30% hydrogen peroxide (Milne et al. 2024).

6.3.3.5 Oxidizing Agents

Biogenic organic substances can hinder the separation of microplastics, particularly during the density separation step. Therefore, for the pure extraction of microplastics, the digestion of organic substances is one of the important steps (Pfeiffer and Fischer 2020). Oxidative digestion methods are less time-consuming and are more efficient (Tuuri et al. 2024). 30% hydrogen peroxide is commonly used as an effective, safer, and selective oxidizing agent due to its lesser effect on the structure and physiology of microplastics. Wet peroxide oxidation (WPO) using the 30% H_2O_2 method was used to digest the GIT, gills, and muscle tissue of various fish (*Silonia silondia, Siganus javus, Agyrosomus amoyensis, Netuma thalassina, Priacanthus* sp., *Sardinella fimbriata, Ameiurus nebulosus, Catostomus commersonii, Micropterus salmoides, Micropterus dolomieu, Ambloplites rupestris, Esox lucius, Macrobrachium rosenbergii, Litopenaeus vannamei, Panulirus cygnus, Portunus armatus*) and zooplankton samples, followed by different separation processes. Besides using the 30% H_2O_2, 15% and 10% H_2O_2 have also been used to digest the tissues of *Mytilus galloprovincialis* and *Penaeus merguiensis* (Mercogliano et al. 2021; Li et al. 2022). The 30% hydrogen peroxide oxidation has a digestion efficiency of 93.69%. For higher organic matter digestion efficacy, hydrogen peroxide can be used in combination with iron sulfate heptahydrate (Fenton's reagent) and an iron catalyst to enhance the oxidative potential of hydrogen peroxide (Prata et al. 2019a, 2019b). This has a digestion efficiency of 97.21% (Yu et al. 2022). The application of sodium dodecyl sulfate (SDS) combined with hydrogen peroxide has been tested for oxidative digestion in planktonic organisms, with no specific degradation of microplastics. Plankton are treated with 2% sodium dodecyl sulfate, followed by the gradual addition of 30% hydrogen peroxide. This does not change the shape of microplastic after its recovery (Tuuri et al. 2024). Sodium hypochlorite in combination with KOH in a 1:1 ratio as an effective organic matter digestive reagent has been tested in fish tissue, with even the diluted form of the mixture having a similar capacity for complete digestion. Therefore, this is considered a more cost-effective working reagent (Enders et al. 2017). Use of 10% sodium hypochlorite at 40–50 °C gave a better outcome as compared to hydrogen peroxide for the digestion of both soft and hard tissues (Pfeiffer et al.2020). Similarly, nitric acid with hydrogen peroxide in a 4:1 ratio by volume at 50 °C with proper stirring can be used for complete digestion of soft tissues of the mussel and the gastrointestinal tract of fish tissues. This method is quicker and more efficient as compared to other digestive methods (Yu et al. 2019). The microwave-assisted oxidation digestion method in combination with hydrogen peroxide has been used to digest the tissue sample of *Penaeus merguiensis* (Li et al. 2022) for quick and efficient breakdown of organic tissue. The ratio of 10% hydrogen peroxide and tissue is to be maintained at 25:1, with sample heating at 550 W inside the microwave for better results.

6.3.3.6 Enzymatic Digestion

Enzymatic digestion is one of the most accepted methods for microplastic analysis due to its nondestructive, eco-friendly nature. Before enzymatic digestion, tissue samples are homogenized using 400 mM Tris-HCl buffer, 60 mM EDTA, 105 mM NaCl, and 1% sodium dodecyl sulfate (SDS) (pH 8) for Protease K and 0.1 M citric acid buffer (pH 5.6) for chitinase (Li et al. 2022). Industrial proteases, Carolase 7089 (AB enzyme), are used for the soft tissue enzymatic digestion for the quantification of microplastics because they are less hazardous, do not require buffer, are in liquid form, and have high recovery without destruction (Catarino et al. 2017). To increase the efficiency of the microplastics purification process, a mixture of enzymes (proteases, cellulases, chitinases, amylases, and lipases) is used (Löder et al. 2017). Karlsson et al. (2017) digested fish tissues using 500 mg ml^{-1} proteinase K with $CaCl_2$ at 50 °C for 2 hrs after which the solution was shaken for 20 mins and incubated at 60 °C for 20 mins. Followed by treatment with H_2O_2 resulted in recovery of almost 97% of microplastics. Sodium dodecyl sulfate along with trishydroxymethyl aminomethane, *Streptomyces griseus* proteases, *Aspergillus niger*, cellulase and Tris(1 M) were used for the analysis of microplastics from edible microalgae (López-Rosales et al. 2022). Fish tissue (gastrointestinal tract) was homogenized using 400 mM Tris buffer (pH 8) and 0.5% SDS. After 60 min of incubation required for protein denaturation, the addition of proteinase K and $CaCl_2$ can be done, followed by incubation at 50 °C for 2 hours, samples shaken at room temperature, and again incubated for 2 hours before the addition of 30% hydrogen peroxide (Karlsson et al. 2017). Enzymatic digestion using proteinase K has also been found to be effective for marine zooplankton species, and it does not show any detrimental effects on microplastics (Cole et al.2014). A combination of proteinase K, pepsin, and trypsin with an oxidative agent such as hydrogen peroxide showed digestion efficiency of around 99%. These were found to be more effective in bivalves and crustaceans. The reactions were carried out at 37 °C for 2 h, followed by the addition of 30% hydrogen peroxide (Yu et al. 2022). Courtene-Jones et al. (2017) used three different enzymes (papain, trypsin, and collagenase) in *Mytilus edulis* for the digestion of mussel soft tissue, and trypsin gave the highest yield without causing any change in microplastic concentration. To increase the digestive efficiency in bivalves' tissue, pancreatic enzymes can also be used. Overnight exposure of *Serripes groenlandicus*, an arctic bivalve species, to pancreatic enzymes resulted in an increased rate of tissue digestion for microplastic extraction, and the method requires little maintenance (Von Friesen et al. 2019).

6.3.3.7 Density Separation

After the digestion of the tissue sample, density separation is carried out to extract microplastics from the digested tissue. High-density salt solutions like NaI (Çevik and Gündoğdu 2018), NaCl (Mercogliano et al. 2021; Hossain et al.2024), $CaCl_2$ (Milne et al.2024), and $ZnCl_2$ (Bornt et al. 2024) give good results. The density separation technique includes the addition of a salt solution, which induces the microplastic to float on the surface.

6.3.3.8 Filtration

The filtration process is carried out before and after the digestion process using different mesh filters (0.2–0.4 μm pore size). The digested samples are filtered on cellulose nitrate membranes of various sizes of 8 μm (Mercogliano et al. 2021) and 0.45 μm (Motivarash et al. 2024). After the digestion process, the supernatant is also filtered using 0.45 μm Whatman membrane filters and vacuum filtered through a glass fiber filter of pore size 1 μm and 45 μm (Evelyn and Pérez-Sabino 2024) (Fig. 6.2) (Table 6.3).

6.4 Visualization and Characterization

There are various methods for the characterization of microplastics after different types of treatment, and the four important techniques include FTIR spectroscopy, Raman spectroscopy, gas chromatography coupled with mass spectrometry (GC/MS), and SEM-EDX analysis for the characterization of microplastics.

6.5 Microscopy

6.5.1 Binocular or Polarized Microscope

This is a low-cost and short-time visualization approach commonly used for the segregation of microplastics from other materials of similar size. It is used for the observation of physical characteristics (like color, shape, and surface texture) and quantification of microplastics (by direct counting) (Alak et al. 2022). This method might create confusion in distinguishing microplastics from other organic and inorganic particles, which could increase the risk of overestimation. Therefore, this method is commonly used for preliminary study and screening (Gniadek and Dąbrowska 2019).

6.5.2 Fluorescence

Fluorescence is a mechanism in which particles absorb short-wavelength light and move to the excited state, which is unstable. Eventually, the excited particles tend to emit the absorbed energy and return to their original state, known as fluorescence. The particles showing fluorescence are known as fluorophores, which can be naturally found in the sample or introduced by fluorescent dyes. A fluorescence microscope can detect the qualitative and quantitative presence of fluorophores in the samples (Zhu et al. 2024). It has high specificity and sensitivity as it can detect fluorescence in low dye concentration. The most commonly used dye for microplastic analysis is Nile red (NR), a lipophilic fluorescent dye, as has been mentioned by many researchers (Sturm et al. 2021; Tarafdar et al. 2022; Ho et al. 2023). The excitation and emission maximum of NR is 549 nm (green) and 628 nm (red),

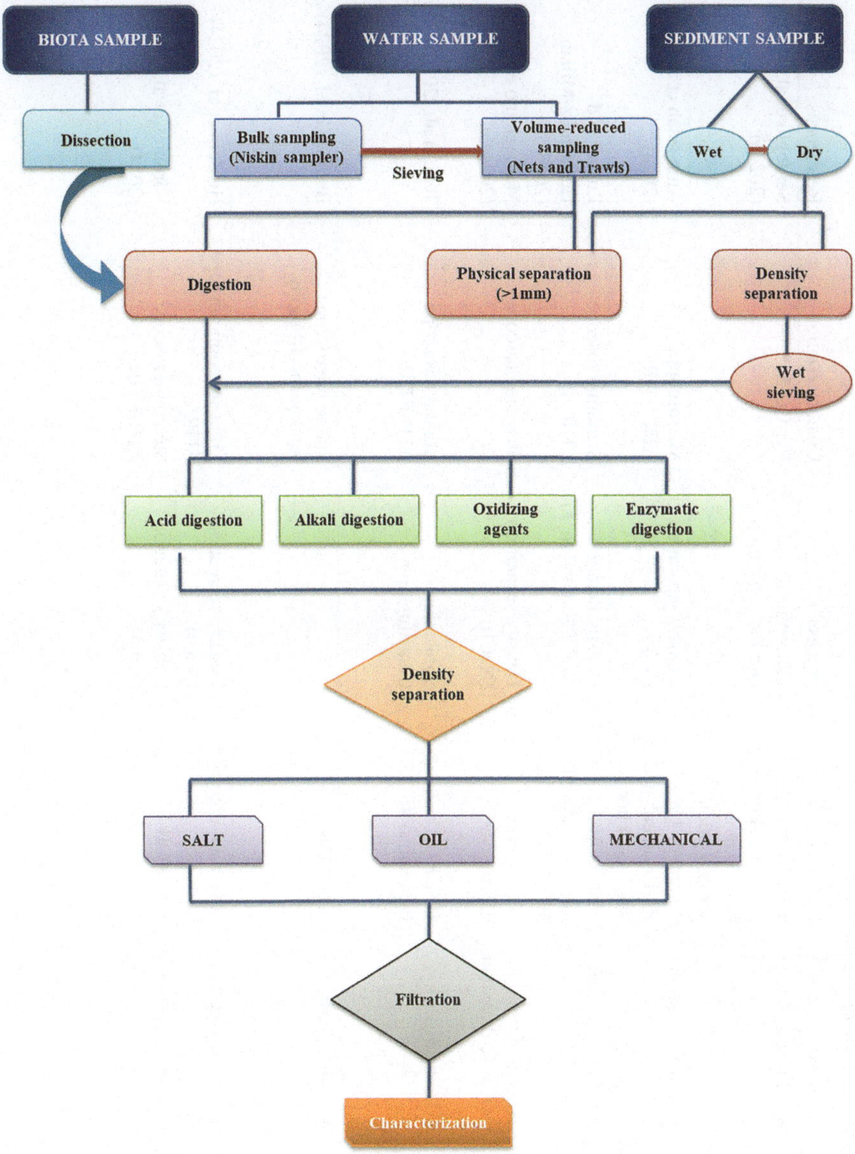

Fig. 6.2 Flowchart showing the extraction processes of biota, water, and sediment samples

respectively (Greenspan and Fowler 1985). According to Liu et al. (2022b), the Nile red staining can be done in two ways, i.e., staining on filter paper and in solution. In the first method, the dye solution will be added to the extracted microplastic particles placed on filter papers. The optimum concentration of NR dye is 1–1000 μg mL^{-1}, and the incubation period is 5 mins–66 hrs, depending upon the dye concentration and sample type (Maes et al. 2017). After incubation, the filter paper can be observed

Table 6.3 Separation of microplastics from biota samples

Species	Digestion process	Chemicals	Process	Characterization	References
Nile tilapia (*Oreochromis niloticus*)—fillet	Alkaline digestion	10% KOH+ NaClO Acid digestion Isopropanol solubilization	Ultrasonication Density separation	Pyr-GC-MS	Sefiloglu et al. (2024a)
Fish GIT and gills	Alkaline digestion	200 ml of 10% KOH—Incubated −60 °C 24 hrs	Density separation (NaI)	Microscopic ATR-FTIR	Amponsah et al. (2024)
Marine vertebrates (mammals, birds, turtles, and fishes)— digestive tract	Alkaline digestion	10% KOH 3:1 ratio—Incubated 60 °C 24 h	Density flotation (NaCl/NaI and SPT)	Stereomicroscope FTIR	Lusher and Hernandez-Milian (2018)
Engraulis encrasicolus (GI tract)	Alkaline digestion Oxidative digestion	10% KOH 250 ml NaCl 15% H_2O_2	Density separation (NaCl)	Stereomicroscope	Mercogliano et al. (2021)
Engraulisencrasicolus *Trachurus trachurus* *Mullus barbatus* *Merlangius merlangus* (GIT)	Alkaline digestion	10% KOH Incubate at 40 °C	Sonication Centrifugation	Light microscope ATR-FTIR	Tepe et al. (2024)
Merluccius merluccius *Sardina pilchardus* *Trisopterus luscus* (heart sample)	Alkaline digestion	10% KOH		Stereo microscope Raman spectroscopy	Barboza et al. (2024)
Silonia silondia (GIT, gills, and muscles)	Oxidative digestion	30% H_2O_2 digestion	Density separation (NaCl)	Light stereomicroscope FTIR	Hossain et al. (2024)
Commercial fish GIT, gills	Alkaline digestion	10% KOH	Density separation (NaCl)	Stereomicroscope ATR-FTIR	Vorobiev et al. (2024)

Siganus javus *Agyrosomus amoyensis* *Netuma thalassina* *Priacanthus* sp. *Sardinella fimbriata*	Oxidative digestion	30% H_2O_2	Density separation (NaCl)	Binocular microscope ATR-FTIR	Dwiyitno et al. (2024)
Pelagic and demersal fish sample	Alkaline digestion	10% KOH		ATR-FTIR Raman analysis	Motivarash et al. (2024)
Oreochromis niloticus *Parachromis managuensis* (digestive tract)	Alkaline digestion	1 Mol l^{-1} NaOH		FTIR-ATR Digital microscope	Evelyn and Pérez-Sabino (2024)
Fish GIT *Ameiurus nebulosus* *Catostomus commersonii* *Micropterus salmoides* *Micropterus dolomieu* *Ambloplites rupestris* *Esox lucius*	Alkaline digestion Oxidative digestion	20% KOH Peroxide oxidation	Density separation ($CaCl_2$)	Dissecting microscope Raman spectrometry	Milne et al. (2024)
Mullus barbatus *Alosa immaculata* (GIT, muscle, brain)	Alkaline digestion	10% KOH	Density separation (NaCl)	Stereomicroscope ATR-FTIR	Atamanalp et al. (2021)
Oryzias latipes (GI tract)	Mixed digestion	10% KOH + 10% HNO_3		Stereozoom microscope SEM/EDS FTIR microspectroscopy Raman microspectroscopy	Wagner et al. (2017)

(continued)

Table 6.3 (continued)

Species	Digestion process	Chemicals	Process	Characterization	References
Bivalves					
Mytilus edulis	Acid digestion	HNO_3 (0%, 9%, 18%, 35%, 50%)		Stereomicroscope FTIR	Catarino et al. (2017)
	Alkaline digestion	NaOH (0.25 M, 0.5 M, 1.0 M, 2.5 M, 5.0 M)			
	Enzymatic digestion	Protease, Corolase			
Mytilus galloprovincialis	Alkaline digestion	10% KOH 250 ml NaCl 15% H_2O_2	Density separation— NaCl	Stereomicroscope	Mercogliano et al. (2021)
Villorita cyprinoides *Crassostrea madrasensis* *Paphia malabarica* *Babylonia spirata* *Perna perna*	Alkaline digestion	10% KOH 1 N HCl for neutralization	Density separation— NaCl	Digital microscope Laser Raman spectroscopy	Abisha et al. (2024)
Donax trunculus	Alkaline digestion	10% KOH	Density separation— NaCl	Raman spectroscopy	Malloggi et al. (2024)
Spondylus spinosus	Alkaline digestion	1:1 KOH:NaClO	Density separation (NaI)	Stereoscopic microscope	Çevik and Gündoğdu (2018)
Crustaceans					
Penaeus merguiensis	Acid digestion	70% HNO_3		Stereomicroscope ATR-FTIR	Li et al. (2022)
	Alkaline digestion	10% KOH			
	Oxidation digestion	10% H_2O_2			
	Microwave-assisted oxidation digestion	10% H_2O_2, boiling			
	Enzymatic digestion	Protease K Chitinase			

Panulirus cygnus *Portunus armatus* (digestive tract)	Alkaline digestion	10% KOH	Density separation ($ZnCl_2$)	Microscope ATR-FTIR	Bornt et al. (2024)
	Acid digestion	70% HNO_3			
	Oxidation digestion	30% H_2O_2			
	Mixed	70% HNO_3 + 30% H_2O_2			
Macrobrachium rosenbergii *Litopenaeus vannamei* (GIT)	Oxidation digestion	30% H_2O_2	Flotation separation (HCOK)	Stereomicroscope FTIR	Reunura and Prommi (2022)
Planktons					
Zooplankton	Acidic method	70% HNO_3 and $NaHCO_3$ for neutralization	Density separation	Confocal microscopy Raman spectroscopy	Tuuri et al. (2024)
	Alkaline method	10% KOH neutralized with citric acid monohydrate (C6H8O7·H2O) 1 M NaOH neutralized with HCl			
	Oxidative method	Sodium dodecyl sulfate 30% hydrogen peroxide			
Copepods	Acid digestion	1 M and 2 M HCl	Ultrasonication	Microscope FTIR	Cole et al. (2014)
	Alkaline digestion	10 M NaOH			
	Enzymatic digestion	Proteinase K			

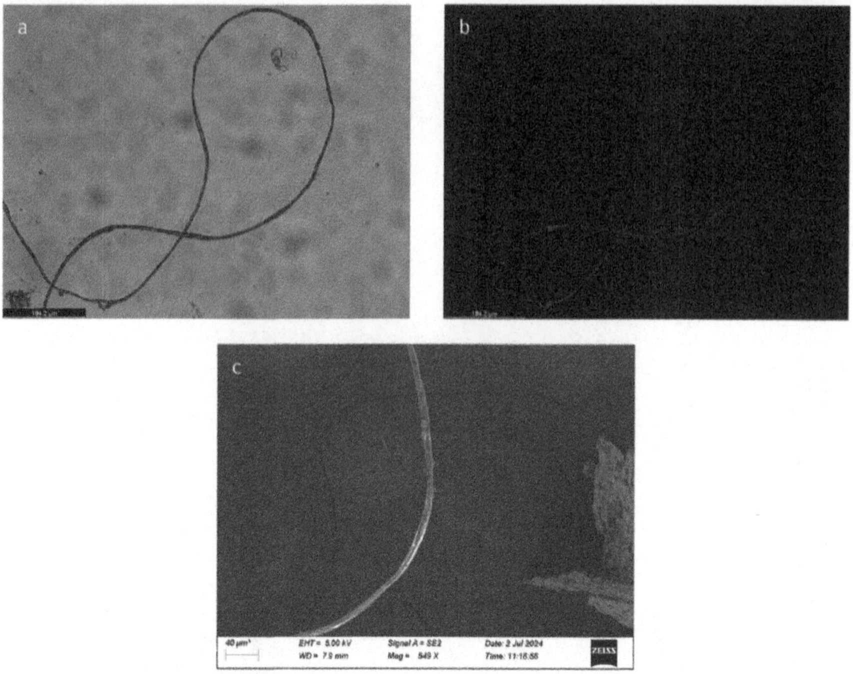

Fig. 6.3 Characterization of microplastics using microscopes. (**a**) Dark field image of fiber, (**b**) Nile red staining of the same fiber, and (**c**) scanning electron microscopic image of fiber

under a fluorescence microscope (Fig. 6.3a and b). The second step is submerging microplastics in a dye solution. For this, Nile red will be dissolved into organic solvents such as chloroform, acetone, methanol, and n-hexane, as NR is hydrophobic (Tamminga 2017). The microplastics suspended in the solution undergo heat and cold treatment to increase the intensity of fluorescence and prevent leaching (Lv et al. 2019). After which, the solution is filtered by vacuum filter and washed using distilled water to remove excess dye (Liu et al. 2022b).

Other types of fluorescent dyes, such as FITC, Safranin T (Lv et al. 2019), 4-chloro-7-nitro-1, 2, 3-benzoxadiazole (Li et al. 2020b), pink and blue reagents, and Rit 93 dye and Kentucky synthetic dyes (Karakolis et al. 2019), can be used to quantify microplastics. Besides various dyes, fluorescence lifetime imaging microscopy (FLIM) can be used to characterize microplastics (Monteleone et al. 2021; Zhou et al. 2022).

6.5.3 SEM Imaging

After the prescreening of microplastic samples in an optical microscope, the selected samples can be observed under SEM. Surface features of microplastics include fractures, cracks, flakes, and granular form and are mainly produced by the weathering process in the natural environment. Using a scanning electron microscope, the surface structure of microplastics can be easily observed. SEM can also be coupled with energy dispersive X-ray spectroscopy (SEM-EDS) for the determination of elemental composition and high-resolution images (Fig. 6.3c).

6.6 Spectroscopy

There are various spectroscopic methods used for microplastic characterization. However, FTIR and Raman spectroscopy have been complementarily used for frequent determination and characterization of microplastics due to strong bands obtained in the IR spectroscopy as well as in Raman spectroscopy. These methods efficiently distinguish crystallinity, chemical modifications, and weathering of microplastics (Gniadek and Dąbrowska 2019).

6.6.1 FTIR (Preparation, Loading, and Observation)

FTIR refers to Fourier transform infrared spectroscopy, which is based on the principle of spectroscopy. As the molecule in an atom has less energy than visible light, it absorbs the infrared (IR) radiation and moves to an excited state. This energy shift changes the vibrational state of the excited molecule, which is then detected by FTIR and depicted in different peaks (Sivasankar and Sunitha 2024). FTIR is generally used to identify different chemical bonds present in the particle (Kavya et al. 2020). FTIR is a reliable method of characterizing microplastics because it can give characteristic spectra of different polymers from which plastics have been derived, such as polyester (PET), polypropylene (PP), and polyethylene (PE). Prior to the FTIR analysis, the environmental samples must undergo an extraction process to bring out microplastic particles without contamination. After extraction, microplastic pellets are prepared using potassium bromide (KBr) and observed under IR spectroscopy (Davis 1962). The samples are exposed to IR of wave number 400–4000 cm^{-1}(Käppler et al. 2016). The spectra obtained for the samples can be compared to the reference spectra to identify the type of polymer (Xu et al. 2019). Natural weathering processes, biofouling, oxidation, and variation in composition—the spectra of environmental samples may vary from commercially available spectrum databases (Sivasankar& Sunitha 2024). Attenuated total reflectance infrared spectroscopy (ATR-FTIR), micro-FTIR (μ-FTIR), and focal plane arrays can be used to optimize the analysis through FTIR. Microplastics of very low particulate size (<20 μm) samples can be identified using micro-FTIR (μ-FTIR) (Song et al. 2014). ATR-FTIR can directly analyze samples, skipping the extraction process with high sensitivity and in less time (de Jesus et al. 2018) (Fig. 6.4).

The extracted microplastics give characteristic peaks of different functional groups present in the microplastics. The type of composition can be deduced from the identified functional groups. For example, the absorption spectra at 2924.95 cm^{-1} and 2850.68 cm^{-1} correspond to C-H asymmetric and symmetric stretching,

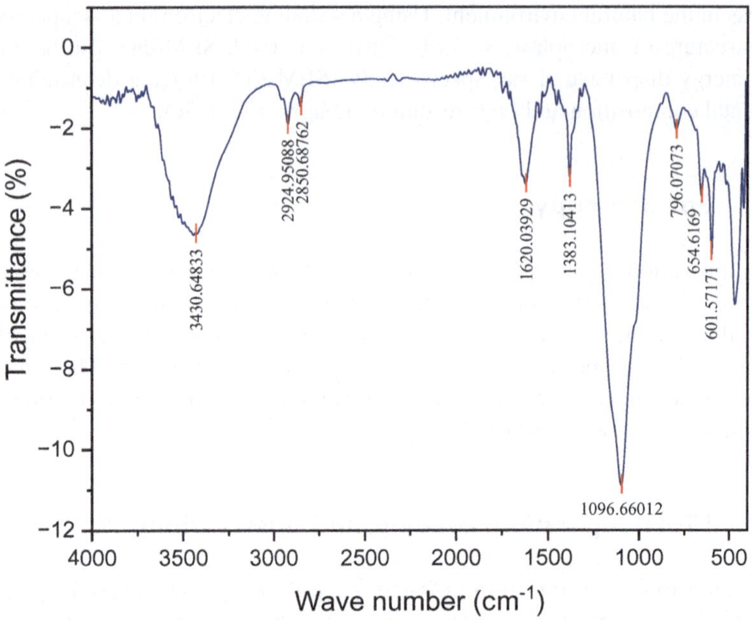

Fig. 6.4 FTIR graph of water sample showing different characteristic peaks of microplastics

respectively, which is present in the maximum form of microplastics as they are polymers of carbon (Jung et al. 2018). The peaks at 3430.64 cm^{-1}, 1620.03 cm^{-1}, 1383.10 cm^{-1}, and 601.57 cm^{-1} are characteristic peaks of N-H stretch, C=O stretch, CH_3 bend, and C-Cl stretch, respectively. Whereas spectra at 1096.66 cm^{-1} (aromatic C-H bending, C stretching), 796.07 cm^{-1} (CH_3 rocking, C-C stretching, CH_2 rocking, C-CH_3 stretching, C-CH stretching), and 654.61 cm^{-1} (N-H bending, C=O bending, aromatic CH out-of-the-plane bend) signify a number of functional groups (Noda et al. 2007; Chércoles Asensio et al. 2009; Mecozzi et al. 2016; Strong 2021; Nandiyanto et al. 2023). Hence, the above analysis depicts the presence of PVC as the graph shows C-C stretching, C-H bending, and C-Cl stretching; PU with C-H stretch and C=O stretch; PE having C=O stretch and aromatic CH out-of-the-plane bend, PS and ABS (Acrylonitrile butadiene styrene) with C-H stretch, and CH out-of-the-plane bend, PP showing C-H stretch, CH_3 rock, C-C stretch, CH_2 rock, C-CH_3 stretch, and C-CH stretch, and cellophane (natural polymer) having C-H stretch and C-C stretch (Beltran and Marcilla 1997; Faraji and Rostami 2020; Mecozzi et al. 2016; Jung et al. 2018).

6.6.2 Raman Spectroscopy

MPs greater than 10 μm and those less than 1 μm can be detected using Raman spectroscopy and micro-Raman spectroscopy, respectively. In this method, microplastics are characterized on the basis of molecular vibration and the chemical nature of the sample. This approach is not used widely because it is expensive and time-consuming compared to FTIR (Campanale et al. 2020).

6.7 Chromatography

Separation of mixtures can be carried out as a result of the interaction with the stationary phase.

6.7.1 GC-MS

Gas chromatography-mass spectrometry is used to quantify and identify volatile particles in the sample. The volatile sample is retained on a selective absorbent, which is then introduced to a chromatographic technique. Separation of complex microplastics can be done using thermal separation, and GC-MS is coupled with Thermal Extraction Desorption Technique (TED-GC-MS) (Sorolla-Rosario et al. 2023; Lee et al. 2023).

6.7.2 Pyr-GC-MS

Pyrolysis gas chromatography linked with a mass spectroscopic detector (Pyr-GC-MS) is prevalent for the recognition of different polymer types and associated organic plastic additives (OPAs). In this method, high-temperature thermal degradation products of microplastics are utilized for deducing structural information related to microplastics (Fries et al. 2013). Pyrolysis temperature varies from 6000–1000 °C depending on the application (Sait 2019). In this technique, pyrolysis is done by various heating techniques like inductive heating, resistive heating, and isothermal furnace (Wirnkor et al. 2019). For the determination of polymer type, various polymer standards are analyzed (Sefiloglu et al. 2024b).

6.8 Data Analysis and Interpretation Using Origin Software Version 2024b

The data generated after FTIR analysis is interpreted using Origin software version 2024b after selecting the X-Y axis with parameters. The plotted data will undergo baseline correction, and required peaks can be marked. The characteristic peaks can be analyzed using standard polymer peaks.

6.9 Quality Assurance/Quality Control (QA/QC)

Airborne contaminants can interfere with the results; hence, precautions should be taken during sampling and processing to minimize contamination. The following methods can be incorporated to avoid cross-contamination: Using glass or metal equipment, avoiding the use of synthetic wear during sampling and processing, cleaning surfaces with ethanol and vials with acid, keeping samples covered, processing in closed vials, operating experiments in clean rooms with less air circulation, or using fume hoods (Prata et al. 2019a, 2019b). The drawback of fluorescence imaging using Nile red is that it can also bind to natural particles, which can result in false positives. In the case of PVC, it shows weak quenching, which can lead to underestimation (Sturm et al. 2021). The staining efficacy also depends upon dye concentration, microplastic type, solvent used, and temperature (Wiggin and Holland 2019). The optimum concentration of Nile red solution should be used, as increased concentration can decrease fluorescence intensity (Lv et al. 2019). Solvent type can affect the solubility of the dyes, and temperature can also affect absorption capacity (Liu et al. 2022b).

There are many factors that affect the quantification of microplastics by interfering with the absorption spectra generated by the particles. The uniformity of distribution of microplastics while making pellets is an important factor in spectra formation (Morgado et al. 2022). The various sizes of microplastics may absorb IR radiation differently, which can interfere with the spectra and result in underestimation (Ivleva 2021). The presence of other unwanted contaminants, like organic or inorganic materials, can affect the identification and quantification of microplastics. Furthermore, as a spectrum of environmental microplastics differs from the commercially available spectra, there is a need for the generation of an updated spectra database (Cai et al. 2019).

6.10 Methods Validated by Authors for the Extraction of Microplastics from Water, Sediment, and Tissue Samples

6.10.1 Processing and Extraction of Microplastics

A. **Water Sample**

1. One thousand ml of water sample was taken and passed through a 1 μm sieve.
2. The residue was collected in a clean, labeled glass beaker using distilled water.
3. Twenty ml of 0.05 M $FeSO_4$. $7H_2O$ solution was added to the beaker and kept at room temperature for 5 minutes, followed by the addition of 20 ml of hydrogen peroxide (H_2O_2). The sample was mixed well and kept at room temperature for 5 min.

4. The solution was boiled at 75 ° C and stirred at 350 rpm for 30 minutes on a magnetic stirrer.
5. Six gm of NaCl was added to the solution and was mixed thoroughly on a magnetic stirrer. The solution was allowed to cool.
6. The cooled solution was passed through a stack of sieves ranging from 5 mm to 1 μm, and the extracted microplastics were collected in different petri plates. The petri plates were transferred to the oven for drying, and the dried samples were collected in closed vials (Fig. 6.5).

B. **Sediment Sample**

1. The sediment samples were collected from the bottom of the estuary and seabed with the help of a Van Veen grab and stored in labeled zip-lock pouches till processing.
2. The sediments were placed in clean, labeled petri dishes and allowed to dry at 60 °C for 96 hrs, and the dried sediments were powdered using a mortar and pestle.
3. Ten gm of sediment sample was taken in a clean beaker of 500 ml capacity, and 100 ml of 4 M NaCl was added. The solution was stirred vigorously for proper mixing.
4. The supernatant was sieved through a 1 μm sieve and collected in a clean beaker and was repeated three times. This was followed by the addition of 20 ml of 0.05 M $FeSO_4.7H_2O$ solution and incubation at room temperature for a period of 5 minutes, and 20 ml of hydrogen peroxide (H_2O_2) was added and incubated at room temperature for a period of 5 minutes.
5. The solution was boiled at 75 ° C and stirred at 350 rpm for 30 minutes on a magnetic stirrer. 3gm of NaCl was added to the beaker and allowed to dissolve completely. The solution was cooled using a cold water bath.

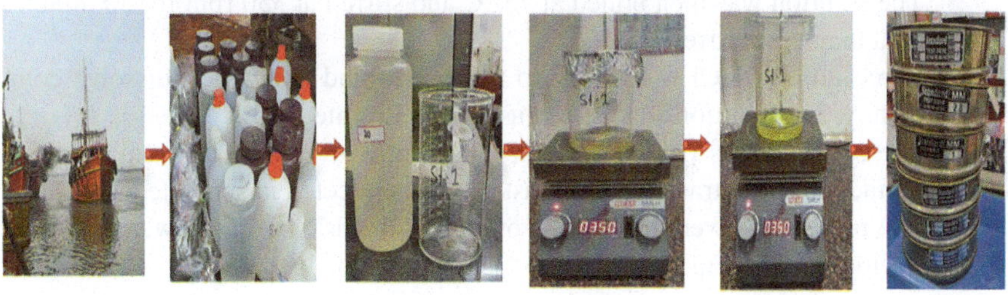

Fig. 6.5 Extraction of microplastics from water samples

Fig. 6.6 Extraction of microplastics from sediment samples

6. The solution was passed through a stack of sieves ranging from 5 mm to 1 μm and extracted microplastics were collected in separate labeled petri plates.
7. The petri plates were dried in an oven, and the dried samples were collected in closed vials (Fig. 6.6).

C. Biota Sample

1. The alimentary canal and muscle tissue of the fish and prawn, the muscle tissue of the crab, and the digestive gland of bivalves were dissected, cleaned, and transferred separately into a glass beaker of 100 ml capacity. This was followed by the addition of 30 ml of 4 M KOH for digestion. For every 5 g of tissue, 30 ml of 4 M KOH was added to the sample.
2. After digestion, 5 ml of hydrogen peroxide (H_2O_2) was added, and the samples were kept overnight. 20 ml of 0.05 M $FeSO_4.7H_2O$ solution was added to the beaker and incubated at room temperature for 5 min. This was followed by the addition of 20 ml of hydrogen peroxide (H_2O_2), and the solution was mixed properly with incubation at room temperature for 5 minutes.
3. The solution was then boiled at 75 °C and stirred at 350 rpm for 30 minutes on a magnetic stirrer.
4. Three gm of NaCl was added to the beaker and allowed to dissolve completely. The solution was cooled using a cold water bath.
5. The solution was passed through a stack of sieves ranging from 5 mm to 1 μm, and the extracted microplastics were collected in separate petri plates.
6. The petri plates were dried in an oven, and the dried samples were collected in closed vials (Fig. 6.7).

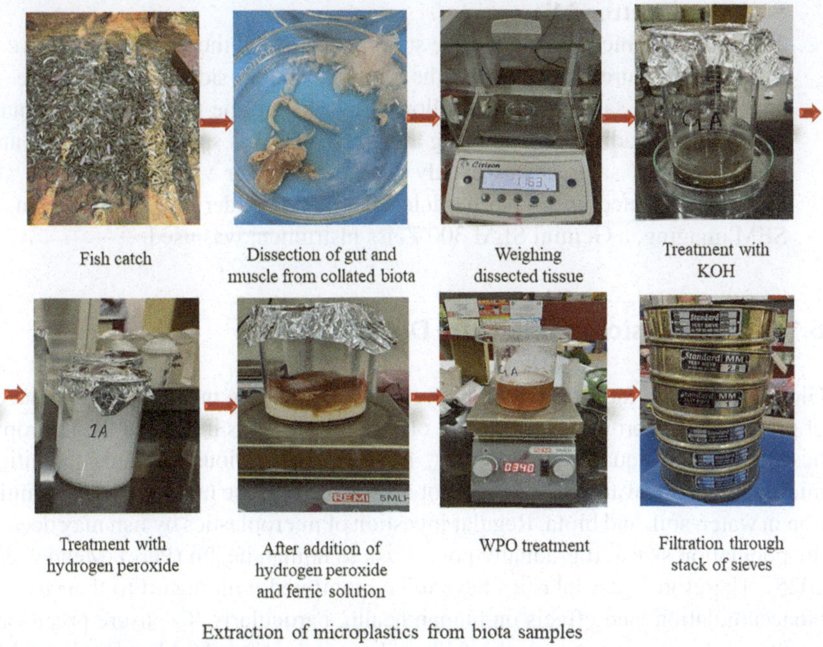

Fig. 6.7 Extraction of microplastics from tissue samples

6.10.2 Characterization of Microplastics

A. **Staining Method**

Nile red is a lipophilic dye (9-diethylamino-5H-benzo[α]phenoxazine-5-one) and is generally used for detecting intracellular lipid droplets and in cytofluorometry (Andrady 2011; Shim et al. 2016; Greenspan and Fowler 1985). A stock solution was prepared by dissolving 1 mg of Nile red in 1 ml of acetone. For a working solution, the concentration used was 10^{-5} mg/ml, and the solution was diluted with acetone. The ratio of Nile red stain and extracted microplastic solution was maintained as 1:1. Three drops of extracted microplastic solution were taken on a clean slide and allowed to dry completely. The working Nile red staining solution was added dropwise using a micropipette to cover the entire dried area and again allowed to dry in the dark. The stained slide was observed under a fluorescence microscope at a magnification of 20x.

B. **Microscopy**

Compound Microscope

The dried slides were observed at 10× magnification under the compound microscope to visualize the type and number of microplastics in the sample. A Nikon microscope equipped with a camera was used to photograph microplastics.

Scanning Electron Microscope

The extracted microplastic sample solution was dried thoroughly to remove any trace of moisture. For loading the sample, double-sided adhesive tape was attached to the sample holder, followed by placing the processed microplastic sample on the adhesive tape using a toothpick. It was spread thoroughly, and a single layer of sample was evenly spread for better visualization. The slide holder was coated with gold particles and placed under an electron beam. For SEM imaging, a Gemini SEM 300 Zeiss instrument was used.

6.11 Conclusion and Future Direction

Globalization and industrialization have led to an increase in the overall production of plastic, and improper management of plastic waste has increased the microplastics load in the aquatic environment. Even though various methods to mitigate microplastics are available, they are not adequate to reduce microplastic contamination in water, soil, and biota. Regular ingestion of microplastics by fish may decrease the population size and gradually pose a risk to human health (Pérez-Méndez et al. 2025). Therefore, microplastics have to be monitored with regard to their toxicity, bioaccumulation, and effects on human health, particularly. To ensure precision in results, replicates of samples should be taken, as each method has its own advantages and disadvantages. Hence, any method to be adopted should be finalized by taking into account the source, composition, and requirements based on the type of ecological environment. There is a need to develop handy techniques and standardize methods for ensuring quick and precision-based extraction of microplastics from different sources and also developing technology for plastic alternatives to save our freshwater bodies and oceans from contamination caused by microplastics and their impact on the health of humans.

References

Abisha C, Kutty R, Gurjar UR et al (2024) Microplastic prevalence, diversity and characteristics in commercially important edible bivalves and gastropods in relation to environmental matrices. J Hazard Mater Adv 13:100392. https://doi.org/10.1016/j.hazadv.2023.100392

Adomat Y, Grischek T (2021) Sampling and processing methods of microplastics in river sediments-a review. Sci Total Environ 758:143691. https://doi.org/10.1016/j.scitotenv.2020.143691

Alak G, Uçar A, Parlak V, Atamanalp M (2022) Identification, characterisation of microplastic and their effects on aquatic organisms. Chem Ecol 38(10):967–987. https://doi.org/10.1080/02757540.2022.2126461

Amponsah AK, Afrifa EA, Essandoh PK, Enyoh CE (2024) Evidence of microplastics accumulation in the gills and gastrointestinal tract of fishes from an estuarine system in Ghana. Heliyon 10(3):e25608. https://doi.org/10.1016/j.heliyon.2024.e25608

Andrady AL (2011) Microplastics in the marine environment. Mar Pollut Bull 62(8):1596–1605. https://doi.org/10.1016/j.marpolbul.2011.05.030

Atamanalp M, Köktürk M, Uçar A et al (2021) Microplastics in tissues (brain, gill, muscle and gastrointestinal) of Mullus barbatus and Alosa immaculata. Arch Environ Contam Toxicol 81:460–469. https://doi.org/10.1007/s00244-021-00885-5

Barboza LGA, Lourenço SC, Aleluia A, Senes GP, Otero XL, Guilhermino L (2024) Are microplastics a new cardiac threat? A pilot study with wild fish from the north East Atlantic Ocean. Environ Res 261:119694. https://doi.org/10.1016/j.envres.2024.119694

Beltran M, Marcilla A (1997) Fourier transform infrared spectroscopy applied to the study of PVC decomposition. Eur Polym J 33(7):1135–1142. https://doi.org/10.1016/S0014-3057(97)00001-3

Bornt K, Linge K, How J, de Lestang S, Hovey R, Langlois T (2024) Microplastic extraction from digestive tracts of large decapods. Mar Pollut Bull 206:116709. https://doi.org/10.1016/j.marpolbul.2024.116709

Bour A, Avio CG, Gorbi S, Regoli F, Hylland K (2018) Presence of microplastics in benthic and epibenthic organisms: influence of habitat, feeding mode and trophic level. Environ Pollut 243:1217–1225. https://doi.org/10.1016/j.envpol.2018.09.115

Cai M, He H, Liu M et al (2018) Lost but can't be neglected: huge quantities of small microplastics hide in the South China Sea. Sci Total Environ 633:1206–1216. https://doi.org/10.1016/j.scitotenv.2018.03.197

Cai H, Du F, Li L, Li B, Li J, Shi H (2019) A practical approach based on FT-IR spectroscopy for identification of semi-synthetic and natural celluloses in microplastic investigation. Sci Total Environ 669:692–701. https://doi.org/10.1016/j.scitotenv.2019.03.124

Campanale C, Savino I, Pojar I, Massarelli C, Uricchio VF (2020) A practical overview of methodologies for sampling and analysis of microplastics in riverine environments. Sustainability 12(17):6755. https://doi.org/10.3390/su12176755

Carson HS, Colbert SL, Kaylor MJ, McDermid KJ (2011) Small plastic debris changes water movement and heat transfer through beach sediments. Mar Pollut Bull 62(8):1708–1713. https://doi.org/10.1016/j.marpolbul.2011.05.032

Catarino AI, Thompson R, Sanderson W, Henry TB (2017) Development and optimization of a standard method for extraction of microplastics in mussels by enzyme digestion of soft tissues. Environ Toxicol Chem 36(4):947–951. https://doi.org/10.1002/etc.3608

Çevik C, Gündoğdu S (2018) Quantity and types of microplastics in the the tissues of the spiny oysters Spondylus spinosus Schreibers, 1793 (Mollusca, Bivalvia) in the Yumurtalık bight (Iskenderun Bay, the northeastern coast of Levatine Sea). Marfresh 1793(October):254–258

Chen YN, Rani A, Chiang CY, Kim H, Pan SY (2023) Monitoring, control and assessment of microplastics in bioenvironmental systems. Environ Technol Innov 32:103250. https://doi.org/10.1016/j.eti.2023.103250

Chércoles Asensio R, San Andrés Moya M, De la Roja JM, Gómez M (2009) Analytical characterization of polymers used in conservation and restoration by ATR-FTIR spectroscopy. Anal Bioanal Chem 395:2081–2096. https://doi.org/10.1007/s00216-009-3201-2

Claessens M, Van Cauwenberghe L, Vandegehuchte MB, Janssen CR (2013) New techniques for the detection of microplastics in sediments and field collected organisms. Mar Pollut Bull 70(1–2):227–233. https://doi.org/10.1016/j.marpolbul.2013.03.009

Cole M, Webb H, Lindeque PK, Fileman ES, Halsband C, Galloway TS (2014) Isolation of microplastics in biota-rich seawater samples and marine organisms. Sci Rep 4(1):4528. https://doi.org/10.1038/srep04528

Constant M, Billon G, Breton N, Alary C (2021) Extraction of microplastics from sediment matrices: experimental comparative analysis. J Hazard Mater 420:126571. https://doi.org/10.1016/j.jhazmat.2021.126571

Coppock RL, Cole M, Lindeque PK, Queirós AM, Galloway TS (2017) A small-scale, portable method for extracting microplastics from marine sediments. Environ Pollut 230:829–837. https://doi.org/10.1016/j.envpol.2017.07.017

Courtene-Jones W, Quinn B, Murphy F, Gary SF, Narayanaswamy BE (2017) Optimisation of enzymatic digestion and validation of specimen preservation methods for the analysis of ingested microplastics. Anal Methods 9(9):1437–1445. https://doi.org/10.1039/C6AY02343F

Cox KD, Covernton GA, Davies HL, Dower JF, Juanes F, Dudas SE (2019) Human consumption of microplastics. Environ Sci Technol 53(12):7068–7074. https://doi.org/10.1021/acs.est.9b01517

Crew A, Gregory-Eaves I, Ricciardi A (2020) Distribution, abundance, and diversity of microplastics in the upper St. Lawrence River. Environ Poll 260:113994. https://doi.org/10.1016/j.envpol.2020.113994

Crichton EM, Noël M, Gies EA, Ross PS (2017) A novel, density-independent and FTIR-compatible approach for the rapid extraction of microplastics from aquatic sediments. Anal Methods 9(9):1419–1428. https://doi.org/10.1039/C6AY02733D

Davis A (1962) Infrared spectra correlation of plastics and resins. In Progress in infrared spectroscopy*: Volume 1* (pp. 17–33). Boston MA: Springer US. doi:https://doi.org/10.1007/978-1-4899-7009-1_3

Davison P, Asch RG (2011) Plastic ingestion by mesopelagic fishes in the North Pacific subtropical gyre. Mar Ecol Prog Ser 432:173–180. https://doi.org/10.3354/meps09142

de Jesus Piñon-Colin T, Rodriguez-Jimenez R, Pastrana-Corral MA, Rogel-Hernandez E, Wakida FT (2018) Microplastics on sandy beaches of the Baja California peninsula, Mexico. Mar Pollut Bull 131:63–71. https://doi.org/10.1016/j.marpolbul.2018.03.055

Dehaut A, Cassone AL, Frère L et al (2016) Microplastics in seafood: benchmark protocol for their extraction and characterization. Environ Pollut 215:223–233. https://doi.org/10.1016/j.envpol.2016.05.018

Di Mauro R, Kupchik MJ, Benfield MC (2017) Abundant plankton-sized microplastic particles in shelf waters of the northern Gulf of Mexico. Environ Pollut 230:798–809. https://doi.org/10.1016/j.envpol.2017.07.030

DiSalvo LH (1988) Observations on the larval and post-metamorphic LIFE of CONCHOLEPAS-CONCHOLEPAS (BRUGUIERE, 1789) in laboratory culture. Veliger 30(4):358–368

Du F, Cai H, Zhang Q, Chen Q, Shi H (2020) Microplastics in take-out food containers. J Hazard Mater 399:122969. https://doi.org/10.1016/j.jhazmat.2020.122969

Dwiyitno D, Barokah GR, Hidayah I, Januar HI, Wibowo S (2024) Microplastic occurrence in different fish organs from two coastal waters in Java Sea, Indonesia. Squalen Bullet Marine Fish Postharvest Biotechnol 19(2):145–153. https://doi.org/10.15578/squalen.941

Enders K, Lenz R, Beer S, Stedmon CA (2017) Extraction of microplastic from biota: recommended acidic digestion destroys common plastic polymers. ICES J Mar Sci 74:326–331. https://doi.org/10.1093/icesjms/fsw173

Eriksen M, Liboiron M, Kiessling T et al (2018) Microplastic sampling with the AVANI trawl compared to two neuston trawls in the bay of Bengal and South Pacific. Environ Pollut 232:430–439. https://doi.org/10.1016/j.envpol.2017.09.058

Evelyn GRQLB, Pérez-Sabino OHJF (2024) Identification of microplastic polymers found in the digestive tract of fish from Lake Amatitlán, Guatemala. Mundo Saúde 48:e15762023. https://doi.org/10.15343/0104-7809.202448e15762023I

Faraji M, Rostami R (2020) Polypyrrole-carbon nanotube-cellophane composite plate with homogeneous network structure for flexible symmetric supercapacitor device. J Mater Sci Mater Electron 31(19):16849–16858. https://doi.org/10.1007/s10854-020-04241-6

Faure F, Demars C, Wieser O, Kunz M, De Alencastro LF (2015) Plastic pollution in Swiss surface waters: nature and concentrations, interaction with pollutants. Environ Chem 12(5):582–591. https://doi.org/10.1071/EN14218

Forsythe K, Egermeier M, Garcia M et al (2024) Viability of elutriation for the extraction of microplastics from environmental soil samples. Environ Sci Adv 3(7):1039–1047. https://doi.org/10.1039/D4VA00087K

Frias JP, Otero V, Sobral P (2014) Evidence of microplastics in samples of zooplankton from Portuguese coastal waters. Mar Environ Res 95:89–95. https://doi.org/10.1016/j.marenvres.2014.01.001

Fries E, Dekiff JH, Willmeyer J, Nuelle MT, Ebert M, Remy D (2013) Identification of polymer types and additives in marine microplastic particles using pyrolysis-GC/MS and scanning electron microscopy. Environ Sci: Processes Impacts 15(10):1949–1956. https://doi.org/10.1039/c3em00214d

Fuller S, Gautam A (2016) A procedure for measuring microplastics using pressurized fluid extraction. Environ Sci Technol 50(11):5774–5780. https://doi.org/10.1021/acs.est.6b00816

Gago J, Filgueiras A, Pedrotti ML, Caetano M, Frias J (2019) Standardised protocol for monitoring microplastics in seawater. Deliverable 4.1. https://doi.org/10.25607/OBP-605

Ghanadi M, Joshi I, Dharmasiri N et al (2024) Quantification and characterization of microplastics in coastal environments: insights from laser direct infrared imaging. Sci Total Environ 912:168835. https://doi.org/10.1016/j.scitotenv.2023.168835

Gniadek M, Dąbrowska A (2019) The marine nano-and microplastics characterisation by SEM-EDX: the potential of the method in comparison with various physical and chemical approaches. Mar Pollut Bull 148:210–216. https://doi.org/10.1016/j.marpolbul.2019.07.067

Goli VSNS, Singh DN (2023) Extraction and characterization of microplastics in landfill-mined-soil-like-fractions: a novel methodology. Chem Eng J 452:139217. https://doi.org/10.1016/j.cej.2022.139217

Grause G, Kuniyasu Y, Chien MF, Inoue C (2022) Separation of microplastic from soil by centrifugation and its application to agricultural soil. Chemosphere 288:132654. https://doi.org/10.1016/j.chemosphere.2021.132654

Green DS, Kregting L, Boots B, Blockley DJ, Brickle P, Da Costa M, Crowley Q (2018) A comparison of sampling methods for seawater microplastics and a first report of the microplastic litter in coastal waters of Ascension and Falkland Islands. Mar Pollut Bull 137:695–701. https://doi.org/10.1016/j.marpolbul.2018.11.004

Greenspan P, Fowler SD (1985) Spectrofluorometric studies of the lipid probe, nile red. J Lipid Res 26(7):781–789. https://doi.org/10.1016/S0022-2275(20)34307-8

Gusmao JB, Diaz O, Gallardo C et al (2022) Water depth and mesoscale oceanography drive neustonic polychaete assemblages in the SE Pacific Ocean. Mar Ecol Prog Ser 682:123–136. https://doi.org/10.3354/meps13929

Han X, Lu X, Vogt RD (2019) An optimized density-based approach for extracting microplastics from soil and sediment samples. Environ Pollut 254:113009. https://doi.org/10.1016/j.envpol.2019.113009

Hanvey JS, Lewis PJ, Lavers JL, Crosbie ND, Pozo K, Clarke BO (2017) A review of analytical techniques for quantifying microplastics in sediments. Anal Methods 9(9):1369–1383. https://doi.org/10.1039/C6AY02707E

He B, Goonetilleke A, Ayoko GA, Rintoul L (2020) Abundance, distribution patterns, and identification of microplastics in Brisbane River sediments, Australia. Sci Total Environ 700:134467. https://doi.org/10.1016/j.scitotenv.2019.134467

He YJ, Qin Y, Zhang TL et al (2021) Migration of (non-) intentionally added substances and microplastics from microwavable plastic food containers. J Hazard Mater 417:126074. https://doi.org/10.1016/j.jhazmat.2021.126074

Hermsen E, Mintenig SM, Besseling E, Koelmans AA (2018) Quality criteria for the analysis of microplastic in biota samples: a critical review. Environ Sci Technol 52(18):10230–10240. https://doi.org/10.1021/acs.est.8b01611

Hidalgo-Ruz V, Gutow L, Thompson RC, Thiel M (2012) Microplastics in the marine environment: a review of the methods used for identification and quantification. Environ Sci Technol 46(6):3060–3075. https://doi.org/10.1021/es2031505

Ho D, Liu S, Wei H, Karthikeyan KG (2023) The glowing potential of Nile red for microplastics identification: science and mechanism of fluorescence staining. Microchem J 109708:109708. https://doi.org/10.1016/j.microc.2023.109708

Holland ER, Mallory ML, Shutler D (2016) Plastics and other anthropogenic debris in freshwater birds from Canada. Sci Total Environ 571:251–258. https://doi.org/10.1016/j.scitotenv.2016.07.158

Hossain MB, Pingki FH, Azad MAS et al (2024) Accumulation, tissue distribution, health hazard of microplastics in a commercially important cat fish, Siloniasilondia from a tropical large-scale estuary. Front Sustain Food Sys 8:1372059. https://doi.org/10.3389/fsufs.2024.1372059

Huppertsberg S, Knepper TP (2018) Instrumental analysis of microplastics—benefits and challenges. Anal Bioanal Chem 410:6343–6352. https://doi.org/10.1007/s00216-018-1210-8

Hurley RR, Lusher AL, Olsen M, Nizzetto L (2018) Validation of a method for extracting microplastics from complex, organic-rich, environmental matrices. Environ Sci Technol 52(13):7409–7417. https://doi.org/10.1021/acs.est.8b01517

Imhof HK, Schmid J, Niessner R, Ivleva NP, Laforsch C (2012) A novel, highly efficient method for the separation and quantification of plastic particles in sediments of aquatic environments. Limnol Oceanogr Methods 10(7):524–537. https://doi.org/10.4319/lom.2012.10.524

Isobe A, Uchida K, Tokai T, Iwasaki S (2015) East Asian seas: a hot spot of pelagic microplastics. Mar Pollut Bull 101(2):618–623. https://doi.org/10.1016/j.marpolbul.2015.10.042

Ivleva NP (2021) Chemical analysis of microplastics and nanoplastics: challenges, advanced methods, and perspectives. Chem Rev 121(19):11886–11936. https://doi.org/10.1021/acs.chemrev.1c00178

Jiang C, Yin L, Li Z et al (2019) Microplastic pollution in the rivers of the Tibet plateau. Environ Pollut 249:91–98. https://doi.org/10.1016/j.envpol.2019.03.022

Jung MR, Horgen FD, Orski SV et al (2018) Validation of ATR FT-IR to identify polymers of plastic marine debris, including those ingested by marine organisms. Mar Pollut Bull 127:704–716. https://doi.org/10.1016/j.marpolbul.2017.12.061

Kapp KJ, Yeatman E (2018) Microplastic hotspots in the Snake and lower Columbia rivers: a journey from the greater Yellowstone ecosystem to the Pacific Ocean. Environ Pollut 241:1082–1090. https://doi.org/10.1016/j.envpol.2018.06.033

Käppler A, Fischer D, Oberbeckmann S, Schernewski G, Labrenz M, Eichhorn KJ, Voit B (2016) Analysis of environmental microplastics by vibrational microspectroscopy: FTIR, Raman or both? Anal Bioanal Chem 408:8377–8391. https://doi.org/10.1007/s00216-016-9956-3

Karakolis EG, Nguyen B, You JB, Rochman CM, Sinton D (2019) Fluorescent dyes for visualizing microplastic particles and fibers in laboratory-based studies. Environ Sci Technol Lett 6(6):334–340. https://doi.org/10.1021/acs.estlett.9b00241

Karami A, Golieskardi A, Choo CK, Romano N, Ho YB, Salamatinia B (2017) A high-performance protocol for extraction of microplastics in fish. Sci Total Environ 578:485–494. https://doi.org/10.1016/j.scitotenv.2016.10.213

Karlsson TM, Vethaak AD, Almroth BC, Ariese F, van Velzen M, Hassellöv M, Leslie HA (2017) Screening for microplastics in sediment, water, marine invertebrates and fish: method development and microplastic accumulation. Mar Pollut Bull 122(1–2):403–408. https://doi.org/10.1016/j.marpolbul.2017.06.081

Kataoka T, Nihei Y, Kudou K, Hinata H (2019) Assessment of the sources and inflow processes of microplastics in the river environments of Japan. Environ Pollut 244:958–965. https://doi.org/10.1016/j.envpol.2018.10.111

Kavya ANL, Sundarrajan S, Ramakrishna S (2020) Identification and characterization of microplastics in the marine environment: a mini review. Mar Pollut Bull 160:111704. https://doi.org/10.1016/j.marpolbul.2020.111704

Klein S, Worch E, Knepper TP (2015) Occurrence and spatial distribution of microplastics in river shore sediments of the Rhine-Main area in Germany. Environ Sci Technol 49(10):6070–6076. https://doi.org/10.1021/acs.est.5b00492

Kononov A, Hishida M, Suzuki K, Harada N (2022) Microplastic extraction from agricultural soils using canola oil and unsaturated sodium chloride solution and evaluation by incineration method. Soil Sys 6(2):54. https://doi.org/10.3390/soilsystems6020054

Lahens L, Strady E, Kieu-Le TC (2018) Macroplastic and microplastic contamination assessment of a tropical river (Saigon River, Vietnam) transversed by a developing megacity. Environ Pollut 236:661–671. https://doi.org/10.1016/j.envpol.2018.02.005

Lee JH, Kim MJ, Kim CS (2023) Detection of microplastic traces in four different types of municipal wastewater treatment plants through FT-IR and TED-GC-MS. Environ Pollut 333:122017. https://doi.org/10.1016/j.envpol.2023.122017

Li Q, Wu J, Zhao X, Gu X, Ji R (2019) Separation and identification of microplastics from soil and sewage sludge. Environ Pollut 254:113076. https://doi.org/10.1016/j.envpol.2019.113076

Li C, Busquets R, Campos LC (2020a) A assessment of microplastics in freshwater systems: a review. Sci Total Environ 707:135578. https://doi.org/10.1016/j.scitotenv.2019.135578

Li L, Luo Y, Peijnenburg WJ, Li R, Yang J, & Zhou Q (2020b)b Confocal measurement of microplastics uptake by plants. MethodsX 7: 100750.doi:https://doi.org/10.1016/j.mex.2019.11.023

Li W, Wufuer R, Duo J, Wang S, Luo Y, Zhang D, & Pan X (2020c)c Microplastics in agricultural soils: extraction and characterization after different periods of polythene film mulching in an arid region. Sci Total Environ 749: 141420.doi:https://doi.org/10.1016/j.scitotenv.2020.141420

Li C, Cui Q, Zhang M, Vogt RD, Lu X (2021) A commonly available and easily assembled device for extraction of bio/non-degradable microplastics from soil by flotation in NaBr solution. Sci Total Environ 759:143482. https://doi.org/10.1016/j.scitotenv.2020.143482

Li JY, Nankervis L, Dawson AL (2022) Digesting the indigestible: microplastic extraction from prawn digestive tracts. Front Environ Chem 3:903314. https://doi.org/10.3389/fenvc.2022.903314

Li Y, Tao L, Wang Q, Wang F, Li G, Song M (2023) Potential health impact of microplastics: a review of environmental distribution, human exposure, and toxic effects. Environ Health 1(4):249–257. https://doi.org/10.1021/envhealth.3c00052

Li G, Pei Z, Li Y et al (2024) A high-precision, effective method for extraction and identification of small-sized microplastics from soil. Talanta 272:125802. https://doi.org/10.1016/j.talanta.2024.125802

Lima ARA, Costa MF, Barletta M (2014) Distribution patterns of microplastics within the plankton of a tropical estuary. Environ Res 132:146–155. https://doi.org/10.1016/j.envres.2014.03.031

Liu M, Song Y, Lu S et al (2019) A method for extracting soil microplastics through circulation of sodium bromide solutions. Sci Total Environ 691:341–347. https://doi.org/10.1016/j.scitotenv.2019.07.144

Liu S, Pan YF, Li HX et al (2022a)a Microplastic pollution in the surface seawater in Zhongsha atoll, South China Sea. Sci Total Environ 822: 153604. doi:https://doi.org/10.1016/j.scitotenv.2022.153604

Liu S, Shang E, Liu J, Wang Y, Bolan N, Kirkham MB, & Li Y (2022b)b What have we known so far for fluorescence staining and quantification of microplastics: a tutorial review. *Frontiers of* Environ Sci Eng 16: 1–14.doi:https://doi.org/10.1007/s11783-021-1442-2

Löder MG, Imhof HK, Ladehoff M et al (2017) Enzymatic purification of microplastics in environmental samples. Environ Sci Technol 51(24):14283–14292. https://doi.org/10.1021/acs.est.7b03055

López-Rosales A, Andrade JM, López-Mahía P, Muniategui-Lorenzo S (2022) Development of an analytical procedure to analyze microplastics in edible macroalgae using an enzymatic-oxidative digestion. Mar Pollut Bull 183:114061. https://doi.org/10.1016/j.marpolbul.2022.114061

Lusher AL, Hernandez-Milian G (2018) Microplastic extraction from marine vertebrate digestive tracts, regurgitates and scats: a protocol for researchers from all experience levels. Bio-protocol 8(22):e3087–e3087. https://doi.org/10.21769/BioProtoc.3087

Lusher AL, Munno K, Hermabessiere L, Carr S (2020a) Isolation and extraction of microplastics from environmental samples: an evaluation of practical approaches and recommendations for further harmonization. Appl Spectrosc 74(9):1049–1065. https://doi.org/10.1177/0003702820938993

Lusher AÁ, Welden NA, Sobral P, Cole M (2020b) Sampling, isolating and identifying microplastics ingested by fish and invertebrates. In: Analysis of nanoplastics and microplastics in food, vol 9. CRC Press, pp 119–148. https://doi.org/10.1039/c6ay02415g

Lv L, Qu J, Yu Z et al (2019) A simple method for detecting and quantifying microplastics utilizing fluorescent dyes-Safranine T, fluorescein isophosphate, Nile red based on thermal expansion and contraction property. Environ Pollut 255:113283. https://doi.org/10.1016/j.envpol.2019.113283

Maes T, Jessop R, Wellner N, Haupt K, Mayes AG (2017) A rapid-screening approach to detect and quantify microplastics based on fluorescent tagging with Nile red. Sci Rep 7(1):44501. https://doi.org/10.1038/srep44501

Mahat S (2017) Separation and quantification of microplastics from beach and sediment samples using the Bauta microplastic-sediment separator Master's thesis, Norwegian University of Life Sciences, Ås

Mai L, Bao LJ, Shi L, Wong CS, Zeng EY (2018) A review of methods for measuring microplastics in aquatic environments. Environ Sci Pollut Res 25:11319–11332. https://doi.org/10.1007/s11356-018-1692-0

Malloggi C, Nalbone L, Bartalena S (2024) The occurrence of microplastics in Donaxtrunculus (Mollusca: Bivalvia) collected along the Tuscany coast (Mediterranean Sea). Animals 14(4):618. https://doi.org/10.3390/ani14040618

Masura J, Baker J, Foster G, & Arthur C (2015) Laboratory methods for the analysis of microplastics in the marine environment: recommendations for quantifying synthetic particles in waters and sediments doi:https://doi.org/10.25607/OBP-604

Mecozzi M, Pietroletti M, Monakhova YB (2016) FTIR spectroscopy supported by statistical techniques for the structural characterization of plastic debris in the marine environment: application to monitoring studies. Mar Pollut Bull 106(1–2):155–161. https://doi.org/10.1016/j.marpolbul.2016.03.012

Mercogliano R, Santonicola S, Raimo G, Gasperi M, Colavita G (2021) Extraction and identification of microplastics from mussels: method development and preliminary results. Italian J Food Safety 10(1). https://doi.org/10.4081/ijfs.2021.9264

Merga LB, Redondo-Hasselerharm PE, Van den Brink PJ, Koelmans AA (2020) Distribution of microplastic and small macroplastic particles across four fish species and sediment in an African lake. Sci Total Environ 741:140527. https://doi.org/10.1016/j.scitotenv.2020.140527

Milne MH, Helm PA, Munno K, Bhavsar SP, Rochman CM (2024) Microplastics and anthropogenic particles in recreationally caught freshwater fish from an urbanized region of the north American Great Lakes. Environ Health Perspect 132(7):077004. https://doi.org/10.1289/EHP13540

Mintenig SM, Int-Veen I, Löder MG, Primpke S, Gerdts G (2017) Identification of microplastic in effluents of waste water treatment plants using focal plane array-based micro-Fourier-transform infrared imaging. Water Res 108:365–372. https://doi.org/10.1016/j.watres.2016.11.015

Möller JN, Heisel I, Satzger A et al (2022) Tackling the challenge of extracting microplastics from soils: a protocol to purify soil samples for spectroscopic analysis. Environ Toxicol Chem 41(4):844–857. https://doi.org/10.1002/etc.5024

Monteleone A, Wenzel F, Langhals H, Dietrich D (2021) New application for the identification and differentiation of microplastics based on fluorescence lifetime imaging microscopy (FLIM). J Environ Chem Eng 9(1):104769. https://doi.org/10.1016/j.jece.2020.104769

Morgado V, Gomes L, da Silva RJB, Palma C (2022) Microplastics contamination in sediments from Portuguese inland waters: physical-chemical characterisation and distribution. Sci Total Environ 832:155053. https://doi.org/10.1016/j.scitotenv.2022.155053

Motivarash Y, Bhatt A, Kardani H (2024) Microplastic (MP) occurrence in pelagic and demersal fishes of Gujarat, northwest coast of India. Environ Sci Pollut Res 31(11):17239–17255. https://doi.org/10.1007/s11356-024-32361-7

Naidoo T, Goordiyal K, Glassom D (2017) Are nitric acid (HNO 3) digestions efficient in isolating microplastics from juvenile fish? Water Air Soil Pollut 228:1–11. https://doi.org/10.1007/s11270-017-3654-4

Nandiyanto ABD, Ragadhita R, Fiandini M (2023) Interpretation of Fourier transform infrared spectra (FTIR): a practical approach in the polymer/plastic thermal decomposition. Indonesian J Sci Technol 8(1):113–126

Nava V, Leoni B (2021) Comparison of different procedures for separating microplastics from sediments. Water 13(20):2854. https://doi.org/10.3390/w13202854

Nikhil VG, Ranjeet K, Varghese GK (2023) Spatio-temporal evaluation and risk assessment of microplastics in nearshore surface waters post-2018 Kerala deluge along the southwest coast of India. Mar Pollut Bull 192:115058. https://doi.org/10.1016/j.marpolbul.2023.115058

Noda I, Dowrey AE, Haynes JL, Marcott C (2007) Group frequency assignments for major infrared bands observed in common synthetic polymers. In: Physical properties of polymers handbook. Springer, New York, NY, pp 395–406. https://doi.org/10.1007/978-0-387-69002-5_22

Nuelle MT, Dekiff JH, Remy D, Fries E (2014) A new analytical approach for monitoring microplastics in marine sediments. Environ Pollut 184:161–169. https://doi.org/10.1016/j.envpol.2013.07.027

Pasquier G, Doyen P, Kazour M, Dehaut A, Diop M, Duflos G, Amara R (2022) Manta net: the golden method for sampling surface water microplastics in aquatic environments. Front Environ Sci 10:811112. https://doi.org/10.3389/fenvs.2022.811112

Patterson J, Jeyasanta KI, Laju RL, Edward JP (2021) Microplastic contamination in Indian edible mussels (*Perna perna* and *Perna viridis*) and their environs. Mar Pollut Bull 171:112678. https://doi.org/10.1016/j.marpolbul.2021.112678

Pérez-Méndez MA, Fraga-Cruz GS, Domínguez-García S, Pérez-Méndez ML, Bocanegra-Díaz CI, Nápoles-Rivera F (2025) Microplastic pollution in soil and water and the potential effects on human health: a review. PRO 13(2):502

Perfetti-Bolaño A, Araneda A, Muñoz K, Barra RO (2022) Occurrence and distribution of microplastics in soils and intertidal sediments at Fildes Bay, Maritime Antarctica. Front Marine Sci 8:774055. https://doi.org/10.3389/fmars.2021.774055

Peters CA, Bratton SP (2016) Urbanization is a major influence on microplastic ingestion by sunfish in the Brazos River basin, Central Texas, USA. Environ Pollut 210:380–387. https://doi.org/10.1016/j.envpol.2016.01.018

Pfeiffer F, Fischer EK (2020) Various digestion protocols within microplastic sample processing—evaluating the resistance of different synthetic polymers and the efficiency of biogenic organic matter destruction. Front Environ Sci 8:572424. https://doi.org/10.3389/fenvs.2020.572424

Pfohl P, Roth C, Wohlleben W (2024) The power of centrifugation: How to extract microplastics from soil with high recovery and matrix removal efficiency. MethodsX 12:102598. https://doi.org/10.1016/j.mex.2024.102598

Poli V, Litti L, Lavagnolo MC (2024) Microplastic pollution in the north-East Atlantic Ocean surface water: How the sampling approach influences the extent of the issue. Sci Total Environ 947:174561. https://doi.org/10.1016/j.scitotenv.2024.174561

Prata JC, Da Costa JP, Duarte AC, Rocha-Santos T (2019a) Methods for sampling and detection of microplastics in water and sediment: a critical review. TrAC Trends Anal Chem 110:150–159. https://doi.org/10.1016/j.trac.2018.10.029

Prata JC, da Costa JP, Girão AV, Lopes I, Duarte AC, Rocha-Santos T (2019b) Identifying a quick and efficient method of removing organic matter without damaging microplastic samples. Sci Total Environ 686:131–139. https://doi.org/10.1016/j.scitotenv.2019.05.456

Prosenc F, Leban P, Šunta U, Bavcon Kralj M (2021) Extraction and identification of a wide range of microplastic polymers in soil and compost. Polymers 13(23):4069. https://doi.org/10.3390/polym13234069

Quinn B, Murphy F, Ewins C (2017) Validation of density separation for the rapid recovery of microplastics from sediment. Anal Methods 9(9):1491–1498. https://doi.org/10.1039/C6AY02542K

Radford F, Zapata-Restrepo LM, Horton AA, Hudson MD, Shaw PJ, Williams ID (2021) Developing a systematic method for extraction of microplastics in soils. Anal Methods 13(14):1695–1705. https://doi.org/10.1039/D0AY02086A

Raguso C, Pedruzzi A, Cetojevic E, Lasagni M, Sanchez-Vidal A, Saliu F (2021) Extraction of microplastic from marine sediments: a comparison between pressurized solvent extraction and density separation. Mar Pollut Bull 168:112436. https://doi.org/10.1016/j.marpolbul.2021.112436

Ramage SJ, Pagaling E, Haghi RK et al (2022) Rapid extraction of high-and low-density microplastics from soil using high-gradient magnetic separation. Sci Total Environ 831:154912. https://doi.org/10.1016/j.scitotenv.2022.154912

Rao Z, Niu S, Zhan N, Wang X, Song X (2020) Microplastics in sediments of river yongfeng from Maanshan City, Anhui Province, China. Bull Environ Contam Toxicol 104:166–172. https://doi.org/10.1007/s00128-019-02771-2

Razeghi N, Hamidian AH, Wu C, Zhang Y, Yang M (2021) Microplastic sampling techniques in freshwaters and sediments: a review. Environ Chem Lett 19(6):4225–4252. https://doi.org/10.1007/s10311-021-01303-x

Reunura T, Prommi TO (2022) Detection of microplastics in *Litopenaeusvannamei* (Penaeidae) and *Macrobrachium rosenbergii* (Palaemonidae) in cultured pond. PeerJ 10:e12916. https://doi.org/10.7717/peerj.12916

Rist S, Vianello A, Winding MHS et al (2020) Quantification of plankton-sized microplastics in a productive coastal Arctic marine ecosystem. Environ Pollut 266:115248. https://doi.org/10.1016/j.envpol.2020.115248

Rowe L, Kubalewski M, Clark R, Statza E, Goyne T, Leach K, Peller J (2018) Detecting microplastics in soil and sediment in an undergraduate environmental chemistry laboratory experiment that promotes skill building and encourages environmental awareness. J Chem Educ 96(2):323–328. https://doi.org/10.1021/acs.jchemed.8b00392

Sait STL (2019) Characterisation of microplastic fibres and their degradation under environmental conditions. Master's thesis, NTNU

Schönlau C, Karlsson TM, Rotander A, Nilsson H, Engwall M, van Bavel B, Kärrman A (2020) Microplastics in sea-surface waters surrounding Sweden sampled by manta trawl and in-situ pump. Mar Pollut Bull 153:111019. https://doi.org/10.1016/j.marpolbul.2020.111019

Schymanski D, Goldbeck C, Humpf HU, Fürst P (2018) Analysis of microplastics in water by micro-Raman spectroscopy: release of plastic particles from different packaging into mineral water. Water Res 129:154–162. https://doi.org/10.1016/j.watres.2017.11.011

Scopetani C, Chelazzi D, Mikola J, Leiniö V, Heikkinen R, Cincinelli A, Pellinen J (2020) Olive oil-based method for the extraction, quantification and identification of microplastics in soil and compost samples. Sci Total Environ 733:139338. https://doi.org/10.1016/j.scitotenv.2020.139338

Sefiloglu FÖ, Brits M, König Kardgar A et al (2024a)a Quantitative analysis of microplastics in Nile tilapia from a recirculating aquaculture system using pyrolysis–gas chromatography–mass spectrometry. *Environmental sciences* Europe 36(1): 172. doi:https://doi.org/10.1186/s12302-024-00987-6

Sefiloglu FÖ, Stratmann CN, Brits M et al (2024b)b Comparative microplastic analysis in urban waters using μ-FTIR and Py-GC-MS: a case study in Amsterdam. Environ Pollut 351: 124088. doi:https://doi.org/10.1016/j.envpol.2024.124088

Setälä O, Norkko J, Lehtiniemi M (2016) Feeding type affects microplastic ingestion in a coastal invertebrate community. Mar Pollut Bull 102(1):95–101. https://doi.org/10.1016/j.marpolbul.2015.11.053

Shim WJ, Song YK, Hong SH, Jang M (2016) Identification and quantification of microplastics using Nile red staining. Mar Pollut Bull 113(1–2):469–476. https://doi.org/10.1016/j.marpolbul.2016.10.049

Silva-Cavalcanti JS, Silva JDB, de França EJ, de Araújo MCB, Gusmão F (2017) Microplastics ingestion by a common tropical freshwater fishing resource. Environ Pollut 221:218–226. https://doi.org/10.1016/j.envpol.2016.11.068

Sivasankar V, Sunitha TG (2024) *Microplastics and pollutants* springer nature pages 129-148. https://doi.org/10.1007/978-3-031-54565-8

Song YK, Hong SH, Jang M, Kang JH, Kwon OY, Han GM, Shim WJ (2014) Large accumulation of micro-sized synthetic polymer particles in the sea surface microlayer. Environ Sci Technol 48(16):9014–9021. https://doi.org/10.1021/es501757s

Sorolla-Rosario D, Llorca-Porcel J, Pérez-Martínez M, Lozano-Castelló D, Bueno-López A (2023) Microplastics' analysis in water: easy handling of samples by a new thermal extraction desorption-gas chromatography-mass spectrometry (TED-GC/MS) methodology. Talanta 253:123829. https://doi.org/10.1016/j.talanta.2022.123829

Stock F, Kochleus C, Bänsch-Baltruschat B, Brennholt N, Reifferscheid G (2019) Sampling techniques and preparation methods for microplastic analyses in the aquatic environment–a review. TrAC Trends Anal Chem 113:84–92. https://doi.org/10.1016/j.trac.2019.01.014

Stoett P, Scrich VM, Elliff CI, Andrade MM, Grilli NDM, Turra A (2024) Global plastic pollution, sustainable development, and plastic justice. World Dev 184:106756

Strong KL (2021) Developing a Fourier-transform infrared (FT-IR) spectroscopy classification tool to identify Common Beach plastics on Hawai'i Island. (Master's thesis,. University of Hawai'i at Hilo

Sturm MT, Horn H, Schuhen K (2021) The potential of fluorescent dyes—comparative study of Nile red and three derivatives for the detection of microplastics. Anal Bioanal Chem 413:1059–1071. https://doi.org/10.1007/s00216-020-03066-w

Tamminga M (2017) Nile red staining as a subsidiary method for microplastic quantifica-tion: a comparison of three solvents and factors influencing application reliability. SDRP J Earth Sci Environ Stud 2(2).doi:10.25177/JESES.2.2.1

Tamminga M, Hengstmann E, Fischer EK (2018) Microplastic analysis in the south Funen archipelago, Baltic Sea, implementing manta trawling and bulk sampling. Mar Pollut Bull 128:601–608. https://doi.org/10.1016/j.marpolbul.2018.01.066

Tan X, Yu X, Cai L, Wang J, Peng J (2019) Microplastics and associated PAHs in surface water from the Feilaixia reservoir in the Beijiang River, China. Chemosphere 221:834–840. https://doi.org/10.1016/j.chemosphere.2019.01.022

Tarafdar A, Choi SH, Kwon JH (2022) Differential staining lowers the false positive detection in a novel volumetric measurement technique of microplastics. J Hazard Mater 432:128755. https://doi.org/10.1016/j.jhazmat.2022.128755

Tepe Y, Aydın H, Ustaoğlu F, Kodat M (2024) Occurrence of microplastics in the gastrointestinal tracts of four most consumed fish species in Giresun, the southeastern Black Sea. Environ Sci Pollution Res:1–10. https://doi.org/10.1007/s11356-024-34814-5

Turra A, Manzano AB, Dias RJS, Mahiques MM, Barbosa L, Balthazar-Silva D, Moreira FT (2014) Three-dimensional distribution of plastic pellets in sandy beaches: shifting paradigms. Sci Reports 4(1):4435. https://doi.org/10.1038/srep04435

Tuuri EM, Gascooke JR, Leterme SC (2024) Efficacy of chemical digestion methods to reveal undamaged microplastics from planktonic samples. Sci Total Environ 947:174279. https://doi.org/10.1016/j.scitotenv.2024.174279

Van Franeker JA, Blaize C, Danielsen J et al (2011) Monitoring plastic ingestion by the northern fulmar *Fulmarus glacialis* in the North Sea. Environ Pollut 159(10):2609–2615. https://doi.org/10.1016/j.envpol.2011.06.008

Vandermeersch G, Van Cauwenberghe L, Janssen CR et al (2015) A critical view on microplastic quantification in aquatic organisms. Environ Res 143:46–55. https://doi.org/10.1016/j.envres.2015.07.016

Vendel AL, Bessa F, Alves VEN, Amorim ALA, Patrício J, Palma ART (2017) Widespread microplastic ingestion by fish assemblages in tropical estuaries subjected to anthropogenic pressures. Mar Pollut Bull 117(1–2):448–455. https://doi.org/10.1016/j.marpolbul.2017.01.081

Vianello A, Boldrin A, Guerriero P, Moschino V, Rella R, Sturaro A, Da Ros L (2013) Microplastic particles in sediments of lagoon of Venice, Italy: first observations on occurrence, spatial patterns and identification. Estuarine Coastal Shelf Sci 130:54–61. https://doi.org/10.1016/j.ecss.2013.03.022

Von Friesen LW, Granberg ME, Hassellöv M, Gabrielsen GW, Magnusson K (2019) An efficient and gentle enzymatic digestion protocol for the extraction of microplastics from bivalve tissue. Mar Pollut Bull 142:129–134. https://doi.org/10.1016/j.marpolbul.2019.03.016

Vorobiev DS, Frank YA, Rakhmatullina SN, Vorobiev ED, Poskrebysheva YR, Oladele O, Trifonov AA (2024) Microplastic ingestion by fish with different feeding habits in the Ob and Yenisei Rivers. https://doi.org/10.17223/19988591/66/13

Wagner J, Wang ZM, Ghosal S, Rochman C, Gassel M, Wall S (2017) Novel method for the extraction and identification of microplastics in ocean trawl and fish gut matrices. Anal Methods 9(9):1479–1490. https://doi.org/10.1039/c6ay02396g

Wang W, Wang J (2018) Investigation of microplastics in aquatic environments: an overview of the methods used, from field sampling to laboratory analysis. TrAC Trends Anal Chem 108:195–202. https://doi.org/10.1016/j.trac.2018.08.026

Wang S, Chen H, Zhou X et al (2020) Microplastic abundance, distribution and composition in the mid-West Pacific Ocean. Environ Pollut 264:114125. https://doi.org/10.1016/j.envpol.2020.114125

Wedemeyer-Strombel KR, Balazs GH, Johnson JB, Peterson TD, Wicksten MK, Plotkin PT (2015) High frequency of occurrence of anthropogenic debris ingestion by sea turtles in the North Pacific Ocean. Mar Biol 162:2079–2091. https://doi.org/10.1007/s00227-015-2738-1

Wen X, Du C, Xu P et al (2018) Microplastic pollution in surface sediments of urban water areas in Changsha, China: abundance, composition, surface textures. Mar Pollut Bull 136:414–423. https://doi.org/10.1016/j.marpolbul.2018.09.043

Wen D, Chen Y, Tong Y, Wang H, Zhang H, Luo Y (2021) Quantification of microplastics in soils using accelerated solvent extraction: comparison with a visual sorting method. Bull Environ Contam Toxicol:1, 770–8. https://doi.org/10.1007/s00128-021-03200-z

Wiggin KJ, Holland EB (2019) Validation and application of cost and time effective methods for the detection of 3–500 μm sized microplastics in the urban marine and estuarine environments surrounding Long Beach, California. Marine Pollut Bullet 143:152–162. https://doi.org/10.1016/j.marpolbul.2019.03.060

Wirnkor VA, Ebere EC, Ngozi VE (2019) Microplastics, an emerging concern: a review of analytical techniques for detecting and quantifying microplatics. Anal Methods Environ Chem J 2:13–30

Wright SL, Kelly FJ (2017) Plastic and human health: a micro issue? Environ Sci Technol 51(12):6634–6647

Xiong X, Wu C, Elser JJ, Mei Z, Hao Y (2019) Occurrence and fate of microplastic debris in middle and lower reaches of the Yangtze River–from inland to the sea. Sci Total Environ 659:66–73. https://doi.org/10.1016/j.scitotenv.2018.12.313

Xu JL, Thomas KV, Luo Z, Gowen AA (2019) FTIR and Raman imaging for microplastics analysis: state of the art, challenges and prospects. TrAC Trends Anal Chem 119:115629. https://doi.org/10.1016/j.trac.2019.115629

Yu Z, Peng B, Liu LY, Wong CS, Zeng EY (2019) Development and validation of an efficient method for processing microplastics in biota samples. Environ Toxicol Chem 38(7):1400–1408. https://doi.org/10.1002/etc.4416

Yu W, Chen J, Zhang S, Zhao Y, Fang M, Deng Y, Zhang Y (2022) Extraction of biodegradable microplastics from tissues of aquatic organisms. Sci Total Environ 838:156396. https://doi.org/10.1016/j.scitotenv.2022.156396

Yuan M, Zhang Y, Guo W, Chen S, Qiu Y, Zhang P (2022) A rapid staged protocol for efficient recovery of microplastics from soil and sediment matrices based on hydrophobic separation. Mar Pollut Bull 182:113978. https://doi.org/10.1016/j.marpolbul.2022.113978

Zhang X, Yu K, Zhang H, Liu Y, He J, Liu X, & Jiang J (2020)A Novel heating-assisted density separation method for extracting microplastics from sediments Chemosphere 256: 127039. doi:https://doi.org/10.1016/j.chemosphere.2020.127039

Zhou F, Wang X, Wang G, Zuo Y (2022) A rapid method for detecting microplastics based on fluorescence lifetime imaging technology (FLIM). Toxics 10(3):118. https://doi.org/10.3390/toxics10030118

Zhu Y, Li Y, Huang J et al (2024) Advanced optical imaging Technologies for Microplastics Identification: progress and challenges. Adv Photon Res 2400038. https://doi.org/10.1002/adpr.202400038

Zobkov M, Esiukova E (2017) Microplastics in Baltic bottom sediments: quantification procedures and first results. Mar Pollut Bull 114(2):724–732. https://doi.org/10.1016/j.marpolbul.2016.10.060

Zobkov MB, Esiukova EE (2018) Microplastics in a marine environment: review of methods for sampling. Process Analyz Microplast Water, Bottom Sediments Coastal Deposits 58:137. https://doi.org/10.1134/S0001437017060169

Advances in Microplastics Research in Freshwater Ecosystems: An Overview of Newly Developed Separation and Identification Methods

7

Ayan Lodh, Mozim Shafi, and Sudha Goel

Abstract

Microplastics (MPs) have become widespread pollutants in freshwater ecosystems, posing a considerable threat to aquatic and terrestrial life, including humans. The absence of standardized extraction and identification methods for MPs is impeding research progress. Current protocols for MP analysis face technical limitations in the context of analysis time, recovery rate, low detection limits, and errors due to manual sample handling, especially when dealing with diminutive-sized (<0.5 mm) particles. There is a necessity for constant improvement in current extraction and identification techniques, leading to the development of new methods for analyzing MPs. This chapter aims to provide an extensive overview of the most recent separation and analytical techniques for MPs, emphasizing their benefits and limitations. We have elaborately described the principles and applications of seven density-independent separation methods, three advanced microscopic imaging techniques, and four spectrometry-based elemental analyses for MP research in freshwater ecosystems. Separation of MPs with the assistance of magnets, electrostatic energy, oils, organic solvents, coag-

A. Lodh
School of Environmental Science and Engineering, Indian Institute of Technology Kharagpur, Kharagpur, West Bengal, India

M. Shafi
Environmental Engineering and Management, Department of Civil Engineering, Indian Institute of Technology Kharagpur, Kharagpur, West Bengal, India

S. Goel (✉)
School of Environmental Science and Engineering, Indian Institute of Technology Kharagpur, Kharagpur, West Bengal, India

Environmental Engineering and Management, Department of Civil Engineering, Indian Institute of Technology Kharagpur, Kharagpur, West Bengal, India
e-mail: sudhagoel@civil.iitkgp.ac.in

K. Bala et al. (eds.), *Occurrence, Detection, and Fate of Microplastics in Freshwater Ecosystems*, https://doi.org/10.1007/978-981-95-0225-7_7

ulants, and frothing agents were described. Characterization using high-definition microscopes, laser-induced spectroscopy, nuclear magnetic resonance, thermal cracking, and inductively coupled mass spectroscopy were addressed. Additionally, we presented algorithmic model-based hybrid approaches, which can be prioritized for future studies to potentially eliminate human error when dealing with minuscule particles. Implementation of the new methods can open up new avenues of MP analysis by providing precise and high-quality research output, although they have been overlooked by the scientific communities. This chapter encourages researchers to select efficient separation and identification methods for MPs research on freshwater ecosystems. Despite advancements, challenges persist in this domain, and this chapter facilitates the necessity for further development of innovative methodologies to enhance the detection and characterization limits, followed by accurate and reliable MP analysis.

Keywords

Microplastics · Novel methods · Extraction · Identification · Automated analysis

7.1 Introduction

The occurrence of small plastic fragments, termed microplastics (MPs), in freshwater ecosystems has become a major scientific concern in recent years. MPs originating from various human activities (mainly from land-based sources) are flushed into fresh and marine water bodies, endangering the whole aquatic habitat (Xu et al. 2020). Rivers play a critical role in transporting MPs, particularly into the marine environment. Past research indicated that runoff from freshwater ecosystems contributes ~80% of marine MPs (Luo et al. 2024a). Although the accumulation of MPs has been concentrated in marine habitats due to their extensive contamination, they have spread from the mountains to the great ocean trenches (Padha et al. 2022; Peng et al. 2018). Evidently, ~15 trillion MP particles, weighing 93,300 tons, are floating in the top foot of seawater and account for 80–85% of waste materials (Bai et al. 2021). Due to their tiny size, MPs can carry toxicants from the environment and eventually create potential hazards to aquatic life (Lodh et al. 2023; Shafi et al. 2024). Aquatic organisms can ingest these MPs, leading to trophic transfer that affects human health primarily via food-chain contamination (Mercogliano et al. 2020). Thus, MPs are regarded as dangerous pollutants, and their existence in freshwater environments requires monitoring.

At present, the extraction, quantification, and identification of plastic particles of various sizes from environmental samples are popular areas of research. However, due to the lack of a standardized protocol for MP analysis, it is difficult to generalize and compare different studies across the world (Bellasi et al. 2021). Most researchers have employed density-mediated separation using saturated salt solutions followed by filtration method to extract MPs from various environmental matrices (Crutchett and Bornt 2024; Cutroneo et al. 2021). For this method, researchers have generally used sodium chloride (NaCl), sodium iodide (NaI), sodium bromide (NaBr), zinc chloride ($ZnCl_2$), and calcium chloride ($CaCl_2$), many of which are highly toxic in nature and expensive. Additionally, recovery rates vary or become low depending on the density and size of MPs, as well as due to the choice of salt solution (Duong et al. 2022). Therefore, these limitations of the conventional separation method can be eliminated by improving or developing new technology. Based on different density-independent features of MP polymer, several new separation methods have been developed. These features are magnetization by surface modification, oleophilic nature, electrical conductivity, and solubility, which have resulted in high separation efficiency even for small particles of <0.5 mm (Fadare et al. 2023; Felsing et al. 2018; Grbic et al. 2019; Mani et al. 2019).

Techniques employed to identify and quantify MPs can be divided into the following categories: visual observation (by the naked eye), microscopic, chromatographic, and spectroscopic analysis. Currently, the most dependable and widely utilized methods for microscopic analysis are stereomicroscopy, fluorescence microscopy, scanning electron microscopy, and transmission electron microscopy. Chemical identification approaches are vibrational spectroscopy, such as Fourier transform infrared (FTIR) and Raman spectroscopy (Kalaronis et al. 2022; Song et al. 2021). Despite offering reliable results, these methods require comparison with a spectrum library and are susceptible to external interference due to weathering. Moreover, vibrational spectroscopy cannot recognize the distinctive signatures of MP samples after surface weathering due to environmental exposure (Lee and Chae 2021). Weathering of MPs induces low-intensity signals of characteristic peaks due to changes in oxidative functional groups on the polymer surface, which leads to a mismatch with reference spectra. In addition, for effective results, the sample surface should be clean and, therefore, pre-treatment is necessary for these methods.

Limitations associated with conventional methods in the analysis of various environmental matrices have led to research on the development of novel methods that offer high sensitivity, rapid analysis, and robustness. These improvements include microscopic techniques such as polarized light microscopy, atomic force microscopy, and hyperspectral imaging to obtain high-resolution images, and laser-induced spectroscopy and thermal cracking to identify the chemical type of MPs polymer (Huang et al. 2021; Liu et al. 2021a, b; Okoffo et al. 2023). In addition, conventional spectroscopy methods have an inability to handle MPs in large quantities, thus consuming more time to analyze. Therefore, the integration of these methods with machine learning (ML) and artificial intelligence (AI)-based platforms offers a promising approach for enhanced identification of MPs with less chance of human error in a short duration (Hu et al. 2024). Computerization can automatically classify microscopic images by segmentation, particle detection, and counting in real environmental samples. Furthermore, ML can extract spectroscopically obtained fingerprints of multiple polymer types for their identification (Jin et al. 2024). In a study, the authors successfully recognized different MP polymers in environmental samples by combining Raman analysis with ML and without any pre-treatment (Coleman 2025). Thus, in spite of relying on conventional methods, future research on MPs in freshwater ecosystems should focus on adopting and improving emerging methods, which will also develop comparability among global studies.

This chapter provides a comprehensive overview of newly developed separation and identification methods for MPs, which emphasize improved sensitivity, reliability, and robustness. A brief idea of the operating principles, applications in real samples, effectiveness in the context of recovery and detection limits of small-sized MPs, with their merits and demerits, is presented. All newly developed methods seem to be transforming the area of MPs research on freshwater, providing appropriate means to solve one of the present world's problems.

7.2 Separation Methods

7.2.1 Magnetic Separation

This method uses nanoparticles with a high specific surface area and magnetic properties. These nanoparticles have a hydrophobic tail, which creates a good interaction and affinity to bind MPs in the sample. Then, magnetized MPs can be separated from the sample using a magnetic field (Fig. 7.1a). Iron-based nanoparticles were frequently used due to their better productivity, high efficiency, and inexpensiveness (Table 7.1). This method was developed and evaluated using Fe-nanoparticles in sediment and surface water samples for various sizes of MPs (~8 mm – <20 μm) (Grbic et al. 2019). The authors achieved an efficiency of ~92% from a seawater sample to recover polyethylene (PE) and polystyrene (PS) beads of <20 μm. The recovery rates of 0.2–1 mm MPs were demonstrated as 84% in freshwater and 78%

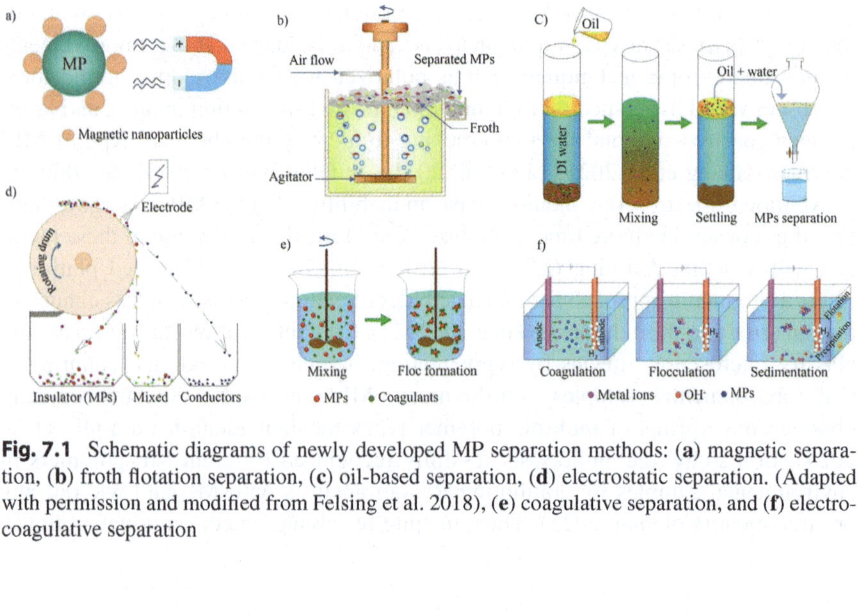

Fig. 7.1 Schematic diagrams of newly developed MP separation methods: (**a**) magnetic separation, (**b**) froth flotation separation, (**c**) oil-based separation, (**d**) electrostatic separation. (Adapted with permission and modified from Felsing et al. 2018), (**e**) coagulative separation, and (**f**) electrocoagulative separation

Table 7.1 Effectiveness of multiple magnetite nanoparticles for MP removal

Nanoparticles type	Matrix type	MPs size (μm)	Efficiency (%)	References
Fe nanoparticles	Seawater	10–20	93	Grbic et al. (2019)
	Freshwater	200–1000	84	
	Sediment	200–1000	78	
C/Fe_3O_4	Drinking water		100	Elmaci (2020)
Mg-hydroxide/Fe_3O_4	Wastewater	≤270	92	Zhang et al. (2021a)
M–CNTs	Water	48	100	Tang et al. (2021)
$Au/Ni/TiO_2$	Water	3	67	Wang et al. (2019)
Fe_2O_3/MnO_2	Water	5-3	10	Ye et al. (2021)
PDA/$Fe_3$04	Wastewater	20–140	100	Zhou et al. (2021a)
Magnetite powder with hexadecyltrimethoxysilane	Sewage sludge	100	80–100	Rhein et al. (2022)

in sediment. To improve the efficiency of this process, two new protocols were developed, known as "magnetic-seeded filtration (MSF)" and "adsorption-magnetic separation (AMS)." In MSF, pre-prepared magnetic material is introduced into the sample, in which they form aggregates with MPs, followed by separation using a magnet (Rhein et al. 2019). The authors employed magnetite Fe_3O_4 (mean dia: 2.75 μm) as a seed for the separation of polymethylmethacrylate (PMMA) and polyvinyl chloride (PVC) from pure water. For this purpose, MPs isolation was performed by creating a hetero-agglomeration of Fe_3O_4 and MPs, resulting in 95% efficiency. In this process, pH and surface charge of the nanoparticles are the key factors, because at higher pH levels, the electrostatic repulsive force acts between the particles, which induces the breakdown of agglomerates and eventually reduce the separation efficiency. Thus, to increase the affinity of seed materials to come closer to MP particles, the concentration of the seed and their rapid movement need to be optimized. For AMS, magnetic carbon nanotubes (M-CNTs) have been used as an adsorbent to isolate MPs from aqueous samples. Polyethylene (PE), polyethylene terephthalate (PET), and polyamide (PA) particles of 48 μm size were successfully coated with M-CNTs, and then the MPs/M-CNT composites were isolated by an external magnetic field with 100% efficiency (Tang et al. 2021). In this case, adsorption of synthesized nanomaterials on MPs was predominant, which is caused by the mechanism of π-π electron conjugation, π-π electron interactions, and hydrogen bond interactions on the polymer surface. Hence, to further improve the attraction for better effectiveness of this method, core-shell superparamagnetic Fe-nanoparticles were prepared by surface alteration process. The authors have combined iron oxides with octadecylphosphonic acid (PAC_{18}) or (12-dodecylphosphonic acid)-N,N-dimethyl-N-octadecyl ammonium bromide ($PAC_{12}NC_{18}$) to modify the surface of Fe-nanoparticles, which were then further used to remove MPs from water samples (Sarcletti et al. 2021). Besides, hydrophobic materials, hydrophilic bare Fe_3O_4 nanoaggregates proved effective for the magnetic separation of some of the most popular MPs (Zandieh and Liu 2022). In this case, hydrogen bonding acts as the primary force for the adsorption of Fe_3O_4. Cao et al. (2024) have reviewed and summarized applications of multiple well-designed Fe-oxide-based composite materials in order to remove MPs from water samples using a magnetic field. In addition, the authors discussed the use of magnetic devices instead of a small piece of magnet to separate MPs. The presence of biogenic substances may hinder the non-specific binding of nanoparticles, which may lead to reduction in the efficiency of the nanoparticles. Thus, this method is appropriate for clean or digested samples.

7.2.2 Electrostatic Separation

Electrostatic separation (ES) is a highly promising method that works on the differentiation of electrical conductivity between MPs and non-plastic particles to enable their separation (Silveira et al. 2018). Briefly, solid matrices like sediment and sand are naturally conductive and acquire a charge when exposed to a high-voltage electric field, whereas polymers are typically non-conductive. This electrostatic difference enable separation when the materials pass through a rotating drum, with the non-conductive materials being isolated (Fig. 7.1d). Additionally, the friction between plastics and non-plastic components transfers electrons, causing plastics to gain a net positive, negative, or neutral charge depending on their triboelectric properties, which facilitates their separation further. By taking advantage of this method, Felsing et al. (2018) separated different common MPs of varying size, shape, and density efficiently. ES has achieved 100% recovery rate for particle sizes ranging from 63 μm to 5 mm. However, Enders et al. 2020 reported comparatively less recovery (80%) for fibers. They also observed only 45% recovery for MPs of 20 μm size, clearly indicating that the efficiency of ES decreases as the size of MPs reduces (Enders et al. 2020). Due to the lightweight of small-sized particles, they are more vulnerable to air turbulence and external interferences, which lead to inconsistent trajectories during separation and ultimately reduce the overall efficiency. Additionally, smaller particles may form agglomerates with other particles, which makes it challenging to separate them effectively. The surface of large MPs can be coated due to the adhesion of small dust particles during the process and therefore obtain high net charges, which leads to a negative effect on their removal (Enders et al. 2020). Thus, this method is suitable for MPs greater than 60 μm (Enders et al. 2020). The advantage of this approach is the ability to handle high-volume samples along with good sample mass reduction capacity, which result reduction of both chemical usages and time. However, while handling large sample volumes, it is advisable to allow density separation after this method, because ES alone is not suitable for enough mass reduction (Kurzweg et al. 2024). This method is less suitable for cohesive samples than sand or sediment. During the process, due to the electrostatic attraction force, fine clay particles are deposited on the rotating drum and form a layer, which eventually reduces the process efficiency as well as the mass reduction rate (Enders et al. 2020). In this regard, oil-based separation can be an alternative for cohesive samples. There is very limited knowledge on the applicability of ES for MP removal from water matrices; hence, extensive research is required in this field for its reliability.

7.2.3 Oil-Based Separation

Oil-based separation has emerged as a novel technique for extracting MPs from environmental samples based on the MP's hydrophobic (oleophilic) nature. It employs the selectivity of MPs for oil over other non-plastic substances that are usually hydrophilic. A small volume of oil (3–10 ml) and Milli-Q water is required to add into the sample, followed by mixing and settling, to enhance the chance of interaction between oil particles and MPs. After settling in the separation unit for some time, MPs move to the oil layer that can be isolated on a filter for identification (Fig. 7.1c). This method was first introduced by Crichton et al. (2017) using canola oil to isolate MPs from sediment samples. Subsequently, other works have used olive oil, castor oil, and sunflower oil effectively on soil, water, and mussel samples (Mani et al. 2019; Scopetani et al. 2020; Song et al. 2022). It was challenging to isolate the oil layer from water and soil because the oil-water boundary layer can contain MPs, and during separation, this layer can be disturbed, which leads to the loss of particles. To overcome this difficulties, a polytetrafluoroethylene cylinder with a removable cap and piston system was used (Scopetani et al. 2020). In this case, after settling, the sample was allowed to be refrigerated at −40 °C for ice formation to facilitate easy separation of the oil phase from the rest of the sample. This approach has benefits, as it is relatively simple, inexpensive, non-hazardous, and suitable in resource-limited conditions. Furthermore, this method has no relation with polymer density; however, the average recovery rate differs from 90% to 95% for low-density to high-density polymers (Scopetani et al. 2020). All the tested oils were efficient; however, the highest efficiency of 99.99% (size 0.3–1 mm) was recorded for castor oil, while canola and olive oils recorded 96.20% (size 0.02–4.4 mm) and 93.50% (size 0.2–2 mm), respectively. This high recovery of MPs by castor oil was mainly caused by the presence of a high amount of long-chain C_{18} fatty acids. These long-chain aliphatic hydrocarbons with high molecular mass lead to strong attraction between the non-polar lipophilic element of the fatty acid molecule and the polymer surface (Mani et al. 2019). Additionally, the high dynamic viscosity ($\eta > 300$ cP) of castor oil enables it to form a thick oil layer over the MPs particles and form oil-MPs cluster. These clusters, therefore, reduce the overall density compared to the water and then easily migrate toward the top surface of the separator unit. Hence, the fatty acid content and viscosity of the oil play a key role in the efficient separation. The separation efficiency for MPs of size ≤360 μm, particularly those with granular, foam, and fiber morphologies, is comparatively low and requires further research for improvement (Lechthaler et al. 2020). This method is suitable for spectroscopic investigation; however, particularly for foam-based or porous polymer, there is a chance of oil adsorption within the pores, which might alter the spectrum (Crichton et al. 2017). For organic-rich samples, an additional digestion step is necessary to achieve adequate results.

7.2.4 Solvent-Based Separation

Different solubility of MPs and other elements in sample matrices makes solvent-based separation an effective alternative. Pressurized fluid extraction (PFE) is a method, that uses selective solvents under subcritical temperature and pressure. In the first phase of the process, semi-volatile organic materials were eliminated by methanol at 100 °C. In the second phase, MPs were isolated using dichloromethane (DCM) at 180 °C and 1500 PSI pressure (Fuller and Gautam 2016). Solvents play a key role, as the effectiveness of this method depends on the selection of a proper organic solvent. Different solvents have been used so far, including hexane, dichloromethane (DCM), ethanol, hexane-DCM composite, and hexane-acetone composite (Table 7.2) (Hong et al. 2017). Fadare et al. (2023) suggested a method to use a water-ethanol mixture as a solvent to separate MPs from surface water and particulate organic matter. They achieved 89–91% removal efficiency for different MPs sizes ranging from 30 μm to 2.5 mm. The process efficiency also relies on the volume of solvent and the separation time. It has several benefits that create great interest among researchers, such as insensitivity to particle size, quick processing, compatibility with spectroscopic analysis, and the need for zero pre-processing steps. Another advantage of PFE is that small MPs can be recovered, which is not always possible with conventional methods. However, this approach is basically destructive in nature, so it is not suitable for the physical characterization of the MPs (Möller et al. 2020). This method poses a significant environmental concern

Table 7.2 Effectiveness of multiple solvents for MP removal

Solvent type	Matrix type	Recovery (%)	References
Dichloromethane	Soil	19–81	Wen et al. (2021)
Tetrahydrofuran	Soil and sediment	83–86	Dierkes et al. (2019)
Dichloromethane	Biosolid	83–95	Okoffo et al. (2020)
Tetrahydrofuran	Sea sand	90–98	Günther and Imhof (2024)
Trifluoroacetic acid and chloroform		12–92	
Trifluoroethanol and chloroform		12–99	
Xylene		27–76	
Ethanol and water	Surface water	89–93	Fadare et al. (2023)
Methanol, hexane, and dichloromethane	Municipal waste	85–94	Fuller and Gautam (2016)
Methanol and dichloromethane	Sea sand and sediment	87–89	Stile et al. (2021)
Decanoic acid, menthol, thymol	Freshwater and saltwater	50–93	Hunter et al. (2023)

due to the use and disposal of toxic solvents. Additionally, the handling of hazardous solvents requires that safety measures be followed. Thus, to eliminate these issues, Hunter et al. (2023) have introduced natural deep eutectic solvents (NADES) to extract MPs, which are classified as an environment friendly green solvent that is derived from natural extracts, such as monocyclic monoterpenoids menthol, thymol, organic acids, and salts. Using NADES, the authors have reported 50–93% extraction efficiency of MPs from freshwater and saltwater.

7.2.5 Coagulative Separation

This method employs coagulative agents to develop the aggregation of MPs into larger flocs, followed by isolation (Fig. 7.1e). The mechanisms involved are adsorption, charge neutralization, sweep flocculation, polymer bridging, and precipitative coagulation (Cui et al. 2024; Girish et al. 2023). The coagulants get attached to the negative or positive surface charge on MPs and form large flocs to neutralize them. Sweep flocculation occurs when the density and size of the aggregates rise because of the adsorption of neighboring MPs by a positively charged coagulant, with the formation of amorphous precipitation. Adsorption comes into action as the MPs adhere to the coagulants via hydrogen bonding, hydrophobic interactions, and electrostatic forces. Recently, the effectiveness of multiple types of coagulants like aluminum (Al)-based, iron (Fe)-based, magnesium (Mg)-based salts, synthetic and natural organic coagulants have been explored (Table 7.3), wherein variation in terms of separation efficiency of MPs has been observed (Girish et al. 2023; Lapointe et al. 2020; Ma et al. 2019a). Fe-based coagulation was adopted for the removal of PE MPs because of their higher density compared to Al-based coagulants. The result showed higher efficiency for the smaller particles (d <0.5 mm) (Ma et al. 2019b). However, overall removal efficiency was <15%, and Al-based salts showed more effectiveness than Fe-based salts in removing PE particles. To enhance the performance of the process, a high dose (15 mg/L) of polyacrylamide (PAM) was added with the coagulant, and a significant increase in removal (~45%) was observed due to the formation of dense floc and high adsorption ability under neutral conditions. However, the application of PAM may result in toxic by-products in treated water and thereby cause an impact on water quality, which requires an additional removal step. The pH of the solution, coagulant dosage, and size of MPs play a key role in this process. With increasing pH, the ability of coagulants to neutralize the surface charge of MPs is reduced. Additionally, the size of the flocs produced depends on the pH of the solution. At alkaline conditions, large-sized flocs are produced, which improve the (Zhou et al. 2021b). Yang et al. (2016) conducted a study where they observed that charge neutralization was predominant in the process, and the removal efficiency increased with the coagulant doses due to the reduction of MP's zeta potential. The optimum coagulant dose can be obtained at zero zeta potential, followed by the highest removal efficiency (Yang et al. 2016). Utilizing a coagulant dose exceeding the appropriate quantity may lead to re-stabilization and charge reversal of the particles (Monira et al. 2021). Moreover, the varying size of

Table 7.3 Effectiveness of multiple coagulants for MP removal

Coagulants type	Matrix type	MPs size (μm)	Removal efficiency (%)	References
Aluminum chloride hexahydrate	Water	?oxy_ delete content=" "?>1-5	98	Xia et al. (2020)
Ferric chloride, polyaluminum chloride, and polyamine	Wastewater	1–6.3	99.4	Rajala et al. (2020)
Aluminum sulphate	Deionized water	3, 6, 25, 45, and 90	96.1	Xue et al. (2021)
Aluminum chlorohydrate and polyacrylamide	Deionized water	0.05–1	98.5	Zhang et al. (2022)
Aluminum chloride hexahydrate and ferric chloride hexahydrate	Deionized water and river water	10–90	33–95.3	Na et al. (2021)
Aluminum sulphate	Water	100	98.9	Li et al. (2021)
Alum coagulant and alum combined with cationic PC sand	Synthetic water	10–100	70.7	Shahi et al. (2020)
PAC, PAM, sodium alginate, and activated silicic acid	Water	100–500	91.45	Zhang et al. (2021a, b)
Magnetic magnesium hydroxide and PAM	Wastewater	≤270	87–92	
Aluminium chloride	Ultrapure water	500	92–100	Lu et al. (2021)

different MPs can differentiate settling efficiency, which affects the effectiveness of the coagulant-based removal. The double-layer effect during charge neutralization makes large MPs more unstable and more likely to form aggregates than small particles. In addition, hydraulic mixing may cause fragmentation of large MPs; thus, high removal efficiency of large MPs can be observed, followed by an increase in the concentration of small MPs (Radityaningrum et al. 2021). For the effective removal of large MPs, the size of the flocs should be similar to or larger than that of the MPs. Meanwhile, a varying size of flocs can remove small MPs sufficiently. Xue et al. (2021) reported a higher recovery rate for 3 μm MPs than for 45 and 90 μm particles. The drawback of this method is the production of a large quantity of sludge.

7.2.6 Electro-Coagulative Separation

Electro-coagulative separation (ECS) is a novel method for separating MPs involving coagulation, flotation, and electrochemical reaction. In ECS, metal electrodes generate metal ions under the influence of voltage differentiation, which act as micro-coagulants (Bhatt et al. 2021). These metal ions dissolve at the anode and immediately hydrolyze to form hydroxides, which polymerize to form amorphous flocculants. These flocculants have a positive-charge surface, which enables them to adsorb negatively charged MPs. Furthermore, hydrogen at the cathode and oxygen at the anode are formed during electrocoagulation together with microbubbles, which lead the flocs to move toward the surface, where they can easily separate by skimming (Núñez et al. 2019) (Fig. 7.1f). It has several benefits, including high efficiency, operational simplicity, and generate less quantity of sludge (Tahreen et al. 2020). Current density, pH, electrolyte concentration, electrolysis time, and MPs concentration are some factors that affect the efficiency of the ECS significantly. The highest removal efficiency (~99%) of PS MPs (size 25–1500 μm) was achieved at pH 4 and a current density of 2.88 mA/cm^2 (Elkhatib et al. 2021). In another study, multiple varieties of MP polymer (polyethylene terephthalate, polyethylene, polypropylene, and polyamide) samples were separated by ECS, where the highest removal of fibers was noticed compared to fragments (Senathirajah et al. 2023). The authors also observed physicochemical changes in MPs, including the deformation of polymer structures and surface disintegration of particles. Hence, ECS increases the chance of the formation of nanoplastics by breaking down MP particles, which has more adverse effect on the environment. Additionally, the production of •OH by electro-oxidation of water can produce an active oxidizing agent, which has the potential to convert toxic substances to non-toxic materials or produce CO_2 gas and subsequently contribute to greenhouse gas emissions (Senathirajah et al. 2023). Akarsu et al. (2021) discovered that using an Al anode in the electrocoagulation process integrated with membrane filtration can attain 100% MP removal efficiency. The Al anode offers greater benefits compared to the Fe anode because the Al anode is capable of producing more microcoagulant compared to the Fe anode in the ECS method.

7.2.7 Froth Flotation Separation

Froth flotation separation involves the addition of surfactants (wetting agents) to facilitate the selective separation of MPs from a liquid sample (Fig. 7.1b). The principle is based on surface wettability: hydrophobic materials tend to float by attaching to bubbles, whereas hydrophilic materials are carried away as settled material. This process uses controlled turbulence to stir the water by introducing dispersed air into the system. Surfactants are added to modify the hydrophobicity of MPs, enhancing their ability to attach to air bubbles and enabling them to float on the surface (Pita and Castilho 2017). Tannic acid, methylcellulose, lignin sulfonate, Triton x-100, sodium lignin sulfonate, calcium lignosulfonate, sorbitan monolaurate, saponin, sodium silicate, and methyl cellulose have been tested as potential wetting agents for the selective flotation of mixed polymers (Pita and Castilho 2017; Wang et al. 2015). Several factors influence the success of this process, including polymer density, pH, dosage of frothing agent, hydrophobicity, flotation time, and particle size. However, polymers with higher density and larger size have more affinity to attach to bubbles, thereby showing improved flotation rate (Zhang et al. 2021d). In addition, high-density polymers have higher collision and adhesion tendencies, which further boost the flotation rate. The effect of floating velocity and process efficiency of two polymers with nearly similar densities but dissimilar surface charges (PVC and PS) was examined (Salerno et al. 2018). The author has revealed that the pH has a significant effect on the floatability of PVC particles. In a study, the authors compared the efficiency of froth flotation in the removal of acrylonitrile butadiene styrene (ABS) and polyvinyl chloride (PVC) MPs from water, differences in the recovery rates were observed. While 97% of PVC MPs were successfully floated, only 58.70% of ABS were removed with adequate terpineol dosage. The findings indicated that common aqueous species like K^+, Na^+, Mg^{2+}, Ca^{2+}, Al^{3+}, and humic acid hinder the flotation of ABS particles, contributing to the variation in recovery rates (Jiang et al. 2022). In natural aquatic environments, plastic debris undergoes slow degradation when exposed to solar radiation and heat, resulting in the development of hydrophilic functional groups on the surface. This alteration can weaken the original hydrophobicity of MPs, thereby diminishing the effectiveness of froth flotation (Zhang et al. 2021c).

7.3 Identification Methods

7.3.1 Polarized Light Microscopy

Polarized light microscopy (PLM) is used in MPs research due to its ability to show the structure of the polymer, which is usually not observed using the standard light microscopy method. This technique changes the phase of light's polarization upon its interaction with MP particles to characterize different polymers in terms of birefringence (anisotropic) properties (Fig. 7.2a) (Zhu et al. 2024). It is more sensitive to microstructure and provides information on the morphology, texture, and the formation of structural imperfections or degradation that may be caused by environmental stress. Those materials, which possess certain types of anisotropy, glow under the polarized light, and isotropic substances appear quite black (Sierra et al. 2020). Liu et al. (2021b) proposed a real-time monitoring method for MPs in water based on the backscattering of the particles to the incident linear polarized light scattered at 120° with a detection limit of 0.01 mg/L. By optimizing the polarization parameters, different physical and chemical properties of MPs can be differentiated. Additionally, it is possible to study changes in the MP's properties due to continuous weathering. This method can be useful to analyze 10 μm-sized particles, and the concentration of these particles can be obtained by measuring the pulse number per unit time. There is a necessity for further enhancement of the method by integrating digital holography (DH) with PLM, which may offer useful information about MPs (Běhal et al. 2022). DH analyze the entire wave field of the object and record both amplitude and phase variations, enabling measurement of morphological as well as optical parameters such as refractive index and optical path difference. Other than that, by changing the polarization state of incident light, selective identification of

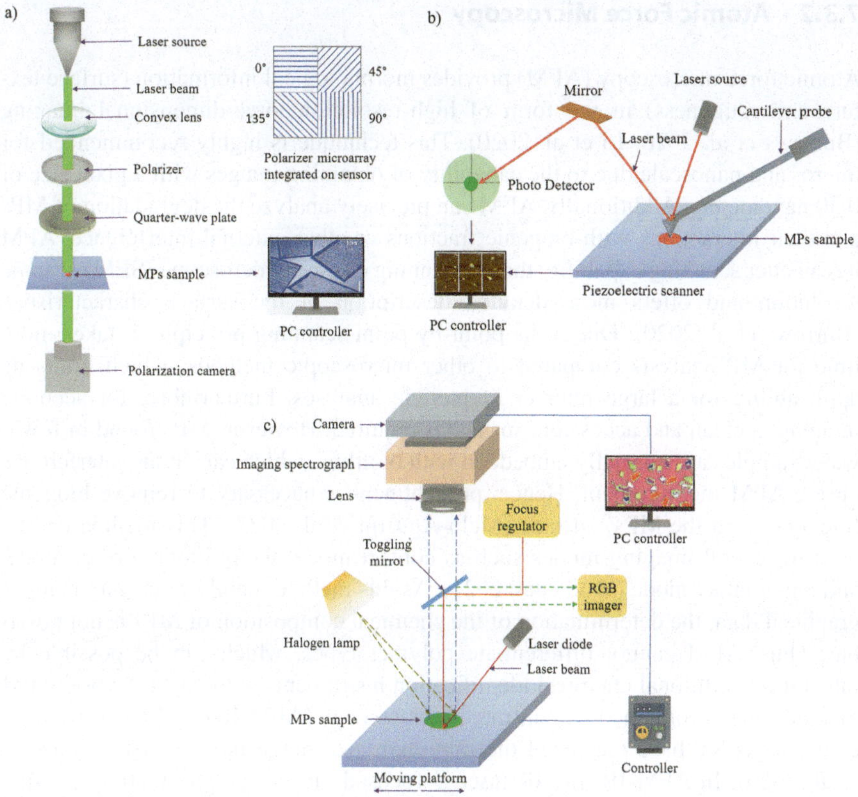

Fig. 7.2 Schematic diagrams of newly developed microscopic methods for MP analysis: (**a**) polarized light microscopy, (**b**) atomic force microscopy, and (**c**) hyperspectral imaging

MPs is possible from other non-plastic fractions based on their different anisotropic nature. In a related study, Valentino et al. (2022) proposed a similar methodology wherein they designed a polarization-sensitive holographic flow cytometer for the detection of microfibers. This novel method helps in characterizing several synthetic (PET, PP, PA) as well as natural microfibers (cotton and wool) of 200 μm size. This depends on the crystal structure of the polymers; the interaction affects and the transference of the polarization light, which can be measured and differs even within the same kind of polymer. The integrated Polarized light microscopy - digital holography (PLM-DH) method requires few optical configurations and sample preparation steps. Thus, this technique is suitable for on-field analysis of MPs (Li et al. 2024). Polymer samples have to be thin enough to allow a proper flow of polarized light through the sample. The variation of color density poses major challenges, whereby samples that have high color intensity or opacity will hinder the transmission of light, making this method unsuitable in some cases.

7.3.2 Atomic Force Microscopy

Atomic force microscopy (AFM) provides morphological information (surface texture and roughness) in the form of high-resolution three-dimensional imaging (Burrows et al. 2020; Li et al. 2020). This technique is highly recommended for micro- and nanoscale due to the capability of obtaining images with a pixel size of 0.30 nanometers. Additionally, AFM can precisely analyze the degradation of MPs and their interactions with biogenic fractions or other external interference. AFM has a better scanning capability than a scanning electron microscope (SEM) at same resolution and offers more detailed description of the surface characteristics (Burrows et al. 2020). Due to its point-by-point scanning principle, it takes more time for MP analysis compared to other microscopic methods, which limits its applicability for a large number of particle analyses. Furthermore, for accurate imaging, a clean and accessible surface is required. However, MPs found in freshwater samples are normally embedded with biofilms, which can create interference during AFM measurement. Hence, pretreatment is necessary to remove biogenic fractions from the MPs' surface (Melo-Agustín et al. 2022). This method can be used in several imaging modes such as contact mode, intermittent-contact mode, and non-contact mode (Woo et al. 2021). As this method mainly relies on the topographical data, the determination of the chemical composition of MPs is not possible. Thus, AFM cannot differentiate polymer types, which can be possible by integrating additional chemical identification instruments with AFM. Hybrid AFM methods, when combined with infrared spectroscopy (AFM-IR) or Raman spectroscopy (AFM-R), have enhanced the minimum size range limit of MPs (Burrows et al. 2020). In AFM-IR, the IR laser is focused on the sample, thereby causing localized heating whenever the sample absorbs the IR radiation at specific wavelengths associated with the chemical bonds. This heating results in thermal expansion and is sensed by the AFM tip as it feels the sample surface deformation due to the absorption (Fig. 7.2b). At the same time, it is possible to capture an image with a resolution of 20–100 nm as well as spectroscopical fingerprints of the polymer's chemical composition (Dazzi and Prater 2017; Woo et al. 2021). This hydride mechanism has been used for the detection and quantification of very small-sized (<20 μm) polymer particles in aqueous matrix (Have et al. 2021). However, the identification of a specific MP particle on which to focus the AFM-IR within an unknown sample can be tricky and time-consuming. The physical contact in AFM can damage the surface of the sample, thereby making it difficult to obtain a realistic image (Mariano et al. 2021). This issue can be eradicated by selecting a non-contact mode analysis, in which the tip is placed a few nanometers away from the sample.

7.3.3 Hyperspectral Imaging

Hyperspectral imaging (HSI) can be defined as a sophisticated analytical method that gathers and analyzes data at different wavelengths, providing high-resolution spectral and spatial data for each pixel in an image (Fig. 7.2c) (Faltynkova et al.

2021). Unlike conventional imaging, which captures only three bands of color (red, green, and blue), HSI divides the spectral range into numerous narrow bands to enable identification of material due to its spectral feature (Huang et al. 2021). HSI generally focuses only on specific wavelength, which includes the visible light spectrum (400–700 nm) and near-infrared NIR-HSI (700–2500 nm) (Huang et al. 2021). NIR-HSI has been determined to be a viable tool for rapid, non-destructive, and automated characterization of MPs, with a particle detection limit of 80 μm (Piarulli et al. 2020). HSI can rapidly scan large area or volume of the sample and deliver high-resolution images and chemical composition, which enables MPs distribution mapping on the sample surface with high variability and is capable of detecting a broad spectrum of polymer (Primpke et al. 2020). Using a modified NIR-HSI device, Zhu et al. (2024) found a detection limit of 100 μm to distinguish 11 types of polymers. Due to rapid development in HSI, model-based methods are emerging as more desirable because they enable accurate assessment of a large set of data, many of which contain spectra with low signal-to-noise ratios. A hybrid NIR-HSI model (Supervised soft independent modeling of class analogy: SIMCA) was used to determine the polymer types of MPs (Vidal and Pasquini 2021). The performance of the SIMCA model was highly sensitive and over 99% for PP, PE, PET, PA, and PS, allowing direct measurement of MPs in the range of 150–300 μm. In a similar study, authors demonstrated the performance of the HSI method based on two different resolutions (150 and 30 μm/pixel), two different ranges of wavelength (1000–1700 and 1000–2500 nm), and three different models (partial least squares-discriminant analysis, PLS-DA; Error correction output coding-support vector machine, ECOC-SVM; neural network pattern recognition, NNPR) to characterize HDPE, PS, and PP MPs (Serranti et al. 2024). As per their recommendation, employing resolution of 150 μm/pixel and wavelength of 1000–1700 nm, followed by PLS-DA model for MPs of >250 μm size, while in the case of <250 μm MPs, a high resolution of 30 μm/pixel with a broad spectrum range of 1000–2500 nm followed by an ECOC-SVM model is useful. The diversity of these results suggests that a broad spectrum range improves the performance of HSI than a short spectrum range. It is important to note that, the quality of the output data for MP analysis using model-based hyperspectral imaging can be affected by multiple factors such as the presence of biofilms, weathering, and chemical additives, which have not been studied enough, particularly when dealing with higher resolution (Knaeps et al. 2021). Thus, future research needs to focus on the applicability of this approach for MP analysis in real water samples. Black or highly pigmented particles capture NIR light from the source sensor, and therefore getting an accurate analysis becomes difficult (Vidal and Pasquini 2021).

7.3.4 Laser-Induced Breakdown Spectroscopy

A high-energy laser pulse is utilized in the process of laser-induced breakdown spectroscopy (LIBS), which is a strong tool in the field of analytical chemistry. This technique is used to vaporize MP samples followed by plasma formation. When

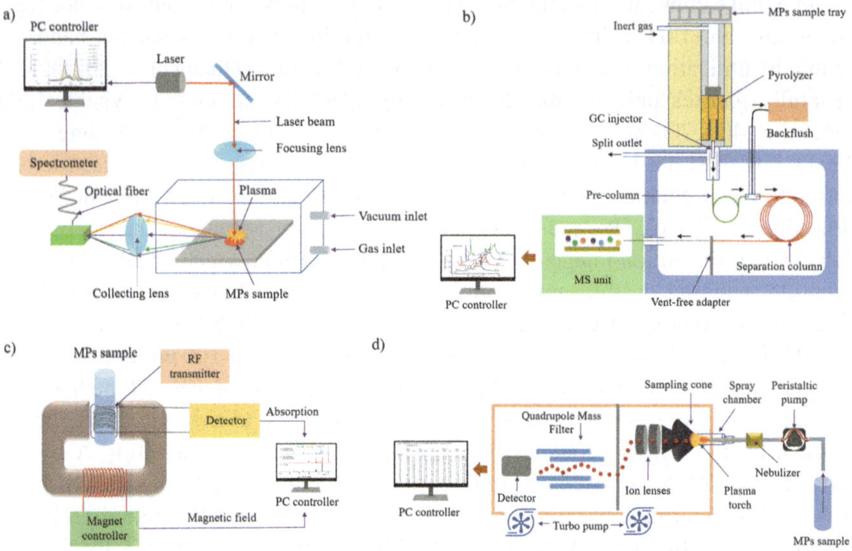

Fig. 7.3 Schematic diagrams of newly developed MP identification methods: (**a**) laser-induced breakdown spectroscopy; (**b**) pyrolysis gas chromatography-mass spectrometry (Adapted with permission and modified from Matsui et al. 2020); (**c**) nuclear magnetic resonance (Adapted with permission and modified from Zia et al. 2019); and (**d**) inductively coupled plasma mass spectrometry (Adapted with modification from Mazarakioti et al. 2022)

plasma reverts to its normal condition, the ionized elements and atoms of the sample release light at a certain wavelength. This light has a correlation with the elemental composition of the sample, which enables LIBS to determine the elemental structure of MPs (Fig. 7.3a) (Brunnbauer et al. 2023; Sommer et al. 2021). In this procedure, the sample can be analyzed in any physical state, and there is no requirement for the sample to be pre-treated beforehand. LIBS can analyze incredibly tiny sample quantities, and the whole process takes milliseconds to complete (Chen et al. 2020). Nevertheless, its application in the area of MPs' research is still insufficiently investigated. LIBS can be applied directly on individual MPs particles to detect the presence of heavy metals. Current research has proved LIBS as a potential tool for polymer classification. By this method, more than 80% precision was achieved to classify between PE and PP pellets (Giugliano et al. 2022). An investigation of a similar nature was carried out by Pořízka et al. (2023), which showed that LIBS has the capability to correctly differentiate between different types of MPs (23–455 μm size), irrespective of weathering, even when they are covered with biofilm layers on their surface. Additionally, neither the color of the sample nor its thickness affects LIBS. A drawback is that LIBS is an elemental analysis, and spectra obtained from different polymers look similar (Giugliano et al. 2022). Therefore, LIBS can only help to detect the presence of MPs but does not provide information about the polymer type (Andoh et al. 2024). For this reason, LIBS are generally employed in combination with other analytical methods such as Raman spectroscopy, HSI, and X-ray fluorescence (XRF) in order to get a more accurate evaluation of MP study (Ferreira et al. 2024). Gajarska et al. (2021) overcame this constraint through the development of a chemometric technique. This strategy involved the modification and optimization of two essential components, mainly the analytical conditions and the spectral variables. The researchers were able to distinguish 20 various kinds of polymer with the help of the LIBS system, which had its discrimination ability enhanced by these upgrades (Gajarska et al. 2021). In another study, authors have proposed a single system combining LIBS with Raman spectroscopy (Vaisakh et al. 2023). This integrated LIBS-Raman system can identify MPs (1–5 mm) from complex environmental samples, including polymer type and attached contaminants on the surface. In addition, due to the capability of multicomponent analysis from a single MP sample without any sample preparation steps, this integrated system can perform real-time MPs monitoring (Ferreira et al. 2024).

7.3.5 Nuclear Magnetic Resonance

Nuclear magnetic resonance (NMR) is an analytical tool that employs the magnetic feature of the atomic nuclei resonating at a specified frequency when subjected to an external magnetic field; most often, the hydrogen nuclei are referred to as ^{1}H-NMR (Fig. 7.3c). From these resonances, it is possible to get data of the molecular structure and the MPs composition of the sample. Besides, ^{1}H-NMR gives highly intense signals through which MPs can be evaluated directly based on their concentration in a sample (Papini et al. 2024). The use of quantitative nuclear magnetic resonance (qNMR) spectroscopy has been identified to be of particular utility for the quantitative assessment of MPs (Peez et al. 2019). Some of the key benefits include cost-effectiveness, rapid analysis, size independence, high sensitivity, precision, easy handling, and the ability to achieve an accuracy of >98% (Peez et al. 2019). A potential limitation may be the loss of shape, size, and color as dissolution of MPs particles in a suitable solvent for analysis makes this method reliant on MPs' physical properties (Peez and Imhof 2020). Peez et al. (2019) established this method, which is generally known as the integration method, together with the calibration curve technique (INT-qNMR). They characterized and quantified PE granules, PET fibers, and PS beads in model samples. During the analysis, toluene-d^8 for PE, mixture of trifluoroacetic acid chloroform-D for PET, and chloroform-D for PS were chosen as a deuterated solvents. In this study, the detection limit was between 19 to 21 μg/mL, while the quantification limit was between 74 to 85 μg/mL. Next, they have analyzed three additional types of MPs:PVC, ABS, and PA—all of which were found to be detectable and measurable using qNMR spectroscopy. This was done using either the integration method (INT) or the peak fitting (PF) method (Peez and Imhof 2020). The application of this method on environmental samples were performed for the first time for the identification and concentration determination of PS and ABS (1–1000 μm) (Papini et al. 2024). The result indicated that PS can be successfully quantified and detected in all samples in the range between 12.3 to 64.6 μg/L, and there were no proper signals for ABS. To test and optimize qNMR spectroscopy for accurate lowest possible concentration limits for different polymer types, Schmidt et al. (2024) conducted a study, where the authors investigated the use of qNMR to quantify six different MP polymers (polystyrene, polyvinyl chloride, polyisoprene-cis, polybutadiene-cis, polylactic acid, and polyurethane). All six polymers were successfully detected within the concentration limits of 0.2 to 10 µg/L. Systematic errors, such as signal overlapping due to matrix effects, require non-plastic elements to be totally excluded from the sample. For this reason, there is always a need for the pre-processing step for environmental samples to get a fair and unbiased output (Papini et al. 2024; Peez et al. 2019).

7.3.6 Thermal Cracking Aided Analysis

The thermal cracking method identifies MPs through controlled depolymerization under high temperature (500–1400 °C) in an inert medium. This technique involves the determination of gaseous compositions that are produced during the process by mass spectroscopy. Therefore, it is a completely destructive method and does not depend on the shape or size of the MPs. The thermal analysis includes pyrolysis gas chromatography-mass spectrometry (Py-GC-MS), thermogravimetric analysis (TGA), differential scanning calorimetry (DSC), and thermal extraction desorption-gas chromatography-mass spectrometry (TED-GC-MS). This method not only indicates the polymer type but also identifies the presence of additives, contaminants, and plasticizers, which provide an in-depth characterization of MPs. There is a limitation in terms of dealing with extremely small size (50–100 µm) MPs because it is difficult to transfer particles manually into the reaction chamber by the operator (Peñalver et al. 2020).

Py-GC-MS is a chromatography-based method used for the determination of the structure of complex substances, particularly for polymers in environmental samples (Fischer and Scholz-Böttcher 2019; Gomiero et al. 2019). It includes pyrolysis (molecular cracking), gas chromatography (isolation of decomposed fragments), and mass spectrometry (identification of chemical structure) (Fig. 7.3b). Two modes of operation of pyrolysis can be classified: micro-furnace pyrolyzer (constant mode), Curie-point pyrolyzer and heated filament pyrolyzer (pulse mode) (Ainali et al. 2021). Presently, micro-furnace pyrolyzers are predominantly employed, which consist of a quartz tube provided with quick heating and cooling facilities and several temperature sensors to monitor heating parameters. Fischer and Scholz-Böttcher (2019) compared micro-furnace (MF) pyrolyzers with Curie-point (CP) pyrolyzers and stated that, MF pyrolyzers are more sensitive and reliable in terms of polymer data. Several operational modes of Py-GC-MS have been documented according to the analytical objectives. These encompass single-shot analysis, double-shot analysis, evolved gas analysis, heart-cut analysis, and reactive pyrolysis-GC-MS (Picó and Barceló 2020). Okoffo et al. (2023) and Santos et al. (2023) were able to identify 5 and 12 different types of polymers, respectively, using the double-shot feature of multi-shot micro-furnace pyrolyzers.

Thermogravimetric analysis (TGA) preliminary relies on the determination of the mass loss profile of the sample with time and temperature. Recently, the TGA-based approach (combination of TGA with other setups) has gained interest for the quick measurement of MPs (Becker et al. 2020; Goedecke et al. 2020). The methods TGA-FTIR and TGA-GC-MS have the potential to be an effective tool for the characterization of polymers (Liu et al. 2021a). Another different approach is differential scanning calorimetry (DSC); it is used to identify the physical properties of polymers by revealing changes in dissolution, transition temperature, crystallization, entropy and enthalpy. However, the DSC has limitations, such as peaks overlapping during the detection of MPs having similar melting points (Woo et al. 2021). The combination of thermogravimetry and thermal desorption (TED-GC-MS) uses a solid-phase adsorbent for identifying MPs. TED-GC-MS was devised to eliminate limitations of Py-GC-MS by separating thermal phase from GC-MS (Duemichen et al. 2019). Compared to Py-GC-MS, this method provides shorter analysis time and able to handle slightly larger sample mass. Applicability of this method for environmental samples was examined, and a blend of polymers (PE, PP, and PS) were successfully identified without any peak overlapping (Table 7.4).

7.3.7 Inductively Coupled Plasma Mass Spectroscopy

Inductively coupled plasma mass spectrometry (ICP-MS) applied in single-particle mode (SP-ICP-MS) provides a new way toward advanced MPs research (Harycki and Gundlach-Graham 2023; Hendriks and Mitrano 2023). In SP-ICP-MS, every single particle is atomized by entering into heated plasma, where the components are ionized and lead to the formation of ion clouds. These ion clouds are isolated from the plasma and distinguished via short pulses and analyzed using a mass spectrometer (Fig. 7.3d) (Gonzalez de Vega et al. 2021). ICP-MS is capable of analyzing

Table 7.4 Difference between Py-GC-MS and TGA-based MP identification methods

Features	Py-GC/MS	TGA-based method (TGA-FTIR/ GC-MS/DSC)
Principle	Pyrolyze molecular fragment, isolation by GC, identification by MS	Works on mass loss against heating, identification by FTIR, GC-MS, thermal transition in DSC
Sample pre-treatment	Not required	Minimal treatment required based on matrix type
Polymer identification	Possible	Possible
Organic additives identification	Possible	Possible
Sample mass requirement	5–200 μg	200 times more than Py-GC/MS
Analysis pattern	Automatic	Automatic
Sensitivity	Very high	Moderate
Mixed MP sample	Effective	Not effective

large quantities of particles (1–2.5 μm size) in a very short duration (≤100 μs) and provides accurate information of the chemical structure, particle sizes, particle number density, and particle mass concentrations (Bolea-Fernandez et al. 2020). Recently, advancements in SP-ICP-MS have been focusing on MP identification by analyzing 13C isotope for improved precision (Bolea-Fernandez et al. 2020; Gonzalez de Vega et al. 2021). Development of this protocol was first established to quantify and identify spherical PS microspheres. Later, the application of an ICP-time-of-flight mass spectrometer (ICP-TOFMS) in single particle mode was studied for the assessment of MPs (Hendriks and Mitrano 2023). During the process, presence of carbon-rich elements such as cells, organic carbons, and other natural particles may create pulse signals similar to those of MPs. ICP-TOFMS allows the identification of different particle compositions by evaluating the elemental spectrum. Despite this method's performance, its effectiveness in complex aqueous samples needs to be tested.

7.3.8 Algorithmic Model-Based Analysis

The development and implementation of machine learning (ML) driven by artificial intelligence (AI) has created a revolution in the area of advanced MPs research (Bertoldi et al. 2021). Inspired by the human brain, deep learning (DL), a segment of ML that follows neural networks, has progressively been applied for MP analysis to develop computational ability and algorithmic skills. In addition, AI-based models significantly reduce the probability of error due to human judgment (Hu et al. 2024). In the field of MPs research, conjugation of AI and ML with conventional analytical methods can increase the accuracy, efficiency, and reliability of MP identification (Enyoh and Wang 2024; Le Han et al. 2023). Multiple applications of ML algorithms were used for spectrum identification, image recognition, image segmentation, and spectral image identification (Su et al. 2023). A number of ML models, including artificial neural networks (ANN), support vector machines (SVM), decision trees (DT), random forest (RF), and principal component analysis (PCA), were frequently used for MP analysis (Lin et al. 2022). Each ML algorithm possesses its own data set of distinctive properties. For further improvement in AI-based monitoring of MPs pollution, a close-range remote sensing technique using satellite or drone-based sensors was developed, in which the obtained dataset can be evaluated using AI models (Gnann et al. 2022). Gonçalves et al. (2020) have used a remote sensing method for MPs quantification in a beach and compared three object-oriented ML models, including SVM, KNN, and RF, which showed nearly similar results (F-score) for detecting colored MPs. However, varied optimization parameters with respect to each model can differentiate the best ML model. In addition, ML can draw quick future predictions on the basis of a small amount of known data. Hence, this approach can be used to predict the fate, ecological impact, and risk factors of MPs in freshwater ecosystems. Nevertheless, a lack of large datasets and incomparable output of different studies are the challenges that urgently need to be overcome by implementing big projects and international cooperation.

7.3.8.1 Artificial Neural Networks (ANN)-Based MP Analysis

Presently, ANN algorithms are widely used in MP research, in which a convolutional neural network (CNN) is frequently adopted as a type of ANN. In a recent research, Luo et al. (2024b) proposed a method for quantitative analysis of MPs in water matrix by integrating CNN with Raman spectroscopy. The results indicated that CNN can effectively provide important information of the Raman spectrum, which can be used to establish correlations among different variables. The R^2 value of CNN for identifying different sizes of PE MPs in varying water samples can reach 0.9972. In a similar study, authors have compared the effectiveness of CNN with the conventional Raman spectram library to evaluate the robustness of this model (Lee et al. 2023). The CNN model showed a higher accuracy of 99.54% than the Raman spectrum library (31.44%), which confirmed that CNN-based DL can identify MPs by analyzing Raman signals effectively. In a different study, the ANN model was used to predict the sorption capacity of Pb, Cd, Cu, Cr, and Zn on the MPs surface in freshwater and saline water samples (Guo and Wang 2021). Based on the prediction, the author reported that MPs in freshwater systems have higher sorption capacity than in saline water. Additionally, weathered MPs have shown better affinity to MPs than virgin MPs. Further development of this model is required to identify the MPs' pollution source and their environmental impact assessment.

7.3.8.2 Support Vector Machines (SVM) and Principal Component Analysis (PCA)-Based MP Analysis

In the context of linearly non-separable data points, such as those of MPs due to their varying physical and chemical nature, SVM can provide very accurate results of MP identification. In comparison, PCA is generally used to separate MPs from the background spectra, which will help to reduce the data for further analysis steps. However, it has no objective function to differentiate polymer types. Wander et al. (2020) eliminated 94% spectrum to be analyzed in the next steps. These separated MP spectra can then be used for other models. SVM model with pre-treated PCA can significantly increase MP identification accuracy from 21% to 92% (Michel et al. 2020). Tan et al. (2023) used SVM in conjunction with quaternion (Q)-PCA and FTIR spectrum to analyze four different spiked MP samples. The authors have reported that compared to other traditional algorithms, SVM-QPCA can provide the highest accuracy and precision. In another study, authors have investigated ML algorithms with FTIR and reported that SVM is well suited to overcome the technical limitations of FTIR, such as handling wide data sets and automated analysis of MPs (Back et al. 2022). However, comparatively less precision was obtained for small (<1 mm) black particles (Shan et al. 2018).

7.3.8.3 Decision Tree (DT) and Random Forest (RF)-Based MP Analysis

Over the past few years, DT and RF have been steadily employed in MPs analysis. DT and RF-based models can be used for rapid identification of MPs based on the obtained spectrum. The results indicated that RF was able to provide more cross-validation accuracy than DT, and after further optimization of process parameters, the accuracy can reach up to 97.24% (Yang et al. 2021). Liu et al. (2023) analyzed one-dimensional (1D) and two-dimensional (2D) CNN, DT, and RF models fitted with FTIR data. They achieved an accuracy of 96.43% for a short data set and 97.44% for a large data set. In another study, Kannankai et al. (2022) used Apriori and RF models to depict the distribution of MPs in global marine ecosystems. Mao et al. (2024) analyzed the RF model to study the influencing factors of MP removal from drinking water using a coagulant. The authors were able to conclude that the shape of MPs was the dominant factor, followed by coagulant type, coagulant dosage, MPs concentration, size, polymer type, and pH.

7.4 Conclusions

Modern research on MPs in freshwater ecosystems has progressed to improve the efficiency of separation and identification methods. These improvements have enhanced our ability to isolate MPs from various complex environmental matrices. Traditional density-based separation methods, though reliable, were time-consuming and had poor recovery rates in terms of small-sized MPs. As a result, more improved techniques were explored to address these limitations. Density-independent methods such as magnetic separation, oil-based extraction, and electrochemical techniques have revolutionized the isolation of <100 μm size MPs by enabling speedy and efficient recovery of diverse polymer types from environmental samples.

Parallel developments in the characterization of MPs have evolved and are still progressing to accurately identify isolated MPs. The majority of the studies have used traditional methods like FTIR and Raman spectroscopy; however, these methods use a subset of MPs from the sample due to instrumental limitations to scan the entire filter. Furthermore, for these well-known identification methods, pretreatment of the sample is highly required for unbiased results. Therefore, to improve the analytical robustness, advanced microscopic methods have evolved, offering high-resolution imaging and three-dimensional topographical mapping of the MP sample. Recently, there have been a lot of advancements observed in terms of MP identification by mass spectroscopy and chromatography. In this regard, Py-GC-MS is attracting researchers due to its simplicity and ability to determine the polymer types as well as other contaminants and additives attached to the MPs. In addition, a combination of LIBS, NMR, and ICP-MS with other traditional methods can analyze MPs in a very short duration. The integration of AI and ML algorithms with other spectroscopic methods have been also explored to exclude human errors and improve accuracy while handling a large number of data sets.

Despite these significant advancements, the current challenge that still persists is the handling of small MPs (<20 μm), which are mainly persistent in the environment. In addition, the rapid advancement in analytical techniques has significantly improved detection efficiencies; however, it has made things complex, particularly related to methodological standardization, which limits cross-study comparisons. Variations in sample collection, preparation, extraction, and identification methods result in huge differences in reported MPs concentrations and characteristics, complicating efforts to establish baseline conditions that are pivotal in developing evidence-based regulatory frameworks and management strategies. Addressing these limitations needs collaborative efforts to establish consensus-based protocols that balance methodological rigor with practical feasibility. Such protocols should include the entire analytical workflow, from field sampling to data reporting, and should be adaptable to diverse environmental contexts while maintaining core standardized elements. Furthermore, the development of certified reference materials specifically designed for MP analysis would significantly enhance quality assurance and method validation efforts across the research community.

References

Ainali NM, Kalaronis D, Kontogiannis A, Evgenidou E, Kyzas GZ, Yang X, Bikiaris DN, Lambropoulou DA (2021) Microplastics in the environment: sampling, pretreatment, analysis and occurrence based on current and newly-exploited chromatographic approaches. Sci Total Environ 794:148725. https://doi.org/10.1016/j.scitotenv.2021.148725

Akarsu C, Kumbur H, Kideys AE (2021) Removal of microplastics from wastewater through electrocoagulation-electroflotation and membrane filtration processes. Water Sci Technol 84(7):1648–1662. https://doi.org/10.2166/wst.2021.356

Andoh CN, Attiogbe F, Bonsu Ackerson NO, Antwi M, Adu-Boahen K (2024) Fourier transform infrared spectroscopy: an analytical technique for microplastic identification and quantification. Infrared Phys Technol 136:105070. https://doi.org/10.1016/j.infrared.2023.105070

Bai Z, Wang N, Wang M (2021) Effects of microplastics on marine copepods. Ecotoxicol Environ Saf 217:112243. https://doi.org/10.1016/j.ecoenv.2021.112243

Becker R, Altmann K, Sommerfeld T, Braun U (2020) Quantification of microplastics in a freshwater suspended organic matter using different thermoanalytical methods – outcome of an interlaboratory comparison. J Anal Appl Pyrolysis 148:104829. https://doi.org/10.1016/j.jaap.2020.104829

Běhal J, Valentino M, Miccio L, Bianco V, Itri S, Mossotti R, Dalla Fontana G, Stella E, Ferraro P (2022) Toward an all-optical fingerprint of synthetic and natural microplastic fibers by polarization-sensitive holographic microscopy. ACS Photonics 9(2):694–705. https://doi.org/10.1021/acsphotonics.1c01781

Bellasi A, Binda G, Pozzi A, Boldrocchi G, Bettinetti R (2021) The extraction of microplastics from sediments: an overview of existing methods and the proposal of a new and green alternative. Chemosphere 278:130357. https://doi.org/10.1016/j.chemosphere.2021.130357

Bertoldi C, Lara LZ, Gomes AA, Fernandes AN (2021) Microplastic abundance quantification via a computer-vision-based chemometrics-assisted approach. Microchem J 160:105690. https://doi.org/10.1016/j.microc.2020.105690

Bhatt P, Pathak VM, Bagheri AR, Bilal M (2021) Microplastic contaminants in the aqueous environment, fate, toxicity consequences, and remediation strategies. Environ Res 200:111762. https://doi.org/10.1016/j.envres.2021.111762

Bolea-Fernandez E, Rua-Ibarz A, Velimirovic M, Tirez K, Vanhaecke F (2020) Detection of microplastics using inductively coupled plasma-mass spectrometry (ICP-MS) operated in single-event mode. J Anal At Spectrom 35(3):455–460. https://doi.org/10.1039/c9ja00379g

Brunnbauer L, Gajarska Z, Lohninger H, Limbeck A (2023) A critical review of recent trends in sample classification using laser-induced breakdown spectroscopy (LIBS). TrAC Trends Anal Chem 159:116859. https://doi.org/10.1016/j.trac.2022.116859

Burrows SD, Frustaci S, Thomas KV, Galloway T (2020) Expanding exploration of dynamic microplastic surface characteristics and interactions. TrAC Trends Anal Chem 130:115993. https://doi.org/10.1016/j.trac.2020.115993

Cao Y, Sathish CI, Guan X, Wang S, Palanisami T, Vinu A, Yi J (2024) Advances in magnetic materials for microplastic separation and degradation. J Hazard Mater 461:132537. https://doi.org/10.1016/j.jhazmat.2023.132537

Chen D, Wang T, Ma Y, Wang G, Kong Q, Zhang P, Li R (2020) Rapid characterization of heavy metals in single microplastics by laser induced breakdown spectroscopy. Sci Total Environ 743:140850. https://doi.org/10.1016/j.scitotenv.2020.140850

Coleman BR (2025) An introduction to machine learning tools for the analysis of microplastics in complex matrices. Environ Sci: Processes Impacts 27(1):10–23. https://doi.org/10.1039/d4em00605d

Crichton EM, Noël M, Gies EA, Ross PS (2017) A novel, density-independent and FTIR-compatible approach for the rapid extraction of microplastics from aquatic sediments. Anal Methods 9(9):1419–1428. https://doi.org/10.1039/c6ay02733d

Crutchett TW, Bornt KR (2024) A simple overflow density separation method that recovers >95% of dense microplastics from sediment. MethodsX 12:102638. https://doi.org/10.1016/j.mex.2024.102638

Cui B, Rong H, Tian T, Guo D, Duan L, Nkinahamira F, Ndagijimana P, Yan W, Naidu R (2024) Chemical methods to remove microplastics from wastewater: a review. Environ Res 249:118416. https://doi.org/10.1016/j.envres.2024.118416

Cutroneo L, Reboa A, Geneselli I, Capello M (2021) Considerations on salts used for density separation in the extraction of microplastics from sediments. Mar Pollut Bull 166:112216. https://doi.org/10.1016/j.marpolbul.2021.112216

Dazzi A, Prater CB (2017) AFM-IR: technology and applications in nanoscale infrared spectroscopy and chemical imaging. Chem Rev 117(7):5146–5173. https://doi.org/10.1021/acs.chemrev.6b00448

Back HM, Vargas Junior EC, Alarcon OE, Pottmaier D (2022) Training and evaluating machine learning algorithms for ocean microplastics classification through vibrational spectroscopy. Chemosphere 287:131903. https://doi.org/10.1016/j.chemosphere.2021.131903

Dierkes G, Lauschke T, Becher S, Schumacher H, Földi C, Ternes T (2019) Quantification of microplastics in environmental samples via pressurized liquid extraction and pyrolysis-gas chromatography. Anal Bioanal Chem 411(26):6959–6968. https://doi.org/10.1007/s00216-019-02066-9

Duemichen E, Eisentraut P, Celina M, Braun U (2019) Automated thermal extraction-desorption gas chromatography mass spectrometry: a multifunctional tool for comprehensive characterization of polymers and their degradation products. J Chromatogr A 1592:133–142. https://doi.org/10.1016/j.chroma.2019.01.033

Duong TT, Le PT, Nguyen TNH, Hoang TQ, Ngo HM, Doan TO, Le TPQ, Bui HT, Bui MH, Trinh VT, Nguyen TL, Da Le N, Vu TM, Tran TKC, Ho TC, Phuong NN, Strady E (2022) Selection of a density separation solution to study microplastics in tropical riverine sediment. Environ Monit Assess 194(2):1–17. https://doi.org/10.1007/S10661-021-09664-0

Elkhatib D, Oyanedel-Craver V, Carissimi E (2021) Electrocoagulation applied for the removal of microplastics from wastewater treatment facilities. Sep Purif Technol 276:118877. https://doi.org/10.1016/j.seppur.2021.118877

Elmaci G (2020) Microwave-assisted rapid synthesis of C@Fe3O4 composite for removal of microplastics from drinking water. Adıyaman Üniversitesi Fen Bilimleri Dergisi 10(1):207–217. https://doi.org/10.37094/adyujsci.739599

Enders K, Tagg AS, Labrenz M (2020) Evaluation of electrostatic separation of microplastics from mineral-rich environmental samples. Front Environ Sci 8:549811. https://doi.org/10.3389/fenvs.2020.00112

Enyoh CE, Wang Q (2024) Automated classification of Undegraded and aged polyethylene terephthalate microplastics from ATR-FTIR spectroscopy using machine learning algorithms. J Polym Environ 32(9):4143–4158. https://doi.org/10.1007/s10924-024-03199-4

Fadare OO, Martin L, Lascelles N, Myers JT, Kaiser K, Xu W, Conkle JL (2023) Binary solvent extraction of microplastics from a complex environmental matrix. Limnol Oceanogr Methods 21(7):414–420. https://doi.org/10.1002/lom3.10554

Faltynkova A, Johnsen G, Wagner M (2021) Hyperspectral imaging as an emerging tool to analyze microplastics: a systematic review and recommendations for future development. Microplastics Nanoplastics 1(1):1–19. https://doi.org/10.1186/s43591-021-00014-y

Felsing S, Kochleus C, Buchinger S, Brennholt N, Stock F, Reifferscheid G (2018) A new approach in separating microplastics from environmental samples based on their electrostatic behavior. Environ Pollut 234:20–28. https://doi.org/10.1016/j.envpol.2017.11.013

Ferreira DS, Babos DV, Lima-Filho MH, Castello HF, Olivieri AC, Verbi Pereira FM, Pereira-Filho ER (2024) Laser-induced breakdown spectroscopy (LIBS): calibration challenges, combination with other techniques, and spectral analysis using data science. J Anal At Spectrom 39(12):2949–2973. https://doi.org/10.1039/d4ja00250d

Fischer M, Scholz-Böttcher BM (2019) Microplastics analysis in environmental samples – recent pyrolysis-gas chromatography-mass spectrometry method improvements to increase the reliability of mass-related data. Anal Methods 11(18):2489–2497. https://doi.org/10.1039/c9ay00600a

Fuller S, Gautam A (2016) A procedure for measuring microplastics using pressurized fluid extraction. Environ Sci Technol 50(11):5774–5780. https://doi.org/10.1021/acs.est.6b00816

Gajarska Z, Brunnbauer L, Lohninger H, Limbeck A (2021) Identification of 20 polymer types by means of laser-induced breakdown spectroscopy (LIBS) and chemometrics. Anal Bioanal Chem 413(26):6581–6594. https://doi.org/10.1007/s00216-021-03622-y

Girish N, Parashar N, Hait S (2023) Coagulative removal of microplastics from aqueous matrices: recent progresses and future perspectives. Sci Total Environ 899:165723. https://doi.org/10.1016/j.scitotenv.2023.165723

Giugliano R, Cocciaro B, Poggialini F, Legnaioli S, Palleschi V, Locritani M, Merlino S (2022) Rapid identification of beached marine plastics pellets using laser-induced breakdown spectroscopy: a promising tool for the quantification of coastal pollution. Sensors 22(18):6910. https://doi.org/10.3390/s22186910

Gnann N, Baschek B, Ternes TA (2022) Close-range remote sensing-based detection and identification of macroplastics on water assisted by artificial intelligence: a review. Water Res 222:118902. https://doi.org/10.1016/j.watres.2022.118902

Goedecke C, Dittmann D, Eisentraut P, Wiesner Y, Schartel B, Klack P, Braun U (2020) Evaluation of thermoanalytical methods equipped with evolved gas analysis for the detection of microplastic in environmental samples. J Anal Appl Pyrolysis 152:104961. https://doi.org/10.1016/j.jaap.2020.104961

Gomiero A, Øysæd KB, Agustsson T, van Hoytema N, van Thiel T, Grati F (2019) First record of characterization, concentration and distribution of microplastics in coastal sediments of an urban fjord in south West Norway using a thermal degradation method. Chemosphere 227:705–714. https://doi.org/10.1016/j.chemosphere.2019.04.096

Gonçalves G, Andriolo U, Gonçalves L, Sobral P, Bessa F (2020) Quantifying marine macro litter abundance on a Sandy Beach using unmanned aerial systems and object-oriented machine learning methods. Remote Sens 12(16):2599. https://doi.org/10.3390/RS12162599

Gonzalez de Vega R, Goyen S, Lockwood TE, Doble PA, Camp EF, Clases D (2021) Characterisation of microplastics and unicellular algae in seawater by targeting carbon via single particle and single cell ICP-MS. Anal Chim Acta 1174:338737. https://doi.org/10.1016/j.aca.2021.338737

Grbic J, Nguyen B, Guo E, You JB, Sinton D, Rochman CM (2019) Magnetic extraction of microplastics from environmental samples. Environ Sci Technol Lett 6(2):68–72. https://doi.org/10.1021/acs.estlett.8b00671

Günther M, Imhof W (2024) Highly selective solid–liquid extraction of microplastic mixtures as a pre-preparation tool for quantitative nuclear magnetic resonance spectroscopy studies. Analyst 149(24):5800–5811. https://doi.org/10.1039/d4an00991f

Guo X, Wang J (2021) Projecting the sorption capacity of heavy metal ions onto microplastics in global aquatic environments using artificial neural networks. J Hazard Mater 402:123709. https://doi.org/10.1016/j.jhazmat.2020.123709

Harycki S, Gundlach-Graham A (2023) Single-particle ICP-TOFMS with online microdroplet calibration: a versatile approach for accurate quantification of nanoparticles, submicron particles, and microplastics in seawater. Anal Chem 95(41):15318–15324. https://doi.org/10.1021/acs.analchem.3c02785

Have C et al (2021) Photoinduced force microscopy as an efficient method towards the detection of nanoplastics. Chem Methods 1(5):205–209. https://doi.org/10.1002/cmtd.202100017

Hendriks L, Mitrano DM (2023) Direct measurement of microplastics by carbon detection via single particle ICP-TOFMS in complex aqueous suspensions. Environ Sci Technol 57(18):7263–7272. https://doi.org/10.1021/acs.est.3c01192

Hong SH, Shim WJ, Hong L (2017) Methods of analysing chemicals associated with microplastics: a review. Anal Methods 9(9):1361–1368. https://doi.org/10.1039/c6ay02971j

Hu B, Dai Y, Zhou H, Sun Y, Yu H, Dai Y, Wang M, Ergu D, Zhou P (2024) Using artificial intelligence to rapidly identify microplastics pollution and predict microplastics environmental behaviors. J Hazard Mater 474:134865. https://doi.org/10.1016/j.jhazmat.2024.134865

Huang H, Qureshi JU, Liu S, Sun Z, Zhang C, Wang H (2021) Hyperspectral imaging as a potential online detection method of microplastics. Bull Environ Contam Toxicol 107(4):754–763. https://doi.org/10.1007/s00128-020-02902-0

Hunter JR, Qiao Q, Zhang Y, Shao Q, Crofcheck C, Shi J (2023) Green solvent mediated extraction of micro- and nano-plastic particles from water. Sci Rep 13(1):1–12. https://doi.org/10.1038/s41598-023-37490-6

Jiang H, Zhang Y, Bian K, Wang H, Wang C (2022) Insight into the effect of aqueous species on microplastics removal by froth flotation: kinetics and mechanism. J Environ Chem Eng 10(3):107834. https://doi.org/10.1016/j.jece.2022.107834

Jin H, Kong F, Li X, Shen J (2024) Artificial intelligence in microplastic detection and pollution control. Environ Res 262:119812. https://doi.org/10.1016/j.envres.2024.119812

Kalaronis D, Ainali NM, Evgenidou E, Kyzas GZ, Yang X, Bikiaris DN, Lambropoulou DA (2022) Microscopic techniques as means for the determination of microplastics and nanoplastics in the aquatic environment: a concise review. Green Anal Chem 3:100036. https://doi.org/10.1016/j.greeac.2022.100036

Kannankai MP, Babu AJ, Radhakrishnan A, Alex RK, Borah A, Devipriya SP (2022) Machine learning aided meta-analysis of microplastic polymer composition in global marine environment. J Hazard Mater 440:129801. https://doi.org/10.1016/j.jhazmat.2022.129801

Knaeps E, Sterckx S, Strackx G, Mijnendonckx J, Moshtaghi M, Garaba SP, Meire D (2021) Hyperspectral-reflectance dataset of dry, wet and submerged marine litter. Earth System Science Data 13(2):713–730. https://doi.org/10.5194/essd-13-713-2021

Kurzweg L, Hauffe M, Schirrmeister S, Adomat Y, Socher M, Grischek T, Fery A, Harre K (2024) Microplastic analysis in sediments of the Elbe River by electrostatic separation and differential scanning calorimetry. Sci Total Environ 930:172514. https://doi.org/10.1016/j.scitotenv.2024.172514

Lapointe M, Farner JM, Hernandez LM, Tufenkji N (2020) Understanding and improving microplastic removal during water treatment: impact of coagulation and flocculation. Environ Sci Technol 54(14):8719–8727. https://doi.org/10.1021/acs.est.0c00712

Le Han X, Jiang NJ, Hata T, Choi J, Du YJ, Wang YJ (2023) Deep learning based approach for automated characterization of large marine microplastic particles. Mar Environ Res 183:105829. https://doi.org/10.1016/j.marenvres.2022.105829

Lechthaler S Hildebrandt L, Stauch G, Schüttrumpf H, Schüttrumpf S (2020) Canola oil extraction in conjunction with a plastic free separation unit optimises microplastics monitoring in water and sediment. https://doi.org/10.1039/d0ay01574a

Lee J, Chae KJ (2021) A systematic protocol of microplastics analysis from their identification to quantification in water environment: a comprehensive review. J Hazard Mater 403:124049. https://doi.org/10.1016/j.jhazmat.2020.124049

Lee S, Jeong H, Hong SM, Yun D, Lee J, Kim E, Cho KH (2023) Automatic classification of microplastics and natural organic matter mixtures using a deep learning model. Water Res 246:120710. https://doi.org/10.1016/j.watres.2023.120710

Li W, Luo Y, Pan X (2020) Identification and characterization methods for microplastics basing on spatial imaging in micro-/nanoscales. Handbook Environ Chem 95:25–37. https://doi.org/10.1007/698_2020_446

Li Y, Zhu Y, Huang J, Ho YW, Fang JKH, Lam EY (2024) High-throughput microplastic assessment using polarization holographic imaging. Sci Rep 14(1):1–11. https://doi.org/10.1038/s41598-024-52762-5

Lin J et al (2022) Recent advances in the application of machine learning methods to improve identification of the microplastics in environment. Chemosphere 307:136092. https://doi.org/10.1016/j.chemosphere.2022.136092

Liu Y, Li R, Yu J, Ni F, Sheng Y, Scircle A, Cizdziel JV, Zhou Y (2021a) Separation and identification of microplastics in marine organisms by TGA-FTIR-GC/MS: a case study of mussels from coastal China. Environ Pollut 272:115946. https://doi.org/10.1016/j.envpol.2020.115946

Liu T, Yu S, Zhu X, Liao R, Zhuo Z, He Y, Ma H (2021b)At alkaline conditions, largeIn-situ detection method for microplastics in water by polarized light scattering. Front Mar Sci 8:739683. https://doi.org/10.3389/fmars.2021.739683

Liu Y, Yao W, Qin F, Zhou L, Zheng Y (2023) Spectral classification of large-scale blended (micro) plastics using FT-IR raw spectra and image-based machine learning. Environ Sci Technol 57(16):6656–6663. https://doi.org/10.1021/acs.est.2c08952

Lodh A, Singh RK, Ruj B (2023) Thermo-catalytic pyrolysis of marine debris: a waste to wealth conversion study for energy recovery. Fuel 350:128850. https://doi.org/10.1016/j.fuel.2023.128850

Lu S, Liu L, Yang Q, Demissie H, Jiao R, An G, Wang D (2021) Removal characteristics and mechanism of microplastics and tetracycline composite pollutants by coagulation process. Sci Total Environ 786:147508. https://doi.org/10.1016/j.scitotenv.2021.147508

Luo W, Fu H, Lu Q, Li B, Cao X, Chen S, Liu R, Tang B, Yan X, Zheng J (2024a) Microplastic pollution differences in freshwater river according to stream order: insights from spatial distribution, annual load, and ecological assessment. J Environ Manag 366:121836. https://doi.org/10.1016/j.jenvman.2024.121836

Luo Y, Su W, Rabbi MF, Wan Q, Xu D, Wang Z, Liu S, Xu X, Wu J (2024b) Quantitative analysis of microplastics in water environments based on Raman spectroscopy and convolutional neural network. Sci Total Environ 926:171925. https://doi.org/10.1016/j.scitotenv.2024.171925

Ma B, Xue W, Ding Y, Hu C, Liu H, Qu J (2019a) Removal characteristics of microplastics by Fe-based coagulants during drinking water treatment. J Environ Sci 78:267–275. https://doi.org/10.1016/j.jes.2018.10.006

Ma B, Xue W, Hu C, Liu H, Qu J, Li L (2019b) Characteristics of microplastic removal via coagulation and ultrafiltration during drinking water treatment. Chem Eng J 359:159–167. https://doi.org/10.1016/j.cej.2018.11.155

Mani T, Frehland S, Kalberer A, Burkhardt-Holm P (2019) Using castor oil to separate microplastics from four different environmental matrices. Anal Methods 11(13):1788–1794. https://doi.org/10.1039/c8ay02559b

Mao Y, Hu Z, Li H, Zheng H, Yang S, Yu W, Tang B, Yang H, He R, Guo W, Ye K, Yang A, Zhang S (2024) Recent advances in microplastic removal from drinking water by coagulation: removal mechanisms and influencing factors. Environ Pollut 349:123863. https://doi.org/10.1016/j.envpol.2024.123863

Mariano S, Tacconi S, Fidaleo M, Rossi M, Dini L (2021) Micro and nanoplastics identification: classic methods and innovative detection techniques. Front Toxicol 3:636640. https://doi.org/10.3389/ftox.2021.636640

Matsui K, Ishimura T, Mattonai M, Iwai I, Watanabe A, Teramae N, Ohtani H, Watanabe C (2020) Identification algorithm for polymer mixtures based on Py-GC/MS and its application for microplastic analysis in environmental samples. J Anal Appl Pyrolysis 149:104834. https://doi.org/10.1016/j.jaap.2020.104834

Mazarakioti EC, Zotos A, Thomatou AA, Kontogeorgos A, Patakas A, Ladavos A (2022) Inductively coupled plasma-mass spectrometry (ICP-MS), a useful tool in authenticity of agricultural products' and foods' origin. Foods 11(22):3705. https://doi.org/10.3390/foods11223705

Melo-Agustín P, Kozak ER, de Jesús Perea-Flores M, Mendoza-Pérez JA (2022) Identification of microplastics and associated contaminants using ultra high resolution microscopic and spectroscopic techniques. Sci Total Environ 828:154434. https://doi.org/10.1016/j.scitotenv.2022.154434

Mercogliano R, Avio CG, Regoli F, Anastasio A, Colavita G, Santonicola S (2020) Occurrence of microplastics in commercial seafood under the perspective of the human food chain. A review. J Agric Food Chem 68(19):5296–5301. https://doi.org/10.1021/acs.jafc.0c01209

Michel APM, Morrison AE, Preston VL, Marx CT, Colson BC, White HK (2020) Rapid identification of marine plastic debris via spectroscopic techniques and machine learning classifiers. Environ Sci Technol 54(17):10630–10637. https://doi.org/10.1021/acs.est.0c02099

Möller JN, Löder MGJ, Laforsch C (2020) Finding microplastics in soils: a review of analytical methods. Environ Sci Technol 54(4):2078–2090. https://doi.org/10.1021/acs.est.9b04618

Monira S, Bhuiyan MA, Haque N, Pramanik BK (2021) Assess the performance of chemical coagulation process for microplastics removal from stormwater. Process Saf Environ Prot 155:11–16. https://doi.org/10.1016/j.psep.2021.09.002

Na SH, Kim MJ, Kim JT, Jeong S, Lee S, Chung J, Kim EJ (2021) Microplastic removal in conventional drinking water treatment processes: performance, mechanism, and potential risk. Water Res 202:117417. https://doi.org/10.1016/j.watres.2021.117417

Núñez J, Yeber M, Cisternas N, Thibaut R, Medina P, Carrasco C (2019) Application of electrocoagulation for the efficient pollutants removal to reuse the treated wastewater in the dyeing process of the textile industry. J Hazard Mater 371:705–711. https://doi.org/10.1016/j.jhazmat.2019.03.030

Okoffo ED, Ribeiro F, O'Brien JW, O'Brien S, Tscharke BJ, Gallen M, Samanipour S, Mueller JF, Thomas KV (2020) Identification and quantification of selected plastics in biosolids by pressurized liquid extraction combined with double-shot pyrolysis gas chromatography–mass spectrometry. Sci Total Environ 715:136924. https://doi.org/10.1016/j.scitotenv.2020.136924

Okoffo ED, Rauert C, Thomas KV (2023) Mass quantification of microplastic at wastewater treatment plants by pyrolysis-gas chromatography–mass spectrometry. Sci Total Environ 856:159251. https://doi.org/10.1016/j.scitotenv.2022.159251

Padha S, Kumar R, Dhar A, Sharma P (2022) Microplastic pollution in mountain terrains and foothills: a review on source, extraction, and distribution of microplastics in remote areas. Environ Res 207:112232. https://doi.org/10.1016/j.envres.2021.112232

Papini G, Petrella G, Cicero DO, Boglione C, Rakaj A (2024) Identification and quantification of polystyrene microplastics in marine sediments facing a river mouth through NMR spectroscopy. Mar Pollut Bull 198:115784. https://doi.org/10.1016/j.marpolbul.2023.115784

Peez N, Imhof W (2020) Quantitative 1H-NMR spectroscopy as an efficient method for identification and quantification of PVC. ABS PA Micropart Anal 145(15):5363–5371. https://doi.org/10.1039/d0an00879f

Peez N, Janiska MC, Imhof W (2019) The first application of quantitative 1H NMR spectroscopy as a simple and fast method of identification and quantification of microplastic particles (PE, PET, and PS). Anal Bioanal Chem 411(4):823–833. https://doi.org/10.1007/s00216-018-1510-z

Peñalver R, Arroyo-Manzanares N, López-García I, Hernández-Córdoba M (2020) An overview of microplastics characterization by thermal analysis. Chemosphere 242:125170. https://doi.org/10.1016/j.chemosphere.2019.125170

Peng X, Chen M, Chen S, Dasgupta S, Xu H, Ta K, Du M, Li J, Guo Z, Bai S (2018) Microplastics contaminate the deepest part of the world's ocean. Geochem Perspect Lett 9:1–5. https://doi.org/10.7185/geochemlet.1829

Piarulli S, Sciutto G, Oliveri P, Malegori C, Prati S, Mazzeo R, Airoldi L (2020) Rapid and direct detection of small microplastics in aquatic samples by a new near infrared hyperspectral imaging (NIR-HSI) method. Chemosphere 260:127655. https://doi.org/10.1016/j.chemosphere.2020.127655

Picó Y, Barceló D (2020) Pyrolysis gas chromatography-mass spectrometry in environmental analysis: focus on organic matter and microplastics. TrAC Trends Anal Chem 130:115964. https://doi.org/10.1016/j.trac.2020.115964

Pita F, Castilho A (2017) Separation of plastics by froth flotation. The role of size, shape and density of the particles. Waste Manag 60:91–99. https://doi.org/10.1016/j.wasman.2016.07.041

Pořízka P, Brunnbauer L, Porkert M, Rozman U, Marolt G, Holub D, Kizovský M, Benešová M, Samek O, Limbeck A, Kaiser J, Kalčíková G (2023) Laser-based techniques: novel tools for the identification and characterization of aged microplastics with developed biofilm. Chemosphere 313:137373. https://doi.org/10.1016/j.chemosphere.2022.137373

Primpke S, Godejohann M, Gerdts G (2020) Rapid identification and quantification of microplastics in the environment by quantum Cascade laser-based hyperspectral infrared chemical imaging. Environ Sci Technol 54(24):15893–15903. https://doi.org/10.1021/acs.est.0c05722

Radityaningrum AD, Trihadiningrum Y et al (2021) Microplastic contamination in water supply and the removal efficiencies of the treatment plants: a case of Surabaya City, Indonesia. J Water Process Eng 43:102195. https://doi.org/10.1016/j.jwpe.2021.102195

Rajala K, Grönfors O, Hesampour M, Mikola A (2020) Removal of microplastics from secondary wastewater treatment plant effluent by coagulation/flocculation with iron, aluminum and polyamine-based chemicals. Water Res 183:116045. https://doi.org/10.1016/j.watres.2020.116045

Rhein F, Scholl F, Nirschl H (2019) Magnetic seeded filtration for the separation of fine polymer particles from dilute suspensions: microplastics. Chem Eng Sci 207:1278–1287. https://doi.org/10.1016/j.ces.2019.07.052

Rhein F, Nirschl H, Kaegi R (2022) Separation of microplastic particles from sewage sludge extracts using magnetic seeded filtration. Water Research X 17:100155. https://doi.org/10.1016/j.wroa.2022.100155

Salerno D, Jordão H, La Marca F, Carvalho MT (2018) Using factorial experimental design to evaluate the separation of plastics by froth flotation. Waste Manag 73:62–68. https://doi.org/10.1016/j.wasman.2017.12.001

Santos LHMLM, Insa S, Arxé M, Buttiglieri G, Rodríguez-Mozaz S, Barceló D (2023) Analysis of microplastics in the environment: identification and quantification of trace levels of common types of plastic polymers using pyrolysis-GC/MS. MethodsX 10:102143. https://doi.org/10.1016/j.mex.2023.102143

Sarcletti M, Park H, Wirth J, Englisch S, Eigen A, Drobek D, Vivod D, Friedrich B, Tietze R, Alexiou C, Zahn D, Apeleo Zubiri B, Spiecker E, Halik M (2021) The remediation of nano–/microplastics from water. Mater Today 48:38–46. https://doi.org/10.1016/j.mattod.2021.02.020

Schmidt J, Haave M, Underhaug J, Wang W (2024) Unlocking the potential of NMR spectroscopy for precise and efficient quantification of microplastics. Microplastics Nanoplastics 4(1):1–13. https://doi.org/10.1186/s43591-024-00095-5

Scopetani C, Chelazzi D, Mikola J, Leiniö V, Heikkinen R, Cincinelli A, Pellinen J (2020) Olive oil-based method for the extraction, quantification and identification of microplastics in soil and compost samples. Sci Total Environ 733:139338. https://doi.org/10.1016/j.scitotenv.2020.139338

Senathirajah K, Kandaiah R, Panneerselvan L, Sathish CI, Palanisami T (2023) Fate and transformation of microplastics due to electrocoagulation treatment: impacts of polymer type and shape. Environ Pollut 334:122159. https://doi.org/10.1016/j.envpol.2023.122159

Serranti S, Capobianco G, Cucuzza P, Bonifazi G (2024) Efficient microplastic identification by hyperspectral imaging: a comparative study of spatial resolutions, spectral ranges and classification models to define an optimal analytical protocol. Sci Total Environ 954:176630. https://doi.org/10.1016/j.scitotenv.2024.176630

Shafi M, Lodh A, Khajuria M, Ranjan VP, Gani KM, Chowdhury S, Goel S (2024) Are we underestimating stormwater? Stormwater as a significant source of microplastics in surface waters. J Hazard Mater 465:133445. https://doi.org/10.1016/j.jhazmat.2024.133445

Shahi NK, Maeng M, Kim D, Dockko S (2020) Removal behavior of microplastics using alum coagulant and its enhancement using polyamine-coated sand. Process Saf Environ Prot 141:9–17. https://doi.org/10.1016/j.psep.2020.05.020

Shan J, Zhao J, Liu L, Zhang Y, Wang X, Wu F (2018) A novel way to rapidly monitor microplastics in soil by hyperspectral imaging technology and chemometrics. Environ Pollut 238:121–129. https://doi.org/10.1016/j.envpol.2018.03.026

Sierra I, Chialanza MR, Faccio R, Carrizo D, Fornaro L, Pérez-Parada A (2020) Identification of microplastics in wastewater samples by means of polarized light optical microscopy. Environ Sci Pollut Res 27(7):7409–7419. https://doi.org/10.1007/s11356-019-07011-y

Silveira AVM, Cella M, Tanabe EH, Bertuol DA (2018) Application of tribo-electrostatic separation in the recycling of plastic wastes. Process Saf Environ Prot 114:219–228. https://doi.org/10.1016/j.psep.2017.12.019

Sommer C, Schneider LM, Nguyen J, Prume JA, Lautze K, Koch M (2021) Identifying microplastic litter with laser induced breakdown spectroscopy: a first approach. Mar Pollut Bull 171:112789. https://doi.org/10.1016/j.marpolbul.2021.112789

Song YK, Hong SH, Eo S, Shim WJ (2021) A comparison of spectroscopic analysis methods for microplastics: manual, semi-automated, and automated Fourier transform infrared and Raman techniques. Mar Pollut Bull 173:113101. https://doi.org/10.1016/j.marpolbul.2021.113101

Song X, Wu X, Song X, Zhang Z (2022) Oil extraction following digestion to separate microplastics from mussels. Chemosphere 289:133187. https://doi.org/10.1016/j.chemosphere.2021.133187

Stile N, Raguso C, Pedruzzi A, Cetojevic E, Lasagni M, Sanchez-Vidal A, Saliu F (2021) Extraction of microplastic from marine sediments: a comparison between pressurized solvent extraction and density separation. Mar Pollut Bull 168:112436. https://doi.org/10.1016/j.marpolbul.2021.112436

Su J, Zhang F, Yu C, Zhang Y, Wang J, Wang C, Wang H, Jiang H (2023) Machine learning: next promising trend for microplastics study. J Environ Manag 344:118756. https://doi.org/10.1016/j.jenvman.2023.118756

Tahreen A, Jami MS, Ali F (2020) Role of electrocoagulation in wastewater treatment: a developmental review. J Water Process Eng 37:101440. https://doi.org/10.1016/j.jwpe.2020.101440

Tan A, Zhao J, Zhao Y, Li X, Su H (2023) Determination of microplastics by FTIR spectroscopy based on quaternion parallel feature fusion and support vector machine. Chemom Intell Lab Syst 243:105018. https://doi.org/10.1016/j.chemolab.2023.105018

Tang Y, Zhang S, Su Y, Wu D, Zhao Y, Xie B (2021) Removal of microplastics from aqueous solutions by magnetic carbon nanotubes. Chem Eng J 406:126804. https://doi.org/10.1016/j.cej.2020.126804

Vaisakh PS, Adarsh UK, Amrutha K, Warrier AK, Kartha VB, Unnikrishnan VK (2023) Integrated LIBS-Raman spectroscopy: a comprehensive approach to monitor microplastics and heavy metal contamination in water resources. Environ Res 231:116198. https://doi.org/10.1016/j.envres.2023.116198

Valentino M, Běhal J, Bianco V, Itri S, Mossotti R, Fontana GD, Battistini T, Stella E, Miccio L, Ferraro P (2022) Intelligent polarization-sensitive holographic flow-cytometer: towards specificity in classifying natural and microplastic fibers. Sci Total Environ 815:152708. https://doi.org/10.1016/j.scitotenv.2021.152708

Vidal C, Pasquini C (2021) A comprehensive and fast microplastics identification based on near-infrared hyperspectral imaging (HSI-NIR) and chemometrics. Environ Pollut 285:117251. https://doi.org/10.1016/j.envpol.2021.117251

Wander L, Vianello A, Vollertsen J, Westad F, Braun U, Paul A (2020) Exploratory analysis of hyperspectral FTIR data obtained from environmental microplastics samples. Anal Methods 12(6):781–791. https://doi.org/10.1039/c9ay02483b

Wang C et al (2015) Flotation separation of waste plastics for recycling – a review. Waste Manag 41:28–38. https://doi.org/10.1016/j.wasman.2015.03.027

Wang L, Kaeppler A, Fischer D, Simmchen J (2019) Photocatalytic TiO2 micromotors for removal of microplastics and suspended matter. ACS Appl Mater Interfaces 11(36):32937–32944. https://doi.org/10.1021/acsami.9b06128

Wen D, Chen Y, Tong Y, Wang H, Zhang H, Luo Y (2021) Quantification of microplastics in soils using accelerated solvent extraction: comparison with a visual sorting method. Bull Environ Contam Toxicol 107(4):770–777. https://doi.org/10.1007/s00128-021-03200-z

Woo H, Seo K, Choi Y, Kim J, Tanaka M, Lee KH, Choi J (2021) Methods of analyzing microsized plastics in the environment. Appl Sci 11(22):10640. https://doi.org/10.3390/app112210640

Xia Y, Xiang XM, Dong KY, Gong YY, Li ZJ (2020) Surfactant stealth effect of microplastics in traditional coagulation process observed via 3-D fluorescence imaging. Sci Total Environ 729:138783. https://doi.org/10.1016/j.scitotenv.2020.138783

Xu S, Ma J, Ji R, Pan K, Miao AJ (2020) Microplastics in aquatic environments: occurrence, accumulation, and biological effects. Sci Total Environ 703:134699. https://doi.org/10.1016/j.scitotenv.2019.134699

Xue J, Peldszus S, Van Dyke MI, Huck PM (2021) Removal of polystyrene microplastic spheres by alum-based coagulation-flocculation-sedimentation (CFS) treatment of surface waters. Chem Eng J 422:130023. https://doi.org/10.1016/j.cej.2021.130023

Yang R, Li H, Huang M, Yang H, Li A (2016) A review on chitosan-based flocculants and their applications in water treatment. Water Res 95:59–89. https://doi.org/10.1016/j.watres.2016.02.068

Yang SJ, Feng WW, Cai ZQ, Wang Q (2021) Study on rapid recognition of marine microplastics based on Raman spectroscopy. Spectrosc Spectr Anal 41(8):2469. https://doi.org/10.3964/j.issn.1000-0593(2021)08-2469-05

Ye H, Wang Y, Liu X, Xu D, Yuan H, Sun H, Wang S, Ma X (2021) Magnetically steerable iron oxides-manganese dioxide core–shell micromotors for organic and microplastic removals. J Colloid Interface Sci 588:510–521. https://doi.org/10.1016/j.jcis.2020.12.097

Zandieh M, Liu J (2022) Removal and degradation of microplastics using the magnetic and Nanozyme activities of bare iron oxide nanoaggregates. Angew Chem Int Ed 61(47):e202212013. https://doi.org/10.1002/anie.202212013

Zhang Y, Zhao J, Liu Z, Tian S, Lu J, Mu R, Yuan H (2021a) Coagulation removal of microplastics from wastewater by magnetic magnesium hydroxide and PAM. J Water Process Eng 43:102250. https://doi.org/10.1016/j.jwpe.2021.102250

Zhang Y, Zhou G, Yue J, Xing X, Yang Z, Wang X, Wang Q, Zhang J (2021b) Enhanced removal of polyethylene terephthalate microplastics through polyaluminum chloride coagulation with three typical coagulant aids. Sci Total Environ 800:149589. https://doi.org/10.1016/j.scitotenv.2021.149589

Zhang Y, Jiang H, Bian K, Wang H, Wang C (2021c) A critical review of control and removal strategies for microplastics from aquatic environments. J Environ Chem Eng 9(4):105463. https://doi.org/10.1016/j.jece.2021.105463

Zhang Y, Jiang H, Bian K, Wang H, Wang C (2021d) Is froth flotation a potential scheme for microplastics removal? Analysis on flotation kinetics and surface characteristics. Sci Total Environ 792:148345. https://doi.org/10.1016/j.scitotenv.2021.148345

Zhang Y, Wang X, Li Y, Wang H, Shi Y, Li Y, Zhang Y (2022) Improving nanoplastic removal by coagulation: impact mechanism of particle size and water chemical conditions. J Hazard Mater 425:127962. https://doi.org/10.1016/j.jhazmat.2021.127962

Zhou H, Mayorga-Martinez CC, Pumera M, Zhou H, Mayorga-Martinez CC, Pumera M (2021a) Microplastic removal and degradation by mussel-inspired adhesive magnetic/enzymatic microrobots. Small Methods 5(9):2100230. https://doi.org/10.1002/smtd.202100230

Zhou G, Wang Q, Li J, Li Q, Xu H, Ye Q, Wang Y, Shu S, Zhang J (2021b) Removal of polystyrene and polyethylene microplastics using PAC and FeCl3 coagulation: performance and mechanism. Sci Total Environ 752:141837. https://doi.org/10.1016/j.scitotenv.2020.141837

Zhu Y, Li Y, Huang J, Lam EY (2024) Smart polarization and spectroscopic holography for real-time microplastics identification. Commun Eng 3(1):1–8. https://doi.org/10.1038/s44172-024-00178-4

Zia K, Siddiqui T, Ali S, Farooq I, Zafar MS, Khurshid Z (2019) Nuclear magnetic resonance spectroscopy for medical and dental applications: a comprehensive review. Eur J Dent 13(01):124–128. https://doi.org/10.1055/s-0039-1688654

Li C, Busquets R, Moruzzi R B, Campos L C (2021) Preliminary study on low-density polystyrene microplastics bead removal from drinking water by coagulation-flocculation and sedimentation. J Water Process Eng 44:102346. https://doi.org/10.1016/j.jwpe.2021.102346

Impact of Microplastics on Human Health

8

Ghanshyam Kumar, Bhanu Duggal, and Paromita Banerjee

Abstract

Plastic consumption has experienced a significant increase in the past decade, resulting in the annual production of 400 million tons. These plastics have the potential to break down into microplastics, which are particulates that are less than 5 mm in diameter. Microplastics are intentionally manufactured, such as microbeads found in cosmetics and personal care products, or through plastic degradation. These particles are now ubiquitous, even in remote regions like the Arctic. They can penetrate the human body through three exposure pathways, i.e., airborne inhalation, oral ingestion of contaminated food and water, and dermal absorption through skin contact. Their presence has been found in various body parts such as the brain, heart, and lungs. In vitro and in vivo studies have linked microplastics to liver dysfunction, gut dysbiosis, urinary tract disorders, respiratory and neurological diseases, and reproductive and developmental toxicity. Microplastics can accumulate in cardiomyocytes (cardiac muscle cells), resulting in oxidative stress and may cause cardiac injury and fibrosis. Their presence has also been associated with atherosclerotic plaques in the carotid arteries, increasing the risk of stroke and overall mortality. The presence of microplastics is also associated with atherosclerosis in coronary arteries, and increased concentration is associated with a high risk of coronary artery disease (CAD) severity. This chapter provides an overview of microplastic production and its various exposure pathways penetrating the human body. The toxicity and mechanisms (oxidative stress and DNA damage, inflammatory and immune

Authors Ghanshyam Kumar and Bhanu Duggal have equally contributed to this chapter.

G. Kumar · B. Duggal (✉) · P. Banerjee
Department of Cardiology, All India Institute of Medical Sciences,
Rishikesh, Uttarakhand, India

K. Bala et al. (eds.), *Occurrence, Detection, and Fate of Microplastics in Freshwater Ecosystems*, https://doi.org/10.1007/978-981-95-0225-7_8

responses) through which microplastics cause disease pathogenesis in different organs of the body are discussed in detail, with a particular emphasis on cardiovascular diseases (CVDs), which cause the highest mortality worldwide. The chapter also discusses the studies conducted to decipher the presence of microplastics and the severity of CVDs. However, a causal relationship between microplastic concentration and human disease severity is yet to be established.

Keywords

Microplastics · Human health · Pathways · Risks · Cardiovascular diseases (CVDs)

8.1 Overview of Plastic Consumption and Toxicity

The global demand for plastics has surged, leading to a significant increase in production, reaching over 400 million tons per year, and is estimated to reach a staggering 1.2 billion tons in 2060 (UN Environment Program). The use of plastics varies from packaging materials, food containers, toys, fishing nets, bags, textiles, and personal care products to the medical field (Filho et al. 2021; Song et al. 2024). With the global COVID-19 pandemic, the usage of single-use plastics, such as disposable face masks, PPE kits, and sanitizer bottles, has increased exponentially (Goswami et al. 2024; Osman et al. 2023). Commonly used plastics include polystyrene (PS), polyethylene terephthalate (PET), polypropylene (PP), nylon, polyamide (PA), polyvinyl chloride (PVC), polyethylene vinyl acetate (PEVA), and polyethylene (PE) (Domenech and Marcos 2021; Li et al. 2023). Due to their high durability and slow degradability, plastics accumulate for more extended periods, causing vast amounts of plastic waste, which subsequently gets decomposed into microplastics, potentially causing many toxic effects on the environment, biodiversity, aquatic life, marine and freshwater ecosystems, and human health (Ali et al. 2024b; Bikiaris et al. 2024; Song et al. 2024). Microplastics have been found in various freshwater environments across the globe, with industrial activity near their source, wastewater effluents, and urban runoffs increasing the toxicity of freshwater with microplastics. This will eventually lead to the loss of biodiversity in aquatic ecosystems (Bhardwaj et al. 2024; Ding et al. 2021). These microplastics can bioaccumulate in aquatic/marine organisms, become a part of the food web, and enter the human body through trophic transfer in their biomagnified form (Miller et al. 2020).

Microplastics, the term first introduced in 2004, are defined as plastics (microscopic plastic fragments) smaller than 5 mm and are a direct result of this production, whether intentional (as in personal care products) or through the degradation of larger plastics. Microplastic sources include ocean and land-based sources, with the majority coming from land-based sources (Bikiaris et al. 2024; Osman et al. 2023). Cosmetics, textiles, single-use plastics, and tires are some of the primary sources of microplastics, with synthetic textiles contributing the most (Saha and Saha 2024). Their ubiquity is alarming, with even the most remote places, such as

Antarctica, the Arctic, and Mount Everest, being contaminated, raising serious concerns (Li et al. 2023; Sun and Wang 2023). The small size of these microplastics makes their removal from the environment difficult, resulting in their accumulation in terrestrial and aquatic ecosystems. This facilitates the transfer of microplastics through the food chain, ultimately affecting humans, who are at the top of the food chain (Lee et al. 2023; Oleksiuk et al. 2022; Sun and Wang 2023). Due to its ubiquitous nature, it has profound implications for the health of the ecosystem and the economies of the tourism and fishing industry (Ghosh et al. 2023; Sun and Wang 2023).

The size, morphology, chemical characteristics (chemical nature of plastics from which those particular microplastics have come), surface charges, concentrations, and presence of exogenous toxins (metal ions, fungicides, persistent organic pollutants (POPs)) of these microplastics along with their exposure dose and duration have differential impacts on human health (Domenech and Marcos 2021; Goodman et al. 2022; Lin et al. 2023b; Zhu et al. 2023). Small and randomly shaped fragments are more likely to enter cells than larger and spherical particles and can cause oxidative stress and other harmful effects. When microplastics come in contact with various biomolecules, such as proteins, a biomolecule corona (the accumulation of biomolecules on their surface when exposed to biological fluids) can quickly form, further altering their toxicity (Kopatz et al. 2023; Li et al. 2023).

Microplastics also act as vectors for other environmental pollutants, such as organic and heavy metal pollutants, pathogenic bacteria, and drugs (Amelia et al. 2021; Goswami et al. 2024). When these bioavailable particles reach the bloodstream following exposure, they may move to secondary organs and build up to a point where they may have detrimental cellular consequences and promote a systemic and/or local inflammatory response, resulting in various toxic effects (Prattichizzo et al. 2024). Because microplastics are alien substances found in animal products, they also jeopardize food safety and productivity, endangering food security (Prata and Dias-Pereira 2023). During processing and packaging, the amount of microplastics present in animal products may exceed the original amount of plastics in them, due to microplastics from the processing and packaging process (Prata and Dias-Pereira 2023).

8.2 Definition and Types of Microplastics

Microplastics can be broadly divided into primary and secondary microplastics based on their origin (degradation of large plastics or intentional production) (Fig. 8.1) (Lee 2024). Both primary and secondary microplastics are transported by atmospheric air over long distances due to their inert nature (Goswami et al. 2024; Sun and Wang 2023). These microplastics can take various forms, such as fragments, foams, films, fibers, granules, and filaments (Emenike et al. 2023).

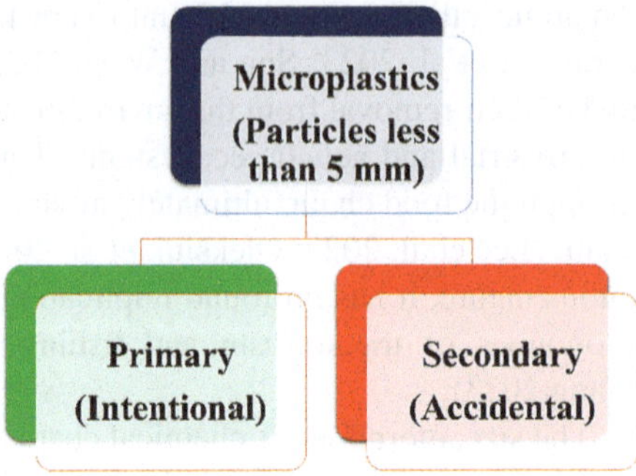

Fig. 8.1 Microplastics and their types

8.2.1 Primary Microplastics (Intentional/Manufactured)

Primary microplastics, intentionally created for use in various industries such as cosmetics, detergents, dyes, resin pellets, powders for injection molding, artificial turf, fishing nets, and abrasive particles, significantly contribute to the microplastic accumulation (Emenike et al. 2023; Ziani et al. 2023). These microplastics are also used in personal care products such as toothpaste, soap, shampoo, and facial gel (Acarer 2023). These microplastics, which can be categorized as microbeads, microfibers, and resin pellets, pose a severe risk to terrestrial and marine ecosystems because of their capacity to adsorb and transfer harmful substances (Emenike et al. 2023; Ziani et al. 2023). These microplastics are mainly discharged into the environment by wastewater stations, potentially making them toxic to exposed ecosystems (Lin et al. 2023b).

8.2.2 Secondary Microplastics (Unintentional/Accidental)

Secondary microplastics, a byproduct of the decomposition, fragmentation, and weathering of larger plastic products such as bottles, packaging materials, bags, and tires, are an unintended consequence of plastic use. Mechanical forces (such as waves and abrasions), temperature fluctuations, sunlight exposure (ultraviolet rays), and chemical and biological reactions contribute to their formation. These microplastics, which can take the form of fragments, fibers, and microbeads, contaminate various ecosystems, including water, air, and soil (Ali et al. 2024b; Emenike et al. 2023; Horvatits et al. 2022; Lin et al. 2023b; Song et al. 2024; Ziani et al. 2023).

Since primary microplastics are produced intentionally, we can control their environmental release. Secondary microplastics are produced unintentionally and in large quantities, so the focus should be more on managing their production (Song et al. 2024).

8.3 Ubiquity of Microplastics in the Environment

Microplastics have infiltrated various ecosystems, from the soil, remote islands, polar regions, groundwater, and oceans to atmospheric environments (Lee et al. 2023; Liu et al. 2024; Sangkham et al. 2023). Their non-biodegradable nature means they can persist for long periods, posing serious environmental and health risks. The detection of microplastics, such as PP, PS, PE, PET, and PEVA, in everyday food items (especially seafood) and water sources indicates that microplastics are now present in almost all spheres of the environment. There is rising concern about microplastics' potential to impact human health by entering the food chain. The severity of this issue is shown by the existence of microplastics in our environment and the food we eat (Emenike et al. 2023; Farady 2019; Ziani et al. 2023). Researchers have found the presence of microplastics (especially polyester fibers) in snow and stream water on Mount Everest, indicating its ubiquity even in remote places with few human settlements (Napper et al. 2020). Microplastics have entered humans through various exposure routes and can accumulate in the body, posing a severe health issue (Liu et al. 2024).

8.4 Exposure Pathways of Microplastics

Humans are exposed to microplastics through various sources, such as industries, air, water, food, and personal care products, and these can enter the body through three possible pathways: airborne inhalation, ingestion (food and water), and dermal contact (Fig. 8.2) (Lee 2024; Li et al. 2023).

8.4.1 Airborne Inhalation

Microplastics have been detected in indoor and outdoor air, providing a pathway for human plastic exposure (Torres-Agullo et al. 2021). PS, PE, and PET particles and fibers are the commonly found constituents of microplastics in the air. Humans are exposed to microplastics through inhalation, as confirmed by microplastics such as PE and PP in the human lungs. Microplastics in the atmosphere are primarily attributed to roads, which contribute approximately 84% of the total (Li et al. 2023). Exposure to microplastics through inhalation is common for people working in waste management, plastics manufacturing, and recycling industries. The microplastics can be transported over long distances by the wind, potentially residing in locations that are remote from the source of contamination. The exposure intensity rate varies with age, gender, and location (urban/rural) (Emenike et al. 2023).

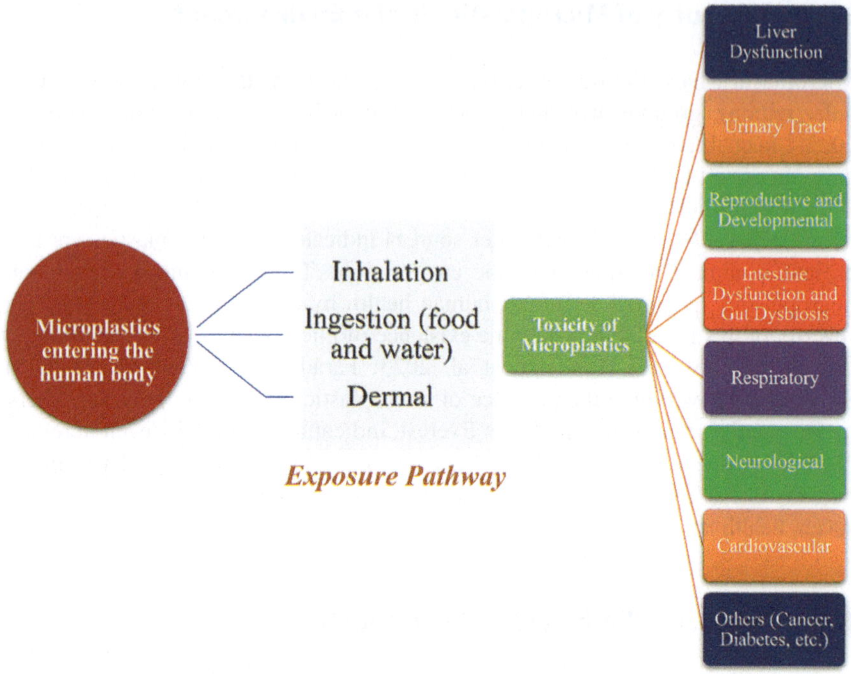

Fig. 8.2 Depiction of the exposure pathway of microplastics into the body and their possible toxicity on various organs of the human body

8.4.2 Oral Ingestion via Food and Water

The most common exposure pathway is exposure through ingestion, especially food diet (Oleksiuk et al. 2022). Humans are exposed to microplastics through food, including seafood (shellfish), sea and table salt, drinking bottles, food containers (usually made up of PP, PS, PE, PET), sugar, drinks sold in plastic bottles, milk, honey, beer, and tea (Emenike et al. 2023; Li et al. 2023; Oleksiuk et al. 2022). The microplastics present in sludge, plastic mulch, and soil can enter the roots, fruits, vegetables, cereals, and legumes, and when consumed, they enter the human body (Li et al. 2023; Mamun et al. 2023). Even infants are exposed to microplastics through the bottle used for feeding, with around 0.66 million microplastic particles reaching their bodies during a year of bottle feeding (Li et al. 2023).

8.4.3 Dermal Absorption Through Skin Contact

Human skin is a barrier against large particles because of the size of skin pores (40–80 μm). However, microplastics still get deposited on the skin, and particles smaller than microplastics (100 nm) can still penetrate the skin (Ageel et al. 2022; Bikiaris et al. 2024; Emenike et al. 2023; Li et al. 2023). Microplastics deposited on human skin pose health concerns. Face creams and cleaners, protective mobile phone cases (which are transferred to the user's hand), and bathing in contaminated water pose a risk of microplastic exposure. The exposure happens through pores in the skin and can lead to skin allergies and other health concerns (Emenike et al. 2023). Even indoor and outdoor dust deposition is a pathway for dermal absorption (Ageel et al. 2022). Some plastic additives, including bisphenols and phthalates (endocrine disrupters), may be absorbed into the skin (Li et al. 2023).

8.5 Mechanisms of Disease Pathogenesis of Microplastics

The disease pathogenesis of microplastics occurs in multiple ways, such as oxidative stress resulting in DNA damage and inflammatory and immune responses (Bikiaris et al. 2024; Yang et al. 2024). Details are discussed below.

8.5.1 Oxidative Stress and DNA Damage

The cytotoxicity of microplastics, a process that causes cell death mainly due to oxidative stress and membrane damage, is a significant health risk associated with microplastic exposure (Li et al. 2023). In vitro tests with human cells and in vivo data with mice indicate that microplastics produce inflammation and induce oxidative stress by increasing the production of reactive oxygen species (ROS), which in turn causes DNA and cell damage (Emenike et al. 2023; Lee et al. 2023; Wu et al. 2023b). Oxidative stress refers to the initial oxidative impacts of particles and the ongoing generation of ROS in cells or tissues exposed to them. This oxidative stress occurs when the antioxidant protective response of cells is overwhelmed by ROS, activating the cell signaling pathway and accelerating the release of pro-inflammatory cytokines, leading to inflammation. When plastic particles are absorbed, the cell membrane's integrity is disrupted, the lipid bilayer changes, pores form, and intracellular ROS generation increases. As a result, the formation of ROS causes mitochondrial dysfunction, the release of pro-inflammatory cytokines, DNA damage (both mitochondrial and nuclear DNA), and impairment of the genetic control of regular DNA repair. Oxidative stress causes DNA damage, including base alterations, base site loss, and strand breaks (Kadac-Czapska et al. 2024; Prata 2018). The DNA damage was validated by increased production of micronuclei and nuclear buds (Ali et al. 2024a). The differential reaction regarding ROS levels depends on cell type, concentration, and exposure time of microplastics (Goodman et al. 2022).

8.5.2 Inflammatory and Immune Responses

The key toxicological mechanism following microplastic exposure is the creation of ROS, as discussed in the above section, which may further drive the production of danger-related molecular patterns and be connected with processes such as toll-like receptor disruption, cytokine synthesis, and inflammatory responses in immune cells (Yang et al. 2022). Microplastic exposure triggers a systemic or local immune response in the body, suppressing innate immunity, increasing pro-inflammatory responses, producing autoimmune disorders, and ultimately causing tissue and organ damage (Bhuyan 2022; Bishop et al. 2024; Yang et al. 2022). PE exposure activates zebrafish's intestinal immune network, leading to mucosal immunoglobulin production. Exposure to PE in mice resulted in increased IL-1α secretion in their serum. Inhaling PS stimulates the expression of inflammatory proteins (TGF-β and TNF-α) in rats' lung tissue. In mice, PS exposure increases inflammation factors (TNF-α, IL-1β, and IFN-γ), leading to intestinal immunological imbalance and increased accumulation of microplastics, resulting in hazardous consequences (Li et al. 2023; Yang et al. 2022). In animal trials, cumulative exposure to microplastics caused chronic inflammation and abnormalities in homeostasis. Human lung cells exposed to microplastics can activate innate immunity by controlling the expression of genes and proteins involved in the immune response (Lee et al. 2023). Human gastric cancer cells exposed to 44 nm PS nanoparticles showed a significant rise in the expression of the body's pro-inflammatory cytokines (IL-6 and IL-8 genes) (Forte et al. 2016). When microplastic particles come into contact with the skin, they can trigger immunological reactions and produce allergies. Exposure to high PP microplastic concentrations (~20 and 25–200 μm) has been demonstrated to increase hypersensitivity in human cells and cause unintended immunological responses (Hwang et al. 2019). A culture of human peripheral blood monocytes with poly(methyl methacrylate) (PMMA) and PS particles enhanced pro-inflammatory cytokines in the supernatant, including IL-6 (Wolff et al. 2023). Microplastics and nanoplastics (MNPs) can interact directly with immune cells and endothelial cells (ECs), resulting in the activation of the endothelium and the promotion of inflammation and atherogenesis (formation of plaques) (Zhu et al. 2023). B and NK cells are believed to be involved in atherosclerotic plaque formation, and the positive correlation between total microplastic content and the quantities of B cells ($CD3^-/CD19^+$ cells) and NK cells ($CD3^-/CD56^+CD16^+$ cells) in acute coronary syndrome patients suggests that microplastics may increase the progression of atherosclerosis by activating immune cells (Yang et al. 2024). Microplastic contents also correlated with these patients' pro-inflammatory cytokine levels (IL-6 and IL-12p70) (Yang et al. 2024). An in vitro study of THP cells with different microplastics, such as PP, PE, PET, and PS, shows that these particles' size (smaller) and nature result in a differential inflammatory response, with size having a more profound effect (Beijer et al. 2022).

8.6 Health Risks Associated with Microplastics

8.6.1 General Toxicological Impacts on Organ Systems

As discussed in Sect. 8.4, microplastics enter the human body through various routes; after entering the body, they can enter the blood circulation and interact with multiple body tissues and organs, which might be damaged (Mamun et al. 2023; Marfella et al. 2024). Microplastics have been found in various human body parts and products, such as the brain, heart, liver, spleen, kidney, intestine, skeletal muscle, feces, blood, arteries, urine, semen, placenta, lower airway, lungs, breast milk, and colon (Lin et al. 2023b; Liu et al. 2024; Marfella et al. 2024; Wang et al. 2024a; Zhao et al. 2024). Microplastics have been found to have toxic effects on various parts of the body, such as the lungs, cardiovascular system, digestive system, urinary tract, and reproductive system, as evidenced by several in vitro studies using human cell lines, in vivo studies using animal models such as rats/mice/zebrafish, and a few studies directly on humans (Fig. 8.2) (Li et al. 2023; Sun and Wang 2023). Various analytical techniques, such as Raman spectroscopy, Fourier transform infrared spectroscopy, and mass spectrometry, have been used to test the presence of microplastics in various biological samples (Osman et al. 2023). Microplastics are also carriers of various toxic compounds, such as carcinogenic pollutants and endocrine disrupters, which can lead to combined toxicity, including cancer of the blood, lungs, breasts, prostate, and ovaries (Goswami et al. 2024; Osman et al. 2023). These toxic compounds are either intentionally incorporated during plastics manufacturing, such as stabilizers, plasticizers, and dyes or acquired through adsorption of coexisting pollutants, such as heavy metals and pesticides, during the environmental transport of microplastics. Due to the diversity in the presence of these contaminants in the various geographical regions, the combined toxicity will also differ across different populations (Lee 2024). Microplastics' toxicity stems from their capacity to enter metabolism, translocate into the circulatory system, and from their difficulty in excretion (Eze et al. 2024).

8.6.2 Liver Dysfunction

The liver is one of the primary organs affected by microplastics, as microplastics get deposited in it (Li et al. 2023). Long-term exposure to microplastics can significantly impact the liver cells and may lead to altered gene expression, structural defects, decreased cell proliferation, and failure (Emenike et al. 2023; Goodman et al. 2022). Exposure to PS microplastics of human hepatocellular (Hep G2) liver cells had increased levels of ROS, and various changes, such as morphology, proliferative, cellular, and metabolic, were observed, indicating the harmful effects of its exposure (Goodman et al. 2022). In an in vitro 3D model, the liver organoids (generated from human pluripotent stem cells) exposed to PS microplastics had altered lipid metabolism, reduced ATP production, and ROS generation (Cheng et al. 2022). Various in vivo studies on mice have also shown that microplastics negatively impact liver function, including fibrogenesis (Ali et al. 2024a). Liver, spleen, and kidney samples of patients without liver disease and cirrhotic liver samples were analyzed, and it was found that elevated microplastic concentrations, such as PS, PVC, PET, PMMA, and PP, were present in patients with liver cirrhosis, indicating that the liver disease is related to microplastics (Horvatits et al. 2022).

8.6.3 Urinary Tract Disorders

Microplastics have been found in various parts of the urinary tract (such as the kidneys and bladder) and urine (O'Callaghan et al. 2024; Zhang et al. 2024). A study found the presence of PEVA, PVC, PP, and PE in male and female urine samples, and this study provided the first preliminary indication of microplastics in human urine (Pironti et al. 2023). Various in vitro studies using various kidney cell lines, such as HEK 293, HEK 293T, and HK-2, have identified the impact of microplastics on cell viability and morphological changes, with potentially harmful effects such as decreased cell viability, increased inflammation, and cancer-inducing pathways (Goodman et al. 2022; O'Callaghan et al. 2024). Urinary tract infections (UTIs) are one of the leading causes of death worldwide. The presence of MNPs in the urinary tract could increase the risk of UTIs. MNPs' rough surfaces create a perfect habitat for bacterial adhesion and biofilm formation, which could lead to more persistent and recurring UTIs. Microplastics can induce urethritis (urethral infection), pyelonephritis (kidney infection), and cystitis (bladder infection) by promoting the intracellular movement of pathogenic bacteria and the creation of the protein corona, which aids in evading the innate immune system (O'Callaghan et al. 2024).

8.6.4 Prenatal Risks: Reproductive and Developmental Toxicity

Microplastic exposure can produce reproductive toxicity, as evidenced by various in vivo studies on mice and zebrafish (Ali et al. 2024a). Exposure results in interference in embryonic development, which leads to spontaneous abortion, preeclampsia, stillbirth, and preterm birth (Paul et al. 2024). Microplastics (PS) exposure negatively impacts female mouse follicle development and oocyte maturation, resulting in lower-quality oocytes. Exposure to PS also significantly affects sperm quality, such as sperm count, motility, and increased sperm deformity rate, affecting male mice's fertility (Ali et al. 2024a; Li et al. 2023). Microplastics can act as carriers of bisphenol A, which can negatively impact reproductive health by interfering with hormonal functions through blockage and mimicking natural hormones (Wang et al. 2024a). Microplastics have also been found in the placental portions (maternal, fetal, and amniochorial membranes) of pregnant women, potentially causing far-reaching toxic effects on the health of offspring (Lee et al. 2023; Ragusa et al. 2021). PS increased microcystin-LR (MCLR) buildup in zebrafish gonads, exacerbating the reproductive harm caused by MCLR. PS also affects the level of sex hormones and the hypothalamic-pituitary-gonadal (HPG) axis, worsening reproductive failure (Emenike et al. 2023; Lin et al. 2023a). Microplastics have also been found in human semen, with PS, PE, and PVC being the most prominent, which can affect sperm motility, subsequently affecting reproductive health (Li et al. 2024). A recent study across various regions of China has found microplastics in semen, with PS, PP, and PE being frequent, with polytetrafluoroethylene (PTFE) exposure significantly associated with decreased total sperm count, sperm concentration, and progressive motility, affecting male reproductive health (Zhang et al. 2024).

Microplastics may also affect offspring's lipid metabolism and reproductive functions. The declining fertility in some parts of the world can also be attributed to the potentially toxic effects of microplastics (Wang et al. 2024a). Exposure to microplastics during the prenatal and postnatal period has also been found to be associated with the development of autism spectrum disorder (Zaheer et al. 2022).

8.6.5 Intestine Dysfunction and Gut Dysbiosis

Gut microbiota has a vital role in maintaining human health (Fackelmann and Sommer 2019). Ingestion of microplastics can serve as a vector for pathogens and pollutants, directly or indirectly impacting or altering gut microbiota composition, resulting in gut dysbiosis (disruption of symbiosis between host and natural gut microbiota) (Emenike et al. 2023; Fackelmann and Sommer 2019). Dysbiosis can lead to pathogenic infections, immune system disruptions, and illnesses such as gastroenteritis and cholera (Fackelmann and Sommer 2019). Various gut microbiota species, such as *Staphylococcus epidermidis, Escherichia coli, and Pseudomonas aeruginosa*, could adhere to the microplastic surface, promoting protective biofilm formation and antibiotic resistance (Tamargo et al. 2022). In fish, exposure to microplastics resulted in the inhibition of digestive enzyme activity (Kim et al. 2022). In humans, microplastic exposure can harm intestinal homeostasis (Tamargo et al. 2022). In the simulated human gastrointestinal system, it was observed that exposure to PS and PE reduced lipid digestion (Tan et al. 2020). Microplastics can induce gut barrier dysfunction; pathogenic bacteria and metabolites in the gut can reach the liver, causing liver damage (Wang et al. 2024c). A study using static and dynamic human gastrointestinal models has reported that PET microplastic exposure (166 mg/intake, which is the estimated daily intake of microplastics) can alter the colon microbial composition and diversity, leading to a decline in essential microbial groups and an increase in pro-inflammatory and disease-related bacteria (Tamargo et al. 2022). Gut microbes may try to adapt to microplastic exposure by degrading these plastics by secreting plastic-degrading enzymes (Nugrahapraja et al. 2022).

8.6.6 Respiratory Diseases

Through inhalation, microplastics, because of their small size, can get deposited in the respiratory tract, including the lungs, and trigger inflammatory responses. Microplastics (mostly fibrous and fragmented) have been found in all the regions of the lungs, with PP and PET being most abundant (Jenner et al. 2022). Microplastics with small size, irregular shape, and positive charge were found to have more toxicity (Vattanasit et al. 2023). Long-term exposure to microplastics in the air could be linked to respiratory diseases, with people having chronic obstructive pulmonary disease or preexisting asthma experiencing exacerbation of their existing condition due to damage to lung tissue (Saha and Saha 2024). Microplastic inhalation can produce oxidative stress in the airways and lungs, resulting in shortness of breath, coughing, fatigue, sneezing, and dizziness (Lee et al. 2023). An in vitro investigation employing human lung epithelial BEAS-2B cells demonstrated that exposure to high concentrations of PS damages human lung cells, raising the risk of chronic obstructive pulmonary disease (Dong et al. 2020; Lee et al. 2023).

8.6.7 Neurological Disorders

Microplastics can result in various neurodegenerative/neurodevelopmental disorders after entering the brain by inducing oxidative stress and neuroinflammation (Prüst et al. 2020). Microplastic exposure inhibits or reduces acetylcholinesterase (AChE) activity, which may impact various activities and functions of the brain, such as neurotransmitter signaling alterations, resulting in behavioral changes and cell damage (Hamed et al. 2022). Chronic exposure to PS damaged the blood-brain barrier of mice, resulting in learning and memory dysfunctions (Li et al. 2023; Prüst et al. 2020). Various other studies have also shown that microplastics can impair the blood-brain barrier and reach the brain through ingestion (Amato-Lourenço et al. 2024; Kopatz et al. 2023). A recent study found evidence of microplastics (PP, nylon, and PEVA) in the deceased human olfactory bulb, suggesting a potential pathway for the translocation of microplastics to the brain (Amato-Lourenço et al. 2024). In marine fish (*D. labrax*), microplastic exposure resulted in lipid peroxidation in the brain and muscle (Barboza et al. 2018). Other studies on larvae, nematodes, crustaceans, and fish have also reported the neurotoxicity of microplastics (Li et al. 2023).

8.7 Impact of Microplastics on Cardiovascular Health

Cardiovascular diseases (CVDs) are the number one killer in the world, with around 20 million deaths worldwide annually (Di Cesare et al. 2024). Hence, it is imperative to understand the impact of microplastics on cardiovascular health and whether it is a risk factor for various CVDs such as coronary artery disease (CAD), arrhythmia, and myocardial infarction (MI) (Persiani et al. 2023; Zhu et al. 2023). Samples taken from the cardiovascular system (atherosclerotic plaques, thrombi, and different cardiac tissues) have yielded a variety of MNP types and have been associated with various cardiovascular risks (Table 8.1). Evidence from animal models such as mice, rats, and zebrafish and in vitro studies on human venous blood and red blood cells have suggested the potentially harmful effects of microplastics on the cardiovascular system, such as hemolysis (RBC destruction), blood coagulation, abnormalities in the atrioventricular valve, and platelet aggregation (Prattichizzo et al. 2024; Zhu et al. 2023). MNPs impact the cardiac system, including abnormal heart rate (tachycardia/bradycardia), impairment of cardiac function (decrease in ventricular contraction frequency), pericardial edema (swelling), myocardial fiber breakage, and myocardial fibrosis (Zhu et al. 2023). Various studies have been conducted to study the toxicity of MNPs on the cardiovascular system, including aquatic fauna such as zebrafish (*Danio rerio*), marine medaka (*Oryzias melastigma*), goldfish (*Carassius auratus*) with exposure to PE and PS, and mammals such as Sprague-Dawley rats, specific pathogen-free (SPF) CD-1® mice, Fischer 344 rats, and Wistar rats, with PS exposure (Persiani et al. 2023; Prattichizzo et al. 2024). Understanding how plastic fragments modify the cardiovascular system in aquatic organisms could help study their impact on human health, as MNPs may ascend the food chain, ultimately affecting humans as apex consumers (Persiani et al. 2023). Microplastics may cause RBC adherence to the vascular endothelium, a critical characteristic of thrombogenesis associated with CVDs (Persiani et al. 2023). POPs with microplastics accumulate in adipose tissue and are released into circulation through lipolysis; they may also be present within the lipid-rich plaque with microplastics because of the plaque's substantial lipid content (Lee 2024).

8.7.1 Accumulation in Cardiomyocytes and Cardiac Injury/Fibrosis

MNPs, acting as environmental stressors, which have an affinity for pathogens, heavy metals, and POPs such as polycyclic aromatic hydrocarbons, can affect ionic conductivity in cardiomyocytes (Persiani et al. 2023). Microplastics resulted in myocardial damage, fibrosis (scarring of cardiac muscle), and impairment of electrophysiological values. When MNPs interact with a developing heart, cardiac function is compromised, resulting in anything from a severe loss of function to total cardiac morphogenesis failure (Persiani et al. 2023). Increased levels of creatine kinase-MB and troponin I in Wister rats' serum indicate apoptosis and structural damage due to exposure to PS. The exposure resulted in the activation of fibrosis-related Wnt/β-catenin signaling pathway, suggesting that exposure leads to cardiovascular toxicity (Li et al. 2020). Microplastics are present in the human heart and its surrounding tissues, with the presence of PMMA in the left atrial appendage, pericardial adipose tissue, and epicardial adipose tissue, which is not related to surgery-related exposure, indicating direct evidence of microplastic presence in the heart (Table 8.1) (Yang et al. 2023).

8.7.2 Myocardial Infarction (MI) Risk and Mortality

Microplastics are associated with an increase in the severity of CAD (Duggal and Kumar 2025). MNPs have the ability to stimulate RBCs and monocytes that interact with ECs, which can lead to atherosclerosis and circulatory diseases (Zhu et al. 2023). Various studies have shown the presence of microplastics in coronary arteries/cardiac tissues and their potential linkage with atherosclerosis and decreased cardiovascular health, especially CAD severity (Table 8.1). Four types of microplastics (PET, PA66, PVC, and PE) with varying shapes and sizes have been significantly detected in human arteries with atherosclerotic plaques, i.e., coronary and

Table 8.1 Details of studies conducted to study the presence of microplastics in human arteries/cardiac tissues

S. no.	Arteries, cardiac tissues, and blood	Sample details	Microplastics present	Key findings	References
1	Carotid artery plaque	257 patients undergoing carotid endarterectomy	PE, PVC	MNPs are possibly a potential risk factor for MI, stroke, or death	Marfella et al. (2024)
2	Coronary and carotid arteries with plaques, aorta without plaques	17 (4 coronary artery, 7 carotid artery, 6 aorta)	PET, PA66, PVC, and PE	Microplastic concentration was significantly higher in atherosclerotic plaques, indicating that these can be associated with atherosclerosis	Liu et al. (2024)
3	Cardiac tissue (left atrial appendages, myocardia, pericardia, epicardial adipose, pericardial adipose)	15 cardiac surgery patients	Various cardiac surgery-related plastics and a cardiac surgery-unrelated plastic, i.e., PMMA	The presence of PMMA in the tissue, which is not related to surgical procedures, provides direct evidence of microplastics in human cardiac tissue	Yang et al. (2023)
4	Aorta, lower extremity arteries	26 (24 arterial dissection, 2 acute arterial embolism of lower extremities)	PE	The presence of microplastics in thrombi indicates their role in thrombosis	Wu et al. (2023a)
5	Blood samples of patients undergoing a coronary angiography	101 (82 acute coronary syndrome cases and 19 controls)	PE, PVC, PS, PP, PA66	Patients with a high risk of CAD have higher concentrations of microplastics, especially in MI patients	Yang et al. (2024)
6	Coronary arteries in the heart, cerebral arteries in the brain, and deep veins in the lower extremities	30 (16 ischemic stroke, 5 MI, and 9 with deep vein thrombosis)	PA66, PVC, PE	A high concentration of microplastics was correlated with a high severity of stroke	Wang et al. (2024b)

carotid arteries, using pyrolysis-gas chromatography/mass spectrometry, indicating that microplastics are associated with atherosclerosis (Liu et al. 2024). A recent study on patients undergoing carotid endarterectomy has shown that those patients who had evidence of MNPs within the carotid plaque had a higher risk (4.5 times) of MI, stroke, or death than patients having no evidence of MNPs in the long-term follow-up of 34 months. However, further studies are required to prove the causality relationship between microplastics and cardiovascular events, as confounding variables, including the health status of patients, can influence the results (Marfella et al. 2024). Another study also confirms the presence of microplastics in the thrombotic aorta and lower extremity arteries, indicating their potential role in thrombosis (Wu et al. 2023a). A recent study on acute coronary syndrome patients found higher concentrations of microplastics to be related to high cardiovascular risks, such as MI. Plastics such as PE, PVC, PS, and PP have been found in the blood samples of these patients, with increased concentrations of microplastics present in more severe cases (Yang et al. 2024). One study has found microplastics in the coronary and cerebral arteries, as well as in deep veins. This study also confirms the direct correlation between the amount of microplastics and disease severity (Wang et al. 2024b). As reported in those studies, possible sources of microplastics given in Table 8.1 are water pipes, food containers, textiles, packaging bags, masks, toothbrushes, toys, agricultural films, medical devices, automobiles, etc.

Other toxic effects of microplastics are their potential linkage to obesity, diabetes, sleep disorders, Parkinson's disease, lipid metabolism disturbances, etc. (Ghosh et al. 2023; Lee et al. 2023).

8.8 Limitations and Future Directions

Microplastics have been found in various organs of the human body, and their presence has been found to have a toxic effect on health; however, there is a lack of human studies on the possible mechanism of impact of microplastics on human health, as most of the studies have been done on mice, zebrafish, or in vitro human cell lines, with a few on ex vivo human samples. There is also a lack of standardized protocols (sampling and detection techniques) for various studies conducted to study the impact of microplastics on human health. The majority of the in vitro and in vivo studies conducted have used only PS. Still, there are other commonly found microplastics in the environment, so future studies should also focus on different types of microplastics (to mimic real-world exposure). It is challenging to ascertain a causal relationship because of limited direct evidence of the health risks of microplastic exposure in the cardiovascular system. A definitive causal relationship still needs to be established on microplastic toxicity in other organs of the human body. A large representative sample, with multiple outcomes and a multidisciplinary approach, is required to decipher the intensity of toxicity of various microplastic exposures on human health.

Considering that there are no clinical trials on removing microplastics from the body, the best way available is to minimize the use of plastics, such as limiting the use of plastic food and beverage containers, along with recycling and proper bioremediation techniques for the environmentally friendly disposal of preexisting microplastics. A potential solution to reduce the toxicity of microplastics is to increase the production and consumption of environmentally friendly biodegradable bioplastics such as polyhydroxyalkanoates. There is also a lack of consumer awareness among different sections of society, especially vulnerable people, such as the poor and the illiterate, regarding the health hazards of microplastics. Therefore, it is necessary to raise awareness about the toxicity of microplastics to encourage behavioral changes that will reduce their consumption.

References

Acarer S (2023) Microplastics in wastewater treatment plants: sources, properties, removal efficiency, removal mechanisms, and interactions with pollutants. Water Sci Technol 87(3):685–710

Ageel HK, Harrad S, Abdallah MA-E (2022) Occurrence, human exposure, and risk of microplastics in the indoor environment. Environ Sci: Processes Impacts 24(1):17–31

Ali N, Katsouli J, Marczylo EL, Gant TW, Wright S, Bernardino de la Serna J (2024a) The potential impacts of micro-and-nano plastics on various organ systems in humans. EBioMedicine 99:104901

Ali SS, Al-Tohamy R, Alsharbaty MHM, Elsamahy T, El-Sapagh S, Lim JW, Sun J (2024b) Microplastics and their ecotoxicological impacts: remediation approaches, challenges and future perspectives – a review. J Clean Prod 452:142153

Amato-Lourenço LF, Dantas KC, Júnior GR, Paes VR, Ando RA, de Oliveira Freitas R, da Costa OMMM, Rabelo RS, Bispo KCS, Carvalho-Oliveira R, Mauad T (2024) Microplastics in the olfactory bulb of the human brain. JAMA Netw Open 7(9):e2440018

Amelia TSM, Khalik WMAWM, Ong MC, Shao YT, Pan H-J, Bhubalan K (2021) Marine microplastics as vectors of major ocean pollutants and its hazards to the marine ecosystem and humans. Prog Earth Planet Sci 8(1):12

Barboza LGA, Vieira LR, Branco V, Figueiredo N, Carvalho F, Carvalho C, Guilhermino L (2018) Microplastics cause neurotoxicity, oxidative damage and energy-related changes and interact with the bioaccumulation of mercury in the European seabass, Dicentrarchus labrax (Linnaeus, 1758). Aquat Toxicol 195:49–57

Beijer NRM, Dehaut A, Carlier MP, Wolter H, Versteegen RM, Pennings JLA, de La Fonteyne L, Niemann H, Janssen HM, Timmermans BG, Mennes W, Cassee FR, Mengelers MJB, Amaral-Zettler LA, Duflos G, Staal YCM (2022) Relationship between particle properties and immunotoxicological effects of environmentally-sourced microplastics. Front Water 4:866732

Bhardwaj LK, Rath P, Yadav P, Gupta U (2024) Microplastic contamination, an emerging threat to the freshwater environment: a systematic review. Environ Syst Res 13(1):8

Bhuyan MS (2022) Effects of microplastics on fish and in human health. Front Environ Sci 10:827289

Bikiaris N, Nikolaidis NF, Barmpalexis P (2024) Microplastics (MPs) in cosmetics: a review on their presence in personal-care, cosmetic, and cleaning products (PCCPs) and sustainable alternatives from biobased and biodegradable polymers. Cosmetics 11(5):145

Bishop CR, Yan K, Nguyen W, Rawle DJ, Tang B, Larcher T, Suhrbier A (2024) Microplastics dysregulate innate immunity in the SARS-CoV-2 infected lung. Front Immunol 15:1382655

Cheng W, Li X, Zhou Y, Yu H, Xie Y, Guo H, Wang H, Li Y, Feng Y, Wang Y (2022) Polystyrene microplastics induce hepatotoxicity and disrupt lipid metabolism in the liver organoids. Sci Total Environ 806:150328

Di Cesare M, Perel P, Taylor S, Kabudula C, Bixby H, Gaziano TA, McGhie DV, Mwangi J, Pervan B, Narula J, Pineiro D, Pinto FJ (2024) The heart of the world. Glob Heart 19(1):11

Ding R, Tong L, Zhang W (2021) Microplastics in freshwater environments: sources, fates and toxicity. Water Air Soil Pollut 232:181

Domenech J, Marcos R (2021) Pathways of human exposure to microplastics, and estimation of the total burden. Curr Opin Food Sci 39:144–151

Dong C-D, Chen C-W, Chen Y-C, Chen H-H, Lee J-S, Lin C-H (2020) Polystyrene microplastic particles: in vitro pulmonary toxicity assessment. J Hazard Mater 385:121575

Duggal B, Kumar G (2025) Cardiotoxicity of microplastics: an emerging cardiovascular risk factor. Curr Cardiol Rev 21(5):1–5

Emenike EC, Okorie CJ, Ojeyemi T, Egbemhenghe A, Iwuozor KO, Saliu OD, Okoro HK, Adeniyi AG (2023) From oceans to dinner plates: the impact of microplastics on human health. Heliyon 9(10):e20440

Eze CG, Nwankwo CE, Dey S, Sundaramurthy S, Okeke ES (2024) Food chain microplastics contamination and impact on human health: a review. Environ Chem Lett 22(4):1889–1927

Fackelmann G, Sommer S (2019) Microplastics and the gut microbiome: how chronically exposed species may suffer from gut dysbiosis. Mar Pollut Bull 143:193–203

Farady SE (2019) Microplastics as a new, ubiquitous pollutant: strategies to anticipate management and advise seafood consumers. Mar Policy 104:103–107

Filho WL, Salvia AL, Bonoli A, Saari UA, Voronova V, Klõga M, Kumbhar SS, Olszewski K, De Quevedo DM, Barbir J (2021) An assessment of attitudes towards plastics and bioplastics in Europe. Sci Total Environ 755:142732

Forte M, Iachetta G, Tussellino M, Carotenuto R, Prisco M, De Falco M, Laforgia V, Valiante S (2016) Polystyrene nanoparticles internalization in human gastric adenocarcinoma cells. Toxicol In Vitro 31:126–136

Ghosh S, Sinha JK, Ghosh S, Vashisth K, Han S, Bhaskar R (2023) Microplastics as an emerging threat to the global environment and human health. Sustainability 15(14):10821

Goodman KE, Hua T, Sang Q-XA (2022) Effects of polystyrene microplastics on human kidney and liver cell morphology, cellular proliferation, and metabolism. ACS Omega 7(38):34136–34153

Goswami S, Adhikary S, Bhattacharya S, Agarwal R, Ganguly A, Nanda S, Rajak P (2024) The alarming link between environmental microplastics and health hazards with special emphasis on cancer. Life Sci 355:122937

Hamed M, Martyniuk CJ, Naguib M, Lee J-S, Sayed AE-DH (2022) Neurotoxic effects of different sizes of plastics (nano, micro, and macro) on juvenile common carp (Cyprinus carpio). Front Mol Neurosci 15:1028364

Horvatits T, Tamminga M, Liu B, Sebode M, Carambia A, Fischer L, Püschel K, Huber S, Fischer EK (2022) Microplastics detected in cirrhotic liver tissue. EBioMedicine 82:104147

Hwang J, Choi D, Han S, Choi J, Hong J (2019) An assessment of the toxicity of polypropylene microplastics in human derived cells. Sci Total Environ 684:657–669

Jenner LC, Rotchell JM, Bennett RT, Cowen M, Tentzeris V, Sadofsky LR (2022) Detection of microplastics in human lung tissue using μFTIR spectroscopy. Sci Total Environ 831:154907

Kadac-Czapska K, Ośko J, Knez E, Grembecka M (2024) Microplastics and oxidative stress—current problems and prospects. Antioxidants 13(5):579

Kim L, Cui R, Kwak JI, An Y-J (2022) Trophic transfer of nanoplastics through a microalgae–crustacean–small yellow croaker food chain: inhibition of digestive enzyme activity in fish. J Hazard Mater 440:129715

Kopatz V, Wen K, Kovács T, Keimowitz AS, Pichler V, Widder J, Vethaak AD, Hollóczki O, Kenner L (2023) Micro-and nanoplastics breach the blood–brain barrier (BBB): biomolecular corona's role revealed. Nano 13(8):1404

Lee D-H (2024) Microplastics and cardiovascular diseases: importance of coexisting environmental pollutants. Circulation 150(12):908–910

Lee Y, Cho J, Sohn J, Kim C (2023) Health effects of microplastic exposures: current issues and perspectives in South Korea. Yonsei Med J 64(5):301–308

Li Z, Zhu S, Liu Q, Wei J, Jin Y, Wang X, Zhang L (2020) Polystyrene microplastics cause cardiac fibrosis by activating Wnt/β-catenin signaling pathway and promoting cardiomyocyte apoptosis in rats. Environ Pollut 265:115025

Li Y, Tao L, Wang Q, Wang F, Li G, Song M (2023) Potential health impact of microplastics: a review of environmental distribution, human exposure, and toxic effects. Environ Health 1(4):249–257

Li N, Yang H, Dong Y, Wei B, Liang L, Yun X, Tian J, Zheng Y, Duan S, Zhang L (2024) Prevalence and implications of microplastic contaminants in general human seminal fluid: a Raman spectroscopic study. Sci Total Environ 937:173522

Lin W, Luo H, Wu J, Liu X, Cao B, Liu Y, Yang P, Yang J (2023a) Polystyrene microplastics enhance the microcystin-LR-induced gonadal damage and reproductive endocrine disruption in zebrafish. Sci Total Environ 876:162664

Lin Y-D, Huang P-H, Chen Y-W, Hsieh C-W, Tain Y-L, Lee B-H, Hou C-Y, Shih M-K (2023b) Sources, degradation, ingestion and effects of microplastics on humans: a review. Toxics 11(9):747

Liu S, Wang C, Yang Y, Du Z, Li L, Zhang M, Ni S, Yue Z, Yang K, Wang Y, Li X, Yang Y, Qin Y, Li J, Yang Y, Zhang M (2024) Microplastics in three types of human arteries detected by pyrolysis-gas chromatography/mass spectrometry (Py-GC/MS). J Hazard Mater 469:133855

Mamun AA, Prasetya TAE, Dewi IR, Ahmad M (2023) Microplastics in human food chains: food becoming a threat to health safety. Sci Total Environ 858:159834

Marfella R, Prattichizzo F, Sardu C, Fulgenzi G, Graciotti L, Spadoni T, D'Onofrio N, Scisciola L, La Grotta R, Frigé C, Pellegrini V, Municinò M, Siniscalchi M, Spinetti F, Vigliotti G, Vecchione C, Carrizzo A, Accarino G, Squillante A, Spaziano G, Mirra D, Esposito R, Altieri S, Falco G, Fenti A, Galoppo S, Canzano S, Sasso FC, Matacchione G, Olivieri F, Ferraraccio F, Panarese I, Paolisso P, Barbato E, Lubritto C, Balestrieri ML, Mauro C, Caballero AE, Rajagopalan S, Ceriello A, D'Agostino B, Iovino P, Paolisso G (2024) Microplastics and nanoplastics in atheromas and cardiovascular events. N Engl J Med 390(10):900–910

Miller ME, Hamann M, Kroon FJ (2020) Bioaccumulation and biomagnification of microplastics in marine organisms: a review and meta-analysis of current data. PLoS One 15(10):e0240792

Napper IE, Davies BFR, Clifford H, Elvin S, Koldewey HJ, Mayewski PA, Miner KR, Potocki M, Elmore AC, Gajurel AP, Thompson RC (2020) Reaching new heights in plastic pollution—preliminary findings of microplastics on Mount Everest. One Earth 3(5):621–630

Nugrahapraja H, Sugiyo PWW, Putri BQ, Ni'matuzahroh, Fatimah, Huang L, Hafza N, Götz F, Santoso H, Wibowo AT, Luqman A (2022) Effects of microplastic on human gut microbiome: detection of plastic-degrading genes in human gut exposed to microplastics—preliminary study. Environments 9(11):140

O'Callaghan L, Olsen M, Tajouri L, Beaver D, Hudson C, Alghafri R, McKirdy S, Goldsworthy A (2024) Plastic induced urinary tract disease and dysfunction: a scoping review. J Expo Sci Environ Epidemiol:1–15. https://doi.org/10.1038/s41370-024-00709-3

Oleksiuk K, Krupa-Kotara K, Wypych-Ślusarska A, Głogowska-Ligus J, Spychała A, Słowiński J (2022) Microplastic in food and water: current knowledge and awareness of consumers. Nutrients 14(22):4857

Osman AI, Hosny M, Eltaweil AS, Omar S, Elgarahy AM, Farghali M, Yap P-S, Wu Y-S, Nagandran S, Batumalaie K, Gopinath SCB, John OD, Sekar M, Saikia T, Karunanithi P, Hatta MHM, Akinyede KA (2023) Microplastic sources, formation, toxicity and remediation: a review. Environ Chem Lett 21(4):2129–2169

Paul I, Mondal P, Haldar D, Halder G (2024) Beyond the cradle–amidst microplastics and the ongoing peril during pregnancy and neonatal stages: a holistic review. J Hazard Mater 469:133963

Persiani E, Cecchettini A, Ceccherini E, Gisone I, Morales MA, Vozzi F (2023) Microplastics: a matter of the heart (and vascular system). Biomedicine 11(2):264

Pironti C, Notarstefano V, Ricciardi M, Motta O, Giorgini E, Montano L (2023) First evidence of microplastics in human urine, a preliminary study of intake in the human body. Toxics 11(1):40

Prata JC (2018) Airborne microplastics: consequences to human health? Environ Pollut 234:115–126

Prata JC, Dias-Pereira P (2023) Microplastics in terrestrial domestic animals and human health: implications for food security and food safety and their role as sentinels. Animals 13(4):661

Prattichizzo F, Ceriello A, Pellegrini V, La Grotta R, Graciotti L, Olivieri F, Paolisso P, D'Agostino B, Iovino P, Balestrieri ML, Rajagopalan S, Landrigan PJ, Marfella R, Paolisso G (2024) Micro-nanoplastics and cardiovascular diseases: evidence and perspectives. Eur Heart J 45(38):4099–4110

Prüst M, Meijer J, Westerink RHS (2020) The plastic brain: neurotoxicity of micro-and nanoplastics. Part Fibre Toxicol 17(1):24

Ragusa A, Svelato A, Santacroce C, Catalano P, Notarstefano V, Carnevali O, Papa F, Rongioletti MCA, Baiocco F, Draghi S, D'Amore E, Rinaldo D, Matta M, Giorgini E (2021) Plasticenta: first evidence of microplastics in human placenta. Environ Int 146:106274

Saha SC, Saha G (2024) Effect of microplastics deposition on human lung airways: a review with computational benefits and challenges. Heliyon 10(2):e24355

Sangkham S, Islam MA, Adhikari S, Kumar R, Sharma P, Sakunkoo P, Bhattacharya P, Tiwari A (2023) Evidence of microplastics in groundwater: a growing risk for human health. Groundw Sustain Dev 23:100981

Song J, Wang C, Li G (2024) Defining primary and secondary microplastics: a connotation analysis. ACS ES&T Water 4(6):2330–2332

Sun A, Wang W-X (2023) Human exposure to microplastics and its associated health risks. Environ Health 1(3):139–149

Tamargo A, Molinero N, Reinosa JJ, Alcolea-Rodriguez V, Portela R, Bañares MA, Fernández JF, Moreno-Arribas MV (2022) PET microplastics affect human gut microbiota communities during simulated gastrointestinal digestion, first evidence of plausible polymer biodegradation during human digestion. Sci Rep 12(1):528

Tan H, Yue T, Xu Y, Zhao J, Xing B (2020) Microplastics reduce lipid digestion in simulated human gastrointestinal system. Environ Sci Technol 54(19):12285–12294

Torres-Agullo A, Karanasiou A, Moreno T, Lacorte S (2021) Overview on the occurrence of microplastics in air and implications from the use of face masks during the COVID-19 pandemic. Sci Total Environ 800:149555

Vattanasit U, Kongpran J, Ikeda A (2023) Airborne microplastics: A narrative review of potential effects on the human respiratory system. Sci Total Environ 904:166745

Wang M, Wu Y, Li G, Xiong Y, Zhang Y, Zhang M (2024a) The hidden threat: Unraveling the impact of microplastics on reproductive health. Sci Total Environ 935:173177

Wang T, Yi Z, Liu X, Cai Y, Huang X, Fang J, Shen R, Lu W, Xiao Y, Zhuang W, Guo S (2024b) Multimodal detection and analysis of microplastics in human thrombi from multiple anatomically distinct sites. EBioMedicine 103:105118

Wang X, Deng K, Zhang P, Chen Q, Magnuson JT, Qiu W, Zhou Y (2024c) Microplastic-mediated new mechanism of liver damage: from the perspective of the gut-liver axis. Sci Total Environ 919:170962

Wolff CM, Singer D, Schmidt A, Bekeschus S (2023) Immune and inflammatory responses of human macrophages, dendritic cells, and T-cells in presence of micro-and nanoplastic of different types and sizes. J Hazard Mater 459:132194

Wu D, Feng Y, Wang R, Jiang J, Guan Q, Yang X, Wei H, Xia Y, Luo Y (2023a) Pigment microparticles and microplastics found in human thrombi based on Raman spectral evidence. J Adv Res 49:141–150

Wu H, Liu Q, Yang N, Xu S (2023b) Polystyrene-microplastics and DEHP co-exposure induced DNA damage, cell cycle arrest and necroptosis of ovarian granulosa cells in mice by promoting ROS production. Sci Total Environ 871:161962

Yang W, Jannatun N, Zeng Y, Liu T, Zhang G, Chen C, Li Y (2022) Impacts of microplastics on immunity. Front Toxicol 4:956885

Yang Y, Xie E, Du Z, Peng Z, Han Z, Li L, Zhao R, Qin Y, Xue M, Li F, Hua K, Yang X (2023) Detection of various microplastics in patients undergoing cardiac surgery. Environ Sci Technol 57(30):10911–10918

Yang Y, Zhang F, Jiang Z, Du Z, Liu S, Zhang M, Jin Y, Qin Y, Yang X, Wang C, Gao H (2024) Microplastics are associated with elevated atherosclerotic risk and increased vascular complexity in acute coronary syndrome patients. Part Fibre Toxicol 21(1):34

Zaheer J, Kim H, Ko IO, Jo E-K, Choi E-J, Lee H-J, Shim I, Woo H-J, Choi J, Kim G-H, Kim JS (2022) Pre/post-natal exposure to microplastic as a potential risk factor for autism spectrum disorder. Environ Int 161:107121

Zhang C, Zhang G, Sun K, Ren J, Zhou J, Liu X, Lin F, Yang H, Cao J, Nie L, Zhang P, Zhang L, Wang Z, Guo H, Lin X, Duan S, Cao J, Huang H (2024) Association of mixed exposure to microplastics with sperm dysfunction: a multi-site study in China. EBioMedicine 108:105369

Zhao B, Rehati P, Yang Z, Cai Z, Guo C, Li Y (2024) The potential toxicity of microplastics on human health. Sci Total Environ 912:168946

Zhu X, Wang C, Duan X, Liang B, Xu EG, Huang Z (2023) Micro-and nanoplastics: A new cardiovascular risk factor? Environ Int 171:107662

Ziani K, Ioniță-Mîndrican C-B, Mititelu M, Neacşu SM, Negrei C, Moroşan E, Drăgănescu D, Preda O-T (2023) Microplastics: a real global threat for environment and food safety: A state of the art review. Nutrients 15(3):617

Microplastics and Aquatic Limnology: Conception of Biogeochemical Processes

9

Siddhant Dash, Payal Mazumder, and Manish Kumar

Abstract

Microplastics (MPs) are emerging as contaminants posing serious environmental threats. Various types of MPs are being discharged into our water bodies, affecting the natural biogeochemical processes in the aquatic ecosystem, including modifications in carbon and nutrient cycling and imparting severe toxicity to flora and fauna. While several pieces of literature are available on the sources and types of MPs being discharged into various global water bodies, deep insights into the various biogeochemical processes affecting the environmental cycle still remain nascent. The present review highlights the key research avenues elucidating the governing mechanisms for changes in the biogeochemical cycle due to microplastic pollution. This includes different studies carried out on the impacts of MPs on the various environmental (carbon, nitrogen and phosphorus) cycles of the aquatic ecosystems. It also presents the toxicity aspects of MPs on the aquatic flora and fauna in different environments. Finally, we discuss the prospects that researchers need to focus on to provide a more comprehensive picture

S. Dash (✉)
Department of Civil Engineering, School of Engineering and Sciences, SRM University AP, Amaravati, Guntur, Andhra Pradesh, India

Escuela de Ingeniería y Ciencias, Tecnologico de Monterrey, Campus Monterey, Monterrey, Nuevo Leon, Mexico

P. Mazumder
Center for the Environment, Indian institute of Technology Guwahati, Guwahati, Assam, India

M. Kumar
Escuela de Ingeniería y Ciencias, Tecnologico de Monterrey, Campus Monterey, Monterrey, Nuevo Leon, Mexico

Sustainability Cluster, School of Advance Engineering, UPES, Dehradun, Uttarakhand, India

K. Bala et al. (eds.), *Occurrence, Detection, and Fate of Microplastics in Freshwater Ecosystems*, https://doi.org/10.1007/978-981-95-0225-7_9

of the biogeochemical processes involved in MP pollution and to achieve sustainable solutions to abate MP contamination.

Keywords

Microplastics · Aquatic limnology · Toxicity · Biogeochemistry · Environmental cycle

9.1 Introduction

The invention of plastic in the early twentieth century was considered one of the greatest in humankind's history. Little did we know that this revolution would one day turn into one of the most critical pollutants on the globe. With its rigidity, elasticity and, most importantly, affordability, the production of plastic has been shooting up since the 1960s (Wang et al. 2021a). This has resulted in an escalation of the plastic pollution problem in the natural environment since these materials are extremely tough to degrade and decompose. Over 80% of all the plastics ever produced have accumulated in the natural environment in some form or another, interfering with the natural processes of the entire ecosystem (Guo et al. 2020). The problems associated with plastics were identified in the early 1970s (Anderson et al. 2016); however, Carpenter et al. (1972) first provided a comprehensive understanding of the seriousness of the plastic pollution problem. Since then, researchers worldwide have been working on different technologies to reduce plastic pollution. However, in the last few years, the term "microplastics" has come into vogue and has gained significant attention owing to their danger to the ecosystem. In the initial days, the microplastics (MPs) were considered as those synthetic polymers whose size is proportionate to the planktonic organisms, i.e. <5 mm (Frias et al. 2014; Hidalgo-Ruz et al. 2012). However, recent classifications by researchers have further categorised plastics under various criteria, such as size: macro (>1 cm), meso (1–<10 mm), micro (1–<1000 μm) and nano (1–<1000 nm); shape; structure and composition (Hartmann et al. 2019).

Primarily, MPs enter the ecosystem through several anthropogenic sources that are mainstream in commercial and industrial use. For example, spherical polyethylene microbeads of sizes ranging from 5 μm to 2 mm are produced for use in mild abrasives and daily-use cosmetic products (Chang 2015; Fendall and Sewell 2009; Lei et al. 2017; Song et al. 2021; Sun et al. 2020). A 150 mL facial scrub can accommodate up to nearly three million microbeads that get discharged as wastewater effluents daily, making up trillions of MPs in the natural environment (Napper et al. 2015). Similarly, other sources include industrial spillage of pellets, accounting for tonnes of MPs entering our water bodies and soil systems (Zbyszewski et al. 2014); degradation of plastic debris into micro- and nanoscale particles (Amorim and Scott-Fordsmand 2021; Vaksmaa et al. 2021) and formation of particles resulting from the abrasion of tyres in vehicles, fragments of garments and textiles, detergents used in laundries, paints and emulsions, etc. (Arias et al. 2022; Järlskog et al.

2021; Kole et al. 2017; Sait et al. 2021; Yang et al. 2021). These particles eventually enter our water bodies, affecting the aquatic biota. In addition, all land-based plastic wastes, such as those in landfills, plastic mulching, sewage sludges, composts and organic fertilisers, also enter our soil and aquatic biota through different phenomena (El Hayany et al. 2022; Gao et al. 2020; He et al. 2019; Hou et al. 2021; Huang et al. 2020; Manikanda Bharath et al. 2021; Khalid et al. 2023; Mahon et al. 2017; Vithanage et al. 2021; Weithmann et al. 2018; Yuan et al. 2022; Zhang et al., 2020, 2022a, b). Although a sewage treatment plant can remove about 99% of MPs following the tertiary treatment, given the gargantuan volume of sewage and greywater generated daily across the globe, one sewage treatment unit can easily release ~10^4–10^8 MPs daily (Kalčíková et al. 2017; Mason et al. 2016). These particles are transported through various environmental systems, such as surface water run-off or airborne transportation by winds, thereby reaching long distances, even the remotest places, such as polar regions or deep oceans (Bergmann et al. 2019; Enyoh et al. 2019). Furthermore, the presence of MPs in soil systems is primarily attributed to the vertical movement of water through infiltration and soil cracking resulting from wet and dry soil cycles. Hence, it is widely reported that MPs are more abundant in shallow soils compared to deep soils (Li et al. 2020). Additionally, disturbances in the root zones of plants and animal activities also allow MPs to migrate into the soil strata (Wang et al. 2022).

MPs pose a significant threat to the aquatic ecosystems. In aquatic environments, the biota is responsible for the transport, storage and transformation of MPs. Various invertebrate species help in the breakdown of the fine particulate and dissolved organic matter present in the aquatic ecosystem. However, MPs present in significant concentrations hinder the natural processes involved owing to their sizes, which resemble the size of coarse particulate matter and fine and dissolved organic matter (Cummins 1974). In addition, MPs are ingested by many aquatic organisms, including birds, and subsequently enter the aquatic and terrestrial food chains through bioaccumulation (Azevedo-Santos et al. 2019; Ehlers et al. 2019; Windsor et al. 2019). In fact, some species even metabolise the carbon stored in the polymers. Likewise, due to their high specific surface area, there is a considerable movement of heavy metals (HMs), organic contaminants, pharmaceutical pollutants and antibiotics through adsorption and desorption processes (Menéndez-Pedriza and Jaumot 2020). Also, the microorganisms from wastewater streams get retained on the surface of MPs, thereby developing biofilms around them, which further adsorb different hazardous chemicals and leach them into the external environment, resulting in interactions with biotic systems (Boyle et al. 2020; Hahladakis et al. 2018; McCormick et al. 2016; Saha et al. 2021). Figure 9.1 presents a clear picture outlining the different pathways by which these particles reach the aquatic environment. Limnology refers to understanding the physico-chemical and biological processes involved within an aquatic system and their interactions with the surrounding environment. Given the vast prevalence of MPs in aquatic systems through massive anthropogenic interventions, limnologists worldwide should consider them as separate units for analysis rather than mixing them with other inorganics derived from natural phenomena such as weathering, disintegration, high temperatures or

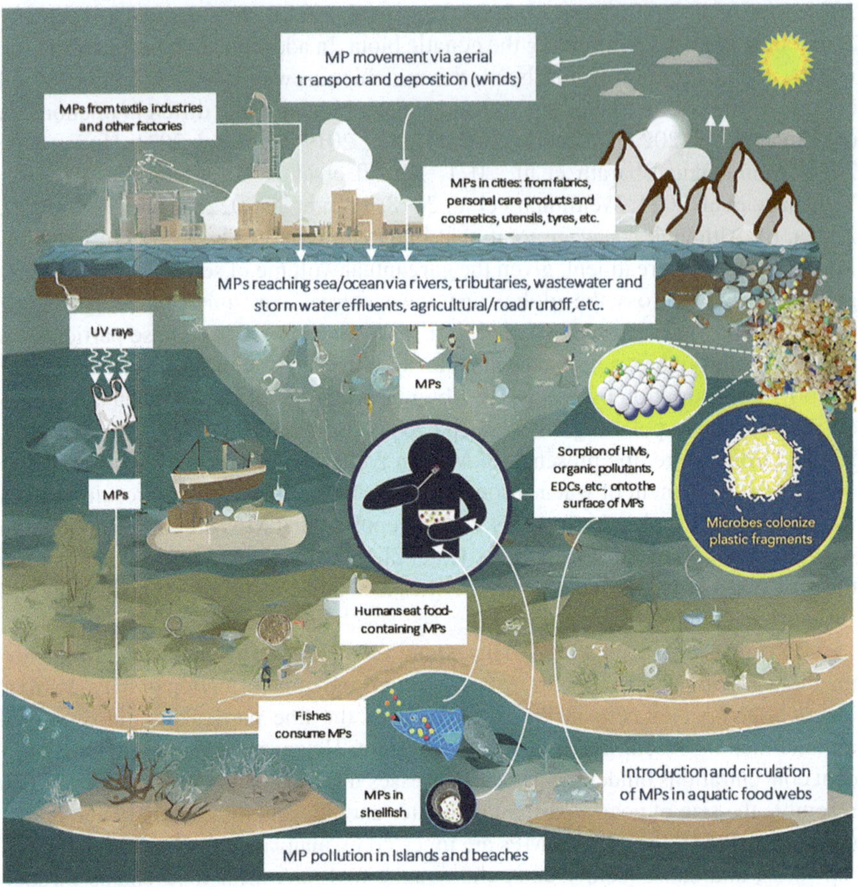

Fig. 9.1 The biogeochemical cycle of microplastics in the environment. MPs: microplastics; HMs: heavy metals; EDCs: endocrine-disrupting chemicals

erosion. Hence, we must integrate standard limnological surveys and water quality assessments with MP concentrations in aquatic environments to better understand the processes involved.

9.2 Global Studies on the Biogeochemistry of MPs

Global studies on understanding the biogeochemistry of MPs in water ecosystems increased significantly in the last few years. Here, we extracted data on the changes in the biogeochemical processes in the aquatic environment due to MPs from the *Scopus* database. We used the keywords *"microplastics" AND "biogeochemical cycle" OR "aquatic limnology" OR "toxicity"* to extract documents from the Scopus database. Figure 9.2 shows the trend of the number of publications in this

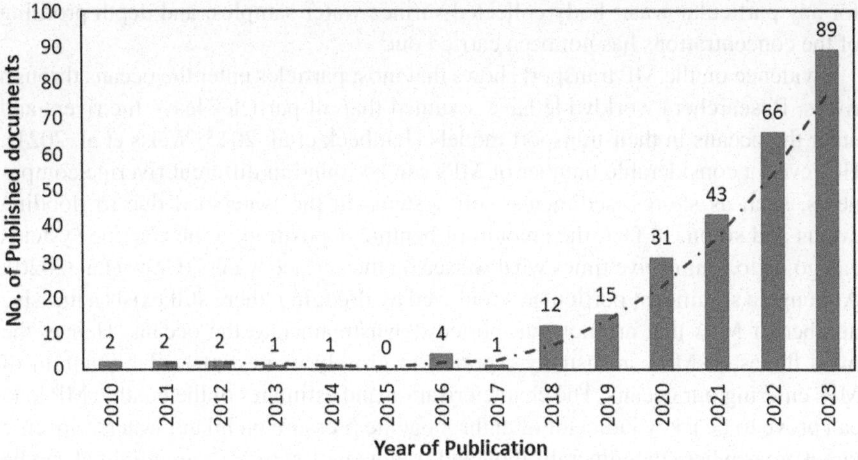

Fig. 9.2 Scientometric analysis showing the number of published articles on understanding the biogeochemistry of MPs in water ecosystems

research domain. As the trendline suggests, in the past dozen years, there has been a significant jump in the number of articles as the years progressed, indicating the growing importance of the research domain among researchers worldwide. It started with lakes and subsequently moved towards rivers and other water bodies. However, the homogeneity of the publications from different geographic locations still needs to be established, as most studies have emanated from Asia and Europe, with China being the major contributor.

When it comes to the aquatic ecosystem, different components show varying concentrations of MPs, depending upon several factors, such as the proximity of the source entry points (sewage discharges, sludge discharge, roads, stormwater run-off, agricultural discharge, etc.), the flow regime of the water body, the rate of precipitation of MPs into the sediment column (depends on the tranquillity of the water) as well as the seasonality of the fluxes. In addition, the MP concentrations can also depend on the season in which the sampling is being carried out and specific hydrological conditions prevailing in the water systems. For example, higher flow regimes are seen to carry more MPs in water bodies, but when we determine their concentrations, they yield lower concentrations compared to the low-flow seasons (Watkins et al. 2019). The hydrological events, such as floods, droughts and anthropogenic control of the waterways (construction of dams, spillways, etc.), alter the transport of MPs in the water bodies, significantly altering their concentrations. Therefore, there is a need to comprehensively monitor the spatial and seasonal fluxes of MPs to better understand their fate and transport and establish a more sophisticated baseline for the concentration ranges in the water bodies. Another critical insight into the MP concentrations is that higher concentrations are usually found on the surface rather than being homogeneous throughout the water column. Hence, MPs collected from the surface are not a good indicator of their concentrations throughout the water column. Most studies that reported MP concentrations for any particular water body collected surface water samples, and depth profiling of the concentrations has not been carried out.

Evidence on the MP transport shows that most particles enter the oceans through rivers. Researchers worldwide have assumed that all particles leave the rivers and enter the oceans in their transport models (Jambeck et al. 2015; Weiss et al. 2021). However, a considerable number of MPs can be found in different riverine components, such as shores, sediments, soil systems in the watershed due to flooding events and so on. In fact, the amount of benthic deposition in the riverine systems can go up to four or five times what we see on the surface waters (Crew et al. 2020). Although a significant portion gets removed by dredging, there still exists a massive number of MPs that are not transported downstream into the oceans. Hence, the mass fluxes of MPs are usually reported to grossly overestimate the quantum of MPs entering our oceans. The characterisation and estimates of the benthic MP load can prove to be a key indicator of anthropogenic pressure on inland waters. Specific target areas where the mineralisation and fragmentation of MPs are minimal can be taken into account. These will act as a function of long-term deposition of the persistent particles in the ecosystem (Hoellein and Rochman 2021). This can be coupled with lake sediment cores, which can be used to track the temporal dynamics of MPs and, thus, prove substantial in recording the palaeolimnological and geogenic records. This is because MPs in sediment layers are ubiquitous and globally distributed, accounting for homogeneity in the sampling strategies.

Aquatic biotas play a significant role in acting as transient reservoirs for MPs. From inland waters to the oceans, all water bodies possess a considerable species diversity, wherein all organisms are symbiotic with each other. These organisms have been reported to take up MPs in various pathways, such as drinking, feeding, swimming, respiration and adsorption. It has been reported that freshwater fishes are more frequently found with MPs in different organs than marine fishes (Covernton et al. 2021). A recent study by Munno et al. (2022) has found the highest ever MP load in the Great Lakes of the USA, necessitating the need for studies on the freshwater ecosystems. In addition, the pathway mechanisms of MPs in freshwater systems need to be better documented. The aquatic biota, however, form the lowest point of the global food chain, where MPs enter the bodies of humans and other higher animals. The retention time of MPs in an aquatic organism's body depends on the metabolism and the complexity of the digestive tract or gill structure of the specific organism. For organisms having higher metabolism and simple digestive tracts, the retention time of MPs is much shorter compared to those organisms that have a complex digestive tract and irregular body structure (Hoang and Felix-Kim 2020; Welden and Cowie 2016). This is because MPs are egested quite quickly. Where the retention time is higher, i.e. at minimum 24–72 h, there is every possibility that the organism is eaten by a higher-order species, thereby ingesting MPs and transferring to the next trophic level. Even if MPs are egested easily, they remain in the water ecosystem and get re-ingested into another organism with a low retention time, and the cycle continues. Hence, even without getting biomagnified, MPs remain in the food chain without having any consequences for their hosts.

Besides the bioaccumulation of MPs in higher biotic species, they also adversely affect biodiversity functions. MPs offer great hosts to many microbial communities that thrive in the ecosystem. These microorganisms tend to form biofilms on the surface of the plastic particles as they offer novel media for their sustenance. These biofilms attract food sources for higher-order organisms, although they do not offer the same food quality as those produced on the natural media (Arias-Andres et al. 2018). As the plastic surfaces offer a durable substrate by design, these microorganisms thrive even under conditions that are not conducive to them. They depend on the carrier vectors to transport them to a new environment suitable for their growth and survival. These include non-native species, pathogens and drug-resistant bacteria (Wang et al. 2021b). The plastic-associated bacteria usually discharged from STPs exhibit higher gene exchange, rendering them resistant to antibiotics and heavy metals (Eckert et al. 2018). In fact, Song et al. (2020) found that there were some strains of antibiotic-resistant *E. coli* that were found and transported by MPs across different environments.

9.3 Influence of MPs on the C-Cycle

It has been determined that MPs, which contain around 80% carbon in their polymeric units, may contribute to the carbon content of soil (Rillig 2018). Carbon dynamics in aquatic environments are influenced by a number of factors, including the composition and structure of MPs and related microbial communities. These dynamics have been studied in a variety of compartments, including water, sediments and biota. Biofilms that form on the surfaces of microplastics act as carbon sources in the water column. As they decompose, they release unsaturated fatty acids that can sustain algae and plankton, ultimately entering the aquatic food web (Rillig 2018; Taipale et al. 2019). By preventing photosynthesis in aquatic organisms, including plants, algae and phytoplankton, MPs can serve as carbon sinks in addition to being a carbon source. In these settings, this inhibition lessens CO_2 fixation (Wieczorek et al. 2019). MPs can also interfere with plankton and algae's metabolic processes, which further reduces the release of CO_2 (Nava and Leoni 2021). MPs have an impact on aquatic sediments as well. There is a favourable association between MP concentrations and the total carbon content of the Brisbane River (He et al. 2019). Although MPs now have a very small role in bottom sediment carbon storage, predictions indicate that the amount of benthic carbon will gradually rise over time. MP-induced changes in sediment structure impede the growth of aquatic plants, which raises CO_2 emissions. MPs have a major impact on altering the biochemistry of sediments. The production of extracellular enzymes that increase CO_2 emissions from sediments is facilitated by enhanced biofilm formation on MP surfaces, which modifies enzyme activity (Gao et al. 2020; Hou et al. 2021). Additionally, because MPs affect the community structure and function of sediment-dwelling microbes, their effects on carbon transformations vary based on their composition and shape (Yi et al. 2021). Collectively, these processes illustrate the complex interactions between MPs and carbon cycling in aquatic ecosystems.

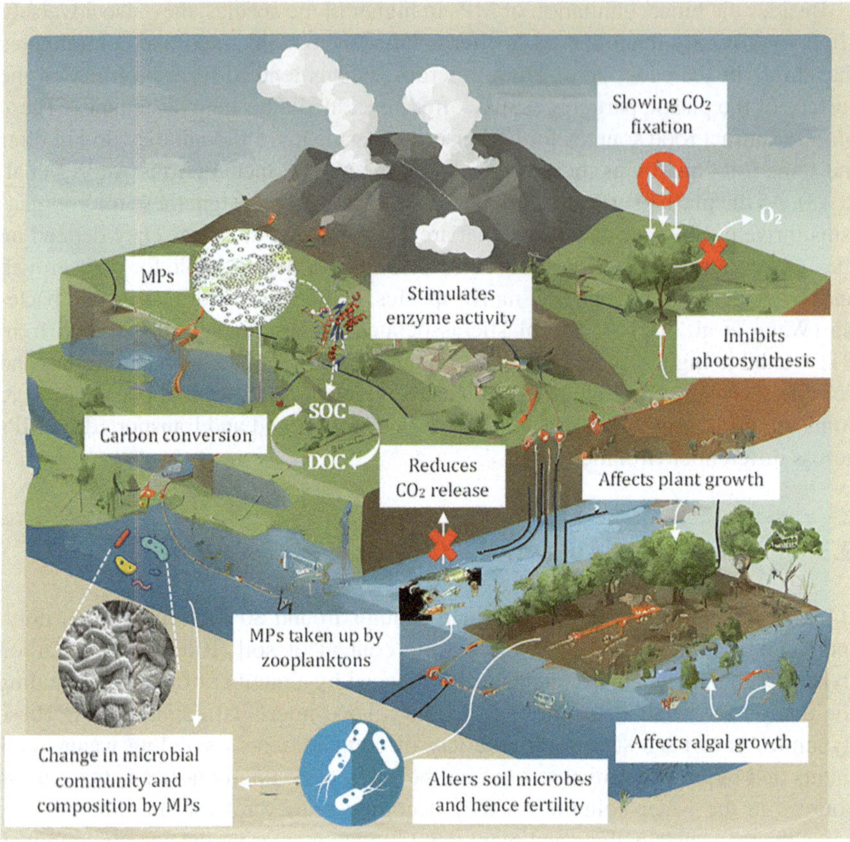

Fig. 9.3 Microplastics and their role in disrupting the environmental carbon cycle

Figure 9.3 illustrates the impact of microplastics on the carbon cycle within aquatic ecosystems.

9.4 Impact of Microplastics on the Nitrogen and Phosphorus Cycles

The nitrogen cycle, crucial for life on Earth, ranks just behind the carbon cycle in importance. It is fundamental to plant growth, which forms the backbone of the human food chain. However, this cycle is increasingly affected by human activities (Widdison and Burt 2008). Nitrogen enters aquatic ecosystems through processes such as nitrogen fixation, ammonification and nitrification (Fig. 9.4), facilitated by diverse microorganisms that maintain its balance in the environment (Sohm et al. 2011; Zehr and Capone 2020; Zhou et al. 2014). In these processes, organic nitrogen is first converted into ammonium (NH_4^+) by saprophytic bacteria. Nitrifying

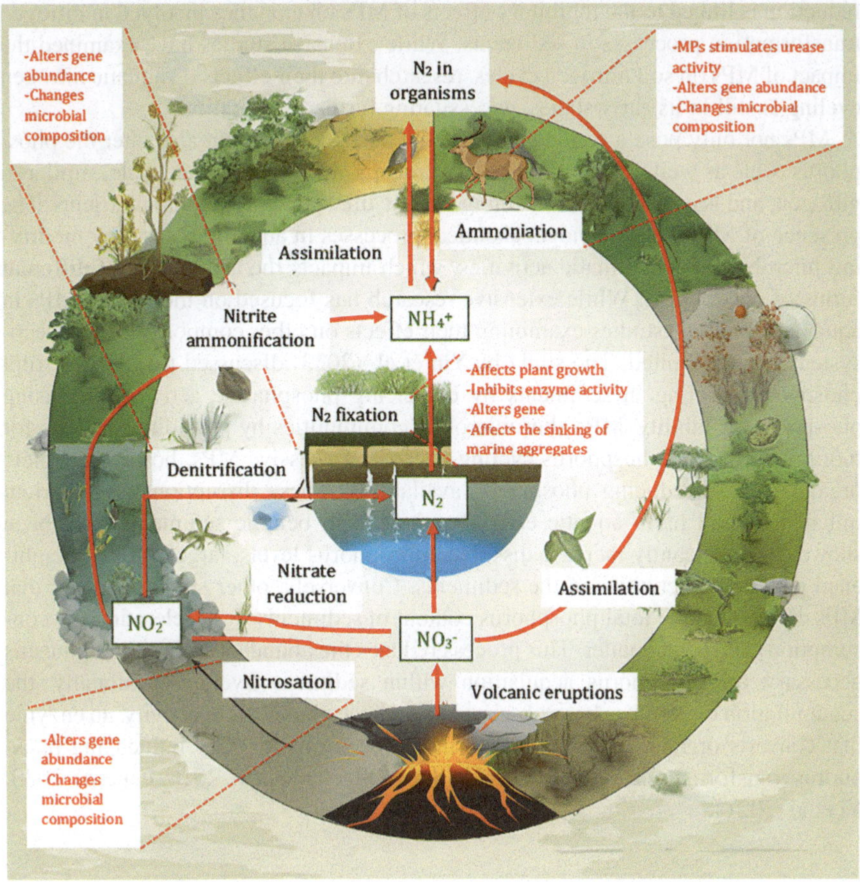

Fig. 9.4 Impacts of microplastics on the nitrogen cycle in the environment

bacteria then transform NH_4^+ into nitrate (NO_3^-), which is absorbed by soil plants and photosynthetic organisms and converted into organic nitrogen. Denitrifying bacteria return NO_3^- to the atmosphere as nitrogen gas (N_2), completing the cycle. Additionally, phytoplanktons consume nitrogen-rich plants, recycling nitrogen through their metabolic processes. Research has shown that biofilms formed on the surfaces of MPs can facilitate the oxidation and denitrification of ammonia nitrogen and nitrite (Shen et al. 2022). However, these biofilms may also hinder nitrogen removal by decreasing the abundance of denitrifying bacteria (Chen et al. 2020). Furthermore, MPs can disrupt the activity of enzymes in nitrogen cycling (Qin et al. 2020) and limit the expression of functional genes responsible for encoding these enzymes (Seeley et al. 2020). Consequently, MP pollution poses a threat to the global nitrogen cycle, potentially threatening the stability of ecosystems worldwide. Yu et al. (2021) found that MPs significantly reduce total nitrogen content in sediment columns, with this effect becoming more pronounced as the plastics age. This reduction is linked to the inhibitory effects of MPs on enzymes involved in nitrogen transformation processes in sediments. While extensive studies have examined the impact of MPs on soil nitrogen cycles, research into their effects on aquatic nitrogen cycling is still in its early stages, necessitating further exploration.

MPs not only pose a serious threat to the nitrogen cycle but also alter the phosphorus cycle as well. Phosphorus plays a crucial role in the nutrient cycle, similar to nitrogen, and is essential for sustaining aquatic life by supplying vital nutrients. The presence of MPs disrupts biogeochemical processes in aquatic systems by modifying microbial and enzymatic activities, which impacts the availability of different forms of phosphorus. While extensive research has focused on the role of MPs in aquatic sediments, studies examining their effects on other components of the ecosystem remain limited. The study by Yin et al. (2023) discussed that MPs disrupt phosphorus cycling in sediments by enhancing phosphatase activity, increasing phosphorus solubility. MPs alter microbial communities by providing surfaces for biofilms, affecting phosphorus-cycling bacteria. Excessive MPs, however, deplete organic matter, reducing phosphorus availability. These disruptions risk nutrient imbalances and harm aquatic ecosystems. MPs in benthic sediments have been shown to significantly increase dissolved phosphorus levels, largely due to heightened enzymatic activities in the sediments. Conversely, other studies suggest that MPs can reduce the total phosphorus content in sediments by accelerating the consumption of organic matter. This process reduces the abundance of microorganisms necessary for phosphorus acquisition within sediment layers. Additionally, the accumulation of MPs has been observed to enhance phosphatase activity, an enzyme that converts organic phosphorus into inorganic forms, thereby influencing phosphorus transformations in sediments (de Souza Machado et al. 2018; Fei et al. 2020; Yi et al. 2021).

9.5 Impact of Microplastics on the Ecotoxicity of Aquatic Systems

The ecotoxicity of MPs is influenced by their high surface area, chemical properties and physical characteristics. Smaller MPs exhibit greater adsorption capabilities than larger ones. Many pollutants adhere to MP surfaces, which are then ingested by aquatic organisms. For instance, Batel et al. (2016) documented the presence of benzo[a]pyrene, a polyaromatic hydrocarbon, on MPs consumed by zebrafish, highlighting the bioaccumulation of toxic substances through the food chain. The combined impact of chemical contaminants and MPs significantly influences contaminant transfer in freshwater fish (Alimi et al. 2018; Fred-Ahmadu et al. 2020). Although limited research exists on the bioaccumulation of contaminants linked to MPs, studies suggest that a wide range of pollutants, particularly pathogens and antibiotic-resistant bacteria, adhere to MP surfaces, causing severe toxic effects such as carcinogenicity, mutagenicity, teratogenicity and endocrine disruption in aquatic organisms (Gross et al. 2025; Jiang et al. 2022). The adsorption and biotransformation of emerging contaminants, including pharmaceuticals, personal care products

(Guo and Wang 2019; Li et al. 2020), PAHs (Bartonitz et al. 2020) and mercury (Graca et al. 2014), depend on ecosystem characteristics like salinity, temperature and pH.

9.6 Concluding Remarks

While the research on understanding the biogeochemistry of MPs in biotic systems is gathering pace, there is still tremendous scope for more in-depth research. Most studies have been carried out for a particular geographical location under a given climatic condition. However, studies on the influence of MPs under varying environmental stresses need to be carried out extensively. For instance, in the case of an extreme hydrological event, such as a flood or cyclone, MPs can get translocated to a new location via wind, erosion, increased discharges of wastewater, enhanced surface run-off or through resuspension and transportation of particles stored in riverbeds or shores. This results in the transportation of the microbial communities, such as bacteria, viruses and other pathogens, habituating on the surfaces of these MPs, to downstream regions and other places that are devoid of MP contamination. This causes significant changes in carbon and nutrient cycling as well as increases the ecotoxicity levels of the ecosystem. It is, therefore, essential that studies on MPs be shifted from laboratory studies to real-world scenarios coupled with complex environmental dynamics, such as changes in precipitation, humidity and organic content. Furthermore, there is limited research on the elemental cycling of MPs. Therefore, substantial attention needs to be paid to the influence of MPs on the aquatic biochemical cycles, such as CH_4 fixation, carbon sequestration and nutrient fixation. In addition, the type and shape of MPs play a pivotal role in the type of microorganisms that get attached to them. This results in enhancing the dynamics and complexity of the processes involved. Hence, a deeper understanding of the structure, composition and concentration of microorganisms can be gained through in-depth study. This can be coupled with the association of microbial communities with MPs at the genetic and protein levels, thereby providing a complete and comprehensive picture of the entire biogeochemical cycle.

Several studies conducted so far have primarily focused on single elemental cycling with respect to MP contamination. However, in the real-world scenario, these elemental cycling processes are synergistic, i.e. they coexist with one another. Additionally, multiple components coexist in an aquatic environment, such as plants, animals and microorganisms. The studies so far have concentrated on the effects of MPs on a single aquatic component. Hence, coupling MPs with multiple elements and components, thereby taking into account their synergistic effects, needs to be deeply explored. Furthermore, there is also a dire need to correlate the influence of MPs on the heavy metals and other contaminants in the water bodies. For example, the adsorption of organic and inorganic pollutants on the surfaces of MPs, which affects the biogeochemical cycles, shall be explored in the future.

MPs are increasingly becoming predominant among the major contaminants in the aquatic ecosystems, together with other contaminants such as nutrients and heavy metals. It is therefore recommended that limnologists develop stricter standardised protocols for MP sampling and analyses in order to understand environmental assessment, the development of stronger regulations and policies to counter plastic pollution, the building of appropriate measures to develop fate and transport mechanisms, and the evaluation of ecotoxicity aspects in natural biota. In addition, public awareness of the MP pollution should be made stronger by promoting eco-designs that do not degrade into fine MPs. Improving our wastewater treatment facilities should also be given due care and consideration to tackle the MP pollution. Hence, there should be comprehensive interdisciplinary research conducted among different sections of researchers to arrive at global sustainability when it comes to understanding the biogeochemical processes involved in MP pollution, thereby achieving viable and sustainable abatement policies.

References

Alimi OS, Farner Budarz J, Hernandez LM, Tufenkji N (2018) Microplastics and nanoplastics in aquatic environments: aggregation, deposition, and enhanced contaminant transport. Environmental science andtechnology, 52(4):1704-1724

Amorim MJB, Scott-Fordsmand JJ (2021) Plastic pollution – a case study with Enchytraeus crypticus – from micro-to nanoplastics. Environ Pollut 271:116363. https://doi.org/10.1016/j.envpol.2020.116363

Anderson JC, Park BJ, Palace VP (2016) Microplastics in aquatic environments: implications for Canadian ecosystems. Environ Pollut 218:269–280. https://doi.org/10.1016/j.envpol.2016.06.074

Arias AH, Alfonso MB, Girones L, Piccolo MC, Marcovecchio JE (2022) Synthetic microfibers and tyre wear particles pollution in aquatic systems: Relevance and mitigation strategies. Environ Pollut 295:118607. https://doi.org/10.1016/j.envpol.2021.118607

Arias-Andres M, Kettner MT, Miki T, Grossart H-P (2018) Microplastics: new substrates for heterotrophic activity contribute to altering organic matter cycles in aquatic ecosystems. Sci Total Environ 635:1152–1159. https://doi.org/10.1016/j.scitotenv.2018.04.199

Azevedo-Santos VM, Gonçalves GRL, Manoel PS, Andrade MC, Lima FP, Pelicice FM (2019) Plastic ingestion by fish: a global assessment. Environ Pollut 255:112994. https://doi.org/10.1016/j.envpol.2019.112994

Bartonitz A, Anyanwu IN, Geist J, Imhof HK, Reichel J, Graßmann J, Beggel S (2020) Modulation of PAH toxicity on the freshwater organism G. roeseli by microparticles. Environmental Pollution, 260:113999

Batel A, Linti F, Scherer M, Erdinger L, Braunbeck T (2016) Transfer of benzo [a] pyrene from microplastics to Artemia nauplii and further to zebrafish via a trophic food web experiment: CYP1A induction andvisual tracking of persistent organic pollutants. Environmental toxicology and chemistry, 35(7):1656-1666

Bergmann M, Mützel S, Primpke S, Tekman MB, Trachsel J, Gerdts G (2019) White and wonderful? Microplastics prevail in snow from the Alps to the Arctic. Sci Adv 5(8). https://doi.org/10.1126/sciadv.aax1157

Boyle D, Catarino AI, Clark NJ, Henry TB (2020) Polyvinyl chloride (PVC) plastic fragments release Pb additives that are bioavailable in zebrafish. Environ Pollut 263:114422. https://doi.org/10.1016/j.envpol.2020.114422

Carpenter EJ, Anderson SJ, Harvey GR, Miklas HP, Peck BB (1972) Polystyrene spherules in coastal waters. Science 178(4062):749–750. https://doi.org/10.1126/science.178.4062.749

Chang M (2015) Reducing microplastics from facial exfoliating cleansers in wastewater through treatment versus consumer product decisions. Mar Pollut Bull 101(1):330–333. https://doi.org/10.1016/j.marpolbul.2015.10.074

Chen G, Feng Q, Wang J (2020) Mini-review of microplastics in the atmosphere and their risks to humans. Science of the Total Environment, 703:135504

Covernton GA, Davies HL, Cox KD, El-Sabaawi R, Juanes F, Dudas SE, Dower JF (2021) A Bayesian analysis of the factors determining microplastics ingestion in fishes. J Hazard Mater 413:125405. https://doi.org/10.1016/j.jhazmat.2021.125405

Crew A, Gregory-Eaves I, Ricciardi A (2020) Distribution, abundance, and diversity of microplastics in the upper St. Lawrence River. Environ Pollut 260:113994. https://doi.org/10.1016/j.envpol.2020.113994

Cummins KW (1974) Structure and function of stream ecosystems. Bioscience 24(11):631–641. https://doi.org/10.2307/1296676

de Souza Machado AA, Lau CW, Till J, Kloas W, Lehmann A, Becker R, Rillig MC (2018) Impacts of microplastics on the soil biophysical environment. Environmental science and technology, 52(17):9656-9665

Eckert EM, Di Cesare A, Kettner MT, Arias-Andres M, Fontaneto D, Grossart H-P, Corno G (2018) Microplastics increase impact of treated wastewater on freshwater microbial community. Environ Pollut 234:495–502. https://doi.org/10.1016/j.envpol.2017.11.070

Ehlers S, Manz W, Koop J (2019) Microplastics of different characteristics are incorporated into the larval cases of the freshwater caddisfly Lepidostoma basale. Aquat Biol 28:67–77. https://doi.org/10.3354/ab00711

El Hayany B, Rumpel C, Hafidi M, El Fels L (2022) Occurrence, analysis of microplastics in sewage sludge and their fate during composting: a literature review. J Environ Manag 317:115364. https://doi.org/10.1016/j.jenvman.2022.115364

Enyoh CE, Verla AW, Verla EN, Ibe FC, Amaobi CE (2019) Airborne microplastics: a review study on method for analysis, occurrence, movement and risks. Environ Monit Assess 191(11):668. https://doi.org/10.1007/s10661-019-7842-0

Fei Y, Huang S, Zhang H, Tong Y, Wen D, Xia X, Barceló, D (2020) Response of soil enzyme activities and bacterial communities to the accumulation of microplastics in an acid cropped soil. Science of theTotal Environment, 707:135634

Fendall LS, Sewell MA (2009) Contributing to marine pollution by washing your face: microplastics in facial cleansers. Mar Pollut Bull 58(8):1225–1228. https://doi.org/10.1016/j.marpolbul.2009.04.025

Fred-Ahmadu OH, Bhagwat G, Oluyoye I, Benson NU, Ayejuyo OO, Palanisami T (2020) Interaction of chemical contaminants with microplastics: Principles and perspectives. Science of the TotalEnvironment, 706:13597834

Frias JPGL, Otero V, Sobral P (2014) Evidence of microplastics in samples of zooplankton from Portuguese coastal waters. Mar Environ Res 95:89–95. https://doi.org/10.1016/j.marenvres.2014.01.001

Gao D, Li X, Liu H (2020) Source, occurrence, migration and potential environmental risk of microplastics in sewage sludge and during sludge amendment to soil. Sci Total Environ 742:140355. https://doi.org/10.1016/j.scitotenv.2020.140355

Graca B, Bełdowska M, Wrzesień P, Zgrundo A (2014) Styrofoam debris as a potential carrier of mercury within ecosystems. Environmental Science and Pollution Research, 21(3):2263-2271

Gross N, Muhvich J, Ching C, Gomez B, Horvath E, Nahum Y, Zaman MH (2025) Effects of microplastic concentration, composition, and size on Escherichia coli biofilm-associated antimicrobial resistance. Appl Environ Microbiol. https://doi.org/10.1128/aem.02282-24

Guo X, Wang J (2019) Sorption of antibiotics onto aged microplastics in freshwater and seawater. Marine Pollution Bulletin, 149:110511

Guo J-J, Huang X-P, Xiang L, Wang Y-Z, Li Y-W, Li H, Cai Q-Y, Mo C-H, Wong M-H (2020) Source, migration and toxicology of microplastics in soil. Environ Int 137:105263. https://doi.org/10.1016/j.envint.2019.105263

Hahladakis JN, Velis CA, Weber R, Iacovidou E, Purnell P (2018) An overview of chemical additives present in plastics: migration, release, fate and environmental impact during their use, disposal and recycling. J Hazard Mater 344:179–199. https://doi.org/10.1016/j.jhazmat.2017.10.014

Hartmann NB, Hüffer T, Thompson RC, Hassellöv M, Verschoor A, Daugaard AE, Rist S, Karlsson T, Brennholt N, Cole M, Herrling MP, Hess MC, Ivleva NP, Lusher AL, Wagner M (2019) Are we speaking the same language? Recommendations for a definition and categorization framework for plastic debris. Environ Sci Technol 53(3):1039–1047. https://doi.org/10.1021/acs.est.8b05297

He P, Chen L, Shao L, Zhang H, Lü F (2019) Municipal solid waste (MSW) landfill: a source of microplastics? -Evidence of microplastics in landfill leachate. Water Res 159:38–45. https://doi.org/10.1016/j.watres.2019.04.060

Hidalgo-Ruz V, Gutow L, Thompson RC, Thiel M (2012) Microplastics in the marine environment: a review of the methods used for identification and quantification. Environ Sci Technol 46(6):3060–3075. https://doi.org/10.1021/es2031505

Hoang TC, Felix-Kim M (2020) Microplastic consumption and excretion by fathead minnows (Pimephales promelas): influence of particles size and body shape of fish. Sci Total Environ 704:135433. https://doi.org/10.1016/j.scitotenv.2019.135433

Hoellein TJ, Rochman CM (2021) The "plastic cycle": a watershed-scale model of plastic pools and fluxes. Front Ecol Environ 19(3):176–183. https://doi.org/10.1002/fee.2294

Hou L, Kumar D, Yoo CG, Gitsov I, Majumder EL-W (2021) Conversion and removal strategies for microplastics in wastewater treatment plants and landfills. Chem Eng J 406:126715. https://doi.org/10.1016/j.cej.2020.126715

Huang Y, Liu Q, Jia W, Yan C, Wang J (2020) Agricultural plastic mulching as a source of microplastics in the terrestrial environment. Environ Pollut 260:114096. https://doi.org/10.1016/j.envpol.2020.114096

Jambeck JR, Geyer R, Wilcox C, Siegler TR, Perryman M, Andrady A, Narayan R, Law KL (2015) Plastic waste inputs from land into the ocean. Science 347(6223):768–771. https://doi.org/10.1126/science.1260352

Järlskog I, Strömvall A-M, Magnusson K, Galfi H, Björklund K, Polukarova M, Garção R, Markiewicz A, Aronsson M, Gustafsson M, Norin M, Blom L, Andersson-Sköld Y (2021) Traffic-related microplastic particles, metals, and organic pollutants in an urban area under reconstruction. Sci Total Environ 774:145503. https://doi.org/10.1016/j.scitotenv.2021.145503

Jiang J, Zhou H, Zhang T, Yao C, Du D, Zhao L, Cai W, Che L, Cao Z, Wu XE (2022) Machine learning to predict dynamic changes of pathogenic Vibrio spp. abundance on microplastics in marine environment. Environ Pollut 305:119257. https://doi.org/10.1016/j.envpol.2022.119257

Kalčíková G, Alič B, Skalar T, Bundschuh M, Gotvajn AŽ (2017) Wastewater treatment plant effluents as source of cosmetic polyethylene microbeads to freshwater. Chemosphere 188:25–31. https://doi.org/10.1016/j.chemosphere.2017.08.131

Khalid N, Aqeel M, Noman A, Fatima Rizvi Z (2023) Impact of plastic mulching as a major source of microplastics in agroecosystems. J Hazard Mater 445:130455. https://doi.org/10.1016/j.jhazmat.2022.130455

Kole PJ, Löhr AJ, Van Belleghem F, Ragas A (2017) Wear and tear of tyres: a stealthy source of microplastics in the environment. Int J Environ Res Public Health 14(10):1265. https://doi.org/10.3390/ijerph14101265

Lei K, Qiao F, Liu Q, Wei Z, Qi H, Cui S, Yue X, Deng Y, An L (2017) Microplastics releasing from personal care and cosmetic products in China. Mar Pollut Bull 123(1–2):122–126. https://doi.org/10.1016/j.marpolbul.2017.09.016

Li J, Song Y, Cai Y (2020) Focus topics on microplastics in soil: analytical methods, occurrence, transport, and ecological risks. Environ Pollut 257:113570. https://doi.org/10.1016/j.envpol.2019.113570

Liu C, Du Y, Yin H, Fan C, Chen K, Zhong J, Gu X (2019) Exchanges of nitrogen and phosphorus across the sediment-water interface influenced by the external suspended particulate matter andthe residual matter after dredging. Environmental Pollution, 246:207-21

Mahon AM, O'Connell B, Healy MG, O'Connor I, Officer R, Nash R, Morrison L (2017) Microplastics in sewage sludge: effects of treatment. Environ Sci Technol 51(2):810–818. https://doi.org/10.1021/acs.est.6b04048

Manikanda Bharath K, Natesan U, Vaikunth R, Praveen Kumar R, Ruthra R, Srinivasalu S (2021) Spatial distribution of microplastic concentration around landfill sites and its potential risk on groundwater. Chemosphere 277:130263. https://doi.org/10.1016/j.chemosphere.2021.130263

Mason SA, Garneau D, Sutton R, Chu Y, Ehmann K, Barnes J, Fink P, Papazissimos D, Rogers DL (2016) Microplastic pollution is widely detected in US municipal wastewater treatment plant effluent. Environ Pollut 218:1045–1054. https://doi.org/10.1016/j.envpol.2016.08.056

McCormick AR, Hoellein TJ, London MG, Hittie J, Scott JW, Kelly JJ (2016) Microplastic in surface waters of urban rivers: concentration, sources, and associated bacterial assemblages. Ecosphere 7(11). https://doi.org/10.1002/ecs2.1556

Menéndez-Pedriza A, Jaumot J (2020) Interaction of environmental pollutants with microplastics: a critical review of sorption factors, bioaccumulation and ecotoxicological effects. Toxics 8(2):40. https://doi.org/10.3390/toxics8020040

Munno K, Helm PA, Rochman C, George T, Jackson DA (2022) Microplastic contamination in Great Lakes fish. Conserv Biol 36(1). https://doi.org/10.1111/cobi.13794

Napper IE, Bakir A, Rowland SJ, Thompson RC (2015) Characterisation, quantity and sorptive properties of microplastics extracted from cosmetics. Mar Pollut Bull 99(1–2):178–185. https://doi.org/10.1016/j.marpolbul.2015.07.029

Nava V, Leoni B (2021) A critical review of interactions between microplastics, microalgae and aquaticecosystem function. Water research, 188:116476

Qin R, Su C, Liu W, Tang L, Li X, Deng X, Chen Z (2020) Effects of exposure to polyether sulfone microplastic on the nitrifying process and microbial community structure in aerobic granular sludge. Bioresourcetechnology, 302:122827

Rillig MC (2018) Microplastic disguising as soil carbon storage

Saha M, Naik A, Desai A, Nanajkar M, Rathore C, Kumar M, Gupta P (2021) Microplastics in seafood as an emerging threat to marine environment: a case study in Goa, west coast of India. Chemosphere, 270:129359

Sait STL, Sørensen L, Kubowicz S, Vike-Jonas K, Gonzalez SV, Asimakopoulos AG, Booth AM (2021) Microplastic fibres from synthetic textiles: environmental degradation and additive chemical content. Environ Pollut 268:115745. https://doi.org/10.1016/j.envpol.2020.115745

Seeley ME, Song B, Passie R, Hale RC (2020). Microplastics affect sedimentary microbial communities and nitrogen cycling. Nature communications, 11(1):2372

Shen M, Song B, Zhou C, Almatrafi E, Hu T, Zeng G, Zhang Y (2022) Recent advances in impacts of microplastics on nitrogen cycling in the environment: a review. Science of the Total Environment, 815:152740

Sohm JA, Webb EA, Capone DG (2011) Emerging patterns of marine nitrogen fixation. Nature Reviews Microbiology, 9(7):499-508

Song J, Jongmans-Hochschulz E, Mauder N, Imirzalioglu C, Wichels A, Gerdts G (2020) The travelling particles: investigating microplastics as possible transport vectors for multidrug resistant E. coli in the Weser estuary (Germany). Sci Total Environ 720:137603. https://doi.org/10.1016/j.scitotenv.2020.137603

Song K, Ding R, Sun C, Yao L, Zhang W (2021) Microparticles and microplastics released from daily use of plastic feeding and water bottles and plastic injectors: potential risks to infants and children in China. Environ Sci Pollut Res 28(42):59813–59820. https://doi.org/10.1007/s11356-021-14939-7

Sun Q, Ren S-Y, Ni H-G (2020) Incidence of microplastics in personal care products: an appreciable part of plastic pollution. Sci Total Environ 742:140218. https://doi.org/10.1016/j.scitotenv.2020.140218

Taipale SJ, Peltomaa E, Kukkonen JVK, Kainz MJ, Kautonen P, Tiirola M (2019) Tracing the fate of microplastic carbon in the aquatic food web by compound-specific isotope analysis. Scientific reports, 9(1):19894

Vaksmaa A, Hernando-Morales V, Zeghal E, Niemann H (2021) Microbial degradation of marine plastics: current state and future prospects. In: Biotechnology for sustainable environment. Springer, Singapore, pp 111–154. https://doi.org/10.1007/978-981-16-1955-7_5

Vithanage M, Ramanayaka S, Hasinthara S, Navaratne A (2021) Compost as a carrier for microplastics and plastic-bound toxic metals into agroecosystems. Curr Opin Environ Sci Health 24:100297. https://doi.org/10.1016/j.coesh.2021.100297

Wang C, Liu Y, Chen W, Zhu B, Qu S, Xu M (2021a) Critical review of global plastics stock and flow data. J Ind Ecol 25(5):1300–1317. https://doi.org/10.1111/jiec.13125

Wang J, Peng C, Li H, Zhang P, Liu X (2021b) The impact of microplastic-microbe interactions on animal health and biogeochemical cycles: a mini-review. Sci Total Environ 773:145697. https://doi.org/10.1016/j.scitotenv.2021.145697

Wang X, Xing Y, Lv M, Zhang T, Ya H, Jiang B (2022) Recent advances on the effects of microplastics on elements cycling in the environment. Sci Total Environ 849:157884. https://doi.org/10.1016/j.scitotenv.2022.157884

Watkins L, Sullivan PJ, Walter MT (2019) A case study investigating temporal factors that influence microplastic concentration in streams under different treatment regimes. Environ Sci Pollut Res 26(21):21797–21807. https://doi.org/10.1007/s11356-019-04663-8

Weiss L, Ludwig W, Heussner S, Canals M, Ghiglione J-F, Estournel C, Constant M, Kerhervé P (2021) The missing ocean plastic sink: gone with the rivers. Science 373(6550):107–111. https://doi.org/10.1126/science.abe0290

Weithmann N, Möller JN, Löder MGJ, Piehl S, Laforsch C, Freitag R (2018) Organic fertilizer as a vehicle for the entry of microplastic into the environment. Sci Adv 4(4). https://doi.org/10.1126/sciadv.aap8060

Welden NAC, Cowie PR (2016) Environment and gut morphology influence microplastic retention in langoustine, Nephrops norvegicus. Environ Pollut 214:859–865. https://doi.org/10.1016/j.envpol.2016.03.067

Widdison PE, Burt TP (2008) Nitrogen cycle. Global Ecology, 2526-2533

Wieczorek AM, Croot PL, Lombard F, Sheahan JN, Doyle TK (2019) Microplastic ingestion by gelatinous zooplankton may lower efficiency of the biological pump. Environmental science and technology, 53(9):5387-5395

Windsor FM, Tilley RM, Tyler CR, Ormerod SJ (2019) Microplastic ingestion by riverine macroinvertebrates. Sci Total Environ 646:68–74. https://doi.org/10.1016/j.scitotenv.2018.07.271

Yang T, Luo J, Nowack B (2021) Characterization of nanoplastics, fibrils, and microplastics released during washing and abrasion of polyester textiles. Environ Sci Technol 55(23):15873–15881. https://doi.org/10.1021/acs.est.1c04826

Yi M, Zhou S, Zhang L, Ding S (2021) The effects of three different microplastics on enzyme activitiesand microbial communities in soil. Water Environment Research, 93(1):24-32

Yin M, Yan B, Wang H, Wu Y, Wang X, Wang J, Fu, C (2023) Effects of microplastics on nitrogen and phosphorus cycles and microbial communities in sediments. Environmental Pollution, 318:120852

Yu H, Zhang Z, Zhang Y, Song Q, Fan P, Xi B, Tan W (2021) Effects of microplastics on soil organic carbon and greenhouse gas emissions in the context of straw incorporation: A comparison with different types ofsoil. Environmental Pollution, 288:117733

Yuan Y, Qin Y, Wang M, Xu W, Chen Y, Zheng L, Chen W, Luo T (2022) Microplastics from agricultural plastic mulch films: a mini-review of their impacts on the animal reproductive system. Ecotoxicol Environ Saf 244:114030. https://doi.org/10.1016/j.ecoenv.2022.114030

Zbyszewski M, Corcoran PL, Hockin A (2014) Comparison of the distribution and degradation of plastic debris along shorelines of the Great Lakes, North America. J Great Lakes Res 40(2):288–299. https://doi.org/10.1016/j.jglr.2014.02.012

Zehr JP, Capone DG (2020) Changing perspectives in marine nitrogen fixation. Science, 368(6492):eaay9514

Zhang L, Xie Y, Liu J, Zhong S, Qian Y, Gao P (2020) An overlooked entry pathway of microplastics into agricultural soils from application of sludge-based fertilizers. Environ Sci Technol 54(7):4248–4255. https://doi.org/10.1021/acs.est.9b07905

Zhang Z, Peng W, Duan C, Zhu X, Wu H, Zhang X, Fang L (2022a) Microplastics pollution from different plastic mulching years accentuate soil microbial nutrient limitations. Gondwana Res 108:91–101. https://doi.org/10.1016/j.gr.2021.07.028

Zhang J, Wang X, Xue W, Xu L, Ding W, Zhao M, Liu S, Zou G, Chen Y (2022b) Microplastics pollution in soil increases dramatically with long-term application of organic composts in a wheat–maize rotation. J Clean Prod 356:131889. https://doi.org/10.1016/j.jclepro.2022.131889

Zhou L, Zhou X, Zhang B, Lu M, Luo Y, Liu L, Li B (2014) Different responses of soil respiration and its components to nitrogen addition among biomes: a meta-analysis. Global change biology, 20(7):2332-2343

Microplastic Occurrence in Sediments and Associated Ecological Risk: A Study of Saurashtra Coastline, Gujarat, India

10

Irédon Adjama, Pavan Hirenbhai Pandya, and Hemen Dave

Abstract

The persistence of microplastics (MPs) in all components of the environment has become a significant global concern. MP contamination has led to the toxicity and bioaccumulation of MPs in living organisms, including humans, and has been proven in various scientific research works. To mitigate the spread of MPs in the environment, governments and policymakers need to be aware of these issues to implement appropriate measures that contribute to environmental and public health protection. Intending to shed light on microplastic pollution in the environment, the current research study focused on the estimation of MPs and associated types of polymers in sediments along the Saurashtra (Gujarat) coastline and assessed the ecological risk posed by these microplastics. The findings revealed an average of 6700 MPs/kg in the sediments, including 2500 fibers/kg, 3000 fragments/kg, 1000 films/kg, and 200 beads/kg. MPs are found in various colors, such as black, white, blue, red, and brown, with fiber shapes and smaller sizes (0.3–1.0 mm) as major ones. Polyethylene (PE) emerged as the predominant polymer. Moreover, the study identified a significant ecological risk due to a considerable abundance of MPs and the diversity in MP polymers. This study provides essential baseline data for pollution monitoring and for the development of strategies to mitigate MP impacts on marine ecosystems along the Saurashtra coastline.

I. Adjama
School of Doctoral Studies & Research (SDSR), National Forensic Sciences University, Gandhinagar, Gujarat, India
e-mail: iredon.phdph21@nfsu.ac.in

P. H. Pandya · H. Dave (✉)
School of Pharmacy, National Forensic Sciences University, Gandhinagar, Gujarat, India
e-mail: pavanhp78@gmail.com; hemen.dave@nfsu.ac.in

K. Bala et al. (eds.), *Occurrence, Detection, and Fate of Microplastics in Freshwater Ecosystems*, https://doi.org/10.1007/978-981-95-0225-7_10

Keywords

Microplastic pollution · Coastal sediments · Ecological risk · Marine biodiversity

10.1 Introduction

Plastic materials are well-known and used in various human activities because of their strength, moisture resistance, adaptability, durability, and cost-effectiveness (Nayanathara Thathsarani Pilapitiya and Ratnayake 2024; Plastics Europe 2024; Thompson et al. 2009). Global plastic production was approximately 390 million metric tons in 2021, driven by various industries that capitalize on plastics (Plastics Europe 2024). However, plastics' durability and imperishable nature also make them persistent pollutants, leading to their extensive spread into ecosystems, especially in marine and coastal ecosystems (Castillo et al. 2016). Microplastics (MPs) are particles of plastics (size <5 mm) and have become a worldwide environmental concern. MPs come from two main sources, which are classified as primary and secondary. The primary sources of MPs are described as those intentionally manufactured in the industry for use in cosmetics, baby care products, and other applications. The secondary sources of MPs originate from the fragmentation of plastic materials (e.g., packaging material and fishing gear) during use or after being discarded into the environment as waste. Over time, these larger plastics degrade into smaller particles, contributing to MP pollution (Ghanadi et al. 2024; Lippiatt et al. 2013; Okoffo et al. 2024).

MPs reach marine ecosystems through several routes, such as runoff from cities and industries, leachate runoff from waste mismanagement facilities, river flow, and air deposits, with about 80% coming from wastewater and farming, while the other 20% comes from sea activities like fishing and shipping (Adjama et al. 2024; Adjama and Dave 2025; Ghanadi et al. 2024; Jambeck et al. 2015; Yang et al. 2024). Because of their small size, low density, and persistent nature, MPs are transported over large distances in aquatic environments, eventually settling in coastal and marine sediments, where they can stay for long periods (Gupta et al. 2024; Xu et al. 2022). These sediments act as long-term reservoirs for MPs, serving as crucial indicators of historical and ongoing plastic pollution (Da Costa Filho et al. 2021; Ghanadi et al. 2024; Lloret et al. 2021). In environmental conditions, MPs are subjected to various reactions, such as enzymatic degradation by microbial activity, photochemical degradation by sunlight exposure, and oxidative or hydrolytic processes. These processes lead to the degradation or breakdown of MPs and result in the release of embedded chemical additives from MPs (Joo et al. 2021). Apart from that, due to their small size, MPs have a large surface area, enhancing their interaction with persistent organic pollutants, heavy metals, and other pollutants on their surface. Therefore, MPs can act as carriers or vectors for transporting these pollutants within the ecosystems (Boiteux et al. 2017; Sabri et al. 2025).

Accumulation of MPs in living organisms through ingestion and other portals of entry results in toxicity due to the physical-chemical properties of MPs, release of harmful chemicals and environmental contaminants absorbed on the MP surface and can disrupt vital biological activities (Adeleye et al. 2024; Fu et al. 2021; Li et al. 2020a; Okoye et al. 2022). In the context of MP contamination in the aquatic environment, several studies have reported the presence of MPs in aquatic organisms and, indirectly, the exposure of human beings to MP pollution via seafood contaminated with MPs (Akoueson et al. 2020; Danopoulos et al. 2020; Hassoun et al. 2023).

Severe plastic pollution is reported along the Indian coastline (Rafey and Siddiqui 2023). As per the report of the United Nations Environment Programme (UNEP), every year about 8–10 million tons of plastic waste are dumped into the oceans worldwide. Due to the large-scale dumping of plastic waste in oceans, the Indian sea bed is found to have a contamination of 4 billion microplastics per square kilometer, the majority of which are in fiber shape (Eriksen et al. 2014; Rabari et al. 2022; Woodall et al. 2014). Similarly, in various studies conducted worldwide, it is reported that an average of 5.25 trillion MP particles are found on the oceans' surfaces (Woodall et al. 2014). The Saurashtra coastline in Gujarat is a particular area of concern due to its heavy industrialization, tourism, fishing activities, and strategic location with 49 ports. The biggest ship-breaking yard is located in the Gujarat district. The Gujarat coastline accounts for one-fifth of India's total coastal length and is home to a thriving fishing industry and major industrial hubs, including the chemical and pharmaceutical sectors. Industrial discharges, along with waste from shipping and urban runoff, contribute significantly to plastic pollution in this region (Rabari et al. 2022).

While research on MPs in the areas of Gujarat's coast has focused primarily on sandy beaches in the Kutch Gulf (Rabari et al. 2022), only limited research has been conducted on the estimation of MPs along the Saurashtra coast and the Gulf of Khambhat, especially in muddy shorelines where industrial effluents from rivers such as the Sabarmati, Narmada, Tapi, and Mahi enter the marine ecosystem (Cole et al. 2011; Rabari et al. 2022; Senapati and Gupta 2014). These estuarine and coastal areas are susceptible to MP accumulation by the influence of hydrodynamic forces that can concentrate plastic particles and make these areas potential hotspots for pollution (Amelia et al. 2021; Rafey and Siddiqui 2023). With the blend of ecological risk due to the probable presence of MPs and the industrial significance of the area, the Saurashtra coastline is an essential site for investigating the broader impacts of MPs on both marine ecosystems and the human populations that depend on them (Rabari et al. 2022). The objective of the current research work was to provide a detailed overview of MP pollution in sediments across the Saurashtra coastline. Representative sediment samples from 12 sampling locations, i.e., Pipavav, Diu, Somnath, Veraval, Adri, Chorwad, Shil, Madhavpur, Porbandar, Dwarka, and Shivrajpu (from east to west), were selected, and the MPs extraction, quantification, and identification were carried out. Also, the ecological risk associated with MP contamination in the coastal sediments was carried out based on the method provided in the previous works (Bilal et al. 2023; Castillo et al. 2024;

Hakanson 1980; Lithner et al. 2011; Ranjbar Jafarabadi et al. 2017). The results of the present research are certainly a valuable addition to the knowledge about the spread of MPs in the Saurashtra coastal region and provide insight into the implications of improper waste management practices, especially plastic waste.

10.2 Materials and Methods

10.2.1 Materials

In this study, high-quality materials and contamination-free laboratory equipment were used to ensure the accuracy and reliability of the results. For collecting the sediment samples, a cylindrical stainless-steel sampler (10 cm depth, 20 cm diameter) was used, and sediments were collected from the mean high-tide line. Sodium chloride (AR grade, Central Drug House Pvt. Ltd.) was used for density separation of MPs, followed by filtration using glass fiber paper (pore size = 2μm and diameter = 47 mm, Macherey-Nagel) with a vacuum filtration assembly. For precise weight measurement, a SHIMADZU electronic weight balance was used throughout the experimental work, a META-LAB hot air oven for drying the samples, and the observation of MPs was carried out using a Stemi 2000-C stereomicroscope. For polymer identification, attenuated total reflectance Fourier transform infrared spectroscopy (ATR-FTIR) (Jasco-4700) was utilized. Milli-Q water was used throughout the experimental work to prevent contamination, ensuring quality control at every stage.

10.2.2 Description of Study and Sampling Process

Gujarat, located on India's western coast, has a large and environmentally diverse 1600-kilometer coastline divided into three separate regions: the Gulf of Khambhat, the Gulf of Kutch, and the Saurashtra coast. The Saurashtra coast, located in the center of Gujarat's coastline along the Arabian Sea, is notable for its different habitats, which include sandy and rocky beaches, estuaries, mudflats, mangroves, and a thriving biodiversity that supports numerous marine and land species. The Saurashtra coastline is a hub for activities related to economic importance, such as fishing, tourism, and industrial activities, and also has places that are important as religious pilgrimages (Mukhopadhyay and Karisiddaiah 2014; Rabari et al. 2023). On the Saurashtra coastline, moving from east to west, 12 locations—such as Pipavav, Diu, Somnath, Veraval, Adri, Chorwad, Mangrol, Shil, Madhavpur, Porbandar, Dwarka, and Shivrajpur—were selected for sampling of coastal sediments (Karthik et al. 2018; Mukhopadhyay and Karisiddaiah 2014; Robin et al. 2020; Senapati and Gupta 2014). The selected sampling sites provide representative sediment types and human activity levels, from highly trafficked ports and tourist beaches to quieter fishing communities. Field visits were conducted at each location to collect sediment samples, measure microplastic levels, identify pollution sources, and assess

Fig. 10.1 Sampling locations along the Saurashtra coastline, mapped using Google Earth based on the imported geographic coordinates of each site."

the risks these plastics pose to marine and coastal ecosystems. GPS coordinates were noted for each sampling site (Fig. 10.1 and Table 10.1).

To examine MPs contamination across these diverse coastal environments, six coastal sediment samples were collected from each sampling area at intervals of 50 meters. The sampling focused on coastal sediments accumulated along the mean high-tide line, where sediment buildup is higher due to intense wave action (Swash) (Alvarez-Zeferino et al. 2020). The sample collection took place over 6 months, starting in November 2023. A cylindrical stainless-steel sampler (10 cm depth, 20 cm diameter) was used to collect the samples. The sampler was pushed into the sediment, after which a stainless-steel case was inserted beneath the sampler to ensure uniformity in all samples. The sediments were sealed in labeled aluminum foil and transported to the laboratory for MP analysis.

10.2.3 Experimental Process

10.2.3.1 Quality Control, Sample Pretreatment, and Storage

Only equipment and tools made of glass and steel were used to avoid external MP contamination during sample pretreatment, processing, or analysis. Throughout the analysis of MPs, Milli-Q water was used for dilution purposes. Additionally, the glass fiber filter used for the extraction of MPs was transferred to glass Petri dishes

Table 10.1 On-site-collected geographic coordinates and major human activities at sampling sites along the Saurashtra coastline

Sr. No.	Location	Latitude	Longitude	Major human activities
A	Pipavav	20°56′20.57″N	71°31′56.13″E	Port activities, industrial shipping, trade
B	Diu	20°42′39.04″N	70°56′6.12″E	Tourism, fishing, hospitality
C	Somnath	20°53′3.88″N	70°24′20.66″E	Religious tourism, temple visits, local handicrafts
D	Veraval	20°54′36.05″N	70°21′8.13″E	Fishing, seafood processing, port activities
E	Adri	20°57′34.87″N	70°16′41.38″E	Small-scale fishing, agriculture
F	Chorwad	20°59′59.18″N	70°13′36.58″E	Tourism, fishing, hospitality
G	Mangrol	21°6′12.78″N	70°6′24.73″E	Fishing, agriculture
H	Shil	21°10′32.65″N	70°1′51.33″E	Small-scale fishing, agriculture
I	Madhavpur	21°15′20.16″N	69°57′21.76″E	Tourism, fishing, agriculture
J	Porbandar	21°37′38.73″N	69°36′46.34″E	Port activities, trade, fishing, tourism
K	Dwarka	22°14′14.25″N	68°57′44.79″E	Religious tourism, fishing, handicrafts
L	Shivrajpur	22°20′5.65″N	68°57′16.47″E	Beach tourism, conservation activities

and carefully observed with the stereomicroscope to ensure cleanliness and promote accurate counting and identification of MPs.

The coastal sediment samples were weighed first and then dried in the oven set at 60 °C (hot air) to remove moisture and make the sieving process easier. The samples remained in the oven until their weight stabilized, indicating that all moisture had been removed. This drying process was crucial, as humidity can interfere with the proper sieving. Once thoroughly dried, the samples were passed through a 5 mm mesh filter, collected, transferred to a beaker, and stored for further analysis.

10.2.3.2 Extraction of Microplastics

The extraction of MPs from sediment samples started after they were dried in an oven set at 60 °C for a duration of 24 h. The dried sediments of each sampling location were weighed in equal quantities and combined; after being homogenized, 100 g of the representative sediment sample was derived from the combined sample using the quartering and coning method. The derived 100 g sediment was put in a beaker and mixed with 400 mL of a saturated solution of sodium chloride (NaCl) (Abidli et al. 2018; Allami et al. 2023; Alves and Figueiredo 2019; Dobaradaran et al. 2018; Martins and Sobral 2011; Rabari et al. 2022). The sample solution was gently stirred with a magnetic stirrer. After 2 h of stirring, the sample solution was kept aside and left undisturbed to allow the density separation of MPs. After 20 h, the supernatant was clear enough to be separated from the sediment. The supernatant was transferred to another beaker and filtered through a glass fiber filter (characteristics: diameter = 47 mm, pore size = 2μm) using vacuum filtration to capture the MPs. The MPs on top of the filter were dried at 60 °C for 1 h. Extraction was performed three times for each representative sediment sample to obtain accurate results.

10.2.3.3 Counting and Physical and Chemical Characterization of Microplastics

MPs counting, followed by physical and chemical characterization, was carried out using a stereomicroscope, ImageJ software, and ATR-FTIR spectroscopy. Dried filters were observed using the stereomicroscope to count the number of MPs present. Along with this counting, for the physical characterization of MPs, the images of each MP were recorded for shape, size, and color identification using ImageJ software. From the surface of the filter paper, each MP was carefully picked and transferred for chemical analysis using ATR-FTIR. The spectra were recorded (64 scans, resolution of 4 cm^{-1}) from 500 to 4000 cm^{-1}. The noise in the spectra obtained was reduced using software. The matching was performed by comparing and correlating the obtained spectra with the available reference from the literature to identify functional groups and polymer types (Jung et al. 2018; Veerasingam et al. 2021).

10.2.4 Calculation Procedure of Ecological Risk Indices

The ecological risk of MPs along the coastline was assessed by following the available procedure in the literature (Castillo et al. 2024; Ranjbar Jafarabadi et al. 2017). The assessment was based on two key indices, the Pollution Load Index (PLI) and Potential Ecological Risk Index (PERI) (Castillo et al. 2024; Haegerbaeumer et al. 2019; Li et al. 2021; Tomlinson et al. 1980).

10.2.4.1 Risk Assessment Standard

The standard assessment procedure of ecological risk associated with MPs was used based on Håkanson's method (1980). This method takes into consideration the individual and cumulative impact of detected MPs based on polymer type. The assessment function is based on three different parameters: the contamination factor (C^f_i) and the reference concentration (C^n_i), which serve as the baseline, and the toxicity coefficient (T^r_i), which is accountable for each MP polymer (Lithner et al. 2011; Rabari et al. 2022).

10.2.4.2 Calculation Procedure for the Microplastic Pollution Load Index

MPs-PLI is calculated to quantify MPs' pollution load level for each sampling location. Therefore, the contamination factor C^f_i for each polymer type is calculated as the ratio of the counted number of MPs in the sample sediment C_i to the reference contamination C^n_i, which is the background level of MPs in Gujarat's coastline sediment, equal to 1.4 MPs/kg as reported in work of Rabari et al. (2022). The cumulative PLI is calculated by taking the nth root of the contamination factors C^f_i for each MP polymer. The interpretation of PLI values was done based on the guidelines given in the works published by Hakanson (1980), Li et al. (2021), Lithner et al. (2011), and Bilal et al. (2023).

10.2.4.3 Calculation Procedure of Potential Ecological Risk Index (PERI)

The PERI is an ecological risk index used to assess the combined environmental risk caused by each MP polymer, which is defined as E^T_i (Bilal et al. 2023; Gupta et al. 2024; Li et al. 2021; Lithner et al. 2011; Ranjbar Jafarabadi et al. 2017). E^T_i represents the ecological risk factor associated with each MP polymer and is calculated as the product of the toxicity coefficient T^r_i and the contamination factor C^f_i. To determine the total ecological risk at each sampling location, the individual E^T_i values calculated for MPs identified are summed, resulting in an overall PERI score. The criteria used for the interpretation of the PERI values are defined in the works of Hakanson (1980), Ranjbar Jafarabadi et al. (2017), and Castillo et al. (2024).

10.3 Results and Discussion

10.3.1 Microplastics Quantification and Abundance Comparison to the Previous Research Works

The abundance of MPs along the Saurashtra coastline shows considerable variation across sampling locations, with significantly high levels of MPs in several locations (refer to Fig. 10.2a). The distribution of MP abundance along the sediment coastline shows a clear relationship with the intensity and type of human activities at each location. The Chorwad area, with the highest concentration of MPs at 963.33 ± 15.28 particles/kg and other high-abundance areas like Somnath and Veraval with 913.33 ± 90.18 and 866.67 ± 20.82 particles/kg, respectively, are hubs of tourism, fishing, industrial, and port activities. These activities contribute significantly to MP pollution, often involving heavy foot traffic, coastal trade, and fishing waste. Moderate-abundance locations such as Madhavpur with 840.00 ± 45.83 MPs/kg, Diu with 820.00 ± 55.68 MPs/kg, and Porbandar with 776.67 ± 20.82 MPs/kg are areas with varied activities, including tourism, fishing, and, in the case of Porbandar, substantial port activities. These sites likely experience moderate pollution inputs from local tourism and maritime activities, with ocean currents possibly aiding microplastic transport. In contrast, low-abundance areas like Shivrajpur, with 173.33 ± 15.28 MPs/kg, Shil, with 423.33 ± 15.28 MPs/kg, and Mangrol, with 466.67 ± 58.59 MPs/kg, are less exposed to sources of MPs. Shivrajpur, for example, benefits from conservation activities that likely limit pollution, while Shil and Mangrol, primarily agricultural and small-scale fishing areas, have limited industrial or tourism impacts.

The abundance of MPs in sediments of the Saurashtra coastline reflects a considerable distribution influenced by human activities. In comparison, the current finding aligns with other local and international investigations of MP contamination along the coastlines (refer to Table 10.2). At the local level, in India, MP abundance

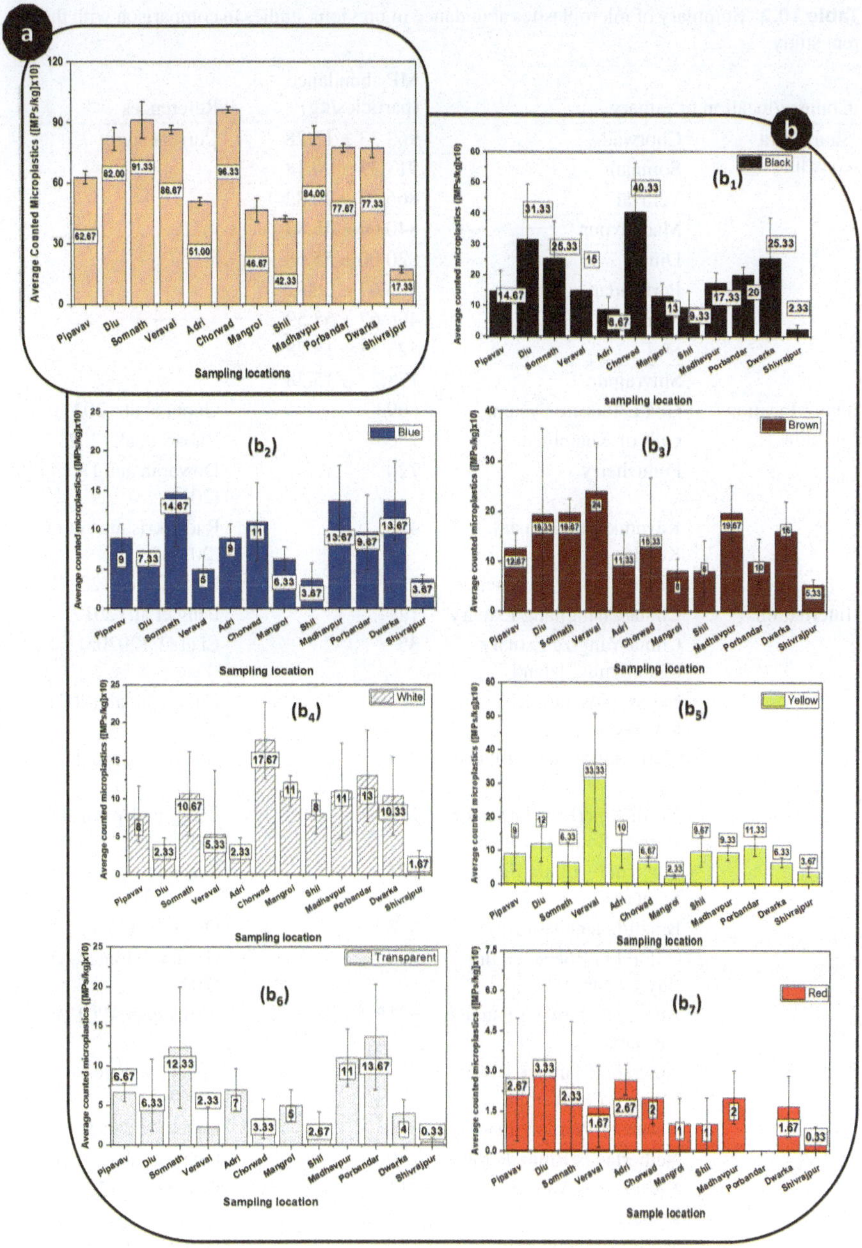

Fig. 10.2 (**a**) Number distribution of microplastics across various sampling locations; (**b**) color distribution of microplastics: (*b1*) black MPs, (*b2*) blue MPs, (*b3*) brown MPs, (*b4*) white MPs, (*b5*) yellow MPs, (*b6*) transparent MPs, and (*b7*) red MPs

Table 10.2 Summary of microplastics abundance in previous studies in comparison with the current study

Country/location or estuary		MP abundance (particles/kg)	References
Saurashtra coastline	Chorwad	963.33 ± 15.28	Current study
	Somnath	913.33 ± 90.18	
	Veraval	866.67 ± 20.82	
	Madhavpur	840.00 ± 45.83	
	Diu	820.00 ± 55.68	
	Porbandar	776.67 ± 20.82	
	Mangrol	466.67 ± 58.59	
	Shil	423.33 ± 15.28	
	Shivrajpur	173.33 ± 15.28	
Other locations in India	Goa Estuarine System	6093	Gupta et al. (2021)
	Gulf of Khambhat	1565	Rabari et al. (2023)
	Puducherry	720.3	Dowarah and Devipriya (2019)
	Kayamkulam Estuary, Kerala	433	Radhakrishnan et al. (2021)
	Gujarat Sandy Beaches	13.7	Rabari et al. (2022)
International	China/Changjiang Estuary	121	Peng et al. (2017)
	China/Yangtze Estuary (Chongming Island)	35	Li et al. (2020b)
	Malaysia/Baram River Estuary	694.4	Choong et al. (2021)
	Thailand/Phuket Province	234	Jiwarungrueangkul et al. (2021)
	North America/Charleston Harbor, South Carolina	2189	Leads and Weinstein (2019)
	Mexico/Tecolutla Estuary, Veracruz	121	Sánchez-Hernández et al. (2021)
	Brazil/Guanabara Bay	528	Olivatto et al. (2019)
	Colombia/Buenaventura Bay Estuary	277.3	Galindo Montero et al. (2023)
	Australia/Derwent Estuary, Tasmania	3315	Willis et al. (2017)
	Australia/Clarence River Estuary, Port Yamba	224	Jahan and Strezov (2019)
	Germany/Warnow Estuary	73	Enders et al. (2019)
	Netherlands/Rhine Estuary	3300	Leslie et al. (2017)
	Netherlands/Westerscheldt Estuary	214	Leslie et al. (2017)
	Morocco/Sebou Estuary, Atlantic Coast	155	Haddout et al. (2022)

across various estuaries and coastal areas shows a diverse range of impacts from fishing, industrial, and tourism activities. The highest contamination along the Saurashtra coastline is seen at Chorwad, Somnath, and Veraval, with an average of 963.33 ± 15.28, 913.33 ± 90.18, and 866.67 ± 20.82 MPs/kg, respectively. These MP abundances are lower than those found in the Goa estuarine system sediments, where very high levels of contamination reached 6093 MPs/kg due to high tourism and the industrial influx (Gupta et al. 2021). Comparatively, Puducherry also records significant MP pollution at 720.3 MPs/kg, likely due to similar pressures from tourism and industrial activities (Dowarah and Devipriya 2019). On the other hand, coastal areas with lower anthropogenic activity, such as the sandy beaches of Gujarat with 13.7 MPs/kg, highlight that reduced MP contamination correlates with lesser human impact (Rabari et al. 2022). Other Indian locations like the Gulf of Khambhat, with 1565 MPs/kg, and the Kayamkulam Estuary in Kerala, with 433 MPs/kg, show varied levels, indicating that specific regional activities can substantially influence microplastic accumulation (Rabari et al. 2023; Radhakrishnan et al. 2021).

Compared to other Asian regions, MP levels generally appear higher in India (refer to Table 10.2), especially in less industrially and commercially active estuaries. In China, for example, the Changjiang Estuary has a recorded level of 121 MPs/kg (Peng et al. 2017), and the Yangtze Estuary at Chongming Island shows only 35 MPs/kg (Li et al. 2020b). These levels are lower than the Saurashtra coastline's high-abundance sites, potentially due to differences in coastal activity intensity and waste management practices. However, Malaysia's Baram River estuary records a substantial 694.4 MPs/kg, comparable to some high-concentration areas in Saurashtra, possibly due to similar fishing and port activities (Choong et al. 2021). In Thailand, Phuket province's level of 234 MPs/kg represents a lower impact from tourism than Saurashtra's tourist-dense areas (Jiwarungrueangkul et al. 2021), suggesting that Saurashtra's unique combination of tourism and industrial activities contributes to its higher levels. In North America, certain regions show higher MP levels than Saurashtra, particularly in areas with heavy maritime and industrial activities. Charleston Harbor Estuary in South Carolina, for instance, reports 2189 MPs/kg, significantly higher than any site along the Saurashtra coastline (Leads and Weinstein 2019). This is likely due to intensive shipping and industrial activities introducing considerable plastic waste into the water. However, Mexico's Tecolutla estuary in Veracruz shows a lower level of 121 MPs/kg, aligning more closely with moderate-activity locations in Saurashtra (Sánchez-Hernández et al. 2021). South American coastline sediments are moderately contaminated with MPs compared to Saurashtra's. Similarly, in the case of Guanabara Bay, Brazil (Olivatto et al. 2019) and Colombia Buenaventura Bay Estuary (Galindo Montero et al. 2023), the number of MPs detected was 528 MPs/kg and 277.3 MPs/kg, respectively.

Generally, in Australia, it's found that the number of MPs detected in coastal sediments exceeds the current case at Saurashtra. For instance, the Derwent Estuary in Tasmania is contaminated with 3315 MPs/kg (Willis et al. 2017), while the Clarence River Estuary in Port Yamba has 224 MPs/kg (Jahan and Strezov 2019). On the side of Europe, MP pollution in sediments is not uniform, comparable to the case of the coastline of Saurashtra. Low contamination levels can be found in the

Warnow Estuary in Germany, with a count of 73 MPs/kg (Enders et al. 2019). However, the Rhine Estuary in the Netherlands stands out with a high number of 3300 MPs/kg, exceeding the cases in Saurashtra (Leslie et al. 2017). Other locations, such as the Westerscheldt Estuary in the Netherlands (214 MPs/kg) and the Sebou Estuary on the Atlantic Coast of Morocco (155 MPs/kg), have comparable levels of MP contamination to those observed in Saurashtra (Haddout et al. 2022).

10.3.2 Physical Characterization

10.3.2.1 Color Identification

The assessment shows that MPs extracted from the coastal sediment of the Saurashtra coastline displayed a wide range of colors, with black and brown colored MPs being the most abundant at all the sampling locations (see Fig. 10.2b). Black MPs were particularly abundant in Chorwad (403.33 ± 161.66 MPs/kg) and Diu (313.33 ± 181.75 MPs/kg), while brown MPs were highly concentrated in sites such as Veraval (240.00 ± 96.44 MPs/kg) and Somnath (196.67 ± 28.87 MPs/kg). In comparison to the results reported in other studies, black and blue colored MPs were commonly detected in sediments (Frias et al. 2016; Karthik et al. 2018). The black MPs may have originated from human activities and tire abrasion and been transported through urban runoff, while brown MPs often originate from degraded packaging materials and fishing equipment. Blue MPs, also among the more commonly found colors in this study, were particularly abundant in Somnath (146.67 ± 68.07 MPs/kg) and Madhavpur (136.67 ± 45.09 MPs/kg). Marine species often mistake the brightly colored MPs for food and ingest them (Wright et al. 2013). In addition, yellow MPs, which were concentrated in Veraval (333.33 ± 174.74 MPs/kg), could be attributed to their origins in fishing gear or agricultural plastic waste. Less abundant colors, such as white, red, and transparent MPs, were generally found in lower quantities, such as white MPs in Mangrol (110 ± 20 MPs/kg) and red MPs in Adri (26.67 ± 5.77 MPs/kg). Transparent MPs, while least common, may represent weathered particles subject to sunlight degradation, often originating from food packaging films or single-use plastics. The variation in color among MPs is indicative of diverse pollution sources, as supported by other studies (Enyoh et al. 2021). Black and dark-colored MPs, being more prevalent, likely result from tire dust, road runoff, and industrial waste, whereas vibrant colors like blue, red, and yellow often come from consumer products and fishing nets (Okoffo et al. 2024; Peng et al. 2017). The color distribution highlights the ecological risks associated with MPs, particularly for marine organisms attracted to brightly colored MPs resembling natural food.

10.3.2.2 Shape Distribution of Microplastics

The shape analysis of MPs in coastal sediments along the Saurashtra coastline confirms a varied distribution of MP shapes, with fibers, fragments, beads, and films identified as the primary types (see Fig. 10.3c). Among these, fiber-shaped MPs emerged as the dominant shape across most sampling locations, with incredibly

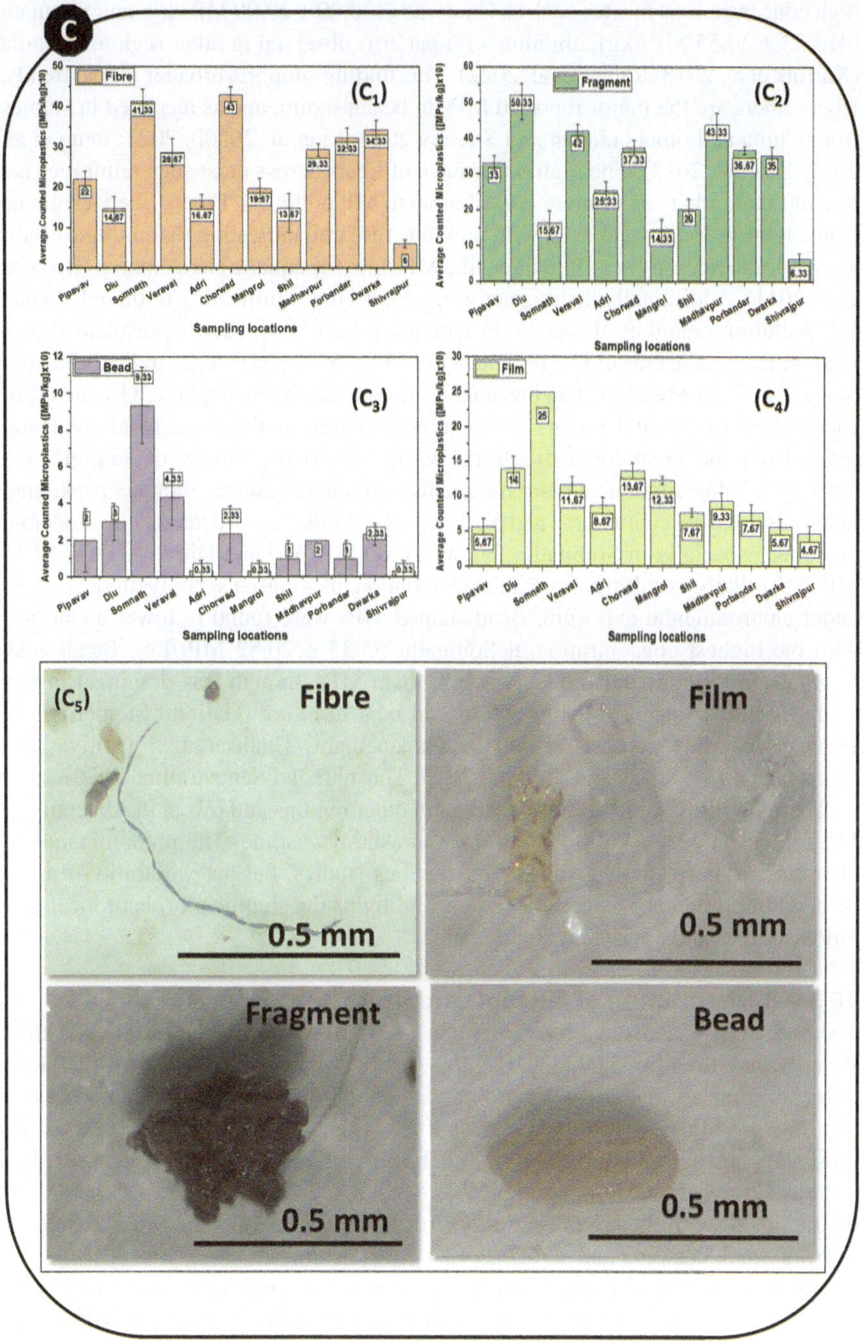

Fig. 10.3 (**c**) Shape distribution of microplastics across various sampling locations: (*c1*) fiber MPs, (*c2*) fragment MPs, (*c3*) beads MPs, (*c4*) film MPs, and (*c5*) example of MP detection

high concentrations in sites such as Chorwad (430.00 ± 20.00 MPs/kg) and Somnath (413.33 ± 30.55 MPs/kg), aligning with patterns observed in other regions of India (Karthik et al. 2018; Rabari et al. 2022). This finding supports broader global trends, where fibers are the major reported MPs in beach sediments, as recorded in studies from China and others (Jahan and Strezov 2019; Li et al. 2020b, 2021; Peng et al. 2017; Xu et al. 2022). The high occurrence of fibers across most sites reinforces the role of textile fibers as primary contributors to MP pollution. Fibers likely originate from clothing and fishing nets, with previous research indicating that a single clothing wash can release over 1900 fiber-like MPs into the aquatic environment (Browne et al. 2011). Additionally, fishing activities contribute significantly to thread-shaped MP pollution (Zhou et al. 2021). Fragments were the next most prevalent shape, with notable concentrations in Diu (503.33 ± 50.33 MPs/kg) and Madhavpur (433.33 ± 37.86 MPs/kg). The presence of fragments can be explained by the influences of environmental factors and human activities, as these fragments typically result from the breakdown of plastic waste due to the effects of sunlight and mechanical degradation. These fragments' common sources include packaging materials, plastic containers, and woven bags. Films were detected in moderate amounts, especially in Somnath (250.00 ± 0.00 MPs/kg) and Diu (140.00 ± 17.32 MPs/kg), likely resulting from the degradation of single-use polyethylene bags under environmental exposure. Bead-shaped MPs were found in lower quantities, with the highest concentration at Somnath (93.33 ± 20.82 MPs/kg). Beads may originate from industrial products, while foam MPs, though less dominant in this study, could be linked to fishing floats and boat furniture (Galindo Montero et al. 2023; Haegerbaeumer et al. 2019; Nayanathara Thathsarani Pilapitiya and Ratnayake 2024; Rafey and Siddiqui 2023; Thushari and Senevirathna 2020).

These findings reflect trends observed in other regions and reveal the diversity of MP sources at each location along the Saurashtra coastline. The predominance of fibers is consistent with findings from various studies, but the variability in fragments, films, and beads among locations highlights the significant role of local pollution sources and environmental dynamics.

10.3.2.3 Distribution of Microplastic Sizes

The size of MPs across various sampling locations along the Saurashtra coastline shows variability (4.0–5.0, 3.0–4.0, 2.0–3.0, 1.0–2.0, and 0.3–1.0 mm), which may reflect differences in sources, environmental conditions, and degradation processes (see Fig. 10.4d). Generally, MPs in the 0.3–1.0 and 1.0–2.0 mm size classes are most abundant across all locations, while larger particles are less prevalent. At sites such as Chorwad and Somnath, the 0.3–1.0 and 1.0–2.0 mm MP size classes are predominant, with mean concentrations of 306.7 ± 11.5 MPs/kg and 270.0 ± 0.0 MPs/kg for Chorwad, and 276.7 ± 40.4 MPs/kg and 263.3 ± 50.8 MPs/kg for Somnath, respectively. Pipavav and Diu exhibit a similar pattern, with small-sized MPs (0.3–1.0 mm) being the most prevalent and a notable presence of MPs in the 2.0–3.0 mm size class. Diu's 2.0–3.0 mm class shows a consistent concentration of 260 MPs/kg with no observed variation. Locations like Veraval and Porbandar also show relatively higher numbers of MPs dominated by small-sized particles. Veraval

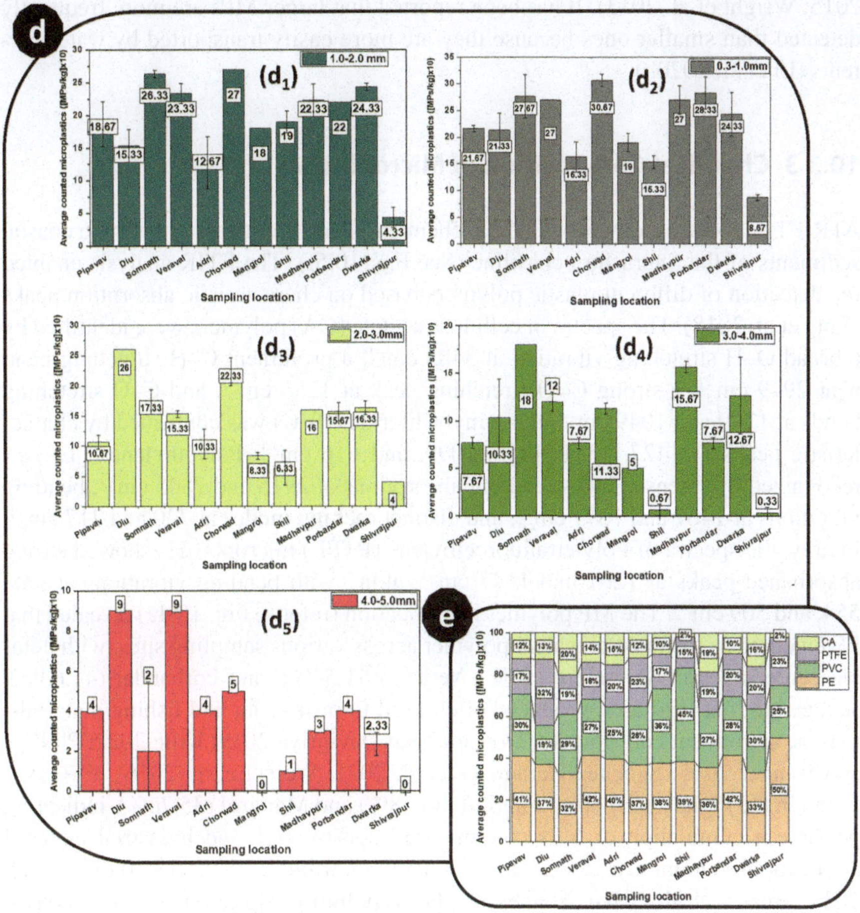

Fig. 10.4 (**d**) Size distribution of microplastics across various sampling locations: (*d1*) 0.3 mm–1.0 mm MPs, (*d2*) 1.0–2.0 mm MPs, (*d3*) 2.0–3.0 mm MPs, (*d4*) 3.0–4.0 mm MPs, and (*d5*) 4.0–5.0 mm MPs; (**e**) chemical composition distribution of MPs

reported a mean of 270 MPs/kg in the 0.3–1.0 mm range, and Porbandar showed a similar MP abundance. However, Shivrajpur reported a notably lower number of MPs across all size classes, especially in those ranging from 3.0 to 5.0 mm. At sites like Mangrol and Shil, the number of MPs significantly decreases for size classes greater than 3.0 mm, which is consistent with the results of other studies. It is reported in several studies that an abundance of smaller-sized MPs is associated with an extensive fragmentation process (Gupta et al. 2021; Karthik et al. 2018; Peng et al. 2017; Rabari et al. 2022; Senapati and Gupta 2014; Woodall et al. 2014). As smaller MPs are likely to be ingested by marine life, their prevalence is of concern, as these MPs can potentially accumulate in the food chain, posing a significant hazard to the ecosystem and human health (Davidson and Dudas 2016; Neves et al.

2015; Wright et al. 2013). It has been reported that larger MPs are more frequently detected than smaller ones because they are more easily transported by water currents (Li et al. 2020b).

10.3.3 Chemical Composition of Microplastics

ATR-FTIR spectroscopy provided the chemical identification of MPs in the coastal sediments of the Saurashtra coastline (See Fig. 10.5f). The FTIR analysis enabled the detection of different plastic polymers based on characteristic absorption peaks (Jung et al. 2018). The spectra of cellulose acetate (CA) polymers were identified by a broad O–H stretching vibration at 3485 cm^{-1}, a prominent C–H stretching band near 2949 cm^{-1}, a strong C=O stretching peak at 1756 cm^{-1}, and C–O stretching bands at 1240 and 1049 cm^{-1}. Polyvinyl Chloride (PVC) was confirmed by characteristic peaks at 1427, 1331, 1255, 1099, and 616 cm^{-1}. Polyethylene (PE) was recognized by intense C–H stretching absorptions at 2915 and 2845 cm^{-1}, bending vibrations at 1467 and 1377 cm^{-1}, and distinct rocking modes at 730 and 717 cm^{-1}. Finally, the spectra of Polytetrafluoroethylene (PTFE) microplastics showed strong absorbance peaks at 1201 and 1147 cm^{-1}, along with bending vibrations at 638, 554, and 509 cm^{-1}. The MP polymer identification (refer to Fig. 10.4e) revealed that PE emerged as the most prevalent polymer across various sampling sites, with notable percentages in Shivrajpur (50%), Veraval (41.54%), and Porbandar (41.63%), suggesting that PE-based products, often used in packaging and fishing materials, may be significant contributors (Dowarah and Devipriya 2019; Kılıç 2022; Phillips and Bonner 2015; Sánchez-Hernández et al. 2021; Wessel et al. 2016). PVC was found in the highest proportions at Shil (44.88%) and Mangrol (35.76%), indicating possible accumulation of MPs coming from construction and industrial material applications commonly associated with PVC (Karthik et al. 2018; Rabari et al. 2023; Wessel et al. 2016). Studies of plastic pollution along other Indian coasts, including Gujarat's sandy beaches, have similarly found PE and PVC as dominant polymers, pointing to the extensive use of these materials in packaging and construction (Karthik et al. 2018). PE's abundance in tourism hotspot locations such as Shivrajpur aligns with findings from other coastal areas, where tourism increases single-use plastic waste, including packaging materials (Ronca 2017). Meanwhile, the elevated PVC presence in Shil and Mangrol may be linked to waste from nearby construction and industrial activities, which use PVC in various applications. PTFE was predominantly found at sites such as Diu (31.71%) and Chorwad (23.18%), where industrial influence is more likely, as PTFE is widely utilized in non-stick coatings and machine components (Crawford and Quinn 2017; Radulovic and Wojcinski 2014). The presence of CA in higher percentages at Somnath (19.71%) and Madhavpur (18.65%) reflects its use in products like cigarette filters and textiles, which are commonly discarded in tourist areas (Epperson et al. 2021). The spatial distribution of these polymers indicates potential sources. For instance, tourist-heavy areas like Shivrajpur and Somnath show a variety of polymers associated with disposable plastics, similar to patterns observed at other tourism-dense

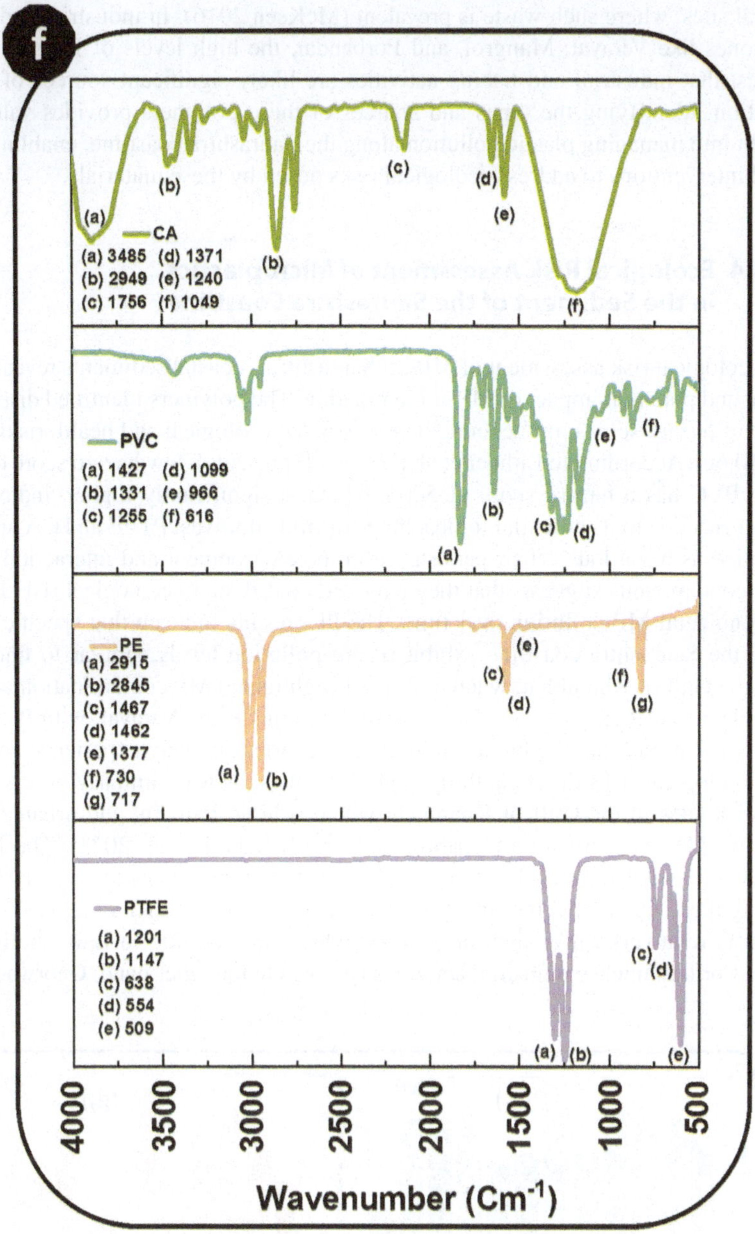

Fig. 10.5 (f) Sample of ATR-FTIR spectra of various polymers of microplastics detected

coastal sites, where such waste is prevalent (McKeen 2016). In industrial and fishing zones like Veraval, Mangrol, and Porbandar, the high levels of PE and PVC suggest that industrial and fishing activities are likely significant sources of MPs pollution. Identifying the types and sources of these polymers provides valuable insight into managing plastic pollution along the Saurashtra coastline, enabling targeted interventions to address ecological risks posed by these materials.

10.3.4 Ecological Risk Assessment of Microplastics in the Sediment of the Saurashtra Coastline

The ecological risk assessment of MPs in Saurashtra's coastal sediments reveals the extent and potential impact of MP contamination. The polymers identified display a range of hazard scores, influencing their associated ecological and health risks (see Fig. 10.6g). According to Lithner et al. (2011), PE presents a low hazard score of 11, while PVC has a hazard score of 5001, which is significantly higher, indicating greater risk due to its potential for leaching harmful additives. PTFE and CA are not classified as hazardous. Their persistence in the environment and interaction with other contaminants suggests that they may still contribute to ecological risk due to environmental MP pollution over time. The PLI results indicate that specific sites along the Saurashtra coastline exhibit severe pollution levels, similar to findings from the Gulf of Khambhat, where a PLI > 1 highlighted MP contamination across multiple areas (Rabari et al. 2023). Chorwad, Somnath, and Veraval, with PLI values above 25, fall into the highest pollution categories and may experience heightened ecological risks due to anthropogenic activities, as was similarly observed in high-risk sites in the Gulf of Khambhat where fishing, tourism, and urbanization intensify MP contamination (Karthik et al. 2018; Rabari et al. 2023). The PERI values across the selected locations differ from one to another (see Fig. 10.6g2). Among these variations, the Shivrajpur site has the lowest PERI value (1377.63), which is relatively lower than other sites, which may be due to fewer pollution sources or less intense industrial activities nearby. On the other hand, Chorwad and

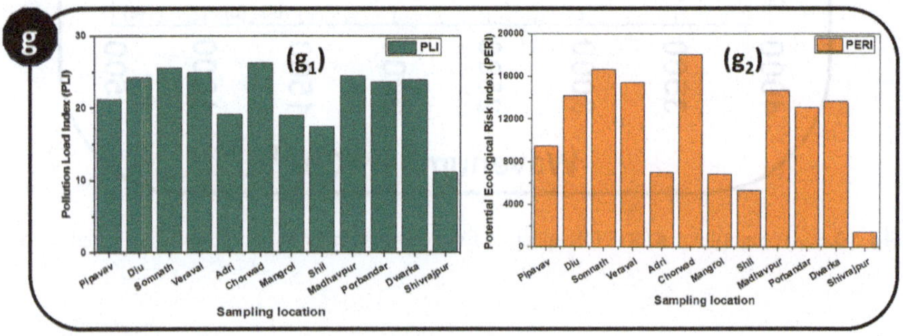

Fig. 10.6 (g) Ecological risk assessment of microplastics: (*g1*) pollution load index, (*g2*) potential ecological risk

Somnath have notably high PERI values of 18049.82 and 16662.95, respectively. This indicates that the ecological risks from MPs are severe in these locations. In summary, the accumulation of MPs along the Saurashtra coastline poses significant ecological risks, particularly in the Chorwad and Somnath sites, where tourism and fishing activities may contribute to the release of MPs. Considerable amounts of particular MPs, such as PE and PVC, suggest common origins, such as industrial components, fishing gear, and plastic packaging.

Furthermore, MPs in coastal sediment environments can be subjected to a variety of environmental processes that will significantly influence their behavior and associated risks. These processes include enzymatic degradation initiated by microbial colonization, photochemical reactions driven by intense sunlight exposure, and oxidative and hydrolytic transformations (Haddout et al. 2022; Karthik et al. 2018; Peng et al. 2017). Such interactions will also progressively fragment MPs into smaller particles and promote the leaching of embedded chemical additives (e.g., plasticizers, flame retardants, and stabilizers) into the surrounding sediment and seawater (Ghobish et al. 2025). The persistence of MPS, along with their small size, high surface area, and inherent chemical reactivity, enhances their ability to adsorb persistent organic pollutants (e.g., pesticides, hydrocarbons, and heavy metals) (Yu et al. 2021). Consequently, MPs will serve as vectors for these toxic compounds, facilitating their transport and accumulation within coastal sediment ecosystems. This accumulation not only increases the toxicological burden on benthic organisms but will also pose a lasting threat to the ecological integrity of coastal environments and the food webs they support (Sfriso et al. 2024). Immediate attention to waste management and pollution control in these coastal areas is essential to reduce MP accumulation and mitigate their ecological impacts. The severe PERI scores underline the urgency of implementing protective measures to preserve the region's marine and coastal ecosystems.

10.4 Conclusion

This study quantifies the abundance and characteristics of MPs in coastal sediments along the Saurashtra coastline, revealing significant variation influenced by human activity and its intensity. High-abundance locations like Chorwad, Somnath, and Veraval recorded MP levels at 963.33 ± 15.28 MPs/kg, 913.33 ± 90.18 MPs/kg, and 866.67 ± 20.82 MPs/kg, respectively, which were attributed to tourism, fishing, and industrial activities. In contrast, conservation-focused and lower-activity areas, such as Shivrajpur, with only 173.33 ± 15.28 MPs/kg, demonstrate the mitigating impact of reduced human intervention on MP pollution. Comparatively, Saurashtra's high levels are still lower than the extreme concentrations reported in Goa's estuarine system (6093 MPs/kg), but they exceed those in less industrialized regions in China, such as the Changjiang Estuary (121 MPs/kg). Color distribution analysis showed that black and brown MPs, linked to urban runoff and fishing equipment, were most prevalent, with notable concentrations in Chorwad (403.33 ± 161.66 MPs/kg) and Veraval (240.00 ± 96.44 MPs/kg). Shape analysis identified fiber MPs, often originating from fishing nets, as the dominant form, with Chorwad and Somnath recording 430.00 ± 20.00 MPs/kg and 413.33 ± 30.55 MPs/kg, respectively. The predominance of smaller-sized MPs (0.3–1.0 mm), found abundantly in sites like Chorwad (306.7 ± 11.5 MPs), suggests higher ecological risks due to their ease of ingestion by marine life, which could lead to bioaccumulation. The chemical analysis revealed PE as the most common polymer, accounting for 50% in Shivrajpur and over 41% in both Veraval and Porbandar. This distribution aligns with the widespread use of PE-based products in fishing and packaging, implicating these industries in coastal MP pollution. The results highlight the urgent need for targeted waste management strategies to reduce the generation of MPs, limit their dispersal into freshwater systems, and ultimately prevent their accumulation in coastal sediments. Implementing strict regulations on the use and disposal of plastic items, particularly in sensitive areas such as beaches along the coastline, is essential to curb the degradation of larger plastic debris into MPs that contaminate coastal environments. In addition, public awareness and conservation initiatives are crucial in reducing the ecological and health risks associated with MPs along the Saurashtra coastline.

Author Contributions The authors thank the National Forensic Sciences University for providing all the resources required to complete this study.

Declaration of Competition Interest The authors declare that they have no known competing financial interests or personal relationships that could have appeared to influence the work reported in this chapter.

Authorship Contribution Statement **Irédon Adjama:** Conceptualization and experimental planning, sampling, experimental work, data analysis and visualization, writing an original draft and editing. **Pavan Hirenbhai Pandya:** Sampling and experimental work. **Hemen Dave:** Conceptualization and supervision, review and editing.

References

Abidli S, Antunes JC, Ferreira JL, Lahbib Y, Sobral P, Trigui El Menif N (2018) Microplastics in sediments from the littoral zone of the north Tunisian coast (Mediterranean Sea). Estuar Coast Shelf Sci 205:1–9. https://doi.org/10.1016/j.ecss.2018.03.006

Adeleye AT, Bahar MM, Megharaj M, Fang C, Rahman MM (2024) The unseen threat of the synergistic effects of microplastics and heavy metals in aquatic environments: a critical Review. Curr Pollut Rep. https://doi.org/10.1007/s40726-024-00298-7

Adjama I, Dave H (2025) Tackling microplastic contamination in sewage sludge: optimizing organic matter degradation, quantifying microplastic presence, and evaluating ecological risks for sustainable agriculture. Sci Total Environ 974:179201. https://doi.org/10.1016/j.scitotenv.2025.179201

Adjama I, Dave H, Amen E (2024) Bibliometric analysis and review of direct factors implicating the impact of nano and microplastics on crop health and development. Plant Nano Biol 9:100083. https://doi.org/10.1016/j.plana.2024.100083

Akoueson F, Sheldon LM, Danopoulos E, Morris S, Hotten J, Chapman E, Li J, Rotchell JM (2020) A preliminary analysis of microplastics in edible versus non-edible tissues from seafood samples. Environ Pollut 263:114452. https://doi.org/10.1016/j.envpol.2020.114452

Allami H, Afzali A, Mirzaei R (2023) Identification of microplastics in coastal sediments of three ports of Dayer, Kangan, and Siraf in Bushehr province, Iran. Reg Stud Mar Sci 63:103007. https://doi.org/10.1016/j.rsma.2023.103007

Alvarez-Zeferino JC, Cruz-Salas AA, Vázquez-Morillas A, Ojeda-Benitez S, Alvarez-Zeferino JC, Cruz-Salas AA, Vázquez-Morillas A, Ojeda-Benitez S (2020) Method for quantifying and characterization of microplastics in sand beaches. Revista Internacional de Contaminación Ambiental 36:151–164. https://doi.org/10.20937/rica.2020.36.53540

Alves VEN, Figueiredo GM (2019) Microplastic in the sediments of a highly eutrophic tropical estuary. Mar Pollut Bull 146:326–335. https://doi.org/10.1016/j.marpolbul.2019.06.042

Amelia TSM, Khalik WMAWM, Ong MC, Shao YT, Pan H-J, Bhubalan K (2021) Marine microplastics as vectors of major ocean pollutants and its hazards to the marine ecosystem and humans. Prog Earth Planet Sci 8:12. https://doi.org/10.1186/s40645-020-00405-4

Bilal M, Taj M, Ul Hassan H, Yaqub A, Shah MIA, Sohail M, Rafiq N, Atique U, Abbas M, Sultana S, Abdali U, Arai T (2023) First report on microplastics quantification in poultry chicken and potential human health risks in Pakistan. Toxics. https://doi.org/10.3390/toxics11070612

Boiteux V, Dauchy X, Bach C, Colin A, Hemard J, Sagres V, Rosin C, Munoz J-F (2017) Concentrations and patterns of perfluoroalkyl and polyfluoroalkyl substances in a river and three drinking water treatment plants near and far from a major production source. Sci Total Environ 583:393–400. https://doi.org/10.1016/j.scitotenv.2017.01.079

Browne MA, Crump P, Niven SJ, Teuten E, Tonkin A, Galloway T, Thompson R (2011) Accumulation of microplastic on shorelines woldwide: sources and sinks. Environ Sci Technol 45:9175–9179. https://doi.org/10.1021/es201811s

Castillo AB, Al-Maslamani I, Obbard JP (2016) Prevalence of microplastics in the marine waters of Qatar. Mar Pollut Bull 111:260–267. https://doi.org/10.1016/j.marpolbul.2016.06.108

Castillo AB, El-Azhary M, Sorino C, LeVay L (2024) Potential ecological risk assessment of microplastics in coastal sediments: their metal accumulation and interaction with sedimentary metal concentration. Sci Total Environ 906:167473. https://doi.org/10.1016/j.scitotenv.2023.167473

Choong WS, Hadibarata T, Yuniarto A, Tang KHD, Abdullah F, Syafrudin M, Al Farraj DA, Al-Mohaimeed AM (2021) Characterization of microplastics in the water and sediment of Baram River estuary, Borneo Island. Mar Pollut Bull 172:112880. https://doi.org/10.1016/j.marpolbul.2021.112880

Cole M, Lindeque P, Halsband C, Galloway TS (2011) Microplastics as contaminants in the marine environment: a review. Mar Pollut Bull 62:2588–2597. https://doi.org/10.1016/j.marpolbul.2011.09.025

Crawford CB, Quinn B (2017) 4 – physiochemical properties and degradation. In: Crawford CB, Quinn B (eds). Elsevier, Microplastic pollutants, pp 57–100. https://doi.org/10.1016/B978-0-12-809406-8.00004-9

Da Costa Filho PA, Andrey D, Eriksen B, Peixoto RP, Carreres BM, Ambühl ME, Descarrega JB, Dubascoux S, Zbinden P, Panchaud A, Poitevin E (2021) Detection and characterization of small-sized microplastics (≥ 5 µm) in milk products. Sci Rep 11. https://doi.org/10.1038/s41598-021-03458-7

Danopoulos E, Jenner LC, Twiddy M, Rotchell JM (2020) Microplastic contamination of seafood intended for human consumption: a systematic review and meta-analysis. Environ Health Perspect 128:126002. https://doi.org/10.1289/EHP7171

Davidson K, Dudas SE (2016) Microplastic ingestion by wild and cultured Manila Clams (Venerupis philippinarum) from Baynes Sound, British Columbia. Arch Environ Contam Toxicol 71:147–156. https://doi.org/10.1007/s00244-016-0286-4

Dobaradaran S, Schmidt TC, Nabipour I, Khajeahmadi N, Tajbakhsh S, Saeedi R, Javad Mohammadi M, Keshtkar M, Khorsand M, Faraji Ghasemi F (2018) Characterization of plastic debris and association of metals with microplastics in coastline sediment along the Persian Gulf. Waste Manag 78:649–658. https://doi.org/10.1016/j.wasman.2018.06.037

Dowarah K, Devipriya SP (2019) Microplastic prevalence in the beaches of Puducherry, India and its correlation with fishing and tourism/recreational activities. Mar Pollut Bull 148:123–133. https://doi.org/10.1016/j.marpolbul.2019.07.066

Enders K, Käppler A, Biniasch O, Feldens P, Stollberg N, Lange X, Fischer D, Eichhorn K-J, Pollehne F, Oberbeckmann S, Labrenz M (2019) Tracing microplastics in aquatic environments based on sediment analogies. Sci Rep 9:15207. https://doi.org/10.1038/s41598-019-50508-2

Enyoh CE, Verla AW, Rakib RJ (2021) Application of index models for assessing freshwater microplastics pollution

Epperson AE, Novotny TE, Halpern-Felsher B (2021) Perceptions about the impact of cigarette filters on the environment and smoking-related behaviors. J Adolesc Health 68:823–826. https://doi.org/10.1016/j.jadohealth.2020.10.022

Eriksen M, Lebreton LCM, Carson HS, Thiel M, Moore CJ, Borerro JC, Galgani F, Ryan PG, Reisser J (2014) Plastic pollution in the world's oceans: more than 5 trillion plastic pieces weighing over 250,000 tons afloat at sea. PLoS One 9:e111913. https://doi.org/10.1371/journal.pone.0111913

Frias JPGL, Gago J, Otero V, Sobral P (2016) Microplastics in coastal sediments from Southern Portuguese shelf waters. Mar Environ Res 114:24–30. https://doi.org/10.1016/j.marenvres.2015.12.006

Fu L, Li J, Wang G, Luan Y, Dai W (2021) Adsorption behavior of organic pollutants on microplastics. Ecotoxicol Environ Saf 217:112207. https://doi.org/10.1016/j.ecoenv.2021.112207

Galindo Montero AA, Costa-Redondo LC, Vasco-Echeverri O, Arana VA (2023) Microplastic pollution in coastal areas of Colombia: review. Mar Environ Res 190:106027. https://doi.org/10.1016/j.marenvres.2023.106027

Ghanadi M, Joshi I, Dharmasiri N, Jaeger JE, Burke M, Bebelman C, Symons B, Padhye LP (2024) Quantification and characterization of microplastics in coastal environments: Insights from laser direct infrared imaging. Sci Total Environ 912:168835. https://doi.org/10.1016/j.scitotenv.2023.168835

Ghobish SA, Motti CA, Bissember AC, Vamvounis G (2025) Microplastics in the marine environment: challenges and the shift towards sustainable plastics and plasticizers. J Hazard Mater 491:137945. https://doi.org/10.1016/j.jhazmat.2025.137945

Gupta P, Saha M, Rathore C, Suneel V, Ray D, Naik A, Unnikrishnan K, Dhivya M, Daga K (2021) Spatial and seasonal variation of microplastics and possible sources in the estuarine system from central west coast of India. Environ Pollut 288:117665. https://doi.org/10.1016/j.envpol.2021.117665

Gupta P, Saha M, Suneel V, Rathore C, Chndrasekhararao AV, Gupta GVM, Junaid CK (2024) Microplastics in the sediments along the eastern Arabian Sea shelf: Distribution, governing factors and risk assessment. Sci Total Environ 910:168629. https://doi.org/10.1016/j.scitotenv.2023.168629

Haddout S, Gimiliani GT, Priya KL, Hoguane AM, Casila JCC, Ljubenkov I (2022) Microplastics in surface waters and sediments in the Sebou Estuary and Atlantic Coast, Morocco. Anal Lett 55:256–268. https://doi.org/10.1080/00032719.2021.1924767

Haegerbaeumer A, Mueller M-T, Fueser H, Traunspurger W (2019) Impacts of micro- and nano-sized plastic particles on benthic invertebrates: a literature review and gap analysis. Front Environ Sci 7. https://doi.org/10.3389/fenvs.2019.00017

Hakanson L (1980) An ecological risk index for aquatic pollution control. A sedimentological approach. Water Res 14:975–1001. https://doi.org/10.1016/0043-1354(80)90143-8

Hassoun A, Pasti L, Chenet T, Rusanova P, Smaoui S, Aït-Kaddour A, Bono G (2023) Detection methods of micro and nanoplastics. In: Özogul F, Department of Seafood Processing Technology, F. of F., Cukurova University, Adana (eds) Adv Food Nutr Res. Academic Press, pp 175–227. https://doi.org/10.1016/bs.afnr.2022.08.002

Jahan S, Strezov V (2019) Assessment of trace elements pollution in the sea ports of New South Wales (NSW), Australia using oysters as bioindicators. Sci Rep 9:1416. https://doi.org/10.1038/s41598-018-38196-w

Jambeck JR, Geyer R, Wilcox C, Siegler TR, Perryman M, Andrady A, Narayan R, Law KL (2015) Plastic waste inputs from land into the ocean. Science 347:768–771. https://doi.org/10.1126/science.1260352

Jiwarungrueangkul T, Phaksopa J, Sompongchaiyakul P, Tipmanee D (2021) Seasonal microplastic variations in estuarine sediments from urban canal on the west coast of Thailand: a case study in Phuket province. Mar Pollut Bull 168:112452. https://doi.org/10.1016/j.marpolbul.2021.112452

Joo SH, Liang Y, Kim M, Byun J, Choi H (2021) Microplastics with adsorbed contaminants: mechanisms and treatment. Environ Challenges 3:100042. https://doi.org/10.1016/j.envc.2021.100042

Jung MR, Horgen FD, Orski SV, Rodriguez CV, Beers KL, Balazs GH, Jones TT, Work TM, Brignac KC, Royer S-J, Hyrenbach KD, Jensen BA, Lynch JM (2018) Validation of ATR FT-IR to identify polymers of plastic marine debris, including those ingested by marine organisms. Mar Pollut Bull 127:704–716. https://doi.org/10.1016/j.marpolbul.2017.12.061

Karthik R, Robin RS, Purvaja R, Ganguly D, Anandavelu I, Raghuraman R, Hariharan G, Ramakrishna A, Ramesh R (2018) Microplastics along the beaches of southeast coast of India. Sci Total Environ 645:1388–1399. https://doi.org/10.1016/j.scitotenv.2018.07.242

Kılıç E (2022) Microplastic ingestion evidence by economically important farmed fish species from Turkey. Mar Pollut Bull 183:114097. https://doi.org/10.1016/j.marpolbul.2022.114097

Leads RR, Weinstein JE (2019) Occurrence of tire wear particles and other microplastics within the tributaries of the Charleston Harbor Estuary, South Carolina, USA. Mar Pollut Bull 145:569–582. https://doi.org/10.1016/j.marpolbul.2019.06.061

Leslie HA, Brandsma SH, van Velzen MJM, Vethaak AD (2017) Microplastics en route: field measurements in the Dutch river delta and Amsterdam canals, wastewater treatment plants, North Sea sediments and biota. Environ Int 101:133–142. https://doi.org/10.1016/j.envint.2017.01.018

Li R, Liu Y, Sheng Y, Xiang Q, Zhou Y, Cizdziel JV (2020a) Effect of prothioconazole on the degradation of microplastics derived from mulching plastic film: apparent change and interaction with heavy metals in soil. Environ Pollut 260. https://doi.org/10.1016/j.envpol.2020.113988

Li Y, Lu Z, Zheng H, Wang J, Chen C (2020b) Microplastics in surface water and sediments of Chongming Island in the Yangtze Estuary, China. Environ Sci Eur 32:15. https://doi.org/10.1186/s12302-020-0297-7

Li Y, Chen G, Xu K, Huang K, Wang J (2021) Microplastics environmental effect and risk assessment on the aquaculture systems from South China. Int J Environ Res Public Health 18:1869. https://doi.org/10.3390/ijerph18041869

Lippiatt S, Opfer S, Arthur C (2013) Marine debris monitoring and assessment: recommendations for monitoring debris trends in the marine environment

Lithner D, Larsson Å, Dave G (2011) Environmental and health hazard ranking and assessment of plastic polymers based on chemical composition. Sci Total Environ 409:3309–3324. https://doi.org/10.1016/j.scitotenv.2011.04.038

Lloret J, Pedrosa-Pamies R, Vandal N, Rorty R, Ritchie M, McGuire C, Chenoweth K, Valiela I (2021) Salt marsh sediments act as sinks for microplastics and reveal effects of current and historical land use changes. Environ Adv 4:100060. https://doi.org/10.1016/j.envadv.2021.100060

Martins J, Sobral P (2011) Plastic marine debris on the Portuguese coastline: a matter of size? Mar Pollut Bull 62:2649–2653. https://doi.org/10.1016/j.marpolbul.2011.09.028

McKeen LW (2016) Fatigue and tribological properties of plastics and elastomers. William Andrew

Mukhopadhyay R, Karisiddaiah SM (2014) The Indian coastline: processes and landforms. In: Kale VS (ed) Landscapes and landforms of India. Springer, Dordrecht, pp 91–101. https://doi.org/10.1007/978-94-017-8029-2_8

Nayanathara Thathsarani Pilapitiya PGC, Ratnayake AS (2024) The world of plastic waste: a review. Cleaner Mater 11:100220. https://doi.org/10.1016/j.clema.2024.100220

Neves D, Sobral P, Ferreira JL, Pereira T (2015) Ingestion of microplastics by commercial fish off the Portuguese coast. Mar Pollut Bull 101:119–126. https://doi.org/10.1016/j.marpolbul.2015.11.008

Okoffo ED, Tan E, Grinham A, Gaddam SMR, Yip JYH, Twomey AJ, Thomas KV, Bostock H (2024) Plastic pollution in Moreton Bay sediments, Southeast Queensland, Australia. Sci Total Environ 920:170987. https://doi.org/10.1016/j.scitotenv.2024.170987

Okoye CO, Addey CI, Oderinde O, Okoro JO, Uwamungu JY, Ikechukwu CK, Okeke ES, Ejeromedoghene O, Odii EC (2022) Toxic chemicals and persistent organic pollutants associated with micro-and nanoplastics pollution. Chem Eng J Adv 11:100310. https://doi.org/10.1016/j.ceja.2022.100310

Olivatto GP, Martins MCT, Montagner CC, Henry TB, Carreira RS (2019) Microplastic contamination in surface waters in Guanabara Bay, Rio de Janeiro, Brazil. Mar Pollut Bull 139:157–162. https://doi.org/10.1016/j.marpolbul.2018.12.042

Peng G, Zhu B, Yang D, Su L, Shi H, Li D (2017) Microplastics in sediments of the Changjiang Estuary, China. Environ Pollut 225:283–290. https://doi.org/10.1016/j.envpol.2016.12.064

Phillips MB, Bonner TH (2015) Occurrence and amount of microplastic ingested by fishes in watersheds of the Gulf of Mexico. Mar Pollut Bull 100:264–269. https://doi.org/10.1016/j.marpolbul.2015.08.041

Plastics Europe (2024) Enabling a sustainable future. WWW document. Plastics Europe. https://plasticseurope.org/. Accessed 10.23.24

Rabari V, Patel K, Patel H, Trivedi J (2022) Quantitative assessment of microplastic in sandy beaches of Gujarat state, India. Mar Pollut Bull 181:113925. https://doi.org/10.1016/j.marpolbul.2022.113925

Rabari V, Patel H, Patel K, Patel A, Bagtharia S, Trivedi J (2023) Quantitative assessment of microplastic contamination in muddy shores of Gulf of Khambhat, India. Mar Pollut Bull 192:115131. https://doi.org/10.1016/j.marpolbul.2023.115131

Radhakrishnan K, Sivapriya V, Rajkumar A, Akramkhan N, Prakasheswar P, Krishnakumar S, Hussain SM (2021) Characterization and distribution of microplastics in estuarine surface sediments, Kayamkulam estuary, southwest coast of India. Mar Pollut Bull 168:112389. https://doi.org/10.1016/j.marpolbul.2021.112389

Radulovic LL, Wojcinski ZW (2014. PTFE (Polytetrafluoroethylene; Teflon®),) In: Wexler P (ed) Encyclopedia of toxicology, 3rd edn. Academic Press, Oxford, pp 1133–1136. https://doi.org/10.1016/B978-0-12-386454-3.00970-2

Rafey A, Siddiqui FZ (2023) A review of plastic waste management in India – challenges and opportunities. Int J Environ Anal Chem 103:3971–3987. https://doi.org/10.1080/03067319.2021.1917560

Ranjbar Jafarabadi A, Riyahi Bakhtiyari A, Shadmehri Toosi A, Jadot C (2017) Spatial distribution, ecological and health risk assessment of heavy metals in marine surface sediments and coastal seawaters of fringing coral reefs of the Persian Gulf, Iran. Chemosphere 185:1090–1111. https://doi.org/10.1016/j.chemosphere.2017.07.110

Robin RS, Karthik R, Purvaja R, Ganguly D, Anandavelu I, Mugilarasan M, Ramesh R (2020) Holistic assessment of microplastics in various coastal environmental matrices, southwest coast of India. Sci Total Environ 703:134947. https://doi.org/10.1016/j.scitotenv.2019.134947

Ronca S (2017) Chapter 10 – polyethylene. In: Gilbert M (ed) Brydson's plastics materials, 8th edn. Butterworth-Heinemann, pp 247–278. https://doi.org/10.1016/B978-0-323-35824-8.00010-4

Sabri NAA, Razak MR, Aris AZ (2025) Fate of microplastics and emerging contaminants: mechanisms of interactions, bioaccumulation and combined toxicity to aquatic organisms. Mar Pollut Bull 214:117822. https://doi.org/10.1016/j.marpolbul.2025.117822

Sánchez-Hernández LJ, Ramírez-Romero P, Rodríguez-González F, Ramos-Sánchez VH, Márquez Montes RA, Romero-Paredes Rubio H, Sujitha SB, Jonathan MP (2021) Seasonal evidences of microplastics in environmental matrices of a tourist dominated urban estuary in Gulf of Mexico, Mexico. Chemosphere 277:130261. https://doi.org/10.1016/j.chemosphere.2021.130261

Senapati S, Gupta V (2014) Climate change and coastal ecosystem in India: Issues in perspectives. Int J Environ Sci

Sfriso AA, Juhmani A-S, Tomio Y, Sfriso A, Rizzolio F, Adeel M, Wahsha M, Munari C, Mistri M (2024) Microplastic accumulation and ecological impacts on benthic invertebrates: insights from a microcosm experiment. Mar Pollut Bull 202:116231. https://doi.org/10.1016/j.marpolbul.2024.116231

Thompson RC, Swan SH, Moore CJ, vom Saal FS (2009) Our plastic age. Philos Trans R Soc B Biol Sci 364:1973–1976. https://doi.org/10.1098/rstb.2009.0054

Thushari GGN, Senevirathna JDM (2020) Plastic pollution in the marine environment. Heliyon 6:e04709. https://doi.org/10.1016/j.heliyon.2020.e04709

Tomlinson DL, Wilson JG, Harris CR, Jeffrey DW (1980) Problems in the assessment of heavy-metal levels in estuaries and the formation of a pollution index. Helgolander Meeresun 33:566–575. https://doi.org/10.1007/BF02414780

Veerasingam S, Ranjani M, Venkatachalapathy R, Bagaev A, Mukhanov V, Litvinyuk D, Mugilarasan M, Gurumoorthi K, Guganathan L, Aboobacker VM, Vethamony P (2021) Contributions of Fourier transform infrared spectroscopy in microplastic pollution research: a review. Crit Rev Environ Sci Technol 51:2681–2743. https://doi.org/10.1080/10643389.2020.1807450

Wessel CC, Lockridge GR, Battiste D, Cebrian J (2016) Abundance and characteristics of microplastics in beach sediments: Insights into microplastic accumulation in northern Gulf of Mexico estuaries. Mar Pollut Bull 109:178–183. https://doi.org/10.1016/j.marpolbul.2016.06.002

Willis KA, Eriksen R, Wilcox C, Hardesty BD (2017) Microplastic distribution at different sediment depths in an urban estuary. Front Mar Sci 4. https://doi.org/10.3389/fmars.2017.00419

Woodall LC, Sanchez-Vidal A, Canals M, Paterson GLJ, Coppock R, Sleight V, Calafat A, Rogers AD, Narayanaswamy BE, Thompson RC (2014) The deep sea is a major sink for microplastic debris. R Soc Open Sci 1:140317. https://doi.org/10.1098/rsos.140317

Wright SL, Rowe D, Thompson RC, Galloway TS (2013) Microplastic ingestion decreases energy reserves in marine worms. Curr Biol 23:R1031–R1033. https://doi.org/10.1016/j.cub.2013.10.068

Xu S, Chen L, Zhang K, Cao Y, Ma Y, Chau HS, Tao D, Wu C, Li C, Lam PKS (2022) Microplastic occurrence in the northern South China Sea, a case for pre and post cyclone analysis. Chemosphere 296:133980. https://doi.org/10.1016/j.chemosphere.2022.133980

Yang S, Lee B-T, Kim S-O, Park S (2024) Review of microplastics in soils: state-of-the-art occurrence, transport, and investigation methods. J Soils Sediments 24:779–792. https://doi.org/10.1007/s11368-023-03689-3

Yu Y, Mo WY, Luukkonen T (2021) Adsorption behaviour and interaction of organic micropollutants with nano and microplastics – a review. Sci Total Environ 797:149140. https://doi.org/10.1016/j.scitotenv.2021.149140

Zhou Z, Zhang P, Zhang G, Wang S, Cai Y, Wang H (2021) Vertical microplastic distribution in sediments of Fuhe River estuary to Baiyangdian Wetland in Northern China. Chemosphere 280:130800. https://doi.org/10.1016/j.chemosphere.2021.130800

11 From Cradle to Contamination: The Life Cycle of Plastics with an Emphasis on Household Waste and Remediation via Biochar

Abhishek Biswas and Gopala Krishna Darbha

Abstract

The widespread use of synthetic polymers has reshaped modern life but has also driven an accelerating environmental crisis. This chapter traces the complete life cycle of plastics, emphasizing the post-consumer phase of plastics and the role of household plastic waste as a key contributor to micro-nanoplastics (MNPs) pollution. It explores the physicochemical transformations and release pathways of plastics under typical domestic conditions, with a focus on their fate in terrestrial and aquatic environments. Factors influencing their mobility and persistence in soil, water, plants, and fauna were explored, alongside sector-wise contributions from packaging, textiles, and consumer goods. Municipal solid waste (MSW) acts as a primary source of plastic pollution in the environment if not treated properly, with comparative assessments from Europe, the USA, and India revealing significant regulatory and infrastructural disparities. While developed nations address legacy contamination, rapidly industrializing regions face mounting consumption and inadequate waste systems, exacerbating pollution risks. To mitigate this, the chapter evaluates the efficacy of affordable, accessible engineered biochar derived from organic residues for remediating plastic and its co-contaminants. Column and batch sorption experiments are required to optimize and remediate MNPs due to biochar's strong affinities for various plastic poly-

A. Biswas
Environmental Nanoscience Laboratory, Department of Earth Sciences, Indian Institute of Science Education and Research Kolkata, Mohanpur, West Bengal, India

G. K. Darbha (✉)
Environmental Nanoscience Laboratory, Department of Earth Sciences, Indian Institute of Science Education and Research Kolkata, Mohanpur, West Bengal, India

Centre for Climate and Environmental Studies, Indian Institute of Science Education and Research Kolkata, Mohanpur, West Bengal, India
e-mail: gkdarbha@iiserkol.ac.in

K. Bala et al. (eds.), *Occurrence, Detection, and Fate of Microplastics in Freshwater Ecosystems*, https://doi.org/10.1007/978-981-95-0225-7_11

mers. The findings highlight the promise of biochar in decentralized plastic waste management and its role in advancing circular, sustainable remediation strategies.

Keywords

Biochar · Pollution · Microplastics · Waste

Abbreviations

LCA	Life cycle assessment
PE	Polyethylene
PP	Polypropylene
PVC	Polyvinyl chloride
PS	Polystyrene
PET	Polyethylene terephthalate
CTC	Cradle-to-contamination
Mt.	Million tons
OECD	Organization for Economic Cooperation and Development
SUPPs	Single-use plastic products
VOCs	Volatile organic compounds
MPs	Microplastics
NPs	Nanoplastics
MNPs	Micro-nanoplastics
AMPs	Airborne microplastics
DOM	Dissolved organic matter
PTEs	Potentially toxic elements
PPCPs	Pharmaceuticals and personal care products
POPs	Persistent organic pollutants
PCPs	Personal care products
WWTPs	Wastewater treatment plants
MSW	Municipal solid waste
BPA	Bisphenol A
BPS	Bisphenol S
Py-GC-MS	Pyrolysis gas chromatography-mass spectrometry
FTIR	Fourier-transform infrared spectroscopy
OPTIR	Optical photothermal infrared spectroscopy
SEM	Scanning electron microscopy
DLS	Dynamic light scattering
MBT	Mechanical biological treatment
CPCB	Central Pollution Control Board
US EPA	United States Environmental Protection Agency
EU REACH	European Union Registration, Evaluation, Authorization, and Restriction of Chemicals
WtE	Waste-to-energy

11.1 Introduction

Plastics, long hailed for their durability, versatility, and low production costs, have now emerged as the major environmental issue of the twenty-first century. A single-use plastic bag and bottle can take 20–450 years to degrade, yet a toothbrush recommended for 3 months can last up to 500 years (Chamas et al. 2020; Chen and Gao 2021; Soni et al. 2023).

Plastic's life cycle begins in the "cradle" with the acquisition of raw ingredients from fossil-based sources such as crude oil and natural gas derivatives via polycondensation and polymerization techniques. These materials are processed using extremely intensive processes to produce monomers such as propylene, ethylene, and styrene, which are then polymerized into a variety of plastics, mostly thermoplastics such as polyethylene (PE), polypropylene (PP), and polystyrene (PS). The use of synthetic materials is becoming increasingly prevalent and unparalleled (Helms and Russell 2016).

Despite the fact that these polymers have numerous advantages in terms of lightweight logistics, product safety and preservation, and cost-effectiveness, plastics pose significant environmental hazards throughout their entire life cycle, particularly at the end of life (EOL). Although global plastic usage has skyrocketed, end-of-life (EOL) management remains inadequate. Only about 10% of plastics are recycled worldwide, while a significant portion leaks into the atmosphere, terrestrial ecosystems, and waterways (OECD 2022). Moreover, greenhouse gas emissions from plastic production and waste, estimated at 1.8 $GtCO_2e$ in 2019, are projected to exceed 2.8 $GtCO_2e$ by 2040 due to escalating production and insufficient waste management (OECD 2024). This cradle-to-contamination (CTC) trajectory underscores systemic inefficiencies in the plastic economy, demanding urgent adoption of sustainable production methods, circular economy frameworks, and robust policy measures to mitigate long-term ecological and health impacts.

11.1.1 Global Production of Plastics and Their Intended Applications

Global plastic production has surged exponentially since the mid-twentieth century, escalating from around 2 million metric tons (Mt) annually in the 1950s to an escalating 500 Mt. by 2025. This exponential increase underscores plastic's pivotal role in modern consumerism and industrial applications, fueled by increasing demand in major sectors such as packaging, construction, textiles, electronics, and automotive manufacturing (OECD 2024).

Global plastic production reveals stark regional disparities (Fig. 11.1), with developing economies, particularly in Asia, dominating output. Asia produces 197.6 Mt., accounting for 37.9% of global production, driven by rapid industrialization, population density, and its central role in manufacturing. The OECD, comprising 38 high-income nations, promotes economic growth and sustainability; its European members, termed OECD Europe, address global issues like plastic pollution

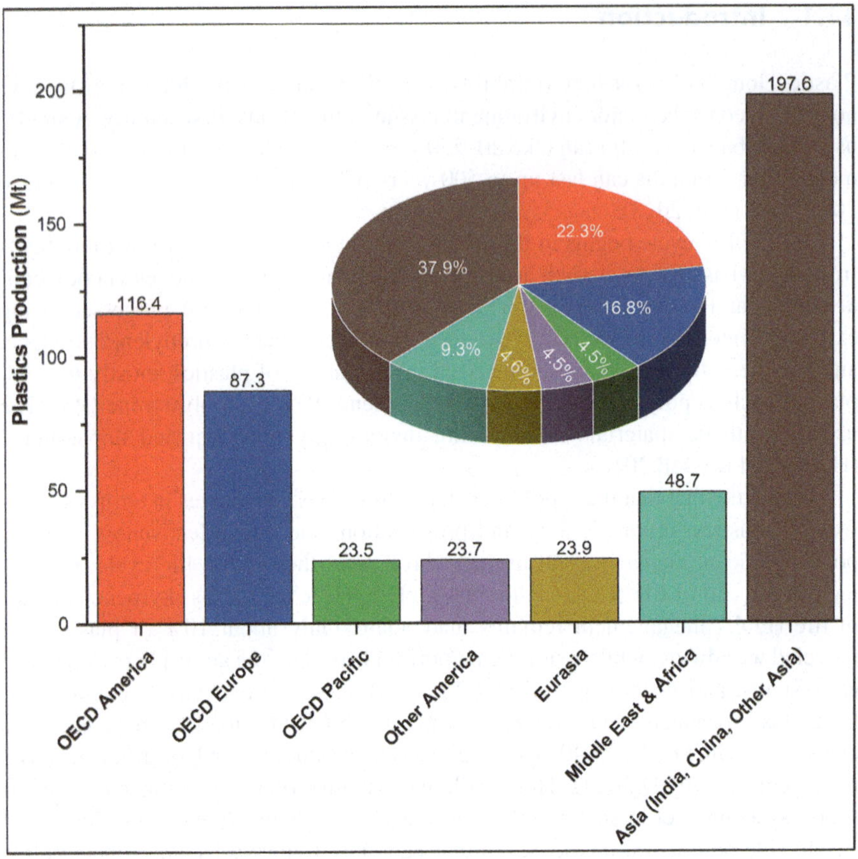

Fig. 11.1 Regional distribution of global plastic production (in million tons, Mt). The inset pie chart illustrates the proportional contributions of each region

collaboratively. In contrast, developed economies in OECD America (116.4 Mt., 22.3%) and OECD Europe (87.3 Mt., 16.8%) contribute significantly less, reflecting a shift in production toward lower-cost regions. Other regions, including the Middle East & Africa (48.7 Mt) and Eurasia (23.9 Mt), play smaller roles.

Asia's (majorly by India and China) outsized production raises environmental concerns, such as higher output increases plastic leakage risks, particularly in regions with inadequate waste management. Without intervention, escalating pollution could undermine ecological and public health. However, systemic reforms across the plastic life cycle by reducing production, replacing them with more natural and biodegradable materials, enhancing circular design, and scaling recycling could cut leakage by 96% by 2040 while boosting recycling rates from 9.5% to 42%. Though such measures may reduce global Gross Domestic Product (GDP) by 0.5% (with Sub-Saharan Africa facing up to 1.5% due to infrastructure costs), the long-term benefits outweigh the economic trade-offs. Addressing these gaps demands targeted investments, technology transfer, and joint cooperation to align production with sustainability goals (OECD 2024). Under the polluter pays principle, individuals in developing economies bear greater responsibility and economic pressure, as increased production leads to higher pollution and associated costs.

11.1.2 Sector-Wise Contribution to Plastic Production, Waste Generation, and Fate

Plastic production and waste generation stem primarily from eight key sectors: packaging (~44%), construction (17%), textiles (13%), consumer goods (~10%), automotive (~5%), electronics (~4%), agriculture (~3%), and other applications. Packaging dominates (~190Mt, 44.2%), contributing significantly more waste than other sectors, with single-use plastic products (SUPPs) driven by e-commerce, food, and beverage industries (Fig. 11.2a, b). In contrast, durable plastics (e.g., in infrastructure and electronics) are designed for long-term use but eventually contribute to end-of-life waste challenges.

Applying the material balance principle, the input to a system should be equal to its output. However, approximately 80–90% of plastic waste contaminates the environment rather than re-entering the production cycle (Fig. 11.2a). Globally, only 9% of plastics are recycled, while the rest are incinerated (19%), landfilled (50%), or mismanaged (22%) (Fig. 11.2c). Consequently, a substantial portion inevitably pollutes terrestrial and aquatic ecosystems. A significant portion is incinerated, releasing hazardous by-products including dioxins, furans, and volatile organic compounds (VOCs). However, the majority of plastic waste, especially SUPPs, accumulates in landfills, water bodies, and terrestrial ecosystems, where it undergoes gradual degradation through photolysis, mechanical abrasion, and microbial interaction (OECD 2022, 2024; Stegmann et al. 2022).

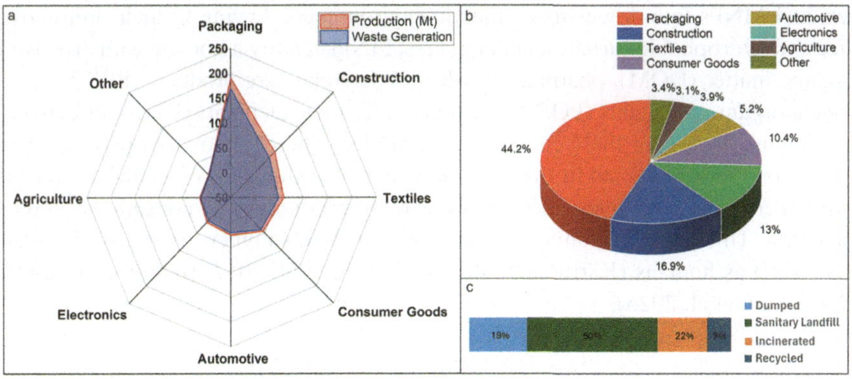

Fig. 11.2 From cradle to grave. (**a**) Comparison of global plastic production (blue) and waste generation (orange) by sector (in million metric tons, Mt). (**b**) Sector-wise contribution to total plastic waste. (**c**) Post-consumption fate of plastic waste

As plastics fragment, they transition into microplastics (MPs, <5 mm) and nanoplastics (NPs, <1000 nm), collectively termed micro-nanoplastics (MNPs), which are now considered emerging global contaminants. These particles persist for decades to centuries, depending on polymer type and environmental conditions, and infiltrate environmental compartments in increasingly alarming concentrations. MNPs come from two main sources: primary microplastics, intentionally produced as small particles (e.g., microbeads in personal care products), and secondary MNPs, which form when larger plastics break down (textiles, packaging, agricultural mulch, pipes and tanks, tire abrasions, and sanitary wastes) (Kumar et al. 2024; Sharma and Sharma 2024). Originating from a diverse array of sources, including household waste, industrial discharges, wastewater effluents, tire abrasions, medical waste, electronic waste, and atmospheric deposition. MNPs are now widely detected across diverse environmental matrices, from Arctic and Himalayan snow, human blood, and inside plants (Bergmann et al. 2022; Kumar et al. 2024; Leslie et al. 2022; Talukdar et al. 2023). Due to their minute size and physicochemical properties, MNPs exhibit high mobility, allowing them to be transported through aquatic systems, infiltrate soils, and disperse via atmospheric currents.

11.2 Micro-nanoplastics (MNP) Pollution

11.2.1 Presence and Transport of MNPs in Aquatic Systems

MNPs are now ubiquitously present in both freshwater and marine environments, entering these ecosystems through various anthropogenic routes, including industrial discharges, treated and untreated wastewater effluents, stormwater runoff, and the direct deposition of litter. Although municipal wastewater treatment facilities exhibit substantial removal efficiencies, they are not designed to fully capture particles in the submicron and nanoscale range, resulting in the continuous release of residual MNPs into downstream aquatic environments. Owing to their diminutive size and hydrophobic surface characteristics, MNPs readily associate with dissolved organic matter (DOM), pharmaceuticals and personal care products (PPCPs), persistent organic pollutants (POPs), and potentially toxic elements (PTEs), effectively serving as carriers for these contaminants. MNPs' inherent buoyancy permits them to stay on the surface and in suspension in aquatic systems for extended periods of time, increasing their bioavailability to primary producers and consumers such as plankton. This leads to biomagnification, which affects higher trophic level organisms such as humans (Kärrman et al. 2006; Ebele et al. 2017; Biswas et al. 2024; Choudhary et al. 2025).

11.2.2 Accumulation and Impact of MNPs in Terrestrial Environments

In terrestrial systems, MNPs contamination is generally generated by agricultural practices, including the decomposition of plastic mulch, the use of wastewater or leachate, and biosolids in the form of compost. They are key sources of an invisible threat, unknowingly releasing these MNPs into agricultural, groundwater, and riverine environments, which can lead to bioaccumulation, biomagnification, and toxicity at varying degrees. Primary producers, such as microalgae and plants, are the first to be harmed, with humans being the last, affecting the entire food chain, including all trophic levels (Liu et al. 2023).

11.2.3 Atmospheric Contamination and Transport of MNPs

MNPs enter the atmosphere via textile fibers, tire abrasions, road dust, and other primary and secondary sources. Open dumping and incineration are major sources of MNPs in the environment. Not only does open burning pollute the environment with MNPs, but it also emits volatile organic compounds (VOCs) (Manikanda Bharath et al. 2023). These partially burned airborne microplastics (AMPs) have the capacity to travel considerable distances; this can influence not just the area of release, but also contaminate far-flung pristine habitats, putting endangered animals and biodiversity at risk. These AMPs can contaminate both aquatic and terrestrial habitats (Chen et al. 2023a, b).

11.2.4 Functional Applications of Plastic Polymers

With over ten polymer types, including copolymers and blends, and more than 10,000 additives, addressing MNP pollution remains a formidable challenge (Pivnenko et al. 2017; Wiesinger et al. 2021). The chemical composition of these materials varies widely depending on functional requirements, regional demographics, climatic conditions, and economic factors. However, most plastic waste is predominantly composed of ten major polymer categories (Table 11.1) (CANSA 2011).

Plastic additives (Table 11.2), including plasticizers such as phthalates, flame retardants, stabilizers, antioxidants, colorants, fillers, and biodegradable additives, have been associated with various forms of toxicity. These substances are often incorporated into plastic formulations to enhance flexibility, thermal resistance, durability, and visual appeal (Goodship 2007; Fred-Ahmadu et al. 2020; Bridson et al. 2023). However, upon environmental exposure or during degradation, they can leach out and exert adverse effects on living organisms. For instance, phthalates are known endocrine-disrupting chemicals (EDCs), while certain flame retardants and stabilizers can be cytotoxic, genotoxic, or even carcinogenic. Antioxidants such as butylated hydroxytoluene (BHT) and Irganox 1010, although added to prevent polymer oxidation, have shown potential for bioaccumulation and can interfere with

Table 11.1 Commonly used plastic polymers

Polymer name	Abbreviation/ code	Origin (Monomer source)	Primary uses
Polyethylene	PE (HDPE 2, LDPE 4)	Ethylene (C_2H_4)	Bags, bottles, packaging, toys
Polypropylene	PP (5)	Propylene (C_3H_6)	Food containers, automotive parts, textiles
Polyvinyl chloride	PVC (3)	Vinyl chloride (C_2H_3cl)	Pipes, flooring, medical tubing (rigid/flexible forms)
Polystyrene	PS (6)	Styrene (C_8H_8)	Foam insulation (EPS), disposable cutlery, CD cases
Polyethylene terephthalate	PET (1)	Ethylene glycol + terephthalic acid	Beverage bottles, synthetic fibers (polyester)
Polyurethane	PUR	Diisocyanate + polyol	Foams (mattresses, insulation), adhesives, elastomers
Polycarbonate	PC	Bisphenol A (BPA) + phosgene	Eyeglass lenses, electronics, bulletproof glass
Acrylonitrile butadiene styrene	ABS	Acrylonitrile, butadiene, styrene	Lego bricks, automotive parts, 3D printing
Polymethyl methacrylate	PMMA	Methyl methacrylate	Acrylic glass (e.g., Plexiglas), dental implants
Nylon (polyamide)	PA	Diamine + dicarboxylic acid	Textiles, fishing nets, mechanical parts
Styrene-butadiene rubber (SBR)	SBR	Styrene + butadiene	Tires, shoe soles, conveyor belts

Table 11.2 Most common plastic additives

Additive type	Example compounds	Purpose	Health/environmental concerns
Plasticizers	Phthalates (DEHP, DINP)	Increase flexibility (e.g., soft PVC)	EDCs, bioaccumulation
Flame retardants	Brominated compounds (PBDEs)	Reduce flammability (e.g., electronics, furniture)	Neurotoxicity, persistent in environment
Stabilizers	Lead/cadmium compounds	Prevent UV/thermal degradation	PTEs toxicity (neuronal dysfunction, cancers)
Antioxidants	BHT, Irganox 1010	Delay oxidation during processing	Potential carcinogenicity
Colorants	Azo dyes, titanium dioxide	Provide color	Some dyes are carcinogenic
Fillers	Calcium carbonate, talc	Reduce cost, modify mechanical properties	Inhalation hazards during manufacturing
Biodegradable additives	Oxo-degradable compounds	Claim to enhance degradation	MNP fragmentation risk

cellular metabolic processes. Colorants and fillers may contain heavy metals or other toxic components, and some biodegradable additives may not fully degrade, instead generating harmful by-products that persist in the environment (Bridson et al. 2023).

However, certain widely used additives, including lead, flame retardants, bisphenol A (BPA), and bisphenol S (BPS), exhibit toxicity and have been linked to endocrine disruption, carcinogenesis, and other adverse health effects. These hazardous compounds raise serious concerns about their widespread use in polymer manufacture and formulation (Manzoor et al. 2022; Rifa and Lavado 2024; Wu et al. 2025).

11.2.5 Detection and Characterization of MNPs

MNPs found in aerial, terrestrial, and aquatic systems are classified based on their physicochemical properties, including size, shape, color, and polymer type. These characteristics aid in determining their origin and utility. Size distribution, fragmentation, thickness, and form all have an impact on their transport dynamics, environmental persistence, and interactions with flora. The composition of polymers offers information about the origin of materials, frequently associating particles with consumer goods, industrial waste, or the breakdown of bigger plastic waste. Thorough profiling of these characteristics is essential for source attribution, fate modeling, and ecological risk evaluation (Geyer et al. 2017; Choudhary et al. 2025).

A variety of sophisticated analytical methods have been created to identify and describe MNPs. Pyrolysis gas chromatography-mass spectrometry (Py-GC-MS) offers highly sensitive, compound-specific identification of polymer types, though it is inherently destructive and precludes subsequent morphological analysis. In contrast, Fourier-transform infrared spectroscopy (FTIR) and Raman spectroscopy are widely adopted for nondestructive chemical analysis, although their resolution decreases significantly for particles smaller than 10 μm due to inherent optical limitations. Optical photothermal infrared spectroscopy (OPTIR) presents a hybrid approach, enhancing spatial resolution for submicron particles (~500 nm). Visual characterization using stereomicroscopy or scanning electron microscopy (SEM) facilitates assessments of particle shape and size but lacks molecular specificity unless combined with complementary spectroscopic techniques. These methodologies, while powerful, are constrained by high operational costs, labor-intensive sample preparation, and reduced reliability when dealing with complex environmental matrices (Jenner et al. 2022; Leslie et al. 2022; Pal et al. 2023; Massardo et al. 2024; Tarafdar et al. 2024).

Although they have applications, these sophisticated methods encounter challenges in establishing standardized protocols, measuring NPs, and dealing with diverse individuals and combinations of synthetic polymers. Environmental heterogeneity complicates the analysis and identification of MNPs. In the future, researchers should focus on enhancing detection limitations, interferences, matrix effects, and analytical biases and employing a multidimensional method to verify and enhance MNP detection reliability and speed. Matrices such as blood, tissues, sewage sludge, leachate, compost, and other organic-rich matrices require centralized standardization in the field of MNP quantification as detection is an important step in the course of remediation (Sujathan et al. 2017; Leslie et al. 2022; Wiesner et al. 2023).

11.2.6 MNPs' Fate in the Environment

In detail, the physicochemical features of MNPs, such as shape, size, density, surface charges, hydrophobicity, crystallinity, weathering, polymer, and additive types, determine their fate in aquatic, terrestrial, and biological systems. These characteristics control aggregation, sorption dynamics, and mobility. Moisture, pH, ionic strength, and organic matter (biofilms/microbes/humics) all contribute to MNP retention and bioavailability in the soil (Singh et al. 2019; Ganie et al. 2024). In aquatic systems, salinity, turbulence, and dissolved organic matter all have an impact on the delivery of important minerals into the roots, as well as cytotoxicity, oxidative stress, and inflammation. Adsorption of other developing pollutants can exacerbate the Trojan horse effect and increase toxicity (Liu et al. 2022; Yu et al. 2024).

Polymer composition influences density, crystallinity, glass transition temperature, and susceptibility to degradation. Common polymers in environments include PE, PP, PS, PVC, Styrene-Butadiene Rubber (SBR), and polyethylene terephthalate (PET). This diverse polymer composition presents a significant challenge for remediation, as no single approach effectively targets all polymer types simultaneously. Surface properties evolve considerably during environmental exposure. Weathering processes, including photo-oxidation, thermal degradation, hydrolysis, mechanical abrasion, and biodegradation, generate surface cracks, pits, and functional groups that enhance reactivity (Chamas et al. 2020). Additionally, the development of biofilms (the "plastisphere") modifies surface chemistry, buoyancy, and aggregation behavior, with implications for both transport and remediation.

11.2.7 Environmental and Health Risks

The widespread presence of MNPs poses a dual threat to ecosystems and human health. In aquatic systems, their ingestion by marine organisms can lead to physiological stress, reduced feeding efficiency, and altered metabolic functions. In terrestrial environments, soil-dwelling organisms, such as earthworms, have shown changes in behavior and reproduction upon MNP exposure. Additionally, humans are exposed to MNPs through contaminated food, water, and air, with potential health consequences including inflammation, oxidative stress, and endocrine disruption (Luqman et al. 2021; Dennis et al. 2025).

11.2.8 Challenges in MNP Removal and Remediation

Traditional wastewater treatment plants (WWTPs) play a significant role in MNP interception; however, studies indicate that a considerable fraction of MNPs bypass conventional filtration systems, eventually entering aquatic environments. Physical removal methods, including membrane filtration, coagulation-flocculation, and sedimentation, have shown varying degrees of success but often pose operational challenges such as membrane fouling and high energy costs. Similarly, chemical and biological treatments remain limited due to the complex nature of MNPs and their interactions with other environmental contaminants (Joseph et al. 2023; Liu et al. 2024).

Given these challenges, adsorption-based techniques have gained attention as an effective and sustainable approach for MNP remediation. The use of adsorbents such as biochar, activated carbon, and other engineered materials presents a promising solution to enhance the removal efficiency of MNPs from contaminated water sources (Wang et al. 2024).

11.3 Life Cycle of Household Plastics

Plastics used in households undergo a complex life cycle, beginning with extraction and production. Synthesized from fossil fuels through polymerization, polymers like PE, PET, PVC, PP, PS, and SBR are molded into durable, lightweight, and cost-effective products such as packaging, containers, toys, and appliances. Household plastics originate from both single-use items, like food wrappers, and durable goods, such as furniture and electronics, making them ubiquitous in daily life. During the use phase, plastics serve multiple functions, from preserving food to insulating electrical components. Post-consumer plastics enter waste streams, with only a fraction being recycled due to contamination or lack of infrastructure; the majority are landfilled, incinerated, or leak into ecosystems, contributing to environmental pollution (Fig. 11.2c) (Civancik-Uslu et al. 2021).

11.3.1 Household Waste as a Major Source of MNP Pollution

Rapid urbanization and exponential population growth have driven increased consumerism and a corresponding rise in waste generation. A significant portion of this waste originates from domestic sources and unauthorized commercial activities, particularly from Micro, Small, and Medium Enterprises (MSMEs), often embedded within residential areas in developing countries. Regulatory authorities unknowingly categorize this household waste as municipal solid or liquid waste. Household waste is a significant contributor to plastic pollution, with municipal solid waste (MSW) and wastewater serving as primary pathways for MNPs release into the environment (He et al. 2019; Kumar et al. 2023).

Liquid waste, commonly termed wastewater, primarily consists of household effluents directed to wastewater treatment plants (WWTPs). In contrast, solid waste, known as municipal solid waste (MSW), is either disposed of in landfills, processed into compost, or incinerated in thermal plants for energy recovery. MNPs stem from diverse anthropogenic activities, with wastewater and MSW among the most critical contributors to environmental contamination. These systems act as persistent entry points, releasing plastic particles into aquatic, terrestrial, and atmospheric compartments. Integrated waste management should adopt a hierarchy (Fig. 11.3) prioritizing prevention, reuse of less deteriorated materials, recycling, recovery, and disposal. In developing nations like India, less deteriorated textile products are often imported and sold at lower prices, but after repeated use, their reusability diminishes, releasing substantial waste into the environment. Public awareness must be raised to discourage purchasing used synthetic clothing for brand appeal and instead promote locally available natural cotton fabrics, as prevention is more effective than post-consumption waste treatment (Srivastava et al. 2020; Alhazmi et al. 2021; Soni et al. 2023).

11.3.2 Knowing the Role of Wastewater and MSW in MNP Pollution

Wastewater and MSW serve as critical indicators for assessing consumerism and MNP pollution, acting as primary pathways linking production to environmental release. These vectors accumulate plastic debris from diverse sources, reflecting consumption patterns while facilitating the transport of MNPs into ecosystems through disposal and treatment processes. Their analysis provides valuable insights

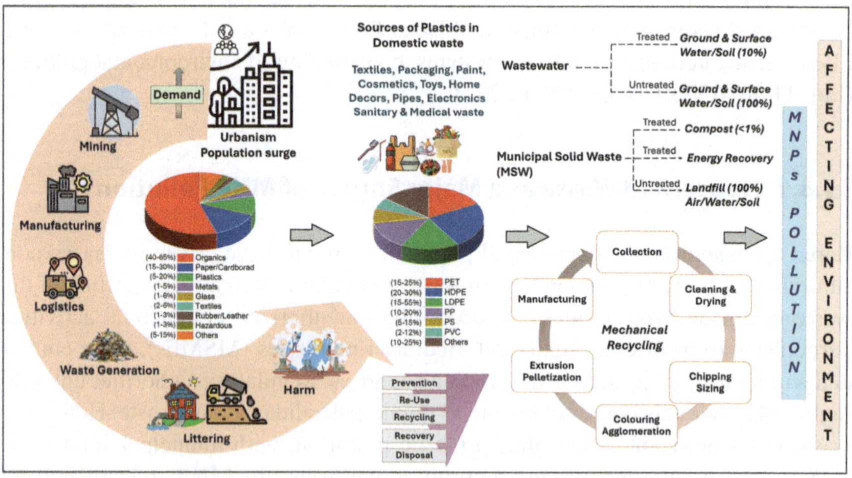

Fig. 11.3 Life cycle of plastics within MSW, highlighting their proportion, sources, polymer types, hierarchy of integrated waste management, and environmental fate

into pollution trends, enabling targeted mitigation strategies to address environmental contamination (Ortega and Cortés-Arriagada 2023; Tumwesigye et al. 2023).

Wastewater is a complex mixture containing textile waste, PPCPs, POPs, PTEs, and other emerging contaminants such as MNPs. Wastewater is broadly classified into greywater (originating from showers, sinks, kitchens, and laundry) and blackwater (toilet waste).

WWTPs are major conduits of MNP pollution, as they receive plastic-rich effluents from households, industries, and stormwater runoff. Common sources of MNPs in wastewater include:

- Synthetic fibers shed from textiles during washing and directly from MSMEs
- Microbeads from personal care products (PCPs)
- Industrial effluents containing plastic additives, resins, and plastic fragments
- Urban runoff carrying degraded plastic particles from roads, packaging waste, and litter

Although modern WWTPs can remove 80–99% of MNPs, a substantial fraction remains in the effluent and sludge. Studies indicate that treated effluent still discharges millions of MNP particles per day into rivers, while sludge disposal on agricultural fields reintroduces MNPs into terrestrial ecosystems (Cristaldi et al. 2020; Carnevale Miino et al. 2024).

MSW comprises household, commercial, and institutional discards, including organics, paper, cardboard, plastics, metals, glass, textiles, rubber/leather, hazardous waste, and others. These pollutants contaminate soil, water, and air, posing significant ecological and health risks due to their persistence and toxicity. Plastics account for 10–15% of global MSW, comprising packaging and synthetic textiles. Over time, plastics from textiles, packaging, SUPPs, PCPs, toys, home decor, pipes, electronics, and sanitary waste directly release MNPs, or released macroplastics degrade into MPs and NPs through mechanical abrasion, UV exposure, and chemical and biological weathering. Improper plastic disposal through landfilling, mechanical recycling, composting, or incineration causes degradation and fragmentation, releasing MNPs (Chen and Gao 2021; He et al. 2019; Manikanda Bharath et al. 2023).

Plastic recycling plants valorize waste by mechanically reprocessing materials through collection, cleaning, shredding, extrusion, and pelletization, extending their life cycle. The recycled output is repurposed for textiles, bottles, and food containers. However, inefficient operations in these facilities release MNPs due to mechanical degradation and poor handling, worsening environmental pollution. The drive to lower production costs often results in the widespread use of low-quality recycled plastics in consumer products, such as disposable food containers and beverage bottles. These materials, while economically attractive, may contain residual additives or degraded polymers, thereby introducing hidden health hazards through direct contact or ingestion (Sánchez et al. 2022; Kaseke et al. 2023).

Landfills serve as long-term reservoirs for MSW, particularly plastics, producing leachates enriched with dissolved organic matter and contaminants, especially in urban areas. When such leachate is used for agricultural irrigation, it introduces MNPs and toxic chemicals that infiltrate groundwater or surface water bodies. These pollutants may be taken up by plants, transferred through trophic levels, and ultimately pose risks to human health (He et al. 2019; Sharma et al. 2019).

11.3.3 Economic Trends in MSW Production and Plastic Contribution

MSW composition varies significantly across high-income, middle-income, and low-income countries, reflecting differences in consumption patterns, economic development, and waste management infrastructure. In high-income countries, MSW is predominantly composed of packaging materials, plastics, paper, and food waste (Table 11.3). A high proportion of single-use plastics and composite packaging materials, such as multilayer films and containers, are generated due to widespread consumerism and convenience-oriented lifestyles (Ncube et al. 2021; De Sousa 2023).

In middle-income countries, the MSW profile mirrors a transitional economy marked by rapid urbanization and industrialization. Organic waste constitutes 40–60% of MSW, while plastics, primarily PET (bottles and textile-derived polyesters), HDPE, LDPE, and PP from insulated clothes, food packaging, shopping bags, and household containers, represent 10–15% and are growing rapidly (Sharma et al. 2019; Georgios et al. 2020; Funk et al. 2020; Soni et al. 2023). Rising demand for consumer goods drives plastic waste generation, yet inadequate waste management systems often lead to environmental leakage. In contrast, middle- and low-income countries exhibit MSW compositions dominated by organic waste (40–60%), with plastics (<10%) mostly comprising LDPE and mixed polyolefins from sachets,

Table 11.3 Economy-wise distribution of MSW

Component	High-income countries (e.g., USA, EU) (%)	Middle-income countries (e.g., India, Brazil) (%)	Low-income countries (e.g., Sub-Saharan Africa) (%)
Organic waste	30–40	40–60	50–70
Paper and cardboard	20–30	10–20	5–15
Plastics	5–15 (US>10, EU < 10)	5–15	4–10
Metals	3–6	1–4	1–3
Glass	4–6	2–5	1–3
Textiles	3–6	2–5	2–6
Rubber/leather	1–3	1–3	1–2
Hazardous waste	1–3	1–4	0.5–2
Other waste	5–10	5–15	5–20

carry bags, and low-cost containers. While constrained industrialization and low packaged goods consumption limit plastic diversity, deficient waste management systems promote open dumping and burning, exacerbating environmental impacts despite reduced plastic generation (Agamuthu and Babel 2023; Idumah and Nwuzor 2019; Pilapitiya and Ratnayake 2024).

Developing nations, including India (Kumar and Agrawal 2020; Soni et al. 2023) and Tanzania (Marshall and Farahbakhsh 2013), face significant challenges in MSW management due to insufficient infrastructure, technological limitations, and the absence of robust regulatory frameworks and monitoring systems. These constraints hinder operational efficiency in waste management processes. Urban pollution stemming from MSW is a pressing concern in these regions, exacerbated by rising waste generation rates. Annually, per capita waste production ranges between 109.2 and 525.6 kg in developing countries, compared to 521.95 and 759.2 kg in industrialized nations, further straining existing waste management capacities (Soni et al. 2023).

11.4 Biochar as a Remediation Material

11.4.1 Production and Properties

Biochar is a carbon-rich substance formed via pyrolysis, the thermochemical degradation of biomass under anoxygenic conditions. Derived from diverse feedstocks such as agricultural residues (wheat, rice, coconut, peanut, walnut, corn, sugarcane, pine, walnut), forestry waste, and urban organic waste, it exhibits recalcitrant properties and a unique physicochemical structure beneficial for environmental applications (Lehmann and Joseph 2015). Biochar produced through pyrolysis at temperatures typically ranging from 300 °C to 900 °C offers a promising approach to mitigate MNP pollution. The physicochemical characteristics of biochar, including surface area, porosity, and functional group composition, are highly tunable based on pyrolysis parameters and feedstock. Sewage sludge biochar can be used as a soil amendment for remediation and increasing soil fertility (Xie et al. 2015; Zornoza et al. 2016).

Biochar's graded pore network (nano, meso, and macropores) has a high surface area (50–1000 m^2/g). With specific surface areas often exceeding 1000 m^2/g when activated, biochar presents high reactivity for pollutant removal (Liang et al. 2019; Ahmad Zamri et al. 2024). Carboxyls and hydroxyls are functional groups that offer hydrophobicity and a negative charge, which aids interactions through complexation, electrostatic forces, and redox reactions (He et al. 2024; Zhang et al. 2025). Moreover, biochar helps in carbon sequestration, and its design from waste biomass into valuable products (banana peel, walnut, peanut, sugarcane bagasse, rice and wheat straw, and coconut shell) supports circular economy principles.

11.4.2 Mechanisms of MNP Adsorption on Biochar

The primary mechanism for MNP adsorption onto biochar is physisorption, which includes hydrophobic interactions, van der Waals forces, and hydrogen bonding (Ganie et al. 2021; Singh et al. 2021). The porous nature of biochar promotes π-π interactions with polymers, while the surface shape allows for mechanical interlocking. Pore capture is important for tiny MNPs. In weathered MNPs, charged functional groups can engage in electrostatic binding with oppositely charged biochar surfaces (Verla et al. 2019). Additionally, hydrogen bonding and metal-bridged surface complexation enhance retention (Kumar et al. 2020; Mota et al. 2025) (Fig. 11.4).

Compared to more capital-intensive remediation technologies (e.g., membrane filtration, electrocoagulation, and advanced oxidation), sorption-based strategies offer scalable, decentralized treatment solutions. Biochar, activated carbon, and functionalized adsorbents have demonstrated high affinity for MNPs across freshwater and saline conditions. Hydrophobic MNPs, particularly polyethylene and polypropylene, are effectively sequestered by biochars with low surface polarity (Shrivastava et al. 2024). However, salinity can attenuate hydrophobic interactions due to salting-out effects. The efficacy of MNP removal via biochar is context-specific, governed by biochar aging, environmental chemistry, and the degree of polymer weathering. Biochars produced at higher temperatures exhibit greater aromaticity and lower polarity, favoring the adsorption of pristine plastics. In contrast, oxygenated biochars exhibit enhanced interactions with aged MNPs due to their polar surfaces (Liu et al. 2019).

Electrostatic interactions become relevant as both MNPs and biochar surfaces acquire charges during weathering. Weathered MNPs exhibit oxygen-containing groups, enhancing hydrogen bonding and surface complexation with biochar. Ash-derived metal ions may further mediate bridging interactions. Biochar demonstrates a high removal efficiency ranging from 70% to 99% for MNPs, indicating its strong

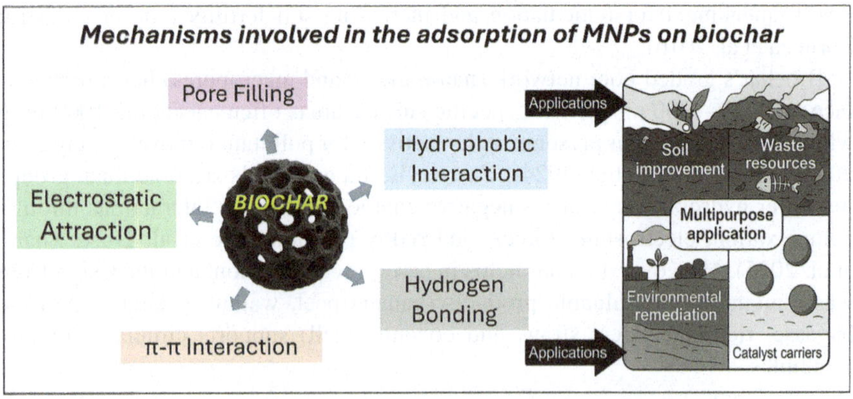

Fig. 11.4 Key mechanisms of MNP adsorption on biochar

potential as an effective and sustainable material for MNP remediation (Table 11.4). Higher pyrolysis temperatures enhance carbon content, aromaticity, surface area, and stability but reduce functional groups and cation exchange capacity; hydrophobic interactions dominate for nonpolar polymers (PE, PP, PS), especially with high-temperature biochars. However, salinity can suppress these effects by reducing the hydrophobic interaction energy (Abuwatfa et al. 2021; Shrivastava et al. 2024).

11.4.3 Factors Affecting Adsorption Efficacy

The adsorption performance of biochar is predominantly controlled by its physicochemical properties, with surface area exhibiting a direct positive correlation with adsorption capacity. Nanoporous biochars generated at elevated temperatures generally demonstrate better efficacy, although pore accessibility can be restricted when the dimensions of MNPs surpass the sizes of the pores (Ahmed et al. 2016). Engineered biochar with surface modification shows a better affinity for removing MNPs. More hydrophobic biochars can effectively adsorb pure MNPs, but oxygenated biochars bind weathered MNPs more easily due to their higher surface polarity. Weathered MNPs have a larger affinity for biochar than immaculate ones due to their enhanced roughness and oxygen functionality (Li et al. 2024).

The physicochemical properties of water have a significant impact on adsorption dynamics; for example, circumneutral pH reduces electrostatic repulsion, encouraging adsorption competence. Ionic strength can increase aggregation by electrostatic contact, whereas DOM competes with adsorption for binding sites, reducing effectiveness (Singh et al. 2019; Kumar et al. 2023). Hydraulic factors such as flow rate, residence period, and flow rate can also influence the cleanup. A high flow rate can reduce contact time and thus lower removal, whereas a low flow rate can reverse this. To achieve the best remediation results, these parameters must be optimized and balanced (Kumar et al. 2021; Chen et al. 2023a, b).

11.4.4 Application of Biochar in Aquatic and Terrestrial MNP Remediation

The use of biochar to remediate MNPs in both aquatic and terrestrial environments has produced promising results. Before use in constructed wetlands or industrial water units. Lab-based column and batch sorption investigations are critical for gaining a clear understanding of dosage optimization, simulating the environment, and increasing productivity and remediation. Batch sorption experiments are critical for understanding the physicochemical interactions between MNPs and biochar in controlled settings. However, realistic environmental circumstances are variable and nonuniform. Nowadays, it is critical to evaluate such biochars in real-world situations, such as with leachate. In these batch sorting investigations, the inherent properties of biochar, particularly its high surface area and porosity, enable efficient MNP adsorption via hydrogen bonding, hydrophobic, and electrostatic interactions.

Table 11.4 Summary of recent studies evaluating biochar-mediated removal of MNPs under batch and column experiments

Biochar feedstock	MNP type (polymer/size)	Experiment type	Interaction mechanism	Removal efficiency	Key findings	References
Palm kernel Coconut shell	PE (146–293 μm)	Column	Hydrophobic interaction	94.13% 92.27%	Efficiently removed MPs from the PCPs	Hanif et al. (2023)
Sugarcane bagasse biochar	PS (<500 nm)	Column	Electrostatic attractions	99%	BC750 showed more removal	Ganie et al. (2021)
Cellulose (400 °C)	PS (20, 200, 2000 nm)	Column	Electrostatic attractions	68.5%, 37.5%, 36.9%	Efficient removal of small-sized MNPs	Tong et al. (2020)
Corn straw and hardwood feedstock Biochar (500 °C)	PS (10 μm)	Column	Entangling, trapping, and sticking	>95%	Potentially remove microbeads from WWTPs	Wang et al. (2020)
Steam activated pine (475 °C) and spruce bark biochar (800 °C)	Spherical PE microbeads (10 μm), cylindrical PE pieces (2–3 mm), and fleece shirt fibers	Column	Pore entrapment	~100% for cylindrical PE pieces and fleece fibers Microbead removal was insufficient	Valorization of wood waste	Siipola et al. (2020)
Corncob (550 °C, 700 °C, 900 °C)	PS (50 nm)	Batch	Hydrophobicity, pore filling	BC900 > BC700 > BC 550	Increased surface area, hydrophobicity, and aromaticity	Abdoul Magid et al. (2021)
Filamentous algae (400 °C, 500 °C, and 600 °C)	PS (1 μm)	Batch	Entanglement, adhesion, encapsulation	176.99 mg/g (5 times)	Promotes valorization	Li et al. (2024)

Rice straw (500 °C)	PS (300 nm)	Batch	Electrostatic attraction, surface complexation, hydrogen bonding, and π-π interactions	99.96–90% (4 cycles) 32.87 mg/g	Displays higher efficiency of magnetic biochar	Wu et al. (2022)
Sewage sludge (480 °C)	PS (80 nm)	Batch and column	Electrostatic interaction, hydrogen bonding, and π-π conjugation	360 mg/g	Most effective way of valorization	Sun et al. (2023)
Banana peel (650 °C)	PS (<75 μm, 75 to 150 μm, and 150 to 300 μm)	Column	Pore filling, hydrophobic interactions	100–91.53% (4 cycles)	Valorization	Subair et al. (2024)

The dynamics of adsorption are greatly modified by feedstock source, pyrolysis temperature, particle size, and other factors. Following that, data are analyzed using isotherm models, primarily the Freundlich and Langmuir (Ganie et al. 2021; Kumar et al. 2023).

Column tests will be more focused on industrial and commercial applications, using optimized and highly efficient biochar with continuous hydraulic flow. Sorptive retention capacity and breakthrough curves are required for performance evaluation in these dynamic settings (Lehmann and Joseph 2015). Building a wetland system using biochars creates a multipurpose system for pollutant removal, filtration, sedimentation, and biota-driven deterioration. One of the most promising remediation strategies is the use of soil amendments. More research is needed to understand scalability, ecological implications, and the full potential of biochar.

Biochar offers multifaceted environmental benefits across soil and water systems while supporting circular economy goals. In soil, it enhances fertility and plant productivity by increasing cation exchange capacity (CEC), improving structure, and boosting water retention. Its stable carbon matrix contributes to long-term carbon sequestration, making it a valuable tool for climate mitigation (Fig. 11.5). In aquatic environments, biochar functions as an efficient sorbent, enabling the removal of pollutants and retention of essential nutrients, thereby improving water quality. Produced from organic waste through pyrolysis, biochar exemplifies resource valorization by converting biomass into a high-value product, reducing urban waste

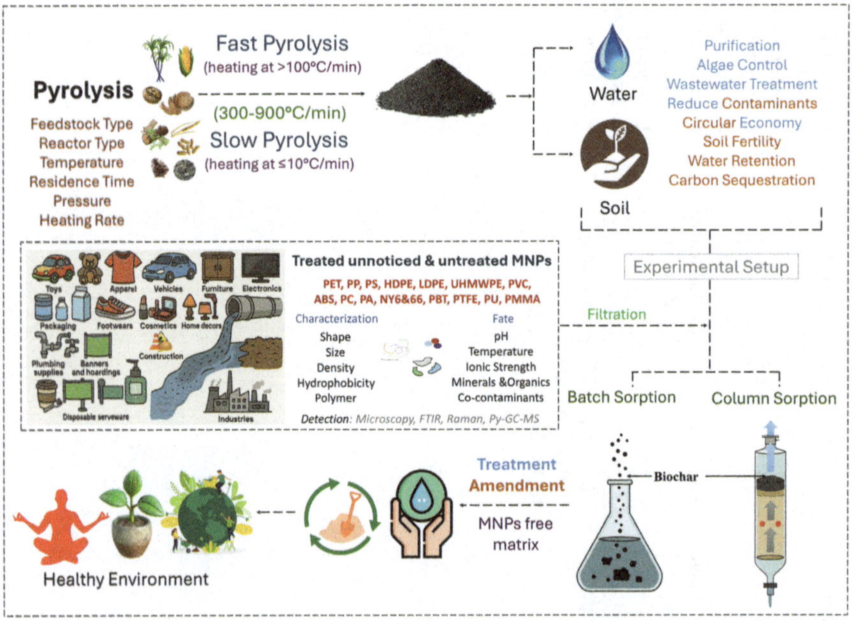

Fig. 11.5 Pyrolytic conversion of biomass waste into biochar and its application in removing MNPs from the environment and its role in promoting sustainability

streams, and reinforcing sustainable waste management practices (Bano et al. 2025; Singh et al. 2022).

11.4.5 Other Adsorbing Alternatives for MNP Remediation

Beyond biochar, several adsorbents are employed for MNPs, each with distinct advantages and limitations. Activated carbon, with its high surface area (500–3000 m^2/g) and microporosity, effectively removes contaminants via physisorption, though its efficacy diminishes for larger particles and in the presence of natural organic matter, compounded by high production costs (Jjagwe et al. 2021; Lawtae and Tangsathitkulchai 2021). Clay minerals, such as kaolinite and gibbsite, leverage electrostatic interactions and cation exchange but face challenges like swelling and clogging, prompting the development of organo-modified and composite variants (Choudhary et al. 2022). Metal-organic frameworks (MOFs) offer exceptional porosity (>7000 m^2/g) and tunable binding mechanisms, yet their environmental application is hindered by water sensitivity and cost, driving research toward stable composites (Ihsanullah 2022). Magnetic composites, incorporating iron oxides for easy recovery, achieve high removal and recovery efficiencies but require optimization to mitigate leaching risks (Cao et al. 2024). Each technology presents trade-offs between performance, cost, and practicality, necessitating context-specific solutions.

11.5 Global Plastic Use: Future Projections and Interventions

Global plastic production has exceeded 8.3 billion Mt. since the 1950s (Geyer et al. 2017), dominated by packaging, construction, and textiles. MNPs arise from primary (e.g., microbeads, industrial pellets) and secondary sources (fragmentation of larger plastics), with around 11 Mt. entering into ocean systems annually (Matavos-Aramyan 2024). Terrestrial contamination, particularly in agricultural soils, exceeds marine levels by a factor of 4–23 (Ren et al. 2021).

By 2050, nearly two-thirds of the global population is projected to live in urban areas, generating approximately 3.40 billion tons of MSW annually, an increase of 70% compared to the current 2.01 billion tons (Hamdan et al. 2025).

By 2050, plastic production in Asia, driven predominantly by China, India, and other Asian nations, is projected to grow substantially faster than in regions such as OECD America, OECD Europe, and the OECD Pacific. China and India, the two largest plastic producers, could significantly contribute to plastic leaching if waste is not properly managed. India, in particular, is expected to see plastic production rise nearly fourfold compared to current levels, making it one of the largest future contributors (Fig. 11.6). In contrast, growth in OECD nations is relatively moderate (~3 times), reflecting plateauing consumption and improved waste management policies. The significance lies in the disproportionate rise of production in developing and emerging economies, where waste management infrastructure often lags

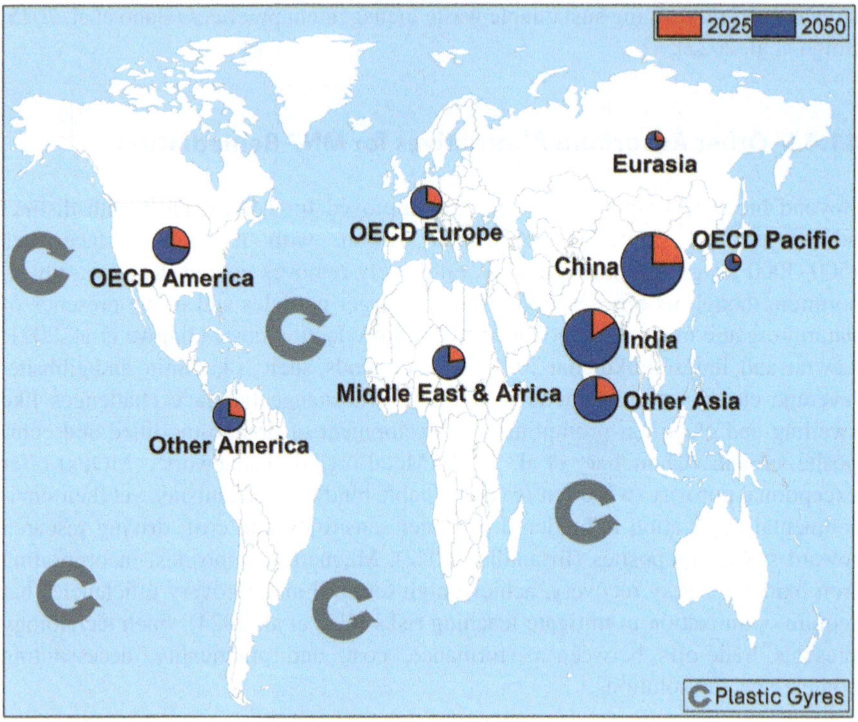

Fig. 11.6 Projected regional plastic production in 2025 (red) and 2050 (blue), with plastic gyres, where mismanaged plastic waste from land-based sources can accumulate through transboundary water transport

behind. This creates a pathway for mismanaged plastic waste to enter river systems, eventually flowing into transboundary waters. From there, currents transport the waste to the five major oceanic plastic gyres, large rotating systems where plastics accumulate over time. This scenario underscores the urgency for global cooperation in plastic waste governance, particularly in rapidly industrializing regions (Andrade et al. 2021).

High-income nations recycle plastics and utilize waste-to-energy (WtE) technologies (Alao et al. 2022); low-income countries rely heavily on open dumping (~90%). India produces 143,449 Mt./day of MSW, with scientific treatment limited to 24.8%. Key initiatives include the Swachh Bharat Mission and Plastic Waste Management Rules (2016, amended 2021), emphasizing segregation, Extended Producer Responsibility (EPR), and bans on select SSUPs (Kumar and Agrawal 2020). India generates approximately 62 Mt. of MSW and 38 billion liters of wastewater annually from urban areas alone. MSW is increasing at a rate of 1–1.33% per year, yet a significant portion remains untreated. This growing urban waste burden is primarily driven by low public awareness and inadequate institutional response, leading to persistent challenges in maintaining clean cities (Ghosh 2016).

In response to rising concerns over plastic contamination in organic waste, Germany has amended its BioWaste Ordinance to enforce stricter impurity thresholds. Effective from May 1, 2025, plastic content in biowaste must not exceed 1%, with penalties up to €2500 for improper disposal, including plastic bags or foliage in biowaste bins. Despite the popularity of compostable bioplastic bags, studies have shown that they do not fully degrade within the standard 12-week composting cycle. Consequently, consumer advocacy groups in Germany have called for transparent and reliable labeling of biodegradable products. While these materials may be certified under EN 13432 (Laycock and Halley 2014), they are often ineffective in real-world composting environments and are discouraged as substitutes for conventional plastics.

The current German Fertilizer Ordinance permits a plastic limit of 0.5% by weight in compost, yet it only accounts for particles larger than 1 mm, thereby excluding microplastics from regulatory oversight. At the EU level, Regulation (EU) 2019/1009, revised in November 2024, sets updated rules for EU fertilizing products but similarly lacks provisions targeting MNPs. Austria's national action plan on microplastics (2022–2025) and the EU's new Packaging Ordinance (2025) underscore a policy shift toward compostability and bio-based packaging, emphasizing upstream control. However, these initiatives primarily focus on prevention, not remediation, reflecting a broader European stance that prioritizes source reduction over technological cleanup strategies (Graça et al. 2023; Wiesner et al. 2023).

The Plastic Waste Management Rules, 2021, marked a significant policy shift, with the nationwide ban on select single-use plastic products (SUPPs) implemented in mid-2022 serving as a catalyst for advancing plastic waste regulation across India (Nøklebye et al. 2023). Policy and Practical Gaps India's 2021 Plastic Waste Management Rules and the subsequent 2022 ban on single-use plastics represent major policy strides. Between 2015 and 2021, treated waste increased from 19% to 50%, while direct landfilling dropped from 54% to 18.4% (CPCB, 2021). Still, daily per capita waste generation in India at around 119.27 g remains lower than in Germany (2.0 kg) or the USA (2.5 kg) (Siddiqi et al. 2021). However, unlike many European policies that prioritize prevention over cleanup, India's reliance on Mechanical Biological Treatment (MBT) systems has led to the unintended misuse of substandard compost. In cities such as Goa and Mumbai, outputs from MBT plants are applied to soil without thorough contaminant screening, increasing the risk of environmental loading with MNPs. These misapplications underscore the importance of adapting technology to regional contexts, a lesson drawn from the United Kingdom's prior missteps with "Compost-Like Output" (Bayard et al. 2010; Judy et al. 2019).

The European Union's Single-Use Plastics Directive (2019/904) bans selected disposable products, while the Packaging and Packaging Waste Regulation fosters circularity (Kiessling et al. 2023); in the United States, the Microbead-Free Waters Act (2015) prohibits the use of microbeads in rinse-off cosmetics (McDevitt et al. 2017); and Canada's Zero Plastic Waste Initiative aims for complete recyclability of plastics by 2030 (Martinez 2022).

11.6 Future Perspectives and Research Gaps

The performance and market behavior of any product are largely driven by consumer demand. Future consumption should prioritize commodities made from natural materials while minimizing the use of synthetic alternatives through reduction and reuse strategies where feasible. A strong, centralized international body is essential to oversee and implement on-the-ground actions, particularly in densely populated developing countries such as India and Indonesia. A cradle-to-contamination life cycle assessment reveals that over 85% of produced plastics eventually enter the environment. These plastics degrade MNPs, which persist in ecosystems, bioaccumulate in organisms, and biomagnify through food chains, posing significant risks to human health. When applied to specific environmental pathways, biochar can significantly reduce MNP contamination, potentially lowering concentrations by 10–20%. However, identifying and targeting these pivotal pathways remains a major research gap.

Post-release, synthetic pollutants should be immobilized using cost-effective biochar. However, the large-scale production of efficient biochar remains challenging and requires standardized, low-cost production methods. Equally critical is the management of spent biochar after pollutant adsorption to prevent secondary environmental risks. With scalable production of multifunctional and economical biochar, MNPs can be effectively trapped in surface soils, thereby preventing their leaching into groundwater, entry into surface waters, or transfer through the food web.

Future research on MNP remediation using biochar must address critical regulatory and methodological gaps to support its environmental deployment. Standardization remains a priority, as variations in biochar synthesis, MNP characteristics, and testing protocols hinder cross-study comparability. Agencies like the CPCB, US EPA, and EU REACH emphasize the need for uniform methods in environmental risk assessments, including controlled parameters such as MNP concentrations, shape, size, pH, and dosage. Similarly, advanced spectroscopic techniques (e.g., FTIR, DLS, SEM-EDS, Raman, and contact angle) are essential for regulatory validation of MNPs-biochar interaction mechanisms. A robust policy framework should also guide life cycle assessment (LCA) and cost-benefit analyses, as advocated by the OECD and United Nations Environment Programme (UNEP), to ensure that biochar's environmental and economic sustainability is benchmarked against conventional treatments like membrane filtration and oxidation processes.

References

Abdoul Magid ASI, Islam MS, Chen Y, Weng L, Li J, Ma J, Li Y (2021) Enhanced adsorption of polystyrene nanoplastics (PSNPs) onto oxidized corncob biochar with high pyrolysis temperature. Sci Total Environ 784:147115. https://doi.org/10.1016/j.scitotenv.2021.147115

Abuwatfa WH, Al-Muqbel D, Al-Othman A, Halalsheh N, Tawalbeh M (2021) Insights into the removal of microplastics from water using biochar in the era of COVID-19: a mini review. Case Stud Chem Environ Eng 4:100151. https://doi.org/10.1016/j.cscee.2021.100151

Agamuthu P, Babel S (2023) Waste management developments in the last five decades: Asian perspective. Waste Manag Res 41:1699–1716. https://doi.org/10.1177/0734242x231199938

Ahmad Zamri MFM, Hassan SHA, Tiong SK, Milano J, Bahru R, Fattah IMR, Mahlia TMI (2024) Progress and challenges of mesoporous catalysts in upgraded pyrolysis of biomass for biofuel production. J Anal Appl Pyrolysis 182:106651. https://doi.org/10.1016/j.jaap.2024.106651

Ahmed MB, Zhou JL, Ngo HH, Guo W, Chen M (2016) Progress in the preparation and application of modified biochar for improved contaminant removal from water and wastewater. Bioresour Technol 214:836–851. https://doi.org/10.1016/j.biortech.2016.05.057

Alao MA, Popoola OM, Ayodele TR (2022) Waste-to-energy nexus: an overview of technologies and implementation for sustainable development. Clean Energy Syst 3:100034. https://doi.org/10.1016/j.cles.2022.100034

Alhazmi H, Almansour FH, Aldhafeeri Z (2021) Plastic waste management: a review of existing life cycle assessment studies. Sustainability 13:5340. https://doi.org/10.3390/su13105340

Andrade H, Glüge J, Herzke D, Ashta NM, Nayagar SM, Scheringer M (2021) Oceanic long-range transport of organic additives present in plastic products: an overview. Environ Sci Eur 33:85. https://doi.org/10.1186/s12302-021-00522-x

Bano A, Aziz MK, Prasad B, Ravi R, Shah MP, Lins PVDS, Meili L, Prasad KS (2025) The multifaceted power of biochar: a review on its role in pollution control, sustainable agriculture, and circular economy. Environ Chem Ecotoxicol 7:286–304. https://doi.org/10.1016/j.enceco.2025.01.004

Bayard R, De Araújo Morais J, Ducom G, Achour F, Rouez M, Gourdon R (2010) Assessment of the effectiveness of an industrial unit of mechanical–biological treatment of municipal solid waste. J Hazard Mater 175:23–32. https://doi.org/10.1016/j.jhazmat.2009.10.049

Bergmann M, Collard F, Fabres J, Gabrielsen GW, Provencher JF, Rochman CM, Van Sebille E, Tekman MB (2022) Plastic pollution in the Arctic. Nat Rev Earth Environ 3:323–337. https://doi.org/10.1038/s43017-022-00279-8

Biswas A, Choudhary A, Darbha GK (2024) From ground to gut: evaluating the human health risk of potentially toxic elements in soil, groundwater, and their uptake by Cocos nucifera in arsenic-contaminated environments. Environ Pollut 344:123342

Bridson JH, Abbel R, Smith DA, Northcott GL, Gaw S (2023) Release of additives and non-intentionally added substances from microplastics under environmentally relevant conditions. Environ Adv 12:100359. https://doi.org/10.1016/j.envadv.2023.100359

CANSA (2011) Plastic-identification-codes. The Cancer Association of South Africa

Cao Y, Sathish C, Guan X, Wang S, Palanisami T, Vinu A, Yi J (2024) Advances in magnetic materials for microplastic separation and degradation. J Hazard Mater 461:132537. https://doi.org/10.1016/j.jhazmat.2023.132537

Carnevale Miino M, Galafassi S, Zullo R, Torretta V, Rada EC (2024) Microplastics removal in wastewater treatment plants: a review of the different approaches to limit their release in the environment. Sci Total Environ 930:172675. https://doi.org/10.1016/j.scitotenv.2024.172675

Chamas A, Moon H, Zheng J, Qiu Y, Tabassum T, Jang JH, Abu-Omar M, Scott SL, Suh S (2020) Degradation rates of plastics in the environment. ACS Sustain Chem Eng 8:3494–3511. https://doi.org/10.1021/acssuschemeng.9b06635

Chen L, Gao M (2021) Formal or informal recycling sectors? Household solid waste recycling behavior based on multi-agent simulation. J Environ Manag 294:113006. https://doi.org/10.1016/j.jenvman.2021.113006

Chen X, Lian X, Wang Y, Chen S, Sun Y, Tao G, Tan Q, Feng J (2023a) Impacts of hydraulic conditions on microplastics biofilm development, shear stresses distribution, and microbial community structures in drinking water distribution pipes. J Environ Manag 325:116510. https://doi.org/10.1016/j.jenvman.2022.116510

Chen Q, Shi G, Revell LE, Zhang J, Zuo C, Wang D, Le Ru EC, Wu G, Mitrano DM (2023b) Long-range atmospheric transport of microplastics across the southern hemisphere. Nat Commun 14:7898. https://doi.org/10.1038/s41467-023-43695-0

Choudhary A, Khandelwal N, Singh N, Tiwari E, Ganie ZA, Darbha GK (2022) Nanoplastics interaction with feldspar and weathering originated secondary minerals (kaolinite and gibbsite) in the riverine environment. Sci Total Environ 818:151831

Choudhary A, George L, Mandal A, Biswas A, Ganie ZA, Darbha GK (2025) Assessment of microplastics and associated ecological risk in the longest river (Godavari) of peninsular India: a comprehensive source-to-sink analysis in water, sediment and fish. Mar Pollut Bull 212:117560. https://doi.org/10.1016/j.marpolbul.2025.117560

Civancik-Uslu D, Nhu TT, Van Gorp B, Kresovic U, Larrain M, Billen P, Ragaert K, De Meester S, Dewulf J, Huysveld S (2021) Moving from linear to circular household plastic packaging in Belgium: prospective life cycle assessment of mechanical and thermochemical recycling. Resour Conserv Recycl 171:105633. https://doi.org/10.1016/j.resconrec.2021.105633

Cristaldi A, Fiore M, Zuccarello P, Oliveri Conti G, Grasso A, Nicolosi I, Copat C, Ferrante M (2020) Efficiency of wastewater treatment plants (WWTPs) for microplastic removal: a systematic review. Int J Environ Res Public Health 17:8014. https://doi.org/10.3390/ijerph17218014

De Sousa FDB (2023) Consumer awareness of plastic: an overview of different research areas. Circ Econ Sust 3:2083–2107. https://doi.org/10.1007/s43615-023-00263-4

Dennis J, Arulraj D, Mistri TK (2025) Unseen toxins: exploring the human health consequences of micro and nanoplastics. Toxicol Rep 14:101955. https://doi.org/10.1016/j.toxrep.2025.101955

Ebele AJ, Abou-Elwafa Abdallah M, Harrad S (2017) Pharmaceuticals and personal care products (PPCPs) in the freshwater aquatic environment. Emerg Contam 3:1–16. https://doi.org/10.1016/j.emcon.2016.12.004

Fred-Ahmadu OH, Bhagwat G, Oluyoye I, Benson NU, Ayejuyo OO, Palanisami T (2020) Interaction of chemical contaminants with microplastics: principles and perspectives. Sci Total Environ 706:135978. https://doi.org/10.1016/j.scitotenv.2019.135978

Funk K, Milford J, Simpkins T (2020) Chapter 19 – waste not, want not: analyzing the economic and environmental viability of waste-to-energy technology for site-specific optimization of renewable energy options. In: Dahiya A (ed) Bioenergy, 2nd edn. Academic Press, pp 385–423. https://doi.org/10.1016/B978-0-12-815497-7.00019-1

Ganie ZA, Khandelwal N, Tiwari E, Singh N, Darbha GK (2021) Biochar-facilitated remediation of nanoplastic contaminated water: effect of pyrolysis temperature induced surface modifications. J Hazard Mater 417:126096. https://doi.org/10.1016/j.jhazmat.2021.126096

Ganie ZA, Mandal A, Arya L et al (2024) Assessment and accumulation of microplastics in the Indian riverine systems: risk assessment and implications of translocation across the water-to-fish continuum. Aquat Toxicol 272:106944. https://doi.org/10.1016/j.aquatox.2024.106944

Georgios B, Batsioula M, Achillas C, Patsios S, Kontogiannopoulos K, Bochtis D, Moussiopoulos N (2020) A life cycle analysis approach for the evaluation of municipal solid waste management practices: the case study of the region of Central Macedonia, Greece. Sustainability 12:8221. https://doi.org/10.3390/su12198221

Geyer R, Jambeck JR, Law KL (2017) Production, use, and fate of all plastics ever made. Sci Adv 3:e1700782. https://doi.org/10.1126/sciadv.1700782

Ghosh SK (2016) Swachhaa Bharat Mission (SBM)—a paradigm shift in waste management and cleanliness in India. Procedia Environ Sci 35:15–27. https://doi.org/10.1016/j.proenv.2016.07.002

Goodship V (2007) Plastic recycling. Sci Prog 90:245–268. https://doi.org/10.3184/003685007X228748

Graça J, Murphy B, Kelleher B (2023) Valorisation alternatives to landfill for organic residues: 2018-RE-MS-15, Online version. ed, EPA Research evidence synthesis report. Environmental Protection Agency, Johnstown Castle, Co. Wexford, Ireland

Hamdan A, Panda S, Jain MS, Raj V, Mathew S (2025) Assessing municipal solid waste in Indian smart cities: a path towards waste-to-energy. Heliyon 11:e42770. https://doi.org/10.1016/j.heliyon.2025.e42770

Hanif MA, Ibrahim N, Dahalan FA et al (2023) Microplastics in facial cleanser: extraction, identification, potential toxicity, and continuous-flow removal using agricultural waste–based biochar. Environ Sci Pollut Res 30:60106–60120. https://doi.org/10.1007/s11356-023-26741-8

He P, Chen L, Shao L, Zhang H, Lü F (2019) Municipal solid waste (MSW) landfill: a source of microplastics? Evidence of microplastics in landfill leachate. Water Res 159:38–45. https://doi.org/10.1016/j.watres.2019.04.060

He R, Hui K, Zhang X, Yao H (2024) Insight into the role of the pore structure and surface functional groups in biochar on the adsorption of sulfamethoxazole from synthetic urine. Appl Sci 14:1715. https://doi.org/10.3390/app14051715

Helms BA, Russell TP (2016) Reaction: polymer chemistries enabling cradle-to-cradle life cycles for plastics. Chem 1:816–818. https://doi.org/10.1016/j.chempr.2016.11.016

Idumah CI, Nwuzor IC (2019) Novel trends in plastic waste management. SN Appl Sci 1. https://doi.org/10.1007/s42452-019-1468-2

Ihsanullah I (2022) Applications of MOFs as adsorbents in water purification: Progress, challenges and outlook. Curr Opin Environ Sci Health 26:100335. https://doi.org/10.1016/j.coesh.2022.100335

Jenner LC, Rotchell JM, Bennett RT, Cowen M, Tentzeris V, Sadofsky LR (2022) Detection of microplastics in human lung tissue using μFTIR spectroscopy. Sci Total Environ 831:154907. https://doi.org/10.1016/j.scitotenv.2022.154907

Jjagwe J, Olupot PW, Menya E, Kalibbala HM (2021) Synthesis and application of granular activated carbon from biomass waste materials for water treatment: a review. J Bioresour Bioprod 6:292–322. https://doi.org/10.1016/j.jobab.2021.03.003

Joseph TM, Al-Hazmi HE, Śniatała B, Esmaeili A, Habibzadeh S (2023) Nanoparticles and nanofiltration for wastewater treatment: from polluted to fresh water. Environ Res 238:117114. https://doi.org/10.1016/j.envres.2023.117114

Judy JD, Williams M, Gregg A, Oliver D, Kumar A, Kookana R, Kirby JK (2019) Microplastics in municipal mixed-waste organic outputs induce minimal short to long-term toxicity in key terrestrial biota. Environ Pollut 252:522–531. https://doi.org/10.1016/j.envpol.2019.05.027

Kärrman A, van Bavel B, Järnberg U, Hardell L, Lindström G (2006) Perfluorinated chemicals in relation to other persistent organic pollutants in human blood. Chemosphere 64:1582–1591. https://doi.org/10.1016/j.chemosphere.2005.11.040

Kaseke T, Lujic T, Cirkovic Velickovic T (2023) Nano- and Microplastics migration from plastic food packaging into dairy products: impact on nutrient digestion, absorption, and metabolism. Foods 12:3043. https://doi.org/10.3390/foods12163043

Kiessling T, Hinzmann M, Mederake L, Dittmann S, Brennecke D, Böhm-Beck M, Knickmeier K, Thiel M (2023) What potential does the EU single-use plastics directive have for reducing plastic pollution at coastlines and riversides? An evaluation based on citizen science data. Waste Manag 164:106–118. https://doi.org/10.1016/j.wasman.2023.03.042

Kumar A, Agrawal A (2020) Recent trends in solid waste management status, challenges, and potential for the future Indian cities – a review. Curr Res Environ Sustain 2:100011. https://doi.org/10.1016/j.crsust.2020.100011

Kumar M, Xiong X, He M, Tsang DCW, Gupta J, Khan E, Harrad S, Hou D, Ok YS, Bolan NS (2020) Microplastics as pollutants in agricultural soils. Environ Pollut 265:114980. https://doi.org/10.1016/j.envpol.2020.114980

Kumar R, Sharma P, Verma A, Jha PK, Singh P, Gupta PK, Chandra R, Prasad PVV (2021) Effect of physical characteristics and hydrodynamic conditions on transport and deposition of microplastics in riverine ecosystem. Water 13:2710. https://doi.org/10.3390/w13192710

Kumar V, Singh E, Singh S, Pandey A, Bhargava PC (2023) Micro- and nano-plastics (MNPs) as emerging pollutant in ground water: environmental impact, potential risks, limitations and way forward towards sustainable management. Chem Eng J 459:141568. https://doi.org/10.1016/j.cej.2023.141568

Kumar D, Biswas JK, Mulla SI, Singh R, Shukla R, Ahanger MA, Shekhawat GS, Verma KK, Siddiqui MW, Seth CS (2024) Micro and nanoplastics pollution: sources, distribution, uptake in plants, toxicological effects, and innovative remediation strategies for environmental sustainability. Plant Physiol Biochem 213:108795. https://doi.org/10.1016/j.plaphy.2024.108795

Lawtae P, Tangsathitkulchai C (2021) The use of high surface area mesoporous-activated carbon from Longan seed biomass for increasing capacity and kinetics of methylene blue adsorption from aqueous solution. Molecules 26:6521. https://doi.org/10.3390/molecules26216521

Laycock BG, Halley PJ (2014) Starch applications. In: Starch polymers. Elsevier, pp 381–419. https://doi.org/10.1016/B978-0-444-53730-0.00026-9

Lehmann J, Joseph S (eds) (2015) Biochar for environmental management: science, technology and implementation. Routledge. https://doi.org/10.4324/9780203762264

Leslie HA, van Velzen MJM, Brandsma SH, Vethaak AD, Garcia-Vallejo JJ, Lamoree MH (2022) Discovery and quantification of plastic particle pollution in human blood. Environ Int 163:107199. https://doi.org/10.1016/j.envint.2022.107199

Li X, Liu W, Zhang J, Wang Z, Guo Z, Ali J, Wang L, Yu Z, Zhang X, Sun Y (2024) Effective removal of microplastics by filamentous algae and its magnetic biochar: performance and mechanism. Chemosphere 358:142152. https://doi.org/10.1016/j.chemosphere.2024.142152

Liang S-X, Duan F-F, Lü Q-F, Yang H (2019) Hierarchical biocarbons with controlled micropores and Mesopores derived from kapok fruit peels for high-performance Supercapacitor electrodes. ACS Omega 4:5991–5999. https://doi.org/10.1021/acsomega.9b00148

Liu G, Zhu Z, Yang Y, Sun Y, Yu F, Ma J (2019) Sorption behavior and mechanism of hydrophilic organic chemicals to virgin and aged microplastics in freshwater and seawater. Environ Pollut 246:26–33. https://doi.org/10.1016/j.envpol.2018.11.100

Liu S, Huang J, Zhang W, Shi L, Yi K, Yu H, Zhang C, Li S, Li J (2022) Microplastics as a vehicle of heavy metals in aquatic environments: a review of adsorption factors, mechanisms, and biological effects. J Environ Manag 302:113995. https://doi.org/10.1016/j.jenvman.2021.113995

Liu Y, Ben Y, Che R, Peng C, Li J, Wang F (2023) Uptake, transport and accumulation of micro- and nano-plastics in terrestrial plants and health risk associated with their transfer to food chain – a mini review. Sci Total Environ 902:166045. https://doi.org/10.1016/j.scitotenv.2023.166045

Liu Z, Liao L, Yang H, Chen Z, Ji H (2024) Impact of micro-nano plastics on microorganisms and thus drinking water treatment effectiveness: a review. Desalin Water Treat 320:100673. https://doi.org/10.1016/j.dwt.2024.100673

Luqman A, Nugrahapraja H, Wahyuono RA, Islami I, Haekal MH, Fardiansyah Y, Putri BQ, Amalludin FI, Rofiqa EA, Götz F, Wibowo AT (2021) Microplastic contamination in human stools, foods, and drinking water associated with Indonesian coastal population. Environments 8:138. https://doi.org/10.3390/environments8120138

Manikanda Bharath K, Muthulakshmi AL, Natesan U (2023) Microplastic contamination around the landfills: distribution, characterization and threats: a review. Curr Opin Environ Sci Health 31:100422. https://doi.org/10.1016/j.coesh.2022.100422

Manzoor MF, Tariq T, Fatima B, Sahar A, Tariq F, Munir S, Khan S, Ranjha MMAN, Sameen A, Zeng X-A, Ibrahim SA (2022) An insight into bisphenol a, food exposure and its adverse effects on health: a review. Front Nutr 9:1047827. https://doi.org/10.3389/fnut.2022.1047827

Marshall RE, Farahbakhsh K (2013) Systems approaches to integrated solid waste management in developing countries. Waste Manag 33:988–1003. https://doi.org/10.1016/j.wasman.2012.12.023

Martinez M (2022) Reduce, reuse, recycle and other plastic rhetoric: examining the influence of environmental rhetoric on plastic governance in Canada. https://doi.org/10.20381/RUOR-28361

Massardo S, Verzola D, Alberti S, Caboni C, Santostefano M, Eugenio Verrina E, Angeletti A, Lugani F, Ghiggeri GM, Bruschi M, Candiano G, Rumeo N, Gentile M, Cravedi P, La Maestra S, Zaza G, Stallone G, Esposito P, Viazzi F, Mancianti N, La Porta E, Artini C (2024) MicroRaman spectroscopy detects the presence of microplastics in human urine and kidney tissue. Environ Int 184:108444. https://doi.org/10.1016/j.envint.2024.108444

Matavos-Aramyan S (2024) Addressing the microplastic crisis: a multifaceted approach to removal and regulation. Environ Adv 17:100579. https://doi.org/10.1016/j.envadv.2024.100579

McDevitt JP, Criddle CS, Morse M, Hale RC, Bott CB, Rochman CM (2017) Addressing the issue of microplastics in the wake of the microbead-free waters act – a new standard can facilitate improved policy. Environ Sci Technol 51:6611–6617. https://doi.org/10.1021/acs.est.6b05812

Mota LSO, De Oliveira PCO, Peixoto BS, Bezerra ES, De Moraes MC (2025) Biochar applications in microplastic and nanoplastic removal: mechanisms and integrated approaches. Environ Sci Water Res Technol 11:222–241. https://doi.org/10.1039/D4EW00709C

Ncube LK, Ude AU, Ogunmuyiwa EN, Zulkifli R, Beas IN (2021) An overview of plastic waste generation and management in food packaging industries. Recycling 6:12. https://doi.org/10.3390/recycling6010012

Nøklebye E, Adam HN, Roy-Basu A, Bharat GK, Steindal EH (2023) Plastic bans in India – addressing the socio-economic and environmental complexities. Environ Sci Pol 139:219–227. https://doi.org/10.1016/j.envsci.2022.11.005

OECD (2022) Global plastics outlook: economic drivers, environmental impacts and policy options. OECD. https://doi.org/10.1787/de747aef-en

OECD (2024) Policy scenarios for eliminating plastic pollution by 2040. OECD. https://doi.org/10.1787/76400890-en

Ortega DE, Cortés-Arriagada D (2023) Atmospheric microplastics and nanoplastics as vectors of primary air pollutants – a theoretical study on the polyethylene terephthalate (PET) case. Environ Pollut 318:120860. https://doi.org/10.1016/j.envpol.2022.120860

Pal SK, Garcés-Sánchez G, Kranert M, Vinu R (2023) Characterization and evaluation of resource recovery potential of beach plastic wastes using analytical Py-GC/MS. J Anal Appl Pyrolysis 172:105996. https://doi.org/10.1016/j.jaap.2023.105996

Pilapitiya PGC, Ratnayake AS (2024) The world of plastic waste: a review. Clean Mater 11:100220. https://doi.org/10.1016/j.clema.2024.100220

Pivnenko K, Granby K, Eriksson E, Astrup TF (2017) Recycling of plastic waste: screening for brominated flame retardants (BFRs). Waste Manag 69:101–109. https://doi.org/10.1016/j.wasman.2017.08.038

Ren S-Y, Kong S-F, Ni H-G (2021) Contribution of mulch film to microplastics in agricultural soil and surface water in China. Environ Pollut 291:118227. https://doi.org/10.1016/j.envpol.2021.118227

Rifa RA, Lavado R (2024) Unveiling the next generation of bisphenol analogs and their impact on human health using *in vitro* methods. Emerg Contam 10:100296. https://doi.org/10.1016/j.emcon.2023.100296

Sánchez A, Rodríguez-Viso P, Domene A, Orozco H, Vélez D, Devesa V (2022) Dietary microplastics: occurrence, exposure and health implications. Environ Res 212:113150. https://doi.org/10.1016/j.envres.2022.113150

Sharma P, Sharma P (2024) Micro(nano)plastics: invisible compounds with a visible impact. F1000Res 13:69. https://doi.org/10.12688/f1000research.142212.2

Sharma A, Ganguly R, Gupta AK (2019) Characterization and energy generation potential of municipal solid waste from nonengineered landfill sites in Himachal Pradesh. India J Hazard Toxic Radioact Waste 23:04019008. https://doi.org/10.1061/(ASCE)HZ.2153-5515.0000442

Shrivastava A, Abhishek K, Gupta AK, Jain H, Kumari M, Patel M, Sharma P (2024) Removal of micro- and nano-plastics from aqueous matrices using modified biochar – a review of synthesis, applications, interaction, and regeneration. J Hazard Mater Adv 16:100518. https://doi.org/10.1016/j.hazadv.2024.100518

Siddiqi MH, Liu X, Hussain MA, Qureshi T, Waqas M, Farooq M, Iqbal T, Nawaz S, Nawaz S (2021) Evolution of kinetic and hydrothermal study of refused derived fuels: Thermogravimetric analysis. Energy Rep 7:1757–1764. https://doi.org/10.1016/j.egyr.2021.03.010

Siipola V, Pflugmacher S, Romar H, Wendling L, Koukkari P (2020) Low-cost biochar adsorbents for water purification including microplastics removal. Appl Sci 10:788. https://doi.org/10.3390/app10030788

Singh N, Tiwari E, Khandelwal N, Darbha GK (2019) Understanding the stability of nanoplastics in aqueous environments: effect of ionic strength, temperature, dissolved organic matter, clay, and heavy metals. Environ Sci Nano 6:2968–2976. https://doi.org/10.1039/C9EN00557A

Singh N, Khandelwal N, Ganie ZA, Tiwari E, Darbha GK (2021) Eco-friendly magnetic biochar: an effective trap for nanoplastics of varying surface functionality and size in the aqueous environment. Chem Eng J 418:129405. https://doi.org/10.1016/j.cej.2021.129405

Singh E, Mishra R, Kumar A, Shukla SK, Lo S-L, Kumar S (2022) Circular economy-based environmental management using biochar: driving towards sustainability. Process Saf Environ Prot 163:585–600. https://doi.org/10.1016/j.psep.2022.05.056

Soni A, Das PK, Kumar P (2023) A review on the municipal solid waste management status, challenges and potential for the future Indian cities. Environ Dev Sustain 25:13755–13803. https://doi.org/10.1007/s10668-022-02688-7

Srivastava V, Vaish B, Singh RP, Singh P (2020) An insight to municipal solid waste management of Varanasi city, India, and appraisal of vermicomposting as its efficient management approach. Environ Monit Assess 192:191. https://doi.org/10.1007/s10661-020-8135-3

Stegmann P, Daioglou V, Londo M, Vuuren D, Junginger M (2022) Plastic futures and their CO2 emissions. Nature 612:272–276. https://doi.org/10.1038/s41586-022-05422-5

Subair A, Krishnamoorthy Lakshmi P, Chellappan S, Chinghakham C (2024) Removal of polystyrene microplastics using biochar-based continuous flow fixed-bed column. Environ Sci Pollut Res 31:13753–13765. https://doi.org/10.1007/s11356-024-32088-5

Sujathan S, Kniggendorf A-K, Kumar A, Roth B, Rosenwinkel K-H, Nogueira R (2017) Heat and bleach: a cost-efficient method for extracting microplastics from return activated sludge. Arch Environ Contam Toxicol 73:641–648. https://doi.org/10.1007/s00244-017-0415-8

Sun M, Wang X-Z, Xiong R-Y, Chen X, Zhai L-F, Wang S (2023) High-performance biochar-loaded MgAl-layered double oxide adsorbents derived from sewage sludge towards nanoplastics removal: mechanism elucidation and QSAR modeling. Sci Total Environ 901:165971. https://doi.org/10.1016/j.scitotenv.2023.165971

Talukdar A, Bhattacharya S, Bandyopadhyay A, Dey A (2023) Microplastic pollution in the Himalayas: occurrence, distribution, accumulation and environmental impacts. Sci Total Environ 874:162495. https://doi.org/10.1016/j.scitotenv.2023.162495

Tarafdar A, Xie J, Gowen A, O'Higgins AC, Xu J-L (2024) Advanced optical photothermal infrared spectroscopy for comprehensive characterization of microplastics from intravenous fluid delivery systems. Sci Total Environ 929:172648. https://doi.org/10.1016/j.scitotenv.2024.172648

Tong M, He L, Rong H, Li M, Kim H (2020) Transport behaviors of plastic particles in saturated quartz sand without and with biochar/Fe3O4-biochar amendment. Water Res 169:115284. https://doi.org/10.1016/j.watres.2019.115284

Tumwesigye E, Felicitas Nnadozie C, Akamagwuna C et al (2023) Microplastics as vectors of chemical contaminants and biological agents in freshwater ecosystems: current knowledge status and future perspectives. Environ Pollut 330:121829. https://doi.org/10.1016/j.envpol.2023.121829

Verla AW, Enyoh CE, Verla EN, Nwarnorh KO (2019) Microplastic–toxic chemical interaction: a review study on quantified levels, mechanism and implication. SN Appl. Sci. 1:1400. https://doi.org/10.1007/s42452-019-1352-0

Wang Z, Sedighi M, Lea-Langton A (2020) Filtration of microplastic spheres by biochar: removal efficiency and immobilisation mechanisms. Water Res 184:116165. https://doi.org/10.1016/j.watres.2020.116165

Wang B, Liu W, Zhang M (2024) Application of carbon-based adsorbents in the remediation of micro- and nanoplastics. J Environ Manag 349:119522. https://doi.org/10.1016/j.jenvman.2023.119522

Wiesinger H, Wang Z, Hellweg S (2021) Deep dive into plastic monomers, additives, and processing aids. Environ Sci Technol 55:9339–9351. https://doi.org/10.1021/acs.est.1c00976

Wiesner Y, Bednarz M, Braun U, Bannick CG, Ricking M, Altmann K (2023) A promising approach to monitor microplastic masses in composts. Front Environ Chem 4:1281558. https://doi.org/10.3389/fenvc.2023.1281558

Wu J, Yang C, Zhao H, Shi J, Liu Z, Li C, Song F (2022) Efficient removal of microplastics from aqueous solution by a novel magnetic biochar: performance, mechanism, and reusability. Environ Sci Pollut Res 30:26914–26928. https://doi.org/10.1007/s11356-022-24130-1

Wu N, He Y, Sun Z, Zhang S, Yang X, Liu QS, Zhou Q, Jiang G (2025) The environmental occurrence, human exposure, and toxicity of novel bisphenol S derivatives: a review. Ecotoxicol Environ Saf 296:118182. https://doi.org/10.1016/j.ecoenv.2025.118182

Xie T, Reddy KR, Wang C, Yargicoglu E, Spokas K (2015) Characteristics and applications of biochar for environmental remediation: a review. Crit Rev Environ Sci Technol 45:939–969. https://doi.org/10.1080/10643389.2014.924180

Yu Z, Xu X, Guo L, Jin R, Lu Y (2024) Uptake and transport of micro/nanoplastics in terrestrial plants: detection, mechanisms, and influencing factors. Sci Total Environ 907:168155. https://doi.org/10.1016/j.scitotenv.2023.168155

Zhang D, Bo W, Wang T, Yılmaz M, Sharma G, Kumar A, Shi H (2025) Multi-mechanism synergistic adsorption of lead and cadmium in water by structure-functionally adapted modified biochar: a review. Desalin Water Treat 101156. https://doi.org/10.1016/j.dwt.2025.101156

Zornoza R, Moreno-Barriga F, Acosta JA, Muñoz MA, Faz A (2016) Stability, nutrient availability and hydrophobicity of biochars derived from manure, crop residues, and municipal solid waste for their use as soil amendments. Chemosphere 144:122–130. https://doi.org/10.1016/j.chemosphere.2015.08.046W

Xu M, [illegible] Y, Sun Z, Zhang S, Yang X, [illegible] Q, Zhou Q, Tang C [illegible] The environmental occurrence, human exposure, and toxicity of novel [illegible]. [illegible] https://doi.org/10.1016/[illegible]

[illegible], Reddy KR, [illegible] Spokas K, [illegible] applications of biochar for environmental remediation [illegible]. Crit Rev Environ Sci Technol [illegible] https://doi.org/10.1080/10643389.2014.[illegible]

[illegible]

[illegible]

Microplastic Pollution: Review of Degradation, Detection, and Treatment Technologies

12

Deborshee Sinha, Adarsh Singh, Ritu Kothari, and Mayur Shirish Jain

Abstract

Microplastics, a pervasive environmental pollutant, pose significant threats to ecosystems and human health due to their persistence, bioaccumulation potential, and complex degradation mechanisms. Originating from primary and secondary sources, these microscopic plastic fragments infiltrate terrestrial, freshwater, and marine environments, raising global concerns. Their slow decomposition releases toxic additives and facilitates the transport of heavy metals and persistent organic pollutants, further amplifying their ecological risks. This chapter provides a comprehensive analysis of microplastic degradation mechanisms, encompassing physical, chemical, and biological pathways, with an emphasis on their efficiency, limitations, and environmental implications. Furthermore, it evaluates advanced detection techniques, including spectroscopic, thermal, and mass-based methodologies, to enhance the accuracy and reliability of microplastic identification in diverse matrices. The study also examines emerging treatment strategies, such as membrane bioreactors, electrocoagulation, advanced oxidation processes, and bioremediation, assessing their scalability, economic feasibility, and effectiveness in mitigating microplastic pollution. Despite technological advancements, challenges persist in large-scale remediation, necessitating an integrated approach combining multiple treatment modalities. The findings underscore the urgent need for innovative policies, cross-disciplinary research, and global collaboration to curb microplastic contamination and develop sustainable mitigation frameworks. Future perspectives highlight the potential of AI-driven detection, biodegradable alternatives, and circular economy principles to revolutionize microplastic management and prevent further environmental deterioration.

D. Sinha · A. Singh · R. Kothari · M. S. Jain (✉)
Department of Civil Engineering, Indian Institute of Technology Indore, Indore, Madhya Pradesh, India

K. Bala et al. (eds.), *Occurrence, Detection, and Fate of Microplastics in Freshwater Ecosystems*, https://doi.org/10.1007/978-981-95-0225-7_12

Keywords

Microplastics degradation · Detection · Treatment methods

Abbreviations

AI	Artificial intelligence
AOPs	Advanced oxidation processes
ATR	Attenuated total reflectance
CD	Cadmium
CO_2	Carbon dioxide
DAF	Dissolved air flotation
DO	Dissolved oxygen
EDS	Energy dispersive spectroscopy
EF-like	Electro-Fenton-like
FTIR	Fourier transform infrared spectroscopy
GAC	Granular activated carbon
Hg	Mercury
ICP-MS	Inductively coupled plasma mass spectrometry
LC-MS	Liquid chromatography-mass spectrometry
LIBS	Laser-induced breakdown spectroscopy
MBR	Membrane bioreactor
MPs	Microplastics
MS	Mass spectrometry
Mt.	Million metric tons
NTA	Nanoparticle tracking analysis
PE	Polyethylene
PET	Polyethylene terephthalate
PHA	Polyhydroxyalkanoates
PLA	Polylactic acid
PCL	Polycaprolactone
POPs	Persistent organic pollutants
PP	Polypropylene
PS	Polystyrene
PVC	Polyvinyl chloride
Py-GC-MS	Pyrolysis-gas chromatography-mass spectrometry
SEM	Scanning electron microscopy
SEM-EDS	Scanning electron microscopy with energy dispersive spectroscopy
TGA-MS	Thermogravimetric analysis-mass spectrometry
TED-GC-MS	Thermal extraction desorption-gas chromatography-mass spectrometry
TiO_2	Titanium dioxide
UF	Ultrafiltration
UV	Ultraviolet
WWTP	Wastewater treatment plant

12.1 Introduction

Plastics have long been the subject of environmental analysis and have expanded to most man-made products. However, there is a lack of unwavering worldwide data, especially about their destiny at the end of their lives. According to scientists, the total quantity of virgin plastics produced until 2017 is 8300 million metric tons (Mt). As of 2015, approximately 9% of the 6300 million tons of plastic waste produced had been recycled, 12% had been incinerated, and 79% had ended up in landfills or the environment (Wojnowska-Baryła et al. 2022). By 2050, over 12,000 Mt. of plastic garbage will be in landfills or the environment if current manufacturing and waste management practices persist (Geyer et al. 2017). The staggering accumulation of plastic waste in our oceans poses a severe environmental crisis. This material litters the water and breaks into minuscule particles known as microplastics. These tiny fragments, often invisible to the naked eye, are becoming pervasive, infiltrating marine ecosystems and threatening the health of aquatic life (Galloway and Lewis 2017). Plastic breaks down into microplastics and nanoplastics under various environmental conditions. Microplastics are pervasive across terrestrial, freshwater, marine, and atmospheric systems (Sudarsan et al. 2024). They can originate from larger plastic items and are easily transported from their source. Ultimately, they may become trapped in specific areas when incorporated into soil and sediments or continuously cycle through habitats influenced by currents and weather patterns (Alice and Simon 2018). Given their remarkably extended degradation cycle, microplastics can easily infiltrate natural ecosystems. These tiny particles can traverse the food chain, accumulating in various organisms over time. This process of bioaccumulation leads to significant pollution concern that endangers the health and stability of entire ecosystems (Zhenyan et al. 2022).

This chapter provides an in-depth exploration of microplastic degradation, a critical challenge in preserving our ecological systems. This chapter examines these diverse types and their origins, highlighting their prevalence in marine and terrestrial ecosystems. Furthermore, this chapter also focuses on various innovative and economically viable degradation techniques, including biological methods utilizing microorganisms, chemical treatments, and advanced filtration systems. The study aims to comprehensively understand the pathways toward mitigating the microplastic crisis by analyzing these approaches' effectiveness, feasibility, and environmental impact.

12.2 Microplastics

According to Frias and Nash (2019) definition of microplastics is *"Microplastics are any synthetic solid particle or polymeric matrix, with regular or irregular shape and with size ranging from 1 μm to 5 mm, of either primary or secondary manufacturing origin, which is insoluble in water"* Microplastics are generally thought to have a diameter of less than 5 mm, while size estimates vary from study to study. They can be classified into two categories: biodegradable microplastics and nonbiodegradable microplastics. Biodegradable plastics, such as polylactic acid (PLA), polyhydroxyalkanoates (PHA), and polycaprolactone (PCL), have various applications in everyday life. They are commonly used in bottles, disposable cups, medical device bags, and fertilizers (Majewsky et al. 2016; González-Pleiter et al. 2019). On the other hand, nonbiodegradable plastics, including polyethylene, polypropylene, polyester, polyethylene terephthalate (PET), and polyurethane, are used in products such as food containers, tires, life jackets, and gaskets. (Zhang et al. 2020; Majewsky et al. 2016; Shah et al. 2008).

12.2.1 Sources of Microplastics

Microplastic originates from various sources, including maritime activities such as lost cargo, discarded fishing and aquaculture gear, and the fragmentation of more significant items that enter through rivers, runoff, tides, winds, and catastrophic events. Additionally, microplastics can enter wastewater treatment systems directly as micrometer-sized particles like garment fibers and cosmetic beads (Law and Thompson 2014). Microplastics are categorized into primary and secondary sources. Primary microplastics are intentionally produced at the micro- or nanoscale and incorporated into paints, coatings, cosmetics, personal care items, and medical supplies. These materials are subsequently released into the environment due to improper disposal. In contrast, secondary microplastics result from the degradation of larger plastic waste through physical abrasion, chemical and biological processes, and photodegradation (Galloway et al. 2017). The most common sources of microplastics are mentioned in Fig. 12.1.

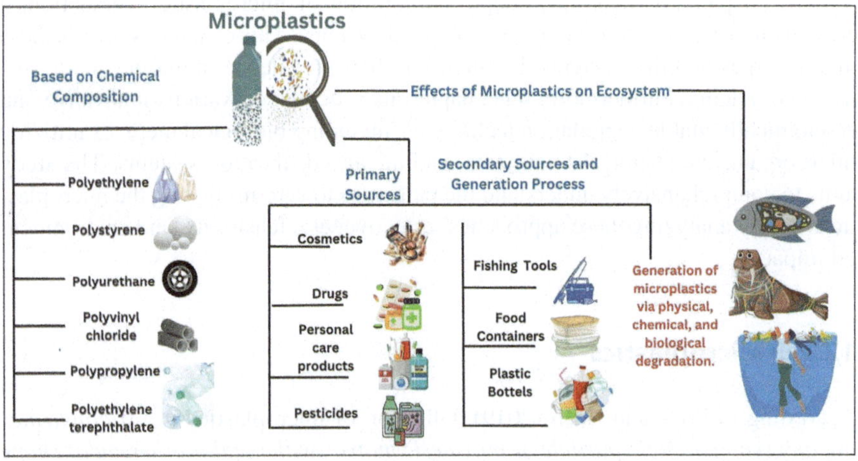

Fig. 12.1 Sources of microplastics in the environment

12.3 Microplastic Degradation Methods

Microplastics decompose very slowly naturally, and during this process, a variety of hazardous materials (including plasticizers, short- and medium-chain chlorinated paraffin, and manufacturing antioxidants) that make up 70% of the plastic weight may be released (Coffin et al. 2019). There are few adequate greens or ways to break down microplastics in the environment. Currently, there are three primary categories into which all microplastic degradation techniques can be divided: physical, chemical, and biological degradation techniques. Table 12.1 describes the details of the three mechanisms for microplastic degradation. Microplastic degradation can be categorized into three primary methods: physical, chemical, and biological, each with distinct mechanisms, advantages, and limitations.

Each degradation method has its strengths and limitations, and an integrated approach combining physical, chemical, and biological processes may enhance the overall effectiveness of microplastic removal (Binbin et al. 2022).

12.4 Analytical Techniques for Microplastic Detection

Detection and quantification of microplastics (MPs) in environmental matrices, such as water, soil, sediment, air, and biological samples, are crucial for evaluating their fate, transport, and ecological effects. Due to their tiny size (1 μm to 5 mm) and complicated polymer structure, detecting microplastics requires sensitive and specific analytical methods to distinguish them from organic and inorganic debris. Analytical techniques for detecting microplastics can be generally categorized into physical (microscopic), spectroscopic, thermal, and mass-based methods, with blends usually yielding the best results (Baruah et al. 2022).

12.4.1 Physical Characterization Techniques (Microscopic Method)

Microscopic techniques like stereo microscopy, scanning electron microscopy (SEM), and polarized light microscopy continue to be core to initial identification, especially in distinguishing MPs by size, shape, and color (Dierkes et al. 2022). Nonetheless, they are plagued by disadvantages like subjectivity, time-consuming, and misidentification owing to organic debris simulating plastic particles (Shim et al. 2017).

12.4.2 Spectroscopic Techniques

In response to these issues, spectroscopic methods like Fourier transform infrared spectroscopy (FTIR) and Raman spectroscopy have become routine for chemical identification (Andoh et al. 2024). FTIR spectroscopy, especially in Attenuated

Table 12.1 Comparison of plastic degradation methods: mechanisms, effectiveness, and applications

Types of degradation	Mechanism	Effectiveness	Types of plastics	References
Physical degradation	Physical degradation occurs through environmental factors such as oxygen, temperature, water, and mechanical forces, leading to polymer chain breakage and the formation of smaller molecules. Pyrolysis, a widely studied physical degradation method, has effectively broken down microplastics at high temperatures	Physical degradation is cost-effective, simple, and fast; its limited effectiveness often necessitates a combination with other methods	PE, PP, and PVC	Ni et al. (2020) Iñiguez et al. (2018)
Chemical degradation	Chemical degradation involves external chemical agents such as peroxides and carbonyl groups, which induce polymer chain scission, reducing molecular weight and altering physical properties. Fenton-like (EF-like) technology using TiO_2/graphite cathodes enhances degradation. This method benefits from using natural light and photocatalysts, allowing the mineralization of microplastics	High cost, inconsistent degradation efficiency, and challenges in preparing high-performance catalysts limit its large-scale application. Future research focuses on developing cost-effective and efficient catalysts for broader industrial use	PVC, PLA, and PE	Miao et al. (2020)
Biodegradation	Biodegradation utilizes microorganisms to break down microplastics into harmless byproducts. Fungal strains capable of degrading polyurethane, achieving an 87% degradation rate in 14 days. Additionally, certain terrestrial snails and bacterial strains have been found to decompose microplastics	This method is environmentally friendly, cost-effective, and sustainable, but its efficiency depends on environmental conditions and microbial strain selection	PU	Álvarez-Barragán et al. (2016)
			PS	Song et al. (2020)
			PE, PET, PP, PS	Auta et al. (2018)

Total Reflectance (ATR) mode, facilitates nondestructive, high-throughput discrimination of polymer types, identifying MPs based on their distinct infrared absorption spectra. Experiments show that FTIR can accurately distinguish microplastics as small as 10 µm and is commonly used together with microscopy to enhance resolution (Andoh et al. 2024). However, FTIR has limitations, including failure to identify nanoplastics and interference from water in aquatic samples, requiring supplementary techniques (Özgenç 2024). Raman spectroscopy, especially Confocal Raman Microscopy, provides better resolution (as low as 1 µm) and the capability to identify smaller particles, which makes it better for nanoplastic identification (1–1000 nm) (Shim et al. 2017). It is, however, constrained by fluorescence interference and the need for large spectral libraries (Ye et al. 2022).

12.4.3 Thermal Analysis Techniques

Thermal methods such as pyrolysis-gas chromatography-mass spectrometry (Py-GC-MS) and thermogravimetric analysis-mass spectrometry (TGA-MS) are commonly used for quantifying polymers, especially in sediment and mixed matrices (Dierkes et al. 2022). Py-GC-MS facilitates exact molecular identification through the thermal breakdown of Mt. into specific volatile compounds, enabling accurate mass quantification. Sophisticated developments in TED-GC-MS (thermal extraction desorption-GC-MS) have enhanced detection sensitivity to quantify MPs down to 1 ng/g concentrations in environmental samples (Ye et al. 2022). However, thermal techniques fail to deliver particle morphology or polymer structure, so they must be combined with spectroscopic methods (Özgenç 2024).

12.4.4 Mass-Based Approaches

Mass-based techniques, including scanning electron microscopy with energy dispersive spectroscopy (SEM-EDS), deliver morphological and elemental analysis, allowing the detection of MPs with incorporated contaminants such as heavy metals and persistent organic pollutants (POPs) (Singh and Kumar 2024). Liquid chromatography in association with mass spectrometry (LC-MS) has also been investigated for polymer-specific molecular fingerprinting, albeit with restricted use owing to economic costs and laborious sample preparation (Singh and Kumar 2024).

12.4.5 Emerging Trends in Microplastic Detection

New technologies involve machine learning-based image analysis for automated counting of microplastics, eliminating human error and enhancing data consistency (Shim et al. 2017). Detection of nanoplastic is still an issue, with most methods having detection limits over 1 µm, leading to the development of super-resolution microscopy, nanoparticle tracking analysis (NTA), and field-flow fractionation with inductively coupled plasma mass spectrometry (ICP-MS) (Dierkes et al. 2022). Improvements in artificial intelligence (AI) and deep learning algorithms are enhancing spectral deconvolution and polymer classification, decreasing dependence on human spectral interpretation (Singh and Kumar 2024). Further, standardization of microplastic detection techniques continues to be necessary since differences in sample processing, density separation, and analytical protocols have resulted in differences in reported concentrations of microplastics (Wirnkor et al. 2019). A worldwide microplastic pollution research meta-analysis revealed that detection efficiency differs by as much as 60% based on the analytical method, highlighting the necessity for harmonized methods (Singh and Kumar 2024). Recent research indicates that combining several methods, including Raman-SEM, Py-GC-MS coupled with FTIR imaging, and laser-induced breakdown spectroscopy (LIBS), offers the most complete characterization of MPs (Özgenç 2024).

12.5 Microplastic Treatment

Microplastic treatment involves various physical, chemical, and biological methods (see detailed treatment methods in Table 12.2) to eliminate or reduce microplastic pollution from wastewater, soil, and air before further degradation into nanoplastics (<100 nm). The treatment process mainly focuses on capturing, filtering, and separating these pollutants to avoid their extensive dispersion. The necessity for microplastic treatment stems from its ubiquitous occurrence in the environment, with an estimated 51 trillion microplastic particles suspended in the world's surface waters (Ahmed et al. 2024). If not treated, Microplastics break down into nanoplastics, enhancing their capacity to bioaccumulate in living organisms and traversing biological barriers like the blood-brain barrier and placental membranes, seriously threatening human health (Na et al. 2021). Research shows that individuals consume about 50,000 to 120,000 microplastic particles annually through food, potable water, and breathing (Cristaldi et al. 2020). In addition, microplastics serve as vectors for long-lived organic pollutants (POPs), heavy metals (Pb, Cd, Hg), and pharmaceutical residues, causing toxic bioaccumulation in aquatic and terrestrial food chains (Wu et al. 2022). Research has also proven that microplastics can take in contaminants at levels as much as 1000 times greater than in the surrounding water, making them more toxic after aquatic organisms consume them (Liu et al. 2021). Synthetic fibers from laundry effluents, microbeads from personal products, tire wear particles, and industrial effluents are principal sources of microplastics in wastewater. Wastewater treatment facilities (WWTPs) have been recognized as significant sources, discharging 5×10^5 to 1.39×10^{10} microplastic particles daily into aquatic environments despite secondary and tertiary treatment (Talvitie et al. 2017). Contamination in soil arises from farming activities like applying sewage sludge as manure, plastic mulching, and industrial effluent discharge. Current research shows that 63,000 to 430,000 tons of microplastics are contributed to European cropland annually through sludge sewage application (Ahmed et al. 2024). Soil persistence

Table 12.2 Microplastic treatment methods

Treatment category	Treatment method	Process description	Steps involved	Implementation feasibility	Efficiency (%)	References
Physical treatment	Membrane bioreactor (MBR)	A combination of membrane filtration and biological treatment for microplastic removal	(i) Biological wastewater treatment (ii) Membrane filtration (iii) Sludge processing	High-used in municipal wastewater treatment plants worldwide	94-99	Talvitie et al. (2017)
Physical treatment	Dissolved air flotation (DAF)	Air bubbles attach to microplastics, lifting them to the surface for removal	(i) Air bubbles injected (ii) Bubbles attach to microplastics (iii) Floating microplastics skimmed off	Moderate-requires integration with existing treatment systems	93.5-95	Ahmed et al. (2024)
Physical treatment	Rapid sand filtration	It uses sand layers to trap microplastics while allowing water to pass	(i) Water flows through sand layers (ii) Microplastics trapped in layers (iii) Periodic backwashing	High-widely used in conventional water treatment plants	45-97	Cristaldi et al. (2020)

(continued)

Table 12.2 (continued)

Treatment category	Treatment method	Process description	Steps involved	Implementation feasibility	Efficiency (%)	References
Physical treatment	Ultrafiltration (UF)	It uses semi-permeable membranes to block particles >0.01 μm	(i) Wastewater forced through membranes (ii) Microplastics retained (iii) Membrane maintenance	High-effective in drinking water and wastewater treatment facilities	98	Sharma et al. (2023)
Physical treatment	Granular activated carbon (GAC)	Adsorbs microplastics onto activated carbon pores	(i) Activated carbon introduced (ii) Microplastics adsorbed (iii) Carbon regeneration/replacement	Moderate-often combined with other filtration techniques for enhanced performance	85	Wu et al. (2022)
Chemical treatment	Electrocoagulation	It uses electric currents to destabilize microplastics and form aggregates	(i) Electrodes placed in wastewater (ii) Electric current generates coagulants (iii) Aggregates removed	Moderate-requires energy input but effective for charged microplastics	90	Liu et al. (2021)
Chemical treatment	Advanced oxidation processes (AOPs)	It uses UV, ozone, or hydroxyl radicals to degrade microplastics into simpler molecules	(i) Oxidants (ozone, UV, etc.) applied (ii) Microplastics undergo chemical breakdown (iii) Residues removed	Moderate-high efficiency but requires monitoring to avoid toxic byproducts	80–95	Na et al. (2021)

Chemical treatment	Coagulation-flocculation	It uses chemical coagulants to aggregate microplastics into larger flocs for removal	(i) Coagulants added (ii) Microplastics form flocs (iii) Settling/filtration removes flocs	High-used in industrial and municipal wastewater treatment plants	90	Iyare et al. (2020)
Chemical treatment	Nanotechnology-based filtration	Engineered nanoparticles selectively trap microplastics for removal	(i) Nanoparticles introduced (ii) Microplastics aggregate (iii) Filtration/ magnetic separation applied	Low-still in experimental stages; future potential is promising	>90	Ahmed et al. (2024)
Biological treatment	Bioremediation	It uses microorganisms to degrade microplastics into CO_2, water, or biodegradable fragments	(i) Plastic-degrading microorganisms introduced (ii) Microplastics broken down (iii) Monitoring of byproducts	Variable-effectiveness depends on microbial species and environmental conditions	50-90	Priya et al. (2023)

of microplastics creates risks for the microbiota of soils, plant integrity, and yields, causing alarms of food web contamination.

Airborne microplastics from tire wear particles, plastic incineration, and synthetic textile shedding have been identified in cities at 9.3–60.4 MP/m^3, and they are responsible for respiratory complications, lung inflammation, and possible neurotoxicity (Ma et al. 2019). Considering these widespread environmental and public health issues, efficient microplastic treatment technologies are vital.

12.6 Cost and Scalability Analysis of Microplastic Treatment Methods

The microplastic pollution challenge has required different treatment technologies with distinct economic and operational features (see Table 12.3). Membrane bioreactors (MBRs) are very effective in microplastic removal (>95% efficiency); however, they are capital and operationally expensive due to membrane fouling and power-demanding processes (Rodríguez-Narvaez et al. 2021). Dissolved air flotation (DAF), on the other hand, has been moderately expensive and efficient, especially in low-density microplastic removal, and thus can be easily integrated into existing treatment plants (Pico et al. 2019). Sand Filtration, being cheap and highly scalable, is extensively used but requires frequent backwashing to ensure efficiency (Barceló and Picó 2019) (Fig. 12.2).

Table 12.3 Cost analysis of different treatment methods

Treatment method	Capital cost	Operational cost	Scalability	Limitations	References
Membrane bioreactors (MBRs)	High	High	Large scale	Expensive membranes, high energy demand	Lares et al. (2018)
Dissolved air flotation (DAF)	Moderate	Moderate	Medium	Requires integration with existing treatment plants	Zhang et al. (2021)
Sand filtration	Low	Low	High	Needs frequent backwashing	Krishnan et al. (2023)
Electrocoagulation	Moderate	High	Medium	Energy-intensive	Perren et al. (2018)
Bioremediation	Low	Low	Scalable	Slow process, dependent on microbial efficiency	Miloloža et al. (2022)
Advanced oxidation processes (AOPs)	High	Moderate	Limited	Risk of toxic byproducts	Kundu et al. (2021)
Nanotechnology	Very high	High	Limited	Still in experimental phase, high cost	Goh et al. (2022)

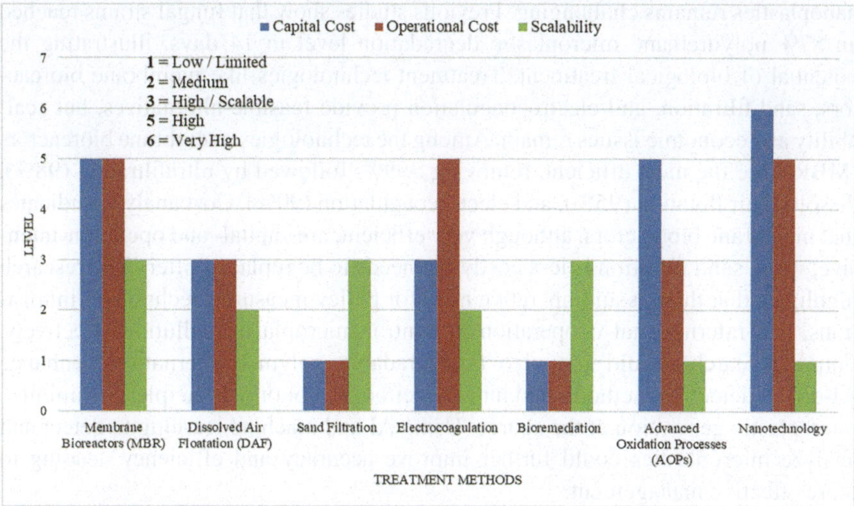

Fig. 12.2 Comparison of cost and scalability of microplastic treatment methods

A low initial capital outlay marks electrocoagulation; however, it is marked by high operating costs due to electricity consumption, making it suitable for medium-scale operations. While this process is highly efficient, its energy requirements impose limitations on its long-term application (Hu et al. 2021). Bioremediation, depending on bacterial and microalgae consortia, is cost-effective and environmentally friendly, but its efficiency is limited by microbial activity efficiency and the slow degradation rate (Das et al. 2023). Advanced oxidation processes (AOPs), including photocatalysis and Fenton-based processes, are a highly efficient but costly methodology, with the additional disadvantage of generating toxic byproducts during the degradation process (Sutradhar 2022). Finally, nanotechnology-based filtration methods are among the most promising but costly alternatives, with research so far limited to experimental phases due to issues of cost, ecological impact, and limited scalability at large scales (Rodríguez-Narvaez et al. 2021).

12.7 Conclusion

The research extensively examines the environmental footprint of microplastics, including their origin, degradation processes, detection methods, and treatment processes. It emphasizes the shocking buildup of microplastics in marine, terrestrial, and atmospheric environments, highlighting their extended degradation process and bioaccumulation characteristics. The study compares physical, chemical, and biological degradation methods, stressing the necessity for integrated systems to raise efficiency. Advanced analytical methods like FTIR, Raman spectroscopy, and Py-GC-MS play important roles in microplastic detection, but detecting nanoplastics remains challenging. Previous studies show that fungal strains reached an 87% polyurethane microplastic degradation level in 14 days, illustrating the potential of biological treatment. Treatment technologies like membrane bioreactors, sand filtration, and electrocoagulation provide feasible alternatives, but scalability and economic issues remain. Among the technologies, membrane bioreactors (MBRs) are the most efficient, removing >99%, followed by ultrafiltration (98%), dissolved air flotation (95%), and electrocoagulation (90%). Cost analysis indicates that membrane bioreactors, although very efficient, are capital- and operation-intensive, while sand filtration is less costly but needs to be replaced often. The research highlights that there is an imperative need for policy measures, technology innovations, and international cooperation to control microplastic pollution effectively. Future research should optimize biodegradable polymer alternatives, enhance AI-driven detection methods, and integrate circular economy principles to minimize microplastic generation at the source. Using AI and machine learning to detect and analyze microplastics could further improve accuracy and efficiency, leading to more effective management.

References

Álvarez-Barragán J, Domínguez-Malfavón L, Vargas-Suárez M, González-Hernández R, Aguilar-Osorio G, Loza-Tavera H (2016) Biodegradative activities of selected environmental fungi on a polyester polyurethane varnish and polyether polyurethane foams. Appl Environ Microbiol 82(17):5225–5235

Auta HS, Emenike CU, Jayanthi B, Fauziah SH (2018) Growth kinetics and biodeterioration of polypropylene microplastics by *Bacillus* sp. and *Rhodococcus* sp. isolated from mangrove sediment. Mar Pollut Bull 127:15–21. https://doi.org/10.1016/j.marpolbul.2017.11.036

Ahmed SF, Islam N, Tasannum N, Mehjabin A, Momtahin A, Chowdhury AA, Almomani F, Mofijur M (2024) Microplastic removal and management strategies for wastewater treatment plants. Chemosphere 347:140648. https://doi.org/10.1016/j.chemosphere.2023.140648

Andoh CN, Attiogbe F, Bonsu Ackerson NO, Antwi M, Adu-Boahen K (2024) Fourier transform infrared spectroscopy: an analytical technique for microplastic identification and quantification. In: Infrared physics and technology, vol 136. Elsevier B.V., p 105070. https://doi.org/10.1016/j.infrared.2023.105070

Baruah A, Sharma A, Sharma S, Nagraik R (2022) An insight into different microplastic detection methods. Int J Environ Sci Technol 19(6):5721–5730). Springer. https://doi.org/10.1007/s13762-021-03384-1

Barceló D, Picó Y (2019) Microplastics in the global aquatic environment: analysis, effects, remediation and policy solutions. J Environ Chem Eng 7(5):103421. https://doi.org/10.1016/j.jece.2019.103421

Coffin S, Huang G-Y, Lee I, Schlenk D (2019) Fish and seabird gut conditions enhance desorption of estrogenic chemicals from commonly ingested plastic items. Environ Sci Technol 53(8):4588–4599. https://doi.org/10.1021/acs.est.8b07140

Cristaldi A, Fiore M, Zuccarello P, Conti GO, Grasso A, Nicolosi I, Copat C, Ferrante M (2020) Efficiency of wastewater treatment plants (Wwtps) for microplastic removal: a systematic review. Int J Environ Res Public Health 17(21):1–23). MDPI AG. https://doi.org/10.3390/ijerph17218014

Dierkes G, Lauschke T, Földi C (2022) Analytical methods for plastic (microplastic) determination in environmental samples. In: Handbook of environmental chemistry, vol 111. Springer, pp 43–67. https://doi.org/10.1007/698_2021_744

Das P, Halder G, Bal M (2023) A critical review on remediation of microplastics using microalgae from aqueous system. In: Science of the Total environment, vol 898. Elsevier B.V., p 166425. https://doi.org/10.1016/j.scitotenv.2023.166425

Frias JPGL, Nash R (2019) Microplastics: finding a consensus on the definition. Mar Pollut Bull 138:145–147. https://doi.org/10.1016/j.marpolbul.2018.11.022

Geyer R, Jambeck JR, Law KL (2017) Production, use, and fate of all plastics ever made. Sci Adv 3(7):e1700782. https://doi.org/10.1126/sciadv.1700782

González-Pleiter M, Tamayo-Belda M, Pulido-Reyes G, Amariei G, Leganés F, Rosal R, Fernández-Piñas F (2019) Secondary nanoplastics released from a biodegradable microplastic severely impact freshwater environments. Environ Sci Nano 6(5):1382–1392

Goh PS, Kang HS, Ismail AF, Khor WH, Quen LK, Higgins D (2022) Nanomaterials for microplastic remediation from aquatic environment: why nano matters? In: Chemosphere, vol 299. Elsevier Ltd. https://doi.org/10.1016/j.chemosphere.2022.134418

Hu K, Tian W, Yang Y, Nie G, Zhou P, Wang Y, Duan X, Wang S (2021) Microplastics remediation in aqueous systems: strategies and technologies. In: Water research, vol 198. Elsevier Ltd. https://doi.org/10.1016/j.watres.2021.117144

Horton AA, Dixon SJ (2018) Microplastics: An introduction to environmental transport processes. WIREs Water. 5:e1268. https://doi.org/10.1002/wat2.1268

Iyare, P. U., Ouki, S. K., & Bond, T. (2020). Microplastics removal in wastewater treatment plants: a critical review. In Environmental science: water research and technology (6, 10, pp. 2664–2675). Royal Society of Chemistry. https://doi.org/10.1039/d0ew00397b

Iñiguez ME, Conesa JA, Fullana A (2018) Recyclability of four types of plastics exposed to UV irradiation in a marine environment. Waste Manag 79:339–345. https://doi.org/10.1016/j.wasman.2018.08.006

Krishnan RY, Manikandan S, Subbaiya R, Karmegam N, Kim W, Govarthanan M (2023) Recent approaches and advanced wastewater treatment technologies for mitigating emerging microplastics contamination—a critical review. In: Science of the total environment, vol 858. Elsevier B.V., p 159681. https://doi.org/10.1016/j.scitotenv.2022.159681

Kundu A, Shetti NP, Basu S, Raghava Reddy K, Nadagouda MN, Aminabhavi TM (2021) Identification and removal of micro- and nano-plastics: efficient and cost-effective methods. Chem Eng J 421:129816. https://doi.org/10.1016/j.cej.2021.129816

Liu W, Zhang J, Liu H, Guo X, Zhang X, Yao X, Cao Z, Zhang T (2021) A review of the removal of microplastics in global wastewater treatment plants: characteristics and mechanisms. In: Environment international, vol 146. Elsevier Ltd. https://doi.org/10.1016/j.envint.2020.106277

Law KL, Thompson RC (2014) Microplastics in the seas. Science 345(6193):144–145. https://doi.org/10.1126/science.1254065

Lares M, Ncibi MC, Sillanpää M, Sillanpää M (2018) Occurrence, identification and removal of microplastic particles and fibers in conventional activated sludge process and advanced MBR technology. Water Res 133:236–246. https://doi.org/10.1016/j.watres.2018.01.049

Ma B, Xue W, Hu C, Liu H, Qu J, Li L (2019) Characteristics of microplastic removal via coagulation and ultrafiltration during drinking water treatment. Chem Eng J 359:159–167). Elsevier B.V. https://doi.org/10.1016/j.cej.2018.11.155

Majewsky M, Bitter H, Eiche E, Horn H (2016) Determination of microplastic polyethylene (PE) and polypropylene (PP) in environmental samples using thermal analysis (TGA-DSC). Sci Total Environ 568:507–511. https://doi.org/10.1016/j.scitotenv.2016.06.017

Miloloža M, Cvetnić M, Kučić Grgić D, Ocelić Bulatović V, Ukić Š, Rogošić M, Dionysiou DD, Kušić H, Bolanča T (2022) Biotreatment strategies for the removal of microplastics from freshwater systems. A review. Environ Chem Lett 20(2):1377–1402). Springer. https://doi.org/10.1007/s10311-021-01370-0

Miao F, Liu Y, Gao M, Yu X, Xiao P, Wang M, Wang S, Wang X (2020) Degradation of polyvinyl chloride microplastics via an electro-Fenton-like system with a TiO_2/graphite cathode. J Hazard Mater 399:123023. https://doi.org/10.1016/j.jhazmat.2020.123023

Na SH, Kim MJ, Kim JT, Jeong S, Lee S, Chung J, Kim EJ (2021) Microplastic removal in conventional drinking water treatment processes: performance, mechanism, and potential risk. Water Res 202:117417. https://doi.org/10.1016/j.watres.2021.117417

Ni B-J, Zhu Z-R, Li W-H, Yan X, Wei W, Xu Q, Xia Z, Dai X, Sun J (2020) Microplastics mitigation in sewage sludge through pyrolysis: the role of pyrolysis temperature. Environ Sci Technol Lett 7(12):961–967. https://doi.org/10.1021/acs.estlett.0c00740

Özgenç E (2024) Advanced analytical techniques for assessing and detecting microplastic pollution in water and wastewater systems. Environ Qual Manag 34(1) Wiley. https://doi.org/10.1002/tqem.22217

Priya A, Anusha G, Thanigaivel S, Karthick A, Mohanavel V, Velmurugan P, Balasubramanian B, Ravichandran M, Kamyab H, Kirpichnikova IM, Chelliapan S (2023) Removing microplastics from wastewater using leading-edge treatment technologies: a solution to microplastic pollution—a review. Bioprocess Biosyst Eng 46(3):309–321). Springer. https://doi.org/10.1007/s00449-022-02715-x

Perren W, Wojtasik A, Cai Q (2018) Removal of microbeads from wastewater using electrocoagulation. ACS Omega 3(3):3357–3364. https://doi.org/10.1021/acsomega.7b02037

Pico Y, Alfarhan A, Barcelo D (2019) Nano- and microplastic analysis: focus on their occurrence in freshwater ecosystems and remediation technologies. In: TrAC – trends in analytical chemistry, vol 113. Elsevier B.V., pp 409–425. https://doi.org/10.1016/j.trac.2018.08.022

Rodríguez-Narvaez OM, Goonetilleke A, Perez L, Bandala ER (2021) Engineered technologies for the separation and degradation of microplastics in water: a review. Chem Eng J 414:128692. https://doi.org/10.1016/j.cej.2021.128692

Sudarsan JS, Dogra K, Kumar R, Raval NP, Leifels M, Mukherjee S et al (2024) Tricks and tracks of prevalence, occurrences, treatment technologies, and challenges of mixtures of emerging contaminants in the environment: with special emphasis on microplastic. J Contam Hydrol 265:104389

Sutradhar M (2022) A review of microplastics pollution and its remediation methods: current scenario and future aspects. Arch Agric Environ Sci 7(2):288–293. https://doi.org/10.26832/24566632.2022.0702019

Sharma A, Kumari S, Chopade RL, Pandit PP, Rai AR, Nagar V, Awasthi G, Singh A, Awasthi KK, Sankhla MS (2023) An assessment of the impact of structure and type of microplastics on ultrafiltration technology for microplastic remediation. Sci Prog 106(2) SAGE. https://doi.org/10.1177/00368504231176399

Shah AA, Hasan F, Hameed A, Ahmed S (2008) Biological degradation of plastics: a comprehensive review. Biotechnol Adv 26(3):246–265. https://doi.org/10.1016/j.biotechadv.2007.12.005

Song Y, Qiu R, Hu J, Li X, Zhang X, Chen Y, Wu W-M, He D (2020) Biodegradation and disintegration of expanded polystyrene by land snails *Achatina fulica*. Sci Total Environ 746:141289. https://doi.org/10.1016/j.scitotenv.2020.141289

Shim WJ, Hong SH, Eo SE (2017) Identification methods in microplastic analysis: a review. Anal Methods 9(9):1384–1391). Royal Society of Chemistry. https://doi.org/10.1039/c6ay02558g

Singh B, Kumar A (2024) Advances in microplastics detection: a comprehensive review of methodologies and their effectiveness. In: TrAC - trends in analytical chemistry, vol 170. Elsevier B.V., p 117440. https://doi.org/10.1016/j.trac.2023.117440

Talvitie J, Mikola A, Koistinen A, Setälä O (2017) Solutions to microplastic pollution—removal of microplastics from wastewater effluent with advanced wastewater treatment technologies. Water Res 123:401–407. https://doi.org/10.1016/j.watres.2017.07.005

Galloway TS, Cole M, Lewis C (2017) Interactions of microplastic debris throughout the marine ecosystem, Nat. Ecol. Evol. 1:116, https://doi.org/10.1038/s41559-017-0116

Wirnkor VA, Ebere EC, Ngozi VE (2019) Microplastics, an emerging concern: a review of analytical techniques for detecting and quantifying microplastics. Anal Methods Environ Chem J 2(2):13–30. https://doi.org/10.24200/amecj.v2.i2.57

Wojnowska-Baryła I, Bernat K, Zaborowska M (2022) Plastic waste degradation in landfill conditions: the problem with microplastics, and their direct and indirect environmental effects. Int J Environ Res Public Health 19(20):13223. https://doi.org/10.3390/ijerph192013223

Wu X, Zhao X, Chen R, Liu P, Liang W, Wang J, Teng M, Wang X, Gao S (2022) Wastewater treatment plants act as essential sources of microplastic formation in aquatic environments: a critical review. Water Res 221:118825. https://doi.org/10.1016/j.watres.2022.118825

Ye Y, Yu K, Zhao Y (2022) The development and application of advanced analytical methods in microplastics contamination detection: a critical review. In: Science of the total environment, vol 818. Elsevier B.V. https://doi.org/10.1016/j.scitotenv.2021.151851

Zhang Y, Jiang H, Bian K, Wang H, Wang C (2021) A critical review of control and removal strategies for microplastics from aquatic environments. J Environ Chem Eng 9(4):105463. https://doi.org/10.1016/j.jece.2021.105463

Zhang J, Gao D, Li Q, Zhao Y, Li L, Lin H, Bi Q, Zhao Y (2020) Biodegradation of polyethylene microplastic particles by the fungus *Aspergillus flavus* from the guts of wax moth *Galleria mellonella*. Sci Total Environ 704:135931. https://doi.org/10.1016/j.scitotenv.2019.135931

Zhenyan Lin, Tuo Jin, Tao Zou, Li Xu, Bin Xi, Dandan Xu, Jianwu He, Lizhi Xiong, Chongjian Tang, Jianwei Peng, Yaoyu Zhou, Jiangchi Fei (2022) Current progress on plastic/microplastic degradation: Fact influences and mechanism. Environmental Pollution 304:119159. https://doi.org/10.1016/j.envpol.2022.119159 ISSN 0269-7491

Microplastic Remediation in Aquatic Environment

13

Dinesh Parida and Regina Nogueira

Abstract

Since microplastics are persistent and bioaccumulative in water, they impose serious environmental hazards. Efficient removal strategies are needed to mitigate their impact. In this chapter, the efficacy of biological degradation by enzyme activity and physical filtration, such as membrane, disk, and sand filtration, is considered. For MPs in the micrometer range, membrane filtration, especially ultrafiltration and nanofiltration, showed excellent removal efficiencies, but disk and sand filtration provided less expensive substitutes for larger particles. Although useful, these processes eventually develop clogging problems and produce secondary waste. On the other hand, by breaking down MPs into biodegradable products, enzymatic degradation by bacterial, fungal, and algal enzymes provides a more sustainable solution. PETase and MHETase are bacterial enzymes. *Pseudomonas* sp. and *Ideonella sakaiensis* are some examples of bacteria that enable polymer degradation. Fungal cultures such as *Aspergillus* and *Fusarium* release hydrolases for polymer degradation. Oxidative enzymes released by algae promote biodegradation. The limitations and advantages of each process are discussed in this chapter without direct comparison. Enzymatic processes provide long-term degradation by breaking down polymers into biode-

D. Parida
Department of Biosciences and Biomedical Engineering, Indian Institute of Technology Indore, Simrol, Madhya Pradesh, India

Institute of Sanitary Engineering and Waste Management, Leibniz University Hannover, Welfengarten, Hannover, Germany
e-mail: phd2101271003@iiti.ac.in

R. Nogueira (✉)
Institute of Sanitary Engineering and Waste Management, Leibniz University Hannover, Welfengarten, Hannover, Germany
e-mail: nogueira@isah.uni-hannover.de

K. Bala et al. (eds.), *Occurrence, Detection, and Fate of Microplastics in Freshwater Ecosystems*, https://doi.org/10.1007/978-981-95-0225-7_13

gradable components, while filtration provides immediate MP removal but can cause secondary waste generation and clogging. Since scalability is a function of further reactor design and enzyme activity, combining these processes offers a promising approach to long-term MP reduction.

Keywords

Filtration · Enzymatic degradation · Microplastics · Removal strategies · Aquatic ecosystems

13.1 Introduction

Among the pertinent concerns about long-term global environmental effects, questions about the widespread distribution and accumulation of microplastics (MPs) in oceanic environments in recent decades have been of particular interest. Since their discovery in 2004, MPs (<5 mm) have triggered a huge public awareness campaign and revived interest in conducting research (Thompson et al. 2004). These MPs are more easily seen because they are commonly found in everyday consumer products (primary) or as a result of degradation from larger polymers (secondary) (Parida et al. 2023). Personal care products, corrosive cleaning products, and synthetic polyester fibers are some examples of primary MPs. Conversely, MPs and nanoplastics are emitted from substances like adhesives, paints, and electronic products. As they persist and accumulate, MPs are also identified as emerging pollutants of global concern with serious environmental impacts (Kiran et al. 2022; Barai et al. 2025). Plastics are extensively used across various industries worldwide due to their versatility, durability, lightweight nature, and low cost. The packaging sector remains the largest consumer, accounting for approximately 40% of total plastic demand globally, primarily for food, beverage, and healthcare products (Geyer et al. 2017). Building and construction industries are the second-largest users, employing plastics for insulation, piping, window frames, and roofing materials, valued for their strength and weather resistance. The automotive sector increasingly relies on plastics to manufacture lightweight vehicle components, thereby improving fuel efficiency and reducing carbon emissions (Real 2023). Similarly, electronics, agriculture, and health care industries utilize plastics for applications ranging from casings and cables to greenhouses and medical devices. As of 2019, global plastic production reached around 368 million tons, with Asia, particularly China, leading production and consumption. Despite the economic benefits, the pervasive use of plastics across industries has contributed to significant environmental concerns (Amadei et al. 2023). Plastic waste accumulation in natural areas threatens human and animal health, imposing risks for livelihoods, food security, and harboring diseases. It also poses economic burdens in fisheries, tourism, and habitats (Prata et al. 2019; Patrício Silva 2021). Plastics are harmful to terrestrial and aquatic ecosystems and adversely affect various natural systems such as soil, water, wastewater, air, and atmosphere. Plastics most often contaminate the aquatic ecosystem and cause various kinds of effects on the ecology and biota (Hartmann et al. 2019; Seeley et al. 2020; Bhatt et al. 2021). MPs pose a significant risk to both public and environmental health due to their persistence, bioaccumulation, and potential for chemical transport. In the environment, MPs can be ingested by a wide range of organisms from planktons and fish to seabirds and mammals, leading to physical blockages, reduced feeding, and impaired reproduction (Wright et al. 2013; Prata 2018). Environmental impacts are also profound, with MPs altering habitat structures, transporting invasive species, and contributing to biodiversity loss in aquatic ecosystems. The ubiquitous presence of MPs in oceans, freshwater systems, soils, and even atmospheric dust highlights the urgent need for removal methods (Prata 2018; Bhatt et al. 2021). In an effort to curb plastic waste and pollution, researchers have developed a number of biological (Ufarté et al. 2015; Kawai et al. 2019; Kim et al. 2020) and technological (Le et al. 2023; González-Camejo et al. 2023; Fu et al. 2023) solutions. Due to affordability and efficacy (Le et al. 2023), physical treatment methods like agglomeration, filtration (Chi et al. 2021), and coagulation (Ma et al. 2019) are used extensively in wastewater treatment for the removal of MPs. These methods, including filtration, surface skimming, and trawling, have been increasingly explored as potential strategies to directly extract MPs from marine systems (Cózar et al. 2014). However, these methods must balance efficiency with minimizing environmental disturbance, leading to ongoing research into optimizing and innovating collection technologies. Various filtration systems, such as sand filters, membrane bioreactors, and fine-mesh sieves, have been tested for MP removal in laboratory and real-world settings (Schwaferts et al. 2019). Apart from environmental factors (Ahmed et al. 2023) that contribute significantly to MPs degradation in the ecosystems, several microbes, such as bacteria (Zadjelovic et al. 2020), fungi (Temporiti et al. 2022), and algae (Manzi et al. 2022), are readily studied to comprehend what role these microbes have in the abatement of MPs. Given the difficulties associated with preventing plastic pollution at its source, there is growing interest in developing and applying physical removal techniques to recover MPs directly from the aquatic environment. As with the progression of technologies, and several reports suggest the potential of microbes, other physical and chemical means are promising in the mitigation of MPs, it has become extremely important to critically assess and evaluate these methods from time to time. So, the main objective of this chapter is to not compare the efficacy of the physical and biological methods of removal, rather carefully understand and assess their efficacy for larger applications in real-life conditions.

13.2 An Overview of MP Removal Using Filtration Methods

MPs are typically removed by conventional methods, such as coagulation, filtration, sedimentation, and clarification, as they impose significant health and ecological risks (Bhatt et al. 2021). Each of these processes has its fair share of advantages and risks, such as pH, density, MPs size, and flow rate of the medium. These parameters largely influence the physical methods in the abatement of MPs (Osman et al. 2023).

Several filtration processes, like sand filtration (Jangid and Sharma 2024), disc filtration (Ahmed et al. 2024), membrane filtration (Jangid and Sharma 2024), biochar or other organic material-mediated filtration (Wang et al. 2020b), are known to produce significant results. This section discusses some of these filtration techniques with considerable detail for the readers.

13.2.1 Rapid Sand Filtration

Recently, rapid sand filtration (RSF) has been proven to be one of the effective methods in removing MPs in WWTPs and from the drinking water. Its negligible environmental footprint, low sensitivity to water quality parameters, and higher flow rate make it extremely viable in MP removal (Talvitie et al. 2017a). As WWTPs contain several treatment processes like pre-sedimentation, coagulation, flocculation, sedimentation, and filtration, RSF is also one of the filtration techniques, along with the above processes, to be swiftly incorporated into WWTPs (Sembiring et al. 2021). The RSF removes MPs by passing it through silica and sand layers, where it actually can interact with these layers. The sand filter is composed of 1 m of gravel with a grain size of 3–5 mm and 0.5 m of quartz with a grain size of 0.1–0.5 mm. In addition to the physical removal of suspended solids, microbial adhesion also removes nutrients and microbes. Most published articles address the effectiveness of this size-based removal process. Sembiring et al. conducted an experiment to determine the effectiveness of RSF in removing MPs. Tire flakes and plastic bags were used to prepare dummy samples of different sizes for the experiment. Bentonite was used to simulate water turbidity. The removal efficiency was 97.7% for plastic flakes and 94.3% for tire flakes. Particle sizes greater than 200 μm were nearly removed in the study (Sembiring et al. 2021). However, in a study conducted by Hidayaturrahman and Lee, membrane disk filters were found to be superior to RSF. In the experiment, effluent from the coagulation WWTP was passed through a 6.8 m deep RSF with 0.8–1.2 mm sand. The removal of microplastics can be improved by rapid sand filtration; however, there will be discrepancies due to diameter variations. The fact that microplastics with a diameter of less than 65 μm can evade the rapid sand filtration system means that they can pass through it (Hidayaturrahman and Lee 2019). However, without the introduction of costly flocculants, this system alone is not very effective and requires frequent maintenance, increasing the overall cost of filtration (Osman et al. 2023).

13.2.2 Membrane Filtration

Membrane filtration technology has been widely used in separation and purification processes due to its high selectivity, continuous operation, and ease of upscaling. It is much more efficient at removing MPs than other technologies (Jangid and Sharma 2024). Compared to other technologies, membrane filtration is highly efficient at removing MPs, which basically acts as a selective barrier. Certain materials, such as ions, molecules, crystals, or particles, such as organic matter, MPs, and clay, are permeable to membranes, and others are not. According to pore sizes, membrane filters are classified as microfiltration (MF) (0.08–2 μm), ultrafiltration (UF) (0.005–0.02 μm), nanofiltration (NF) (0.002 μm), and reverse osmosis filtration, which is used for desalination of brackish or seawater typically (Le et al. 2023; Sangkham et al. 2025). With the advancements of several membrane bioreactors (MBRs), this membrane filtration technique is being incorporated into the mainstream filtration processes gradually, which might open several new doors in the future for the filtration techniques. Critical evaluation made it clear that several physical characteristics, like particle type, size, mass, and form, play a huge role in the separation processes (Poerio et al. 2019). For the removal of polyamide (PA) and polystyrene (PS) plastics, a small-scale home filtration system was developed, and its efficiency was tested. Membranes like polycarbonate (PCA), poly-tetrafluoro ethylene (PTFE), and cellulose acetate were tested for their efficacy in removing the MPs, where cellulose acetate was found to be extremely suitable in the home filtration system. Though it was also observed that mishandling and mechanical rupture of membranes might lead to MP penetration or degradation in the system itself, which imposed as a great challenge for the particular system. Therefore, it was concluded that membrane filtration is indeed a challenging and complex system (Pizzichetti et al. 2021). Due to low energy consumption, compact structure of the plant, greater separation efficiency, and low cost, UF can also be touted as one of the promising separating techniques that produces high-quality drinking water (Molinari et al. 2006). A study conducted by Ma et al. investigated the use of Fe-based coagulant in UF and coagulation to lower the risk of polyethylene (PE) contamination from drinking water. PE, being one of the frequently detected polymers from water, is quite difficult to remove due to its dense structure. However, incorporation of polyacrylamide (PAM) improved the efficiency of coagulation, resulting in the increase in the removal of PE from 13% to 91%. But, after coagulation with PAM and $FeCl_3$. $6H_2O$, the large PE particles reduced membrane fouling, thereby decreasing the flow rate almost by 10% (Ma et al. 2019). Even though UF successfully removes particles, in-depth research is required to determine the effects of MPs on cake formation, fouling characters, etc. Moreover, the removal techniques are also controlled by the composition of the membrane, so the final treatment stages need to be investigated thoroughly to remove different types of MPs in a more effective manner (Talvitie et al. 2017b; Poerio et al. 2019).

13.2.3 Disc Filtration

Disc filtration has a medium to high removal efficiency in removing MPs from wastewater. For improved performance and to achieve consistency, adjustments in the operating conditions are much needed, as discussed by several researchers (Simon et al. 2019; Iyare et al. 2020). In a disc filter, filtration, cake formation, dewatering, and discharge are the main processes, where solid particles are trapped when the discs rotate, forming filter cakes. Compressed air, vacuum, or, in some cases, gravity are used to dewater the system. After one cycle, the filter cake is removed carefully to allow subsequent operations (Hou and Zhang 2012; Chi et al. 2021). A study conducted by Simon et al. revealed the efficacy of disc filtration in removing particles larger than 10 μm. The disc filtration here retained a significant percentage of MPs particles (89.7% and 75.6%), albeit with periodic interruption. Some particles larger than the pore size of the disc escaped, which lowered the efficiency of the system. However, it still showed great potential with a considerable MP removal rate and can be upscaled in the future for larger applications (Simon et al. 2019). Disc filters actually have efficiency from 0.5% to 98.5% as argued by Jani et al., in their review, where they discussed the effectiveness of several filters with a variable mode of operational feature. So, it indicated the need for optimization of this technology for effective removal in WWTPs (Jani et al. 2024). In another study reported by Talvitie et al., several tertiary treatment methods, such as rapid and sand filtration, disc filtration, and air flotation, were scrutinized. Here, they reported membrane bioreactors having the highest removal efficiency of 99.9%, whereas the disc filter came a close second with a random efficiency from 40% to 98.5% (Talvitie et al. 2017b).

13.2.4 Gravity Filtration

Rapid gravity filtration (RGF), or in simpler terms, gravity filtration, has been explored several times for the removal of MPs in WWTPs. A bed of granular media, typically of anthracite or sand, is used in RGF, where water is suspended in a downward motion, where the granular bed entraps the suspended solids and MPs effectively. Even though the efficiency of RGF is backed by several peer-reviewed data and studies (González-Camejo et al. 2023). González-Camejo et al. further investigated the role of RGF and its efficacy in the removal of MPs in WWTPs, where RGF showed great potential in high MP removal from municipal and industrial wastewater (González-Camejo et al. 2023). Another study indicated that albeit gravity filtration is an effective technique in removing the larger particles, several improvements are needed for the removal of smaller MPs (<500 μm). The pore size of the filter media imposed a limitation that allowed smaller particles to pass into the effluent (Mintenig et al. 2017). In a similar study, Carr et al. showcased that primary gravity filtration techniques retained almost 40–70% of the total particles. But, smaller particles, especially <1 mm particles, were missed, but in the subsequent treatment processes, 10–50% of the smaller particles also got removed, leaving behind a small fraction of MPs in the wastewater at the end. The study highlighted the need to improve the WWTPs and optimize all the processes in an effective way where the MPs also can be captured without releasing them into the environment, thereby decreasing the risk of MPs exposure in the ecosystem (Carr et al. 2016).

13.2.5 Biological Filtration

Microorganisms are used in the tertiary biological filtration systems, where organic waste, nutrients, and MPs are effectively removed in WWTPs. The biological filtration process requires the treated water from the primary and secondary treatment steps to pass through a filtration medium with microbes in an attempt to break down the pollutants and nutrients to further simpler constituents (Iyare et al. 2020). The role of WWTPs and tertiary treatment systems is studied widely for MP removal in the Gulf of Finland. In a study conducted at Viikinmaki WWTP, the high levels of MPs from the influent were reduced significantly by the WWTP processes. In the primary sedimentation process, a considerable amount of microfibers was removed; further, biological filtration removed the other MPs. A small number of MPs remained in the effluent (Talvitie et al. 2015).

A list of different filtration methods deployed for the physical removal of MPs from distinguished water sources is given in Table 13.1.

13.2.6 Other Organic Material-Mediated Filtration Systems

Several organic materials have been investigated for microplastic removal by filtration because of their efficacy and environmental sustainability. A study compares the effectiveness of biochar filters produced from corn straw and hardwood at different temperatures for removing 10 μm microplastic spheres. It reported that biochar, particularly from corn straw at 500 °C, demonstrated over 95% removal efficiency, significantly outperforming sand filters. The honeycomb structure and thin chips of biochar contribute to this high capacity. The study suggests that biochar could improve sand filter efficiency for microplastic removal in wastewater treatment (Wang et al. 2020b). Another quintessential study developed an ultralight porous sponge by crosslinking natural corn starch and gelatin, which is easily broken down by enzymes to glucose. With 90% efficiency and exceptional mechanical robustness, the sponge facilitates easy compression of micro- and nanoscale polymers into the pores. Additionally, 70% of the complex materials, such as soil and seawater, can also be removed. Pores of the sponge effectively absorb MPs fractions by a process called heterogeneous diffusion. It is a cost-effective way of removing MPs (Fu et al. 2023). Unlike the traditional filter configurations, these materials offer a less harmful alternative and are often cheaper, so they can be easily replaced, offering safe and long-term solutions for large-scale applications in wastewater and pollution control systems.

13.3 Enzymatic Deterioration of MPs

Enzymatic degradation is readily studied in recent times, and holds promising potential in the abatement of MPs in the environment. Though most of these studies are lab-scale, optimizing these studies in lab-scale conditions is also important in

Table 13.1 List of filtration methods used in the treatment of wastewater and water from several sources, along with their removal efficiency in the removal of MPs

Filtration method	Types of samples	Details of the method	Types of MPs retention	Efficiency	References
Membrane microfiltration	Wastewater from WWTP	Electrocoagulation/electro-flotation-assisted removal	PE and PVC	100%	Akarsu et al. (2021)
Anthracite sand filtration	River surface water	Coagulation with aluminum hydroxide, flocculation, and chlorination assisted removal using a drinking water plant	Polyester and cellulose	50–80%	Cherniak et al. (2022)
Aluminosilicate filtration	Effluent discharge from WWTP	Aluminosilicate filter media was modified by cationic surfactant	PE and PA	96%	Shen et al. (2021)
Rapid sand filtration	Effluent from urban WWTP	Membrane bioreactor was compared with RSF	LDPE, nylon, melamine, and polyvinyl	75%	Bayo et al. (2020)
Hybrid biochar sand filtration	Wastewater	The sand filters were modified using biochar spheres	–	95%	Wang et al. (2020b)
Membrane filtration	–	Polycarbonate, cellulose acetate, and polytetrafluoroethylene membranes were used	PA and PS	94%	Pizzichetti et al. (2021)
Rapid gravity filtration and membrane ultrafiltration	Wastewater from plastic industry	UF membranes were able to remove all microliter particles higher than 25 μm	LDPE	40%	González-Camejo et al. (2023)
Granular activated carbon filtration	Raw and treated drinking water from DWTPs	Coagulation combined with sedimentation	PAM	60%	Wang et al. (2020a)
Membrane bioreactor	Water from municipal wastewater	Both primary and secondary treatment facilities, including disc filter, rapid sand filtration, and dissolved air flotation, were utilized and compared	PE and PES	99%	Talvitie et al. (2017a)
Intermittent filtration	Wastewater from urban WWTP	Three filters with different substrates (silica sand, sand/coke, and coke) were analyzed	PE, PES, PET, and PP	98%	Martín-García et al. (2022)

(continued)

Table 13.1 (continued)

Filtration method	Types of samples	Details of the method	Types of MPs retention	Efficiency	References
Concrete-based filtration	Urban stormwater	A concrete-based filtration system with two comparative fine filtration media (sand and biochar) was built	PE and PP	96%	Pankkonen (2020)
Ceramic microfiltration	Wastewater from WWTP	MPs <20 μm across are not removed effectively	–	45–98%	Takeuchi et al. (2023)
Magnetic filtration	Wastewater from WWTP	Magnetic filter was prepared using an acidification-magnetization method with fly ash as the base material. The filter specifically targeted the removal of 1-μm PSMPs	PS	–	Zhao et al. (2024)

WWTP Wastewater treatment plant, *RSF* Rapid sand filtration, *PE* Polyethylene, *PVC* Polyvinyl chloride, *PS* Polystyrene, *PP* Polypropylene, *PES* Polyester, *PAM* Polyacrylamide, *PA* Polyamide, *LDPE* Low-density polyethylene, *PSMPs* Polystyrene microspheres

conceptualizing a long-term or larger application or pollution mitigation system (Parida et al. 2023). Microbes like bateria and fungi secrete enzymes that are known to depolymerize polymers in controlled conditions, leading the way for enzyme-assisted degradation of MPs in the environment (Dhali et al. 2024). Microbial enzymes, such as laccases, PETase, carboxylases, oxygenases, and MHETases, have shown significant potential in degrading polyethylene terephthalate (PET) and polyurethane (PU) (Yoshida et al. 2021; Amobonye et al. 2021; Buchholz et al. 2022). Unlike traditional treatment processes, biological degradation systems can be an environmentally viable option, which requires the selection and stability of appropriate enzymes, scalability, and optimization of reaction processes to improve the wastewater treatment processes, resulting in environmental cleanup (Othman et al. 2021). PETase and laccases are known to hydrolyze plastics and have been discovered through in vitro processes, but *in silico* approaches such as molecular docking and modeling are aiding the enzyme engineering in improving substrate specificity (Skariyachan et al. 2022; Parida et al. 2024). Additionally, metagenomics is also expanding its horizon in unveiling the unculturable microbes in enzyme-substrate binding, which has been remarkable in identifying novel plastic-hydrolyzing enzymes (Lam et al. 2015; Ufarté et al. 2015). Combining these techniques can accelerate the production of functional enzymatic solutions and produce a feasible solution for large-scale wastewater treatment and ecosystem restoration. Several enzymes have been isolated till now from various microbes and other biogenic sources, which are discussed in detail in the following sub-sections.

13.3.1 Mechanism of Enzyme Action

With the abundance of MPs in the environment, the microbes in those ecosystems have developed significant tolerance for the abundance concentration of MPs. They have been known to withstand extreme conditions caused by this abundant concentration (Oberbeckmann and Labrenz 2020). The stress responses and enzyme activity are highly coordinated, which actively regulate cellular metabolic processes (Othman et al. 2021). Additionally, these enzymes not only regulate cellular function, but they also assist in the degradation of native contaminants such as organic pesticides and MPs. The polymeric structure of MPs is readily broken down by the enzymes into monomers, which can actively be utilized by the microbes as a carbon source for the energy metabolism of the microbes (Winkel 2017; Akarsu et al. 2021). Each enzyme has a unique pattern and mechanism of interaction with respect to the particular MPs. Typically, these enzymes go through two steps in their mechanism. Enzyme hydrolases such as lipases, carboxylesterases, cutinases, and proteases have been known to modify the MP surface (Othman et al. 2021; Dhali et al. 2024). On the other hand, certain MPs hydrolases do not readily deteriorate MPs until they modify their surface predicted by Kawai et al. Thus, these enzymes modify the MP surface by hydrophilic interaction without changing the core material. Elemental spectroscopy chemical analysis (ESCA) revealed significant details regarding the mechanism of the enzymatic action of microbes. In comparison with control and enzyme-treated MPs, significant C–O bond changes were observed in the study of Kawai et al. (2014, 2019). An illustration was made depicting the enzymatic depolymerization and subsequent mineralization of polymers (Fig. 13.1). Enzymes such as

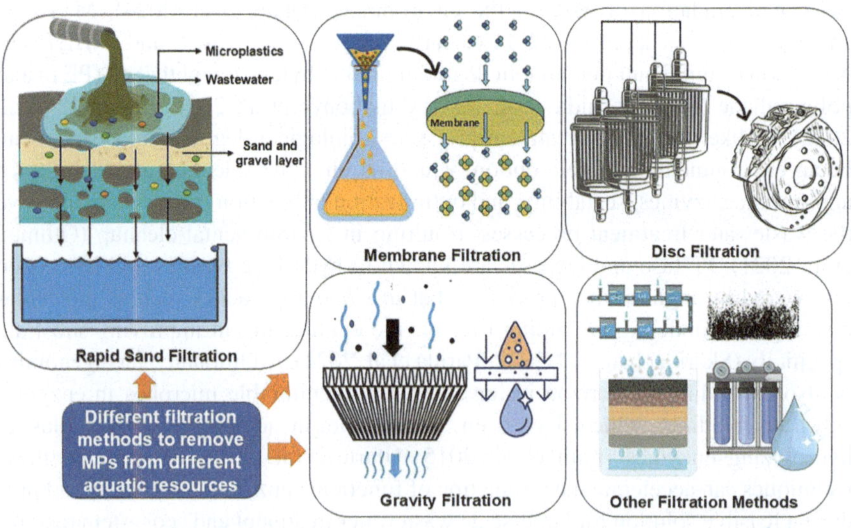

Fig. 13.1 Overview of various filtration methods used for the removal of MPs, including sand filtration, membrane filtration, disc filtration, gravity filtration, and other filtration techniques

cutinases, PETases, and lipases hydrolyze the ester bonds present in polymers like PET, leading to the release of monomers and oligomers. Cutinases and PETases exhibit a high affinity for the amorphous regions of plastics, where polymer chains are less densely packed and thus more accessible (Taniguchi et al. 2019). Meanwhile, for polyolefins like PE and PP, which lack hydrolyzable bonds, oxidative enzymes like laccases, peroxidases, and monooxygenases initiate degradation through oxidation of carbin-carbon backbones (Restrepo-Flórez et al. 2014). These oxidative reactions introduce functional groups (e.g., carbonyls and hydroxyls) that make the plastic surface more hydrophilic and susceptible to further breakdown by microbial attack. Some enzymes act synergistically, for example, in *Ideonella sakaiensis*, PETase first cleaves PET into mono(2-hydroxyethyl) terephthalic acid (MHET), and then MHETase further hydrolyzes MHET into terephthalic acid and ethylene glycol (Yoshida et al. 2021). The efficiency of microbial enzymatic degradation depends on various factors such as crystallinity, molecular weight, surface area, and additives present in the plastics. Although complete mineralization of plastics remains challenging, advancements in protein engineering and synthetic biology have led to the development of more efficient and stable enzyme variants, enhancing the degradation rates under environmental conditions (Tournier et al. 2020) (Fig. 13.2).

13.3.2 Bacterial Enzymes

Bacteria have wide metabolic pathways, therefore used commonly in the bioremediation of several pollutants (Coelho et al. 2015). Several bacterial strains have been reported to have potential biodegradation potential of plastics. As plastics are used as a carbon source for the metabolic processes of microbes, and are also used as an ecological niche for their growth, bacterial remediation systems have been gaining significant attention worldwide by researchers (Gilani et al. 2023). Mineralization, bioassimilation, fragmentation, and biodeterioration are the four main processes that aid plastic degradation through bacteria (Jeyavani et al. 2021). The polymers can be used by bacteria as a carbon source when they release hydrolase groups of enzymes that modify the polymer surface, which results in the mineralization of the polymer chains. In a study by Sivan et al., biodegradation of PE was executed by *Rhodococcus ruber*, which secreted laccase enzymes to oxidize the polymer backbone, leading to depolymerization. They found that after 8 weeks of exposure, the bacteria formed a biofilm on the surface of polyethylene and degraded 7.5% of the material. The most efficient bacteria for plastic biodegradation are biofilm-forming bacteria (Sivan et al. 2006; Santo et al. 2013). Table 13.2 lists some examples of microbial strains (fungi, bacteria, and algae) with enzymes capable of degrading some MPs. These examples show how bacteria modify their genomes to adapt to their environment and secrete the right enzymes to break down PET. The latter is more efficient than other mechanisms for degrading plastics. However, because bacteria are slow at degrading plastics, more research is needed to increase the effectiveness of bacterial strains. In addition, comparatively little gene expression

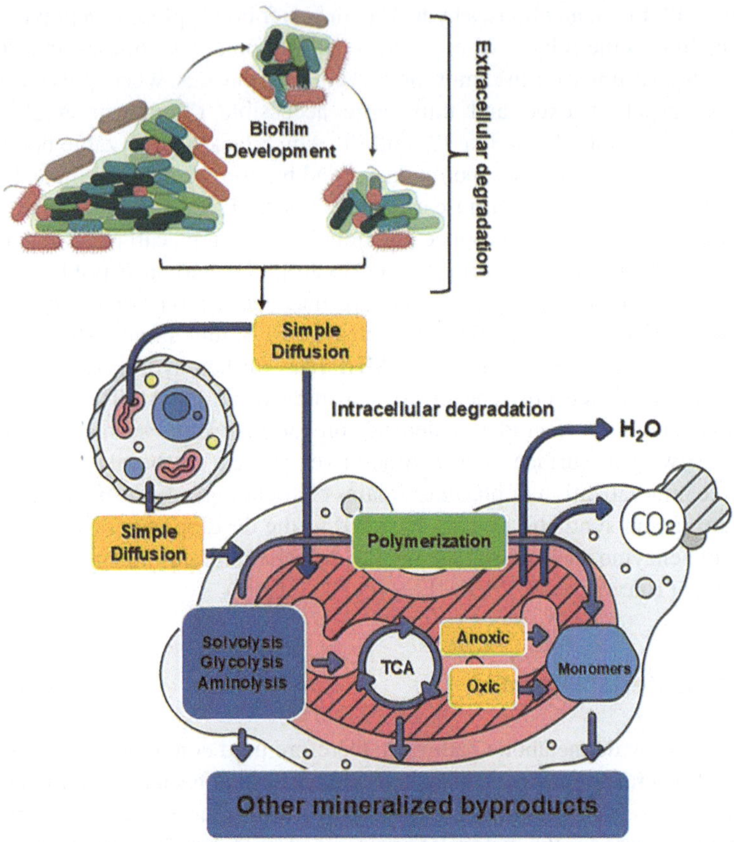

Fig. 13.2 Schematic illustration of intra- and extracellular enzymatic degradation pathways in microbes

research has been done on bacterial biodegradation of plastics. Future research could develop recombinant bacteria with the following favorable characteristics improving their viability under fluctuating pH and temperature and adapting to different environmental conditions (Jeyavani et al. 2021; Dhali et al. 2024).

13.3.3 Fungal Enzymes

During polymer biodegradation, fungi colonize the plastic surface and consume it as a substrate under the regulation of environmental factors such as pH and temperature. Fungi secrete enzymes such as cutinases (Liu et al. 2024), lipases (Safdar et al. 2024), proteases (Brunner et al. 2018), carboxylesterases (Temporiti et al. 2022), esterases (Temporiti et al. 2022), lignocellulolytic enzymes, and certain pro-oxidant ions that are designed to degrade plastics. Oxidation/hydrolysis enzymes convert high molecular weight polymers into low molecular weight ones by making them more hydrophilic. Since high molecular weight molecules are large and cannot pass through the fungal cell membrane, they primarily depolymerize them into small monomers before they can get through (Srikanth et al. 2022). Biodegradation is made more difficult large molecular weight, hydrophobic characteristics, and insolubility of plastics (Wei and Zimmermann 2017). Because of their hyphal apical growth pattern, filamentous fungi can spread mycelial networks across materials (Daccò et al. 2020). Oxidoreductase enzymes, such as laccases and peroxidases, break down PET and PU into smaller molecules (Brunner et al. 2018; Temporiti et al. 2022). To develop effective fungal cell factories for industrial application, further studies need to be done to fully understand the process and genetic background of fungal enzymes with the capability of biodegrading plastic waste.

Table 13.2 List of microbial strains along with the types of enzymes they produce for the degradation of different types of MPs

Microbial strain	Species name	Name of the enzyme	Target MP	Degradation efficiency	References
Bacteria	*Meyerozyma guilliermondii* and *Serratia marcescens*	Multicopper oxidase	PE	15.87%	Lou et al. (2022)
	Acinetobacter baumannii	Multicopper oxidase	PE	–	Zhang et al. (2022)
	Alcanivorax sp. 24	Two cytochrome P450, three alkane monooxygenases, AlkB, and two enzymes involved in the degradation of long-chain alkanes, AlmA, 13 esterases, 17 peroxidases, and a laccase-like protein	PHB, PES	–	Zadjelovic et al. (2020)
	Pseudomonas aeruginosa E7	Alkane hydroxylase	PE	40.8%	Jeon and Kim (2015)
	Bacillus sp. and *Rhodococcus* sp.	Extracellular enzymes—not identified	PP	6.4%, 4%	Auta et al. (2018)
	Comamonas testosteroni NCIMB 8955, Bacillus firmus NCTC 10335, and *Paenibacillus macquariensis NCTC 10419*	Extra- and intracellular enzymes	HDPE	21.81%	Newrick et al. (2025)

(continued)

Table 13.2 (continued)

Microbial strain	Species name	Name of the enzyme	Target MP	Degradation efficiency	References
Fungi	*Aspergillus flavus*	Laccase and laccase-like multicopper oxidases and Manganese Peroxidase	PE (PEDX3)	–	Zhang et al. (2020)
	C. echinulata, *Pestalotiopsis* sp., *H. anthochroum*, *P. lilacinus*, *Aspergillus* sp., and *L. theobromae*	Laccase	PE	–	Sheik et al. (2015)
	Penicillium simplicissimum	Laccase and manganese peroxidase	PE	38%	Sowmya et al. (2015)
	Aspergillus flavus (ITCC 6051)	Esterase	PU	–	Mathur and Prasad (2012)
	Cladosporium cladosporioides, Xepiculopsis graminea, and Penicillium griseofulvum	Oxidases, laccases, and peroxidases	PU	14%	Brunner et al. (2018)
	Cladosporium sp. *P7*	Cutinase-like enzyme CpCut1	PU	–	Liu et al. (2024)
	Aspergillus niger	Lipase	PE, PET, PS	3.8%, 3.6%, 5%	Safdar et al. (2024)
Algae	*Synechocystis* sp.	Laccase	PS	23.2%	Gowthami et al. (2025)
	Marinobacter sp. and *Bacillus* sp.	Esterase	LDPE	99.63%	Khandare et al. (2022)
	Chlamydomonas reinhardtii	PETase	PET		Kim et al. (2020)
	Phaeodactylum tricornutum	PETase	PET	–	Moog et al. (2019)

PE Polyethylene, *PS* Polystyrene, *PET* Polyethylene terephthalate, *LDPE* Low-density polyethylene, *HDPE* High-density polyethylene, *PU* Polyurethane, *PHB* Polyhydroxybutyrate, *PES* Polyester

13.3.4 Algal Enzymes

Microalgae play significant role in responding to the several environmental pollution, so how MPs interact with algae in the environment, whether positively or negatively are needed to be scrutinized (Wang et al. 2022). Different algae respond to different polymers in various ways; they can interact or adhere to the plastic surface by metabolically processing them, or they can protect the MPs from UV degradation and photocatalysis. In addition to settling, the presence of MPs and microalgae in heteroaggregates can alter the bioavailability, fate, and biodegradability of MPs (Leong et al. 2021). Algal bioremediation is the solution for MPs because algae can degrade high-order polymers without releasing endotoxins (Yang et al. 2021). Algae and MPs are fixed and form a chemical bond because carbon is the dominant constituent of the extracellular framework of the molecular forms, e.g., extracellular polymeric substances and extracellular polymers. Understanding how MPs can bind to algae is critical, to avoid disrupting the ecological balance and efficiency of natural aquatic food webs (Khandare et al. 2022; Manzi et al. 2022). Restoration of microbial plastics would be possible through other processes such as mineralization, bio-fragmentation, and biodeterioration under anaerobic and aerobic environments (Carr et al. 2020). Microorganisms would depolymerize the target material and mineralize monomeric molecules by using enzymes that act on specific polymers. Environmental factors (pH, temperature, humidity, and salinity) would provide ideal conditions for enzyme function, and the geometry of the MP would allow adhesion of microbes. UV light and photooxidation are examples of abiotic factors that affect microbial degradation (Gewert et al. 2015; Manzi et al. 2022). MP degradation can be accelerated by using these abiotic factors (light, temperature, nutrient starvation, phytohormones, and salt) to promote algal growth and enzyme activity. However, in order to condition cells with the necessary enzymes, the algal metabolic process must first be fully understood (Manzi et al. 2022).

13.3.5 Enzymes from Other Sources

The scientific community has recently become more interested in the biological breakdown of plastics; although it is commonly known that plastic-degrading enzymes are produced by bacteria, fungi, and algae, new research has found that other organisms also produce these enzymes. Especially, their potential use as a means to mitigate plastics has been investigated. Among all the known lepidopterans, *Galleria mellonella* is certainly one of the most promising of them, as it produces the first polyethylene-degrading enzymes reported (Boschi et al. 2024). Scientists discovered that certain beetles and larvae, members of the Anobiidae, Dermestidae, and Tenebrionidae families, have the peculiar ability to break down plastic wrappings for the first time in the 1950s. The aim then was to identify repellent treatments or resistant materials to keep insects out of packing films, since insects that could demolish plastics were viewed negatively. In contrast, this unusual capacity discovered in some insects is now considered to be a likely new and renewable method of addressing the issue of plastic waste elimination (Riudavets et al. 2007; Boschi et al. 2024). Since then, several researchers have reported the efficiency of insects in the degradation of different types of plastics. A study by Wang and colleagues investigated the in vivo degradation process of PS by plastic-eating insects, *Galleria mellonella.* The study found that although metabolism could be enhanced, PS microplastics were depolymerized and degraded independently by gut bacteria. Styrene oxide-phenylacetaldehyde and 4-methylphenol-4-hydroxybenzaldehyde-4-hydroxybenzoate were suggested as two possible metabolic pathways. Although enzymes that catalyze effective PS biodegradation have been identified, further research is needed (Wang et al. 2022). In another report, the effect of feeding wheat bran or beeswax to larvae on plastic degradation was investigated. After 21 days, the results showed significant loss of plastic mass, depolymerization, and biodegradation. Larval supplementation reduced plastic uptake but increased survival. The diets affected the gut microbiota, with the PS and PE diets strongly correlated with *Bacillus* and *Serratia* (Lou et al. 2020). Other than insects, another study considered the capacity of African smaller mealworms (*Alphitobius diaperinus*) to degrade plastic. The mitogenomic (mitochondrial genomics) structure of the mealworm and its capacity to feed on PS were tested. A novel subspecies was suggested by the results, showing a circular complete mitochondrial genome. The research also discovered that the lesser mealworm contains a bacterial community linked to PS degradation and can live on a PS diet (Ndotono et al. 2024). Research has shown that the microbial population in bovine rumen fluid was capable of degrading polymers. The hydrolytic activity of bovine (*Bos taurus*) rumen on model synthetic polyesters was studied. Rumen fluid hydrolyzes synthetic aromatic polyesters with higher terephthalic acid content, better than hydrolysis by PET. Bio-based polyester poly (ethylene furanoate) (PEF) was also degraded by rumen fluid. Bacteria were the most abundant in the rumen microbiome (98%), followed by Archaea and Eukaryota. Certain taxa, such as *Pseudomonas*, were found to hydrolyze polyester. Another study using the mealworm *Tenebrio molitor* and the superworm *Zophobas morio* showed the symbiotic nature of the microbiota and the host insect. They studied the degradation capacity of mealworms and superworms, with a particular focus on the potential of the gut microbiota to produce enzymes that accelerate plastic degradation (Sanz et al. 2024). By demonstrating the variability of plastic-degrading enzyme sources beyond traditional microbes, this research opens the door to new possibilities for biotechnological interventions to address plastic pollution.

13.4 Conclusion and Future Outlook

MP pollution is a severe threat to aquatic organisms, and there is a requirement for efficient removal techniques. Membrane filtration, disc filtration, and sand filtration are a few of the filtration techniques that can effectively and instantly remove microplastics from water. Their long-term efficiency, however, relies on clogging, maintenance requirements, and production of secondary waste, which emphasizes the necessity of monitoring and optimization of systems. Enzymatic biodegradation, using bacterial, fungal, and algal enzymes, is a more environmentally friendly process that breaks down MPs into biodegradable products. Enzymatic activities show promising laboratory reports, but their widespread application in real environmental settings remains problematic. Further studies are needed to improve the stability, activity, and efficacy of enzymes in mixed ecosystems. A two-pronged approach using filtration for initial removal and enzymatic degradation for continuous degradation may provide a viable solution for microplastic mitigation. Improved design optimization of filtration systems and scale-up of enzymatic processes will be critical to improving the efficiency and feasibility of these processes for effective use. Nevertheless, scaling up enzymatic MP removal from laboratory to field applications still faces challenges related to enzyme stability, cost, and delivery methods in complex aquatic environments. Future research should focus on developing robust, scalable biodegradation platforms that integrate microbial communities, engineered enzymes, and pretreatment methods to accelerate MP breakdown under natural environmental conditions.

References

Ahmed ASS, Billah MM, Ali MM, Bhuiyan MKA, Guo L, Mohinuzzaman M, Hossain MB, Rahman MS, Islam MS, Yan M, Cai W (2023) Microplastics in aquatic environments: a comprehensive review of toxicity, removal, and remediation strategies. Sci Total Environ 876:162414. https://doi.org/10.1016/j.scitotenv.2023.162414

Ahmed SF, Islam N, Tasannum N, Mehjabin A, Momtahin A, Chowdhury AA, Almomani F, Mofijur M (2024) Microplastic removal and management strategies for wastewater treatment plants. Chemosphere 347:140648. https://doi.org/10.1016/j.chemosphere.2023.140648

Akarsu C, Kumbur H, Kideys AE (2021) Removal of microplastics from wastewater through electrocoagulation-electroflotation and membrane filtration processes. Water Sci Technol 84:1648–1662. https://doi.org/10.2166/wst.2021.356

Amadei AM, Rigamonti L, Sala S (2023) Exploring the EU plastic value chain: a material flow analysis. Resour Conserv Recycl 197:107105. https://doi.org/10.1016/j.resconrec.2023.107105

Amobonye A, Bhagwat P, Singh S, Pillai S (2021) Plastic biodegradation: frontline microbes and their enzymes. Sci Total Environ 759:143536. https://doi.org/10.1016/j.scitotenv.2020.143536

Auta HS, Emenike CU, Jayanthi B, Fauziah SH (2018) Growth kinetics and biodeterioration of polypropylene microplastics by Bacillus sp. and Rhodococcus sp. isolated from mangrove sediment. Mar Pollut Bull 127:15–21. https://doi.org/10.1016/j.marpolbul.2017.11.036

Barai DP, Gajbhiye SL, Bhongade YM, Kanhere HS, Kokare DM, Raut NA, Bhanvase BA, Dhoble SJ (2025) Performance evaluation of existing and advanced processes for remediation of microplastics: a comprehensive review. J Environ Chem Eng 13:116194. https://doi.org/10.1016/j.jece.2025.116194

Bayo J, López-Castellanos J, Olmos S (2020) Membrane bioreactor and rapid sand filtration for the removal of microplastics in an urban wastewater treatment plant. Mar Pollut Bull 156:111211. https://doi.org/10.1016/j.marpolbul.2020.111211

Bhatt P, Pathak VM, Bagheri AR, Bilal M (2021) Microplastic contaminants in the aqueous environment, fate, toxicity consequences, and remediation strategies. Environ Res 200:111762. https://doi.org/10.1016/j.envres.2021.111762

Boschi A, Scieuzo C, Salvia R, Arias CF, Perez RP, Bertocchini F, Falabella P (2024) Beyond microbial biodegradation: plastic degradation by Galleria mellonella. J Polym Environ 32:2158–2177. https://doi.org/10.1007/s10924-023-03084-6

Brunner I, Fischer M, Rüthi J, Stierli B, Frey B (2018) Ability of fungi isolated from plastic debris floating in the shoreline of a lake to degrade plastics. PLoS One 13:e0202047. https://doi.org/10.1371/journal.pone.0202047

Buchholz PCF, Feuerriegel G, Zhang H, Perez-Garcia P, Nover L, Chow J, Streit WR, Pleiss J (2022) Plastics degradation by hydrolytic enzymes: the plastics-active enzymes database – PAZy. Proteins Struct Funct Bioinf 90:1443–1456. https://doi.org/10.1002/prot.26325

Carr SA, Liu J, Tesoro AG (2016) Transport and fate of microplastic particles in wastewater treatment plants. Water Res 91:174–182. https://doi.org/10.1016/j.watres.2016.01.002
Carr CM, Clarke DJ, Dobson ADW (2020) Microbial polyethylene terephthalate hydrolases: current and future perspectives. Front Microbiol 11. https://doi.org/10.3389/fmicb.2020.571265
Cherniak SL, Almuhtaram H, McKie MJ, Hermabessiere L, Yuan C, Rochman CM, Andrews RC (2022) Conventional and biological treatment for the removal of microplastics from drinking water. Chemosphere 288:132587. https://doi.org/10.1016/j.chemosphere.2021.132587
Chi Y, Yang P, Ma Z, Wang H, Liu Y, Jiang B, Hu Z (2021) The study on internal flow characteristics of disc filter under different working condition. Appl Sci 11:7715. https://doi.org/10.3390/app11167715
Coelho LM, Rezende HC, Coelho LM, de Sousa PAR, Melo DFO, Coelho NMM (2015) Bioremediation of polluted waters using microorganisms. In: Advances in bioremediation of wastewater and polluted soil. InTech
Cózar A, Echevarría F, González-Gordillo JI, Irigoien X, Úbeda B, Hernández-León S, Palma ÁT, Navarro S, García-de-Lomas J, Ruiz A, Fernández-de-Puelles ML, Duarte CM (2014) Plastic debris in the open ocean. Proc Natl Acad Sci 111:10239–10244. https://doi.org/10.1073/pnas.1314705111
Daccò C, Girometta C, Asemoloye MD, Carpani G, Picco AM, Tosi S (2020) Key fungal degradation patterns, enzymes and their applications for the removal of aliphatic hydrocarbons in polluted soils: a review. Int Biodeterior Biodegradation 147:104866. https://doi.org/10.1016/j.ibiod.2019.104866
Dhali SL, Parida D, Kumar B, Bala K (2024) Recent trends in microbial and enzymatic plastic degradation: a solution for plastic pollution predicaments. Biotechnol Sustain Mater 1:11. https://doi.org/10.1186/s44316-024-00011-0
Fu J, Liu N, Peng Y, Wang G, Wang X, Wang Q, Lv M, Chen L (2023) An ultra-light sustainable sponge for elimination of microplastics and nanoplastics. J Hazard Mater 456:131685. https://doi.org/10.1016/j.jhazmat.2023.131685
Gewert B, Plassmann MM, MacLeod M (2015) Pathways for degradation of plastic polymers floating in the marine environment. Environ Sci Process Impacts 17:1513–1521. https://doi.org/10.1039/C5EM00207A
Geyer R, Jambeck JR, Law KL (2017) Production, use, and fate of all plastics ever made. Sci Adv 3. https://doi.org/10.1126/sciadv.1700782
Gilani IE, Sayadi S, Zouari N, Al-Ghouti MA (2023) Plastic waste impact and biotechnology: exploring polymer degradation, microbial role, and sustainable development implications. Bioresour Technol Rep 24:101606. https://doi.org/10.1016/j.biteb.2023.101606
González-Camejo J, Morales A, Peña-Lamas J, Lafita C, Enguídanos S, Seco A, Martí N (2023) Feasibility of rapid gravity filtration and membrane ultrafiltration for the removal of microplastics and microlitter in sewage and wastewater from plastic industry. J Water Process Eng 51:103452. https://doi.org/10.1016/j.jwpe.2022.103452
Gowthami A, Syed Marjuk M, Santhanam P, Thirumurugan R, Muralisankar T, Perumal P (2025) Marine microalgae – mediated biodegradation of polystyrene microplastics: Insights from enzymatic and molecular docking studies. Chemosphere 370:144024. https://doi.org/10.1016/j.chemosphere.2024.144024
Hartmann NB, Hüffer T, Thompson RC, Hassellöv M, Verschoor A, Daugaard AE, Rist S, Karlsson T, Brennholt N, Cole M, Herrling MP, Hess MC, Ivleva NP, Lusher AL, Wagner M (2019) Are We speaking the same language? Recommendations for a definition and categorization framework for plastic debris. Environ Sci Technol 53:1039–1047. https://doi.org/10.1021/acs.est.8b05297
Hidayaturrahman H, Lee T-G (2019) A study on characteristics of microplastic in wastewater of South Korea: identification, quantification, and fate of microplastics during treatment process. Mar Pollut Bull 146:696–702. https://doi.org/10.1016/j.marpolbul.2019.06.071
Hou JX, Zhang YC (2012) Study on filtration performance of rotary disc filter with different filter discs. Adv Mater Res 610–613:1265–1269. https://doi.org/10.4028/www.scientific.net/AMR.610-613.1265

Iyare PU, Ouki SK, Bond T (2020) Microplastics removal in wastewater treatment plants: a critical review. Environ Sci Water Res Technol 6:2664–2675. https://doi.org/10.1039/D0EW00397B

Jangid NK, Sharma R (2024) Micro plastics. CRC Press, Boca Raton

Jani V, Wu S, Venkiteshwaran K (2024) Advancements and regulatory situation in microplastics removal from wastewater and drinking water: a comprehensive review. Microplastics 3:98–123. https://doi.org/10.3390/microplastics3010007

Jeon HJ, Kim MN (2015) Functional analysis of alkane hydroxylase system derived from Pseudomonas aeruginosa E7 for low molecular weight polyethylene biodegradation. Int Biodeterior Biodegradation 103:141–146. https://doi.org/10.1016/j.ibiod.2015.04.024

Jeyavani J, Sibiya A, Shanthini S, Ravi C, Vijayakumar S, Rajan DK, Vaseeharan B (2021) A review on aquatic impacts of microplastics and its bioremediation aspects. Curr Pollut Rep 7:286–299. https://doi.org/10.1007/s40726-021-00188-2

Kawai F, Oda M, Tamashiro T, Waku T, Tanaka N, Yamamoto M, Mizushima H, Miyakawa T, Tanokura M (2014) A novel Ca2+-activated, thermostabilized polyesterase capable of hydrolyzing polyethylene terephthalate from Saccharomonospora viridis AHK190. Appl Microbiol Biotechnol 98:10053–10064. https://doi.org/10.1007/s00253-014-5860-y

Kawai F, Kawabata T, Oda M (2019) Current knowledge on enzymatic PET degradation and its possible application to waste stream management and other fields. Appl Microbiol Biotechnol 103:4253–4268. https://doi.org/10.1007/s00253-019-09717-y

Khandare SD, Agrawal D, Mehru N, Chaudhary DR (2022) Marine bacterial based enzymatic degradation of low-density polyethylene (LDPE) plastic. J Environ Chem Eng 10:107437. https://doi.org/10.1016/j.jece.2022.107437

Kim JW, Park S-B, Tran Q-G, Cho D-H, Choi D-Y, Lee YJ, Kim H-S (2020) Functional expression of polyethylene terephthalate-degrading enzyme (PETase) in green microalgae. Microb Cell Factories 19:97. https://doi.org/10.1186/s12934-020-01355-8

Kiran BR, Kopperi H, Venkata Mohan S (2022) Micro/nano-plastics occurrence, identification, risk analysis and mitigation: challenges and perspectives. Rev Environ Sci Bio/Technol 21:169–203. https://doi.org/10.1007/s11157-021-09609-6

Lam KN, Cheng J, Engel K, Neufeld JD, Charles TC (2015) Current and future resources for functional metagenomics. Front Microbiol 6. https://doi.org/10.3389/fmicb.2015.01196

Le L-T, Bui X-B, Tran C-S, Chiemchaisri C, Pandey A (2023) Membrane and filtration processes for microplastic removal. In: Current developments in biotechnology and bioengineering. Elsevier, pp 203–220

Leong YK, Chew KW, Chen W-H, Chang J-S, Show PL (2021) Reuniting the biogeochemistry of algae for a low-carbon circular bioeconomy. Trends Plant Sci 26:729–740. https://doi.org/10.1016/j.tplants.2020.12.010

Liu J, Xin K, Zhang T, Wen Y, Li D, Wei R, Zhou J, Cui Z, Dong W, Jiang M (2024) Identification and characterization of a fungal cutinase-like enzyme CpCut1 from Cladosporium sp. P7 for polyurethane degradation. Appl Environ Microbiol 90. https://doi.org/10.1128/aem.01477-23

Lou Y, Ekaterina P, Yang S-S, Lu B, Liu B, Ren N, Corvini PF-X, Xing D (2020) Biodegradation of polyethylene and polystyrene by greater Wax Moth Larvae (Galleria mellonella L.) and the effect of co-diet supplementation on the core gut microbiome. Environ Sci Technol 54:2821–2831. https://doi.org/10.1021/acs.est.9b07044

Lou H, Fu R, Long T, Fan B, Guo C, Li L, Zhang J, Zhang G (2022) Biodegradation of polyethylene by Meyerozyma guilliermondii and Serratia marcescens isolated from the gut of waxworms (larvae of Plodia interpunctella). Sci Total Environ 853:158604. https://doi.org/10.1016/j.scitotenv.2022.158604

Ma B, Xue W, Ding Y, Hu C, Liu H, Qu J (2019) Removal characteristics of microplastics by Fe-based coagulants during drinking water treatment. J Environ Sci 78:267–275. https://doi.org/10.1016/j.jes.2018.10.006

Manzi HP, Abou-Shanab RAI, Jeon B-H, Wang J, Salama E-S (2022) Algae: a frontline photosynthetic organism in the microplastic catastrophe. Trends Plant Sci 27:1159–1172. https://doi.org/10.1016/j.tplants.2022.06.005

Martín-García AP, Egea-Corbacho Á, Franco AA, Albendín G, Arellano JM, Rodríguez-Barroso R, Coello MD, Quiroga JM (2022) Application of intermittent sand and coke filters for the removal of microplastics in wastewater. J Clean Prod 380:134844. https://doi.org/10.1016/j.jclepro.2022.134844

Mathur G, Prasad R (2012) Degradation of polyurethane by Aspergillus flavus (ITCC 6051) isolated from soil. Appl Biochem Biotechnol 167:1595–1602. https://doi.org/10.1007/s12010-012-9572-4

Mintenig SM, Int-Veen I, Löder MGJ, Primpke S, Gerdts G (2017) Identification of microplastic in effluents of waste water treatment plants using focal plane array-based micro-Fourier-transform infrared imaging. Water Res 108:365–372. https://doi.org/10.1016/j.watres.2016.11.015

Molinari R, Argurio P, Poerio T (2006) Ultrafiltration of polymer-metal complexes for metal ion removal from wastewaters. Macromol Symp 235:206–214. https://doi.org/10.1002/masy.200650325

Moog D, Schmitt J, Senger J, Zarzycki J, Rexer K-H, Linne U, Erb T, Maier UG (2019) Using a marine microalga as a chassis for polyethylene terephthalate (PET) degradation. Microb Cell Factories 18:171. https://doi.org/10.1186/s12934-019-1220-z

Ndotono EW, Tanga CM, Kelemu S, Khamis FM (2024) Mitogenomic profiling and gut microbial analysis of the newly identified polystyrene-consuming lesser mealworm in Kenya. Sci Rep 14:21370. https://doi.org/10.1038/s41598-024-72201-9

Newrick BA, Valdés D, Laca A, Laca A, Díaz M (2025) Enhanced biodegradation of high-density polyethylene microplastics: study of bacterial efficiency and process parameters. J Hazard Mater 485:136822. https://doi.org/10.1016/j.jhazmat.2024.136822

Oberbeckmann S, Labrenz M (2020) Marine microbial assemblages on microplastics: diversity, adaptation, and role in degradation. Annu Rev Mar Sci 12:209–232. https://doi.org/10.1146/annurev-marine-010419-010633

Osman AI, Hosny M, Eltaweil AS, Omar S, Elgarahy AM, Farghali M, Yap P-S, Wu Y-S, Nagandran S, Batumalaie K, Gopinath SCB, John OD, Sekar M, Saikia T, Karunanithi P, Hatta MHM, Akinyede KA (2023) Microplastic sources, formation, toxicity and remediation: a review. Environ Chem Lett 21:2129–2169. https://doi.org/10.1007/s10311-023-01593-3

Othman AR, Hasan HA, Muhamad MH, Ismail NI, Abdullah SRS (2021) Microbial degradation of microplastics by enzymatic processes: a review. Environ Chem Lett 19:3057–3073. https://doi.org/10.1007/s10311-021-01197-9

Pankkonen P (2020) Urban stormwater microplastics–characteristics and removal using a developed filtration system. Master thesis

Parida D, Sangtani R, Bala K (2023) Microplastics: the stemming environmental challenge and the quest for the missing mitigation strategies. Int Biodeterior Biodegradation 179:105581. https://doi.org/10.1016/j.ibiod.2023.105581

Parida D, Katare K, Ganguly A, Chakraborty D, Konar O, Nogueira R, Bala K (2024) Molecular docking and metagenomics assisted mitigation of microplastic pollution. Chemosphere 351:141271. https://doi.org/10.1016/j.chemosphere.2024.141271

Patrício Silva AL (2021) New frontiers in remediation of (micro)plastics. Curr Opin Green Sustain Chem 28:100443. https://doi.org/10.1016/j.cogsc.2020.100443

Pizzichetti ARP, Pablos C, Álvarez-Fernández C, Reynolds K, Stanley S, Marugán J (2021) Evaluation of membranes performance for microplastic removal in a simple and low-cost filtration system. Case Stud Chem Environ Eng 3:100075. https://doi.org/10.1016/j.cscee.2020.100075

Poerio T, Piacentini E, Mazzei R (2019) Membrane processes for microplastic removal. Molecules 24:4148. https://doi.org/10.3390/molecules24224148

Prata JC (2018) Airborne microplastics: consequences to human health? Environ Pollut 234:115–126. https://doi.org/10.1016/j.envpol.2017.11.043

Prata JC, Silva ALP, da Costa JP, Mouneyrac C, Walker TR, Duarte AC, Rocha-Santos T (2019) Solutions and integrated strategies for the control and mitigation of plastic and microplastic pollution. Int J Environ Res Public Health 16:2411. https://doi.org/10.3390/ijerph16132411

Real LEP (2023) Plastics statistics: production, recycling, and market data. In: Recycled materials for construction applications. Springer, Cham, pp 103–113

Restrepo-Flórez J-M, Bassi A, Thompson MR (2014) Microbial degradation and deterioration of polyethylene – a review. Int Biodeterior Biodegradation 88:83–90. https://doi.org/10.1016/j.ibiod.2013.12.014

Riudavets J, Salas I, Pons MJ (2007) Damage characteristics produced by insect pests in packaging film. J Stored Prod Res 43:564–570. https://doi.org/10.1016/j.jspr.2007.03.006

Safdar A, Ismail F, Imran M (2024) Biodegradation of synthetic plastics by the extracellular lipase of Aspergillus niger. Environ Adv 17:100563. https://doi.org/10.1016/j.envadv.2024.100563

Sangkham S, Pitakpong A, Kumar R (2025) Advanced approaches to microplastic removal in landfill leachate: advanced oxidation processes (AOPs), biodegradation, and membrane filtration. Case Stud Chem Environ Eng 11:101056. https://doi.org/10.1016/j.cscee.2024.101056

Santo M, Weitsman R, Sivan A (2013) The role of the copper-binding enzyme – laccase – in the biodegradation of polyethylene by the actinomycete Rhodococcus ruber. Int Biodeterior Biodegradation 84:204–210. https://doi.org/10.1016/j.ibiod.2012.03.001

Sanz L, Tran T, Kainer D (2024) Potential of tenebrio molitor and zophobas morio in plastic degradation: mechanisms, microorganisms, and enzymes. https://doi.org/10.5281/zenodo.13621718

Schwaferts C, Niessner R, Elsner M, Ivleva NP (2019) Methods for the analysis of submicrometer- and nanoplastic particles in the environment. TrAC Trends Anal Chem 112:52–65. https://doi.org/10.1016/j.trac.2018.12.014

Seeley ME, Song B, Passie R, Hale RC (2020) Microplastics affect sedimentary microbial communities and nitrogen cycling. Nat Commun 11:2372. https://doi.org/10.1038/s41467-020-16235-3

Sembiring E, Fajar M, Handajani M (2021) Performance of rapid sand filter – single media to remove microplastics. Water Supply 21:2273–2284. https://doi.org/10.2166/ws.2021.060

Sheik S, Chandrashekar KR, Swaroop K, Somashekarappa HM (2015) Biodegradation of gamma irradiated low density polyethylene and polypropylene by endophytic fungi. Int Biodeterior Biodegradation 105:21–29. https://doi.org/10.1016/j.ibiod.2015.08.006

Shen M, Hu T, Huang W, Song B, Zeng G, Zhang Y (2021) Removal of microplastics from wastewater with aluminosilicate filter media and their surfactant-modified products: performance, mechanism and utilization. Chem Eng J 421:129918. https://doi.org/10.1016/j.cej.2021.129918

Simon M, Vianello A, Vollertsen J (2019) Removal of 10 μm microplastic particles from treated wastewater by a disc Filter. Water 11:1935. https://doi.org/10.3390/w11091935

Sivan A, Szanto M, Pavlov V (2006) Biofilm development of the polyethylene-degrading bacterium Rhodococcus ruber. Appl Microbiol Biotechnol 72:346–352. https://doi.org/10.1007/s00253-005-0259-4

Skariyachan S, Taskeen N, Kishore AP, Krishna BV (2022) Recent advances in plastic degradation – from microbial consortia-based methods to data sciences and computational biology driven approaches. J Hazard Mater 426:128086. https://doi.org/10.1016/j.jhazmat.2021.128086

Sowmya HV, Ramalingappa KM, Thippeswamy B (2015) Degradation of polyethylene by Penicillium simplicissimum isolated from local dumpsite of Shivamogga district. Environ Dev Sustain 17:731–745. https://doi.org/10.1007/s10668-014-9571-4

Srikanth M, Sandeep TSRS, Sucharitha K, Godi S (2022) Biodegradation of plastic polymers by fungi: a brief review. Bioresour Bioprocess 9:42. https://doi.org/10.1186/s40643-022-00532-4

Takeuchi H, Tanaka S, Koyuncu CZ, Nakada N (2023) Removal of microplastics in wastewater by ceramic microfiltration. J Water Process Eng 54:104010. https://doi.org/10.1016/j.jwpe.2023.104010

Talvitie J, Heinonen M, Pääkkönen J-P, Vahtera E, Mikola A, Setälä O, Vahala R (2015) Do wastewater treatment plants act as a potential point source of microplastics? Preliminary study in the coastal Gulf of Finland, Baltic Sea. Water Sci Technol 72:1495–1504. https://doi.org/10.2166/wst.2015.360

Talvitie J, Mikola A, Koistinen A, Setälä O (2017a) Solutions to microplastic pollution – removal of microplastics from wastewater effluent with advanced wastewater treatment technologies. Water Res 123:401–407. https://doi.org/10.1016/j.watres.2017.07.005

Talvitie J, Mikola A, Setälä O, Heinonen M, Koistinen A (2017b) How well is microlitter purified from wastewater? – a detailed study on the stepwise removal of microlitter in a tertiary level wastewater treatment plant. Water Res 109:164–172. https://doi.org/10.1016/j.watres.2016.11.046

Taniguchi I, Yoshida S, Hiraga K, Miyamoto K, Kimura Y, Oda K (2019) Biodegradation of PET: current status and application aspects. ACS Catal 9:4089–4105. https://doi.org/10.1021/acscatal.8b05171

Temporiti MEE, Nicola L, Nielsen E, Tosi S (2022) Fungal enzymes involved in plastics biodegradation. Microorganisms 10:1180. https://doi.org/10.3390/microorganisms10061180

Thompson RC, Olsen Y, Mitchell RP, Davis A, Rowland SJ, John AWG, McGonigle D, Russell AE (2004) Lost at sea: where is all the plastic? Science 304(80):838. https://doi.org/10.1126/science.1094559

Tournier V, Topham CM, Gilles A, David B, Folgoas C, Moya-Leclair E, Kamionka E, Desrousseaux M-L, Texier H, Gavalda S, Cot M, Guémard E, Dalibey M, Nomme J, Cioci G, Barbe S, Chateau M, André I, Duquesne S, Marty A (2020) An engineered PET depolymerase to break down and recycle plastic bottles. Nature 580:216–219. https://doi.org/10.1038/s41586-020-2149-4

Ufarté L, Laville É, Duquesne S, Potocki-Veronese G (2015) Metagenomics for the discovery of pollutant degrading enzymes. Biotechnol Adv 33:1845–1854. https://doi.org/10.1016/j.biotechadv.2015.10.009

Wang Z, Lin T, Chen W (2020a) Occurrence and removal of microplastics in an advanced drinking water treatment plant (ADWTP). Sci Total Environ 700:134520. https://doi.org/10.1016/j.scitotenv.2019.134520

Wang Z, Sedighi M, Lea-Langton A (2020b) Filtration of microplastic spheres by biochar: removal efficiency and immobilisation mechanisms. Water Res 184:116165. https://doi.org/10.1016/j.watres.2020.116165

Wang S, Shi W, Huang Z, Zhou N, Xie Y, Tang Y, Hu F, Liu G, Zheng H (2022) Complete digestion/biodegradation of polystyrene microplastics by greater wax moth (Galleria mellonella) larvae: direct in vivo evidence, gut microbiota independence, and potential metabolic pathways. J Hazard Mater 423:127213. https://doi.org/10.1016/j.jhazmat.2021.127213

Wei R, Zimmermann W (2017) Microbial enzymes for the recycling of recalcitrant petroleum-based plastics: how far are we? Microb Biotechnol 10:1308–1322. https://doi.org/10.1111/1751-7915.12710

Winkel BSJ (2017) When an enzyme isn't just an enzyme anymore. J Exp Bot 68:1387–1389. https://doi.org/10.1093/jxb/erx080

Wright SL, Thompson RC, Galloway TS (2013) The physical impacts of microplastics on marine organisms: a review. Environ Pollut 178:483–492. https://doi.org/10.1016/j.envpol.2013.02.031

Yang X, Wang H, Zhang L, Kong L, Chen Y, He Q, Li L, Grossart H-P, Ju F (2021) Marine algae facilitate transfer of microplastics and associated pollutants into food webs. Sci Total Environ 787:147535. https://doi.org/10.1016/j.scitotenv.2021.147535

Yoshida S, Hiraga K, Taniguchi I, Oda K (2021) Ideonella sakaiensis, PETase, and MHETase: from identification of microbial PET degradation to enzyme characterization, pp 187–205

Zadjelovic V, Gibson MI, Dorador C, Christie-Oleza JA (2020) Genome of Alcanivorax sp. 24: a hydrocarbon degrading bacterium isolated from marine plastic debris. Mar Genomics 49:100686. https://doi.org/10.1016/j.margen.2019.05.001

Zhang J, Gao D, Li Q, Zhao Y, Li L, Lin H, Bi Q, Zhao Y (2020) Biodegradation of polyethylene microplastic particles by the fungus Aspergillus flavus from the guts of wax moth Galleria mellonella. Sci Total Environ 704:135931. https://doi.org/10.1016/j.scitotenv.2019.135931

Zhang Y, Lin Y, Gou H, Feng X, Zhang X, Yang L (2022) Screening of polyethylene-degrading bacteria from rhyzopertha dominica and evaluation of its key enzymes degrading polyethylene. Polymers (Basel) 14:5127. https://doi.org/10.3390/polym14235127

Zhao H, Song F, Zhou H, Ji P (2024) Enhanced removal of microplastics from wastewater treatment plants by a novel magnetic filter. Environ Pollut 361:124854. https://doi.org/10.1016/j.envpol.2024.124854

The manufacturer's authorised representative in the EU is Springer Nature Customer Service Centre GmbH, Europaplatz 3, 69115 Heidelberg, Germany. If you have any concerns regarding our products, please contact ProductSafety@springernature.com

Printed and bound by CPI Group (UK) Ltd, Croydon, CR0 4YY
30/09/2026
02225300-0016